AF709759

Advances in Ceramic Matrix Composites IV

Ceramic Transactions

Volume 96

Advances in Ceramic Matrix Composites IV

Edited by

J.P. Singh

Argonne National Laboratory

Narottam P. Bansal

National Aeronautics and Space Administration, Lewis Research Center

Published by

The American Ceramic Society

735 Ceramic Place

Westerville, Ohio 43081

Papers from the Ceramic Matrix Composites Symposium, held at the 100th Annual Meeting of The American Ceramic Society in Cincinnati, Ohio, May 3–6, 1998.

Copyright 1999, The American Ceramic Society. All rights reserved.

No part of this book may be reproduced, stored in a retrieval system, or transmitted in any form or by any means, electronic, mechanical, photocopying, microfilming, recording, or otherwise, without written permission from the publisher.

Permission to photocopy for personal or internal use beyond the limits of Sections 107 and 108 of the U.S. Copyright Law is granted by the American Ceramic Society, provided that the base fee of US$5.00 per copy, plus US$.50 per page, is paid directly to the Copyright Clearance Center, 222 Rosewood Dr., Danvers, MA 01923, USA. The fee code for users of the Transactional Reporting Service for *Ceramic Transactions Volume 96* is 1-57498-059-9/99 $5.00+$.50. This consent does not extend to other kinds of copying, such as copying for general distribution, for advertising or promotional purposes, or for creating new collective works. Requests for special photocopying permission and reprint requests should be directed to the Director of Publications, The American Ceramic Society, 735 Ceramic Place, Westerville OH 43081, USA.

COVER PHOTO: "Fracture surfaces of the 80SN-20BAS two step sintered sample (1800-4hr-1950-3hr)," is courtesy of N. Nagarajan, F. Yu, Y. Fang, and K.W. White, and appears as figure 9c in their paper, "Mechanical Behavior of the Silicon Nitride-Barrium Aluminosilicate Ceramic Matrix Composite with Reduced Liquid Phase," which begins on page 257.

Library of Congress Cataloging-in-Publication Data
A CIP record for this book is available from the Library of Congress.

For information on ordering titles published by The American Ceramic Society, or to request a publications catalog, please call 614-794-5890.

Printed in the United States of America.

1 2 3 4–02 01 00 99

ISSN 1042-1122
ISBN 1-57498-059-9

Contents

Preface v

Fibers, Coatings, and Interfaces

Tensile Strength and Microstructure of Hi-Nicalon Fibers Extracted from Celsian Matrix Composites 3
Narottam P. Bansal

Durability of Mullite/YSZ-Coated SiC in 90% H_2O/O_2 17
Kang N. Lee and Robert A. Miller

Oxidation Resistant Coatings for Ceramic Matrix Composite Components 27
Virginie M. Vaubert, David P. Stinton, and Deidre A. Hirschfeld

Microstructure of Ti_3SiC_2 Coatings Synthesized by CVD 39
E. Pickering, W.J. Lackey, and S. Crain

The Mechanical Response of Sintered Alumina Bundles and Porous Alumina Matrix Composites 51
Paul E. Cantonwine and Haydn N.G. Wadley

Silicon Carbide Laminates with Porous Interlayers Made by EPD 63
Luc J. Vandeperre and Omer O. Van Der Biest

Continuous Fiber Ceramic Composites: Processing and Microstructure/Property Relationships

Hermetic SiC-SiC Composite Tubes 77
Witold Kowbel, Yin Liu, Calvin Bruce, James C. Withers, Lynne E. Kolaya, and Nathan Lewis

Processing and Properties of Nicalon-Reinforced Zirconium Phosphate Composites 89
Barry A. Bender, Todd L. Jessen, and Scott Browning

The Effect of Microstructure on the Mechanical Behavior of CFCMCs under Dynamic and Static Loading Conditions 101
T.L. Jessen and B.A. Bender

SYLRAMIC™ Ceramic-Matrix Composites Processing and Properties 113
T.E. Easler, D.R. Petrak, and A. Szweda

High-Thermal Conductivity SiC-SiC Composites 123
W. Kowbel, C. Bruce, K.T. Tsou, and J.C. Withers

Temperature Dependency of Strength of a Unidirectional SiC Fiber-Reinforced (BaSr)$Al_2Si_2O_8$ Celsian Composite 135
Ö. Ünal and N.P. Bansal

Ceraset™ and Blackglas™ Mini-Tow Composites with Carbon-Coated Nicalon™ Tows 149
S.T. Gonczy and K.T. Faber

Flexure Test Method Investigation for Various Nextel™ Continuous Reinforced Blackglas™ Ceramic Matrix Composites 161
R. Belardinelli and S.L. Pauletti

Nonconventional, Damage Tolerant Oxide/Oxide Fiber-Composites 173
Timothy Radsick, B. Saruhan, and H. Schneider

Nano, Particulate, Whisker, and Self-Reinforced Composites

Pressureless Sintered Alumina/Silicon Carbide Nanocomposites 185
K. Niihara, Y.K. Jeong, T. Sekino, and Y.H. Choa

Mechanical and Electrical Properties of 3Y-TZP/SiC and 8YSZ/SiC Nanocomposites 197
Koichi Niihara, Noriko Bamba, Yong-Ho Choa, and Tohru Sekino

Mechanical and Electrical Properties of ZnO/Ag Nanocomposites 209
Y. Hayashi, Y.H. Choa, J.P. Singh, and K. Niihara

Mechanical Properties and Thermal Shock Resistance of SiO_2-Si_3N_4 Composite Materials. 219
Li Baoshun, Yao Junjie, Zhang Yufeng, and Huang Xueren

Reaction-Bonded Mullite-ZrO_2 Composites with Oxide Additives 231
E. Rocha-Rangel, H. Balmori-Ramírez, and S.D. De la Torre

Preparation, Microstructural, and Mechanical Properties of Ti(C,N) Whisker/Al_2O_3 Composites 237
P. Pettersson, M. Johnsson, and M. Nygren

Microstructural Modification of 70% Si_3N_4-30% BAS Ceramic Matrix Composites by Varying the Particle Size Distribution of Si_3N_4 Starting Powders 245
Feng Yu and Kenneth W. White

Mechanical Behavior of the Silicon Nitride-Barium Aluminosilicate Ceramic Matrix Composites with Reduced Liquid Phase 257
N. Nagarajan, F. Yu, Y. Fang, and K.W. White

Phase Relationships and Microstructures in Composite Systems

Phase Relationships and Microstructural Control of Isothermally Solidified YAG-Spinel-Alumina Composites 271
Shuqiang Wang, Takashi Akatsu, Yasuhiro Tanabe, and Eiichi Yasuda

Precipitation of $YAlO_3$ in YAG-Spinel Eutectic Composites with YAG-Rich Composition 285
Shuqiang Wang, Takashi Akatsu, Yasuhiro Tanabe, and Eiichi Yasuda

Microstructural Characteristics of YAG-Spinel Eutectic Composites with Changed Compositions 295
Shuqiang Wang, Takashi Akatsu, Yasuhiro Tanabe, and Eiichi Yasuda

Grain Boundary Phase in α/β-SiAlON Composites 307
Tzer-Shin Sheu

Eutectic Precipitation in Spinel-YAG Composite with Off-Eutectic Composition 315
Shuqiang Wang, Takashi Akatsu, Yasuhiro Tanabe, and Eiichi Yasuda

Thermodynamics of the Mg-C-N-O System for Evaluation of MgO-C Bricks 329
Siang-Chun Liu

Environmental Effects and Creep Behavior

Stability of BN Interphases and Their Effect on the Deformation of Fiber Reinforced Ceramic Matrix Composite Materials 341
Margaret M. Stackpoole, Rajendra K. Bordia, Charles H. Henager Jr., and Russell H. Jones

Environmentally Induced Failure Mechanism Mapping for Continuous-Fiber, Ceramic Composites 351
C.A. Lewinsohn, C.H. Henager Jr., and R.H. Jones

Compositional Dependence of the Oxidation Behavior of $SiC-ZrO_2$-Mullite Composites 361
Cheng-Yuan Tsai and Chien-Cheng Lin

High-Temperature Mechanical Properties of Advanced Nicalon Fibers and Their Relationship to Composite Failure 371
C.A. Lewinsohn , H. Serizawa, D.E. Johnston, G.E. Youngblood, R.H. Jones, and A. Kohyama

Flexural Creep and Creep Rupture of Monolithic Zircon and Zircon Silicon Carbide Fiber Composites 381
Umashankar Anandakumar and Raj N. Singh

Creep Deformation in Small Grain Size Al_2O_3+TiN Composite Polycrystals 393
Qiu Tai and Alain Mocellin

Damage Evolution, Residual Stress Measurement, and Micromechanical Modeling

Measuring Damage Evolution in Ceramic Matrix Composites Using Thermoelastic Stress Analysis. 407
Thomas J. Mackin and Mark C. Roberts

Study of Residual Stress in Sapphire-Fiber Reinforced Glass Matrix Composite Using Fluorescence Spectroscopy 417
D. Banerjee, Raj N. Singh, H. Rho, and H.E. Jackson

A Novel Approach for the Fiber Bridging Analysis in Ceramic Matrix Composites 429
Yongjian Sun and Raj N. Singh

A 2-D Model for Crack Deflection in Thermoelastically Loaded Inclusion Containing Ceramics. 439
Marc Genevrier and Alain Mocellin

The Effect of Fabric Orientation on the Flexural Behavior of CFCCs: Experiment and Theory 451
N. Miriyala, P.K. Liaw, C.J. McHargue, A. Selvarathinam, and L.L. Snead

Index 463

Preface

Ceramic composites are leading candidate materials for high-temperature structural applications. This proceedings volume contains papers given at a symposium on ceramic matrix composites held during the 100th Annual Meeting and Exposition of The American Ceramic Society in Cincinnati, Ohio, May 3–6, 1998. This symposium provided an international forum for scientists and engineers to discuss all aspects of ceramic composites. A total of 98 papers, including invited talks, oral presentations, and posters, were presented from 16 countries (the United States, Australia, Belgium, Brazil, France, Germany, Italy, Japan, Mexico, Norway, the People's Republic of China, Republic of Korea, Russia, Spain, Sweden, and Taiwan). The speakers represented universities, industry, and government research laboratories.

This volume contains 40 invited and contributed papers on various aspects of ceramic matrix composites, all peer-reviewed according to The American Ceramic Society procedures. The latest developments are covered in the areas of fiber reinforcements, coatings and interfaces, composite processing and characterization, phase relationships and microstructures in composite systems, environmental effects and creep behavior, damage evolution, residual stress measurement, and micromechanical modeling. The result is that the most important aspects necessary for understanding and further development of ceramic composites are discussed.

The organizers are grateful to all participants and session chairs for their time and effort, to authors for their timely submissions and revisions of the manuscripts, and to reviewers for their valuable comments and suggestions; without the contributions of all involved, this volume would not have been possible. Financial support from The American Ceramic Society is gratefully acknowledged. Thanks are due to the staff of the Meetings and Publications Department of The American Ceramic Society for their tireless efforts. Especially, we greatly appreciate the helpful assistance and cooperation of Sarah Godby throughout the production process of this volume. Finally, special thanks are extended to Jae Hong Cheon for his invaluable help during organization of the symposium and preparation of the proceedings.

J.P. Singh
Narottam P. Bansal

Fibers, Coatings, and Interfaces

TENSILE STRENGTH AND MICROSTRUCTURE OF HI-NICALON FIBERS EXTRACTED FROM CELSIAN MATRIX COMPOSITES

Narottam P. Bansal
National Aeronautics and Space Administration
Lewis Research Center
Cleveland, Ohio 44135-3191

ABSTRACT

Tensile strengths of uncoated and BN/SiC coated Hi-Nicalon fibers, extracted from celsian matrix composites by dissolving away the matrix in HF, were determined. Strengths of uncoated and BN/SiC coated Hi-Nicalon fibers before and after HF treatment were also measured. The elemental composition and thickness of the fiber coatings were determined using scanning Auger analysis. Microstructural analyses of the fibers and the coatings were done by SEM and TEM. The average tensile strength of uncoated Hi-Nicalon was 3.19±0.73 GPa with a Weibull modulus of 5.41. The Hi-Nicalon/BN/SiC fibers showed an average strength of 3.04±0.53 GPa and Weibull modulus of 6.66. After HF treatments, the average strengths of the uncoated and BN/SiC coated Hi-Nicalon fibers were 2.69±0.67 GPa and 2.80±0.53 GPa and the Weibull moduli were 4.93 and 5.96, respectively. The BN/SiC coated fibers extracted from the celsian matrix composite exhibited a strength of 2.38±0.40 GPa and a Weibull modulus of 7.15. Strength of the uncoated Hi-Nicalon fibers in the composite was so severely degraded that they disintegrated into small fragments during extraction with HF. The uncoated fibers probably undergo mechanical surface damage during hot pressing of the composites. The BN layer acts as a compliant layer which protects the coated fibers from mechanical damage during composite processing.

1. INTRODUCTION

Hi-Nicalon fiber [1] has a desirable combination of tensile strength, elastic modulus, density, and retention of these properties at temperatures up to ~1200 °C. Thus it is suitable for a variety of aerospace applications as a reinforcement in high performance structural composites with ceramic, metal, and polymer matrices. The Hi-Nicalon fiber surface needs to be protected with appropriate ceramic coating(s) in order to alleviate chemical reactions with ceramic and metal matrices during composite processing and use, and also, to provide a weak fiber-matrix interface for toughened ceramic composites.

In a recent study [2] it was observed that the $0.75BaO\text{-}0.25SrO\text{-}Al_2O_3\text{-}2SiO_2$ (BSAS) celsian matrix composite reinforced with Hi-Nicalon (uncoated) fibers showed low flexure strength along with catastrophic failure. In contrast, the composite reinforced with BN/SiC coated Hi-Nicalon fibers was strong, tough and showed graceful failure. Fiber push-through tests and microscopic examinations indicated a weak fiber-matrix interface in both composites, and no chemical reaction was observed between the fibers and the matrix. It was postulated that the low strength of the Hi-Nicalon (uncoated)/celsian composite was due to degradation of fiber strength during composite processing.

To the extent authorized under the laws of the United States of America, all copyright interests in this publication are the property of The American Ceramic Society. Any duplication, reproduction, or republication of this publication or any part thereof, without the express written consent of The American Ceramic Society or fee paid to the Copyright Clearance Center, is prohibited.

The primary objective of the present research was to verify the postulate of this earlier study. Another objective was to carry out microstructural and chemical analyses of the fibers and the coatings. Room temperature tensile strengths of the as-received Hi-Nicalon fibers and those coated with a dual BN/SiC surface layer were measured. Fibers were also extracted from the celsian matrix composites by leaching away the oxide matrix in 40% HF in order to investigate processing related fiber strength degradation. Strength of fibers treated with 40% HF were also determined. Elemental compositions and thicknesses of the fiber coatings were determined by scanning Auger analysis. Microstructural analyses of the fibers and the coatings were done by scanning electron microscopy (SEM) and transmission electron microscopy (TEM).

2. MATERIALS

Hi-Nicalon fiber from Nippon Carbon Co. was used in the present study. The properties of the as-received fiber are shown in Table I. These fibers contain nanocrystalline (10-20 nm) β-SiC along with 30-40 atomic % excess carbon which is also reflected in its low elastic modulus. A large variation (10-18 μm) in fiber diameter was observed; the manufacturer reports an average value of ~14 μm.

Table I. Properties of Hi-Nicalon fibers[a]

Property	Value
Density, g/cm^3	2.74
Diameter, μm	14
Filaments/tow	500
Denier, g/9000m	1800
Tensile strength, GPa	2.8
Elastic modulus, GPa	269
Thermal expansion coefficient, 10^{-6}/K	3.5 (25-500 °C)
Specific heat, J/g-K	0.67 (25 °C); 1.17 (500 °C)
Thermal conductivity, W/m-K	7.77 (25 °C); 10.1 (500 °C)
Electrical resistivity, Ω-cm	1.4
Chemical composition, wt %	63.7Si, 35.8C, 0.5O
C/Si, atomic ratio	1.3-1.4
Temperature capability, °C	1200

[a]Data from Nippon Carbon Co.

The polyvinyl alcohol (PVA) sizing on the as-produced Hi-Nicalon fibers was burned off in a Bunsen burner flame. Single filaments were carefully separated from the fiber tow for testing. All the coatings on the fibers were applied by 3M Co. with a continuous chemical vapor deposition (CVD) reactor. The BN coating was deposited at ~1000 °C utilizing a proprietary precursor and was amorphous to partly turbostratic in nature. A thin overcoating of SiC was also applied to the BN-coated fibers. The nominal coating thickness was 0.4 μm of BN and 0.3 μm of SiC. Uncoated and BN/SiC coated fibers were extracted from celsian matrix composites by leaching away the matrix in 40% HF for 24 h. The BN/SiC coated and uncoated Hi-Nicalon fibers were also treated with 40% HF for 24 h to determine if HF leaching affected fiber strength.

3. EXPERIMENTAL PROCEDURES

3.1. Electron Microscopy

Fiber surfaces were examined using SEM and TEM. For cross-sectional analysis, fibers were mounted in a high temperature epoxy and polished before examination. Fiber cross-sectional thin foils for TEM were prepared using a procedure developed for ceramic fibers which involves epoxy potting, slicing, polishing, dimple grinding, and Ar ion beam milling. A thin carbon coating was evaporated onto the TEM thin foils and SEM specimens for electrical conductivity prior to analysis. The thin foils were examined in a Philips EM400T operating at 120 keV. SEM was performed using a JEOL JSM-840A operating at 15 keV.

3.2. Scanning Auger Analysis

The elemental compositions of the fiber near the surface and of the fiber surface coatings were analyzed [3] with a scanning Auger microprobe (Fisons Instruments Microlab Model 310-F). The fibers for this analysis were mounted on a stainless steel sample mount by tacking the ends with colloidal graphite. Depth profiling was performed by sequential ion-beam sputtering and Auger analysis. The ion etching was done with 3 keV Argon ions rastered over an ~1 mm^2 area on the specimen. The etch rate in Ta_2O_5 under these conditions was 0.05 nm/s.

Auger electron spectroscopy (AES) analysis was performed using an electron beam current of ~1.5 nA. The beam was rastered over a 2×20 μm^2 area of the fiber with the long axis of the area aligned with the long axis of the fiber. Spectra were acquired in integral (as opposed to derivative) form at a beam energy of 2 keV and depth profiles were created by plotting peak areas against ion etch time. The atomic concentrations were calculated by dividing the peak areas by the spectrometer transmission function and the sensitivity factors for each peak, then scaling the results to total 100%. The sensitivity factors were derived from spectra of ion etched Ag, Si, B, Ti, SiC, BN, and TiO_2 standards. The sensitivity factors used for each element should not be trusted to better than ±20%. The depth scale is from the Ta_2O_5 calibration and no attempt has been made to adjust for the actual etch rate for each material. Only the fibers with a smooth surface coating, rather than those having thick and rough coating morphologies, were used for Auger analysis.

3.3. Tensile Strength Measurement

Room temperature tensile strengths of the individual filaments were measured in ambient atmosphere with a screw driven Micropull fiber test frame equipped with pneumatic grips. A 1000 g load cell, Sensotec model 34, was used. Measurements were carried out at a constant crosshead speed of 1.261 mm/min (0.05 in./min). A single filament was mounted on a 20 lb paper tab using Hardman extra fast setting epoxy. The side portions of the tab were cut with a hot wire just before application of the load producing a fiber gage length of 2.54 cm (1 in.). Twenty filaments of each type of fiber were tested. For the as-received fibers, actual diameter of each filament tested was measured using a Zeiss Axioskop microscope with a Sony DXC-960MD video camera attached to a Boeckeler VIA-150 video measurement system. Magnification ranges from 100X to 800X, depending upon the size of the fiber. Weibull parameters for tensile strength of as-received fibers were calculated using the measured diameters and also an average diameter of 13.5 μm. For all other fibers tested, tensile strengths were calculated based only on an average filament diameter of 13.5 μm.

4. RESULTS AND DISCUSSION

4.1. Microstructure

SEM micrographs showing the surface of as-received, flame desized Hi-Nicalon fibers are presented in Fig. 1. The fiber surface appears to be fairly smooth and featureless. EDS compositional spectra (Fig. 2) taken from the polished cross-section of the fiber indicate the

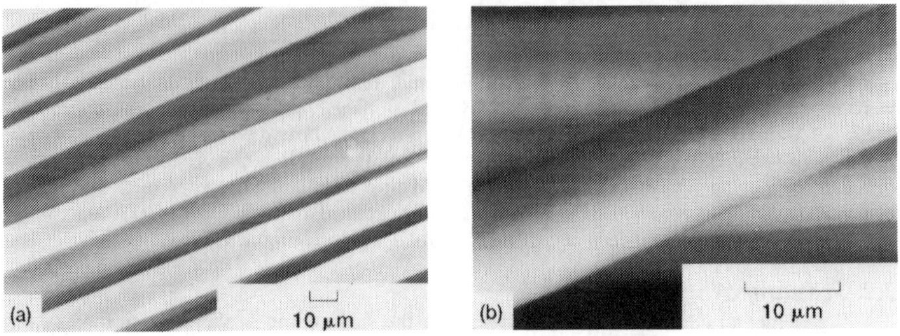

Figure 1.—SEM micrographs showing surface of desized Hi-Nicalon fibers.

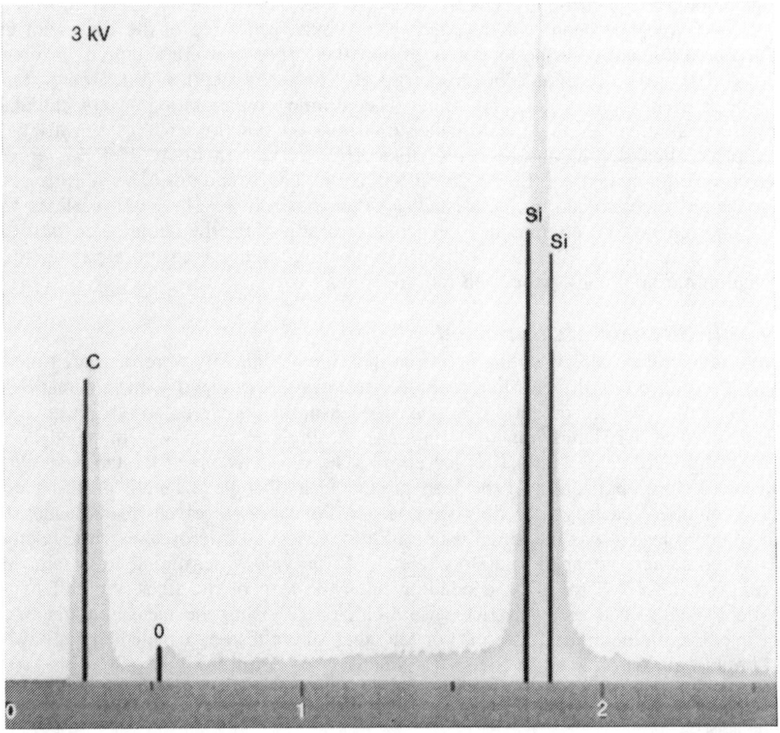

Figure 2.—EDS spectra from the cross-section of desized Hi-Nicalon fiber at 3 kV.

presence of only Si and C along with a small amount of oxygen which is in qualitative agreement with the manufacturer's data (Table I). The surface of HF-treated Hi-Nicalon fibers is also fairly smooth and featureless as seen from the SEM micrograph in Fig. 3. SEM micrographs from the surface and polished cross-section of the BN/SiC coated Hi-Nicalon fibers are given in Figs. 4 (a) and (b), respectively. The coating on most of the fibers is smooth and uniform, but some fibers have a quite granular coating. The BN coating is often nodular as can be seen in Fig. 5. The nodules in the BN layer remain or, often, are enhanced by the subsequent overcoating of SiC. The number density of the nodules varies considerably from filament to filament, but is more consistent along an individual fiber. Within a tow, the fibers on the outside had thicker and more granular coatings than those on the inside. Overall, the coating thickness was reasonably uniform. Generally, the BN coating thickness was 0.2-0.4 μm, but could be much higher at the nodules. The total coating thickness on a fiber could be as high as 7 μm when a granular cluster is attached (Fig. 5). Typically, the SiC coating thickness varied between 0.2-0.3 μm. Sharp contrast is observed between the fiber, BN coating, and SiC coating in polished cross-sections (Fig. 6). The microstructure of these coatings is quite similar to those of BN/SiC coatings on HPZ ceramic fibers [3].

TEM analysis of the loose, BN/SiC coated Hi-Nicalon fibers in cross-section revealed four distinct layers in the BN coating (Fig. 7). While there was no detectable variation in composition between the layers, there was a subtle variation in crystallographic structure. All four layers were turbostratic, but layers 1 and 3 (counting from the fiber outward) were more directional. The basic structural units of the BN are more aligned parallel to the fiber surface, giving sharper cusps in microelectron diffraction patterns (Fig. 8). The SiC layer on the as-coated fibers typically ion milled away before electron transparent thin regions were obtained, but appeared to be adherent to the BN layers.

After treatment with HF, the BN/SiC surface coating was cracked, and spalled off at some places on the fibers, as seen in the SEM micrographs in Fig. 9. SEM micrographs showing surfaces of the uncoated and BN/SiC coated fibers extracted from the celsian composite are given in Figs. 10 and 11, respectively. Matrix particles are still sticking to many of the fibers and cracks in the coating are observed on those fibers having a thick granular coating.

4.2. Scanning Auger Analysis

Elemental composition depth profiles, as obtained from scanning Auger analysis of as-received, but desized, Hi-Nicalon fibers and those having BN/SiC surface coatings are shown in Figs. 12 (a) and (b), respectively. Depth profiles of the uncoated and coated fibers, after treatment with HF, are also presented in Figs. 12 (c) and (d), respectively. The elemental composition (atomic %) of the as-received fiber is ~42% Si and 56% C with C/Si atomic ratio of 1.33. A trace amount of oxygen is also detected. These results are in good agreement with the fiber composition supplied by the manufacturer (Table I). The composition of the HF treated Hi-Nicalon fiber is found to be ~43% Si, ~54% C, and a trace of oxygen. The Hi-Nicalon/BN/SiC fiber has ~130 nm thick layer of Si-rich SiC followed by ~500 nm thick layer of B-rich BN. However, the coating thickness was non-uniform from filament to filament as seen from the SEM micrographs in Figs. 4 and 5. The fibers on the outside of a tow had thicker and more granular coatings than those on the inside. The BN coating is contaminated with oxygen and contains a high concentration of carbon. In fact, the concentration of carbon is larger than those of boron and nitrogen. The results for the HF treated Hi-Nicalon/BN/SiC fiber are quite similar to those for the untreated fiber. However, a trace amount of F is also detected in the BN layer of the treated fiber.

4.3. Tensile Strength

The strength of ceramic fibers is determined by the statistical distribution of flaws in the material. The tensile strength of ceramic fibers is generally analyzed on the basis of the well

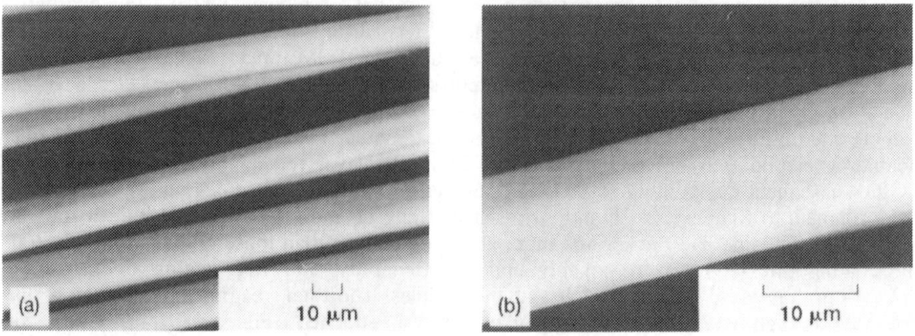

Figure 3.—SEM micrographs showing surface of Hi-Nicalon fiber after treatment with HF.

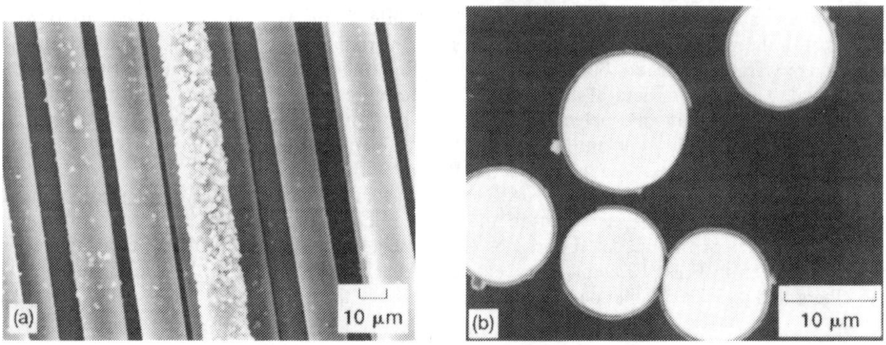

Figure 4.—SEM micrographs. (a) Surface. (b) Polished cross-section of BN/SiC coated Hi-Nicalon fibers.

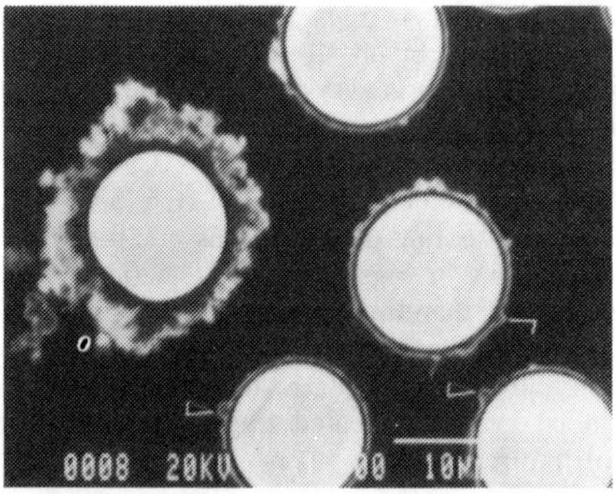

Figure 5.—SEM micrograph from polished cross-section of BN/SiC coated Hi-Nicalon fibers showing nodular coating. Some of the nodules are indicated by arrows.

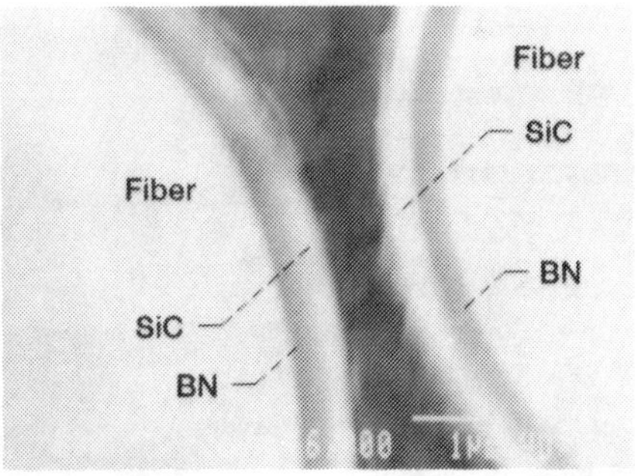

Figure 6.—SEM micrograph at high magnification from polished cross-section of BN/SiC coated Hi-Nicalon fibers showing sharp contrast between the fiber and the coatings.

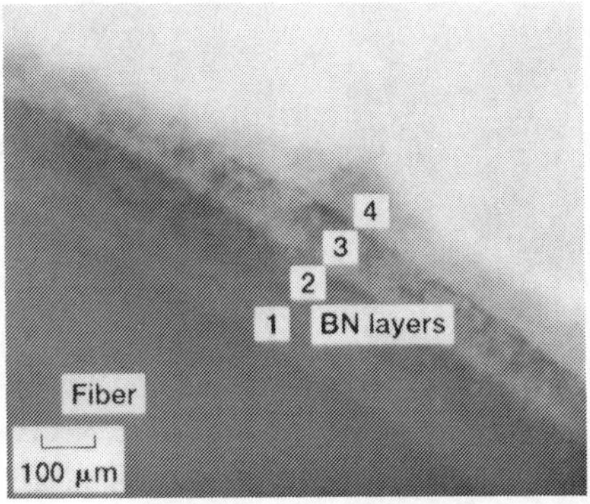

Figure 7.—TEM bright field image showing cross-section of the as-received BN/SiC coated Hi-Nicalon fiber; four distinct sublayers are visible within the BN coating.

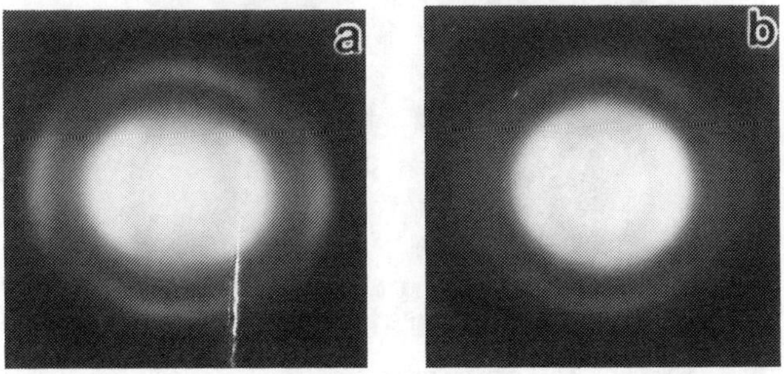

Figure 8.—Microelectron diffraction patterns typical for BN layers in BN/SiC coated Hi-Nicalon fibers. (a) Layers 1 and 3. (b) Layers 2 and 4.

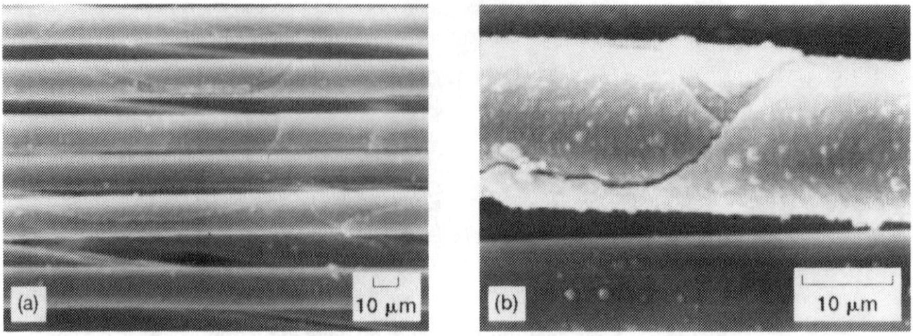

Figure 9.—SEM micrographs showing surface of BN/SiC coated Hi-Nicalon after treatment with HF.

Figure 10.—SEM micrographs showing surface of uncoated Hi-Nicalon fibers extracted from celsian matrix composite by dissolving away the matrix in HF.

Figure 11.—SEM micrograph showing surface of BN/SiC coated Hi-Nicalon fibers extracted from celsian matrix composite by dissolving away the matrix in HF.

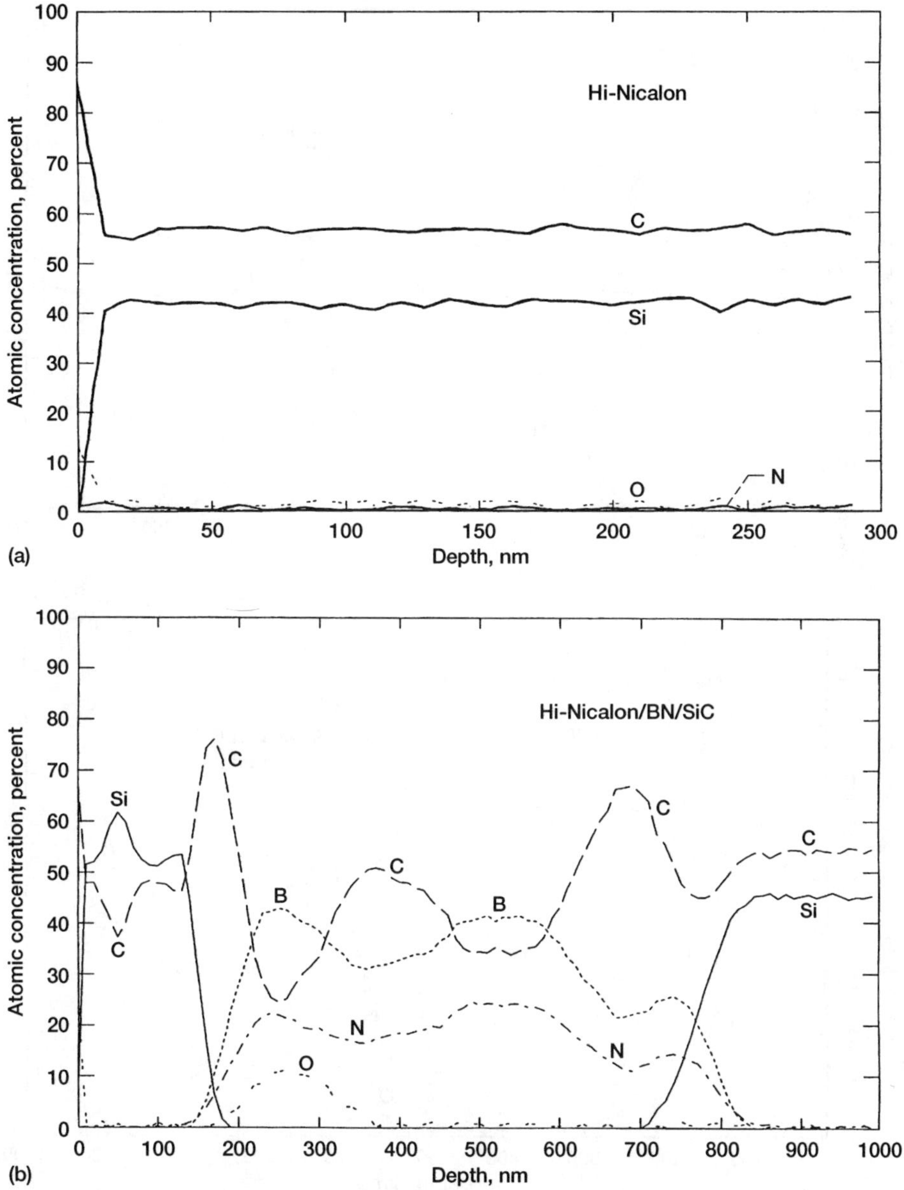

Figure 12.—Surface depth profiles of various elements obtained using scanning Auger microprobe. (a) Hi-Nicalon. (b) Hi-Nicalon/BN/SiC. (c) HF treated Hi-Nicalon. (d) HF treated Hi-Nicalon/BN/SiC fibers.

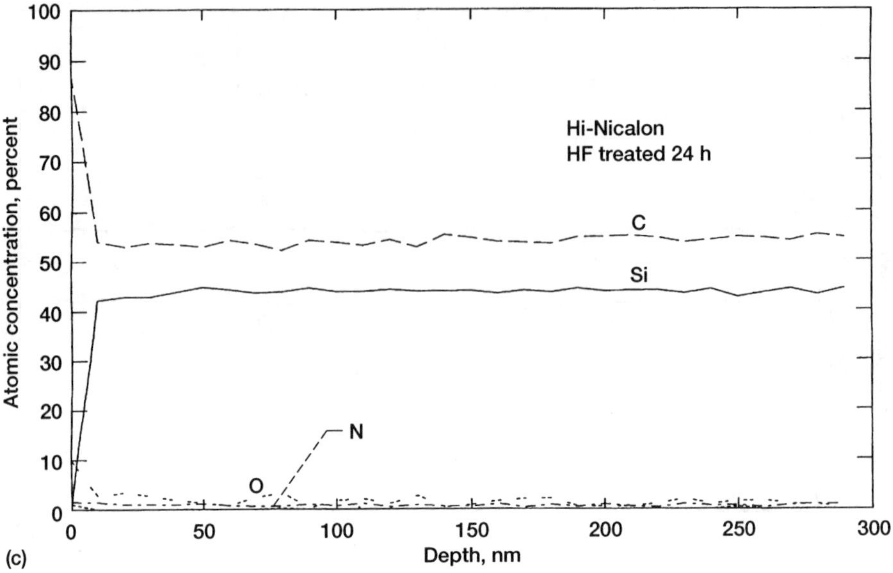

(c)

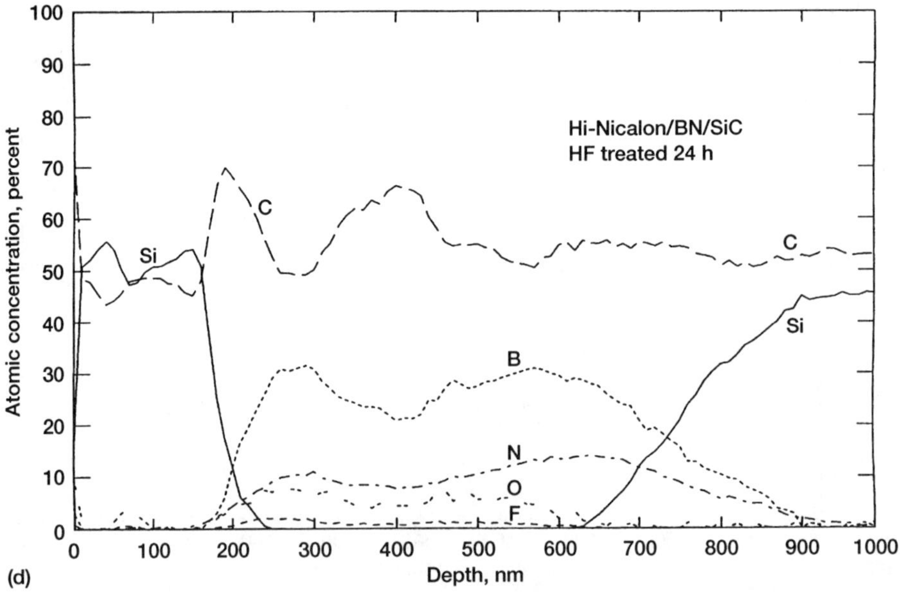

(d)

Figure 12.—Concluded. (c) HF treated Hi-Nicalon. (d) HF treated Hi-Nicalon/BN/SiC fibers.

known Weibull statistics [4]. An excellent review of the subject was provided by Van der Zwaag [5]. The Weibull modulus describes the distribution of strength in materials which fail at defects according to weakest link statistics. The probability of survival of a material is given by the empirical equation:

$$P_s = 1 - P(s) = \exp[-V\{(s - s_u)/s_o\}^m] \quad (1)$$

where P_s is the survival probability ($P(s) = 1 - P_s$ is the failure probability) of an individual fiber at an applied stress of s, V is the fiber volume, s_u is the stress below which failure never occurs, s_o is the scale parameter, and m is the shape or flaw dispersion parameter. For a given material, the Weibull modulus m and s_o are constant. Assuming $s_u = 0$ and uniform fiber diameter along the length, L, corresponding to the volume, V, eq. (1) can be rearranged as:

$$\ln \ln (1/P_s) = \ln \ln [1/(1 - P(s))] = m \ln s + \text{constant} \quad (2)$$

The experimental data are ranked in ascending order of strength values and the cumulative probability $P(s_i)$ is assigned as

$$P(s_i) = i/(1 + N) \quad (3)$$

where i is the rank of the tested fiber in the ranked strength tabulation and N is the total number of fibers tested. A least-squares linear regression analysis is then applied to a plot of $[\ln\ln 1/P_s]$ vs. ln(s) whose slope is the Weibull modulus m.

In has been recently reported [6, 7] that the strengths and Weibull modulus of fibers obtained from measured and average fiber diameters are in good agreement. Therefore, the Weibull parameters for the tensile strengths of fibers tested in the present study were evaluated using an average filament diameter of 13.5 μm. The $\ln \ln (1/P_s)$ vs. ln s Weibull probability plots for room temperature tensile strength of uncoated Hi-Nicalon fibers and those having a dual BN/SiC surface coating are shown in Fig. 13. The values of Weibull parameters obtained from linear regression analysis are given in Table II. The tensile strength of as-received Hi-Nicalon fiber is 3.19±0.73 GPa and the Weibull modulus, m, is 5.41. The manufacturer's information data sheet [1] reports a value of 2.8 GPa for the room temperature tensile strength of these fibers. The BN/SiC coated Hi-Nicalon fibers show an average tensile strength of 3.04±0.53 GPa with m value of 6.66 indicating no strength degradation during fiber coating by CVD.

Weibull plots for the uncoated and BN/SiC coated Hi-Nicalon fibers which have been treated with 40% HF for 24 h and those extracted from the celsian matrix composite are given in Fig. 14 and the values of Weibull parameters are summarized in Table II. After HF treatment, both the as-received and the BN/SiC coated fibers show strength degradation of ~8-15 %. The tensile strength of the BN/SiC coated fibers extracted from the celsian matrix composite by dissolving away the matrix in HF has degraded to 2.38±0.40 GPa with an m value of 7.15. Exposure of the fibers to high temperature during hot pressing of the composites could probably account for this strength degradation. In contrast, the fibers extracted from the Hi-Nicalon (uncoated)/celsian composite were so weak that they disintegrated into small pieces during extraction with HF. The strength of these uncoated fibers was probably lost due to mechanical surface damage during composite hot pressing. The BN layer in the BN/SiC coated fibers acts as a compliant layer protecting the fibers from mechanical damage during composite processing. Uncoated Hi-Nicalon fibers extracted from lithium aluminosilicate (LAS) glass-ceramic matrix composites [8], which were hot pressed at 1360 °C for 40 min,

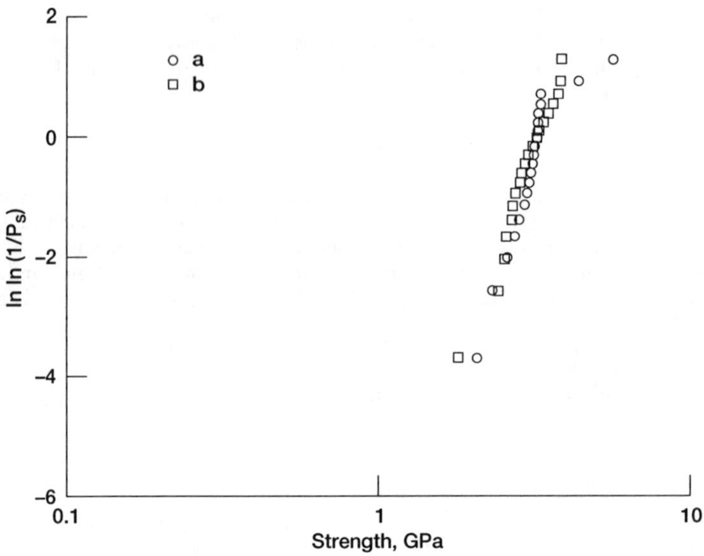

Figure 13.—Weibull probability plots for room temperature tensile strengths of fibers. (a) Hi-Nicalon. (b) BN/SiC coated Hi-Nicalon.

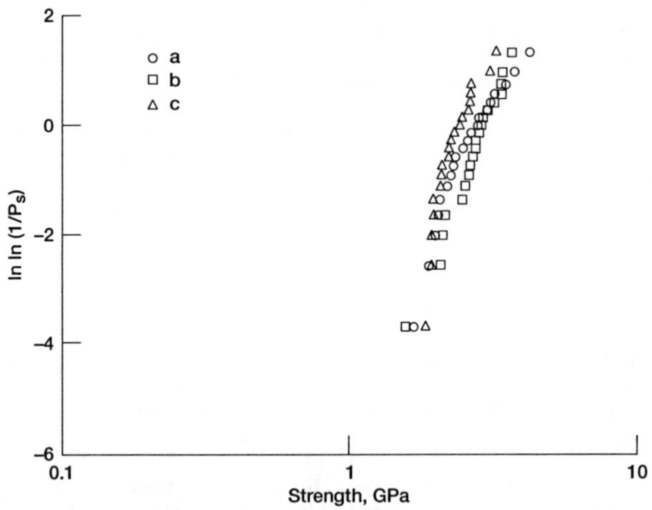

Figure 14.—Weibull probability plots for room temperature tensile strengths of chemically treated fibers. (a) HF treated Hi-Nicalon. (b) BN/SiC coated Hi-Nicalon treated with HF. (c) BN/SiC coated Hi-Nicalon extracted from celsian matrix composite by leaching away matrix in HF.

Table II. Room temperature strength of Hi-Nicalon fibers[a]
[Gage length = 2.54 cm; Crosshead speed = 1.261 mm/min]

Fiber treatment	Average strength, GPa	Std. dev., GPa	Weibull modulus, m
As-received, flame desized	3.19	0.73	5.41
As-received, flame desized; HF treated, 24 h	2.69	0.67	4.93
BN-SiC coated	3.04	0.53	6.66
BN-SiC coated; HF treated, 24 h	2.80	0.53	5.96
Uncoated, extracted from composite[b]	---	---	---
BN-SiC coated, extracted from composite by dissolving away matrix in HF, 24 h	2.38	0.40	7.15

[a]An average fiber diameter of 13.5μm used; 20 filaments tested for each fiber type.
[b]Fragmented into small pieces during extraction with HF; could not measure strength.

showed only 20-25% reduction in tensile strength. This is because the LAS composites were densified for a short time just above the matrix melting point making use of the viscous flow of glass. In contrast, the celsian matrix composites were hot pressed at temperatures much below the matrix melting point.

5. SUMMARY OF RESULTS

The average tensile strength of as-received Hi-Nicalon was 3.19±0.73 GPa with a Weibull modulus of 5.41. The BN/SiC coated Hi-Nicalon fibers showed an average strength of 3.04±0.53 GPa and Weibull modulus of 6.66. After HF treatments, the average strengths of the uncoated and BN/SiC coated Hi-Nicalon fibers were 2.69±0.67 GPa and 2.80±0.53 GPa and the Weibull modulus were 4.93 and 5.96, respectively. The BN/SiC coated fibers extracted from celsian matrix composite, by leaching away the matrix in HF, exhibited a strength of 2.38±0.40 GPa and Weibull modulus of 7.15. The strength of the uncoated Hi-Nicalon fibers in the composite had been so severely degraded that they disintegrated into small pieces during extraction with HF. This strength degradation is probably caused by mechanical surface damage to the uncoated fibers during hot pressing of the composites. The elemental composition and thickness of the fiber coatings were determined using scanning Auger analysis. HF treatment had no effect on composition of the fibers. SEM analyses of the fibers and the coatings gave no evidence that HF treatment damaged the fibers.

6. CONCLUSIONS

A premise of an earlier study was that the low strength of uncoated Hi-Nicalon fiber-reinforced celsian matrix composite could be due to mechanical damage to fibers during hot pressing. In contrast, composites made with BN/SiC coated fibers had strength levels one might expect based on as-produced fiber properties. From this study of fibers leached from composites it is concluded that the premise was correct. The BN coating in the BN/SiC coated Hi-Nicalon fibers acts as a compliant layer protecting the fibers from mechanical damage during composite hot pressing.

Acknowledgments

Thanks are due to Dan Gorican for fiber strength measurements, John Setlock for SEM, Rob Dickerson for TEM, and Darwin Boyd for scanning Auger analysis.

REFERENCES

[1]M. Takeda, J. Sakamoto, A. Saeki, and H. Ichikawa, "High Performance Silicon Carbide Fiber Hi-Nicalon for Ceramic Matrix Composites," *Ceram. Eng. Sci. Proc.*, **16**[4], 37–44 (1995).

[2]N.P. Bansal and J.I. Eldridge, "Hi-Nicalon Fiber-Reinforced Celsian Matrix Composites: Influence of Interface Modification," *J. Mater. Res.*, **13**, 1998.

[3]N.P. Bansal and R.M. Dickerson, "Tensile Strength and Microstructural Characterization of HPZ Ceramic Fibers," *Mater. Sci. Eng.*, **A222** [2], 149–157 (1997).

[4]W.A. Weibull, "A Statistical Distribution Function of Wide Applicability," *J. Appl. Mech.*, **18** [9], 293–297 (1951).

[5]S. Van der Zwaag, "The Concept of Filament Strength and the Weibull Modulus," *J. Testing Eval.*, **17** [5], 292–298 (1989).

[6]M.D. Petry, T.-I. Mah, and R.J. Kerans, "Validity of Using Average Diameter for Determination of Tensile Strength and Weibull Modulus of Ceramic Filaments," *J. Am. Ceram. Soc.*, **80** [10], 2741–2744 (1997).

[7]N.P. Bansal, "Effects of HF Treatments on Tensile Strength of Hi-Nicalon Fibers," NASA TM–1998-206626, February 1998.

[8]W.K. Tredway, "Hi Nicalon SiC Fiber Reinforced Glass and Glass-Ceramic Matrix Composites," *Ceram. Eng. Sci. Proc.*, **17**[4], 291–298 (1996).

DURABILITY OF MULLITE/YSZ-COATED SiC IN 90% H_2O/O_2

Kang N. Lee*
Chemical Engineering Department
Cleveland State University
Cleveland, OH 44115

Robert A. Miller
NASA Lewis Research Center
Cleveland, OH 44135

ABSTRACT

Plasma-sprayed mullite ($3Al_2O_3 \cdot 2SiO_2$) and mullite/yttria-stabilized-zirconia (YSZ) dual layer coatings have been developed to protect silicon-based ceramics from environmental attack. Mullite-based coating systems show excellent long-term durability under thermal cycling in air. These systems, however, suffer from water-vapor enhanced oxidation of SiC in high water vapor environment. Water vapor transports through the cracks in the mullite-based coatings and attacks the SiC along the interface. In an effort to improve the durability of mullite-based coating systems in water vapor, the mullite/SiC interface was modified by applying a thin layer (1-5 μm) of Mo which was subsequently annealed in inert atmosphere to form molybdenum silicide. Some areas of the mullite/SiC interface, where the molybdenum silicide was intact after water vapor exposure, showed significantly improved durability, suggesting the potential of interface modification for improving the water vapor resistance of mullite-based coating systems.

INTRODUCTION

Silicon-based ceramics are prime candidates for hot section structural components of heat engines and heat exchanger tubes for industrial furnaces. One

* Resident Research Associate at NASA Lewis Research Center

To the extent authorized under the laws of the United States of America, all copyright interests in this publication are the property of The American Ceramic Society. Any duplication, reproduction, or republication of this publication or any part thereof, without the express written consent of The American Ceramic Society or fee paid to the Copyright Clearance Center, is prohibited.

potential barrier to such applications is their environmental durability. The excellent oxidation resistance of silicon-based ceramics at high temperatures in dry oxygen is due to the formation of a protective external silica scale. However, the normally protective silica scale can be degraded by reaction with impurities, such as alkali salts [1,2] or water vapor [3,4].

Molten Na_2SO_4 can deposit in gas turbine engines operating near marine environments or by contaminants from the fuel [5]. The Na_2SO_4 then partially decomposes and dissolves the protective silica scale, leading to accelerated degradation of SiC [1]. In coal-fired combustion environments, combustion gases contain low levels of alkali salts because of naturally occurring minor alkali components in the coal. These alkali salts can dissolve in the silica scale and enhance the transport of oxygen, leading to drastically enhanced oxidation [2]. In heat engines, substantial amounts of water vapor, typically about 10%, is produced from burning hydrocarbon fuels in air [6]. The water vapor reacts with silica, forming gaseous $Si(OH)_4$ [3]. In high pressure combustion environments, the higher water vapor pressure generates even higher levels of $Si(OH)_4$, resulting in linear volatilization of silica. The linear volatilization of silica coupled with the parabolic oxidation of SiC results in overall paralinear kinetics [4], causing a rapid degradation of SiC. Therefore, the realization of the full potential of silicon-based ceramics in high temperature structural applications depends on the development of environmental protection schemes.

Plasma-sprayed mullite ($3Al_2O_3 \cdot 2SiO_2$) and mullite/yttria-stabilized-zirconia (YSZ) dual layer coatings have been developed to protect silicon-based ceramics from the environmental attack [7-8]. Mullite coatings provide excellent protection in air [9,10] and molten salt environment [11]. They, however, suffer from selective vaporization of silica in the presence of water vapor [12]. In the mullite/YSZ coating system, the mullite coating provides the bonding while the YSZ coating provides the resistance to water vapor. This paper will discuss the durability of mullite/YSZ-coated SiC in water vapor and our recent efforts to improve the water-vapor resistance of mullite-based coating systems through interface modification.

EXPERIMENTAL

Mullite and YSZ coatings were applied by atmospheric pressure plasma spraying onto 2.5 x 0.6 x 0.15 cm sintered α-SiC coupons (Hexoloy^{TM},

Carborundum, Niagara Falls, NY). The SiC substrates were roughened (R_a ** = 5-6 μm) by etching in Na_2CO_3 to achieve a good mechanical bond [7]. Fused mullite powder with the particle size of 44-74 μm was used (Cerac, Inc., Milwaukee, WI). Typical coating thickness was 100-200 μm for the mullite coating and 50 μm for the YSZ coating. Details of the coating parameters are described elsewhere [7].

To modify the mullite/SiC interface, a thin flash layer (1-5 μm) of Ni, Mo, Y, Mg, or Zr was sputter coated on SiC and annealed in Ar-5% H_2 to promote the reaction between the flash layer and SiC, prior to the application of mullite coating. The annealing was performed at 1200 - 1350°C for 5 - 100h.

Coated coupons were exposed to 2h thermal cycling in air or 90% H_2O/O_2 at 1 atm. Thermal cycling was performed using an automated thermal cycling furnace. Each thermal cycle consisted of 2h at temperature, rapid cooling to low temperature, and 20 min at the low temperature (<100°C). Typically, samples reached the high temperature within 2 min and the low temperature within 5 min in each cycle.

Tested samples were mounted in epoxy, polished to 1 μm using diamond suspension, and examined using Scanning Electron Microscopy (SEM) and Energy Dispersive Spectroscopy (EDS).

RESULTS

Degradation of Mullite/YSZ-Coated SiC in Water Vapor

Mullite/YSZ coatings on SiC exhibit excellent long-term durability under thermal cycling in dry air. Figure 1 is the cross-section of mullite/YSZ-coated SiC after 500h at 1300°C with 2h cycles in air, showing excellent oxidation protection. They also provide excellent protection from water vapor attack in short-term exposures. Figure 2 is the cross-section of mullite/YSZ-coated SiC after 100h at 1300°C with 2h cycles in 90% H_2O/O_2. However, mullite/YSZ coatings fail to provide the protection from water vapor in extended exposures (>100h). Figure 3 shows the cross-section of mullite/YSZ-coated SiC after 200h at 1300°C with 2h cycles in 90% H_2O/O_2. Note the formation of thick porous silica scale on the SiC along the mullite/SiC interface. The thick porous silica scale is typical of the scale formed on uncoated SiC in high water vapor environment.

** Average distance from the roughness profile to the mean line

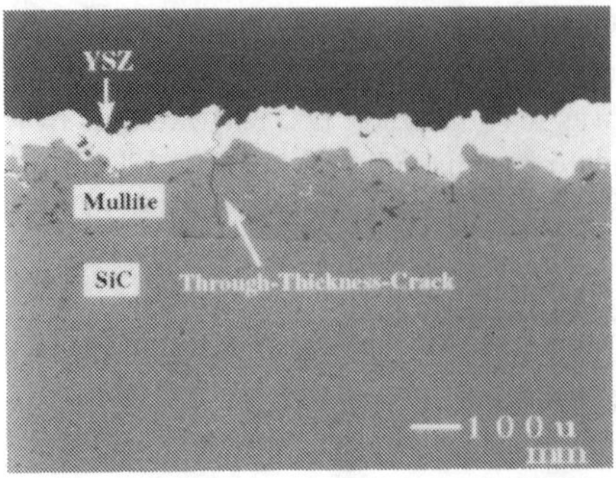

Fig. 1. Mullite/YSZ-coated SiC after 500h with 2h cycles at 1300°C in air, showing excellent oxidation resistance.

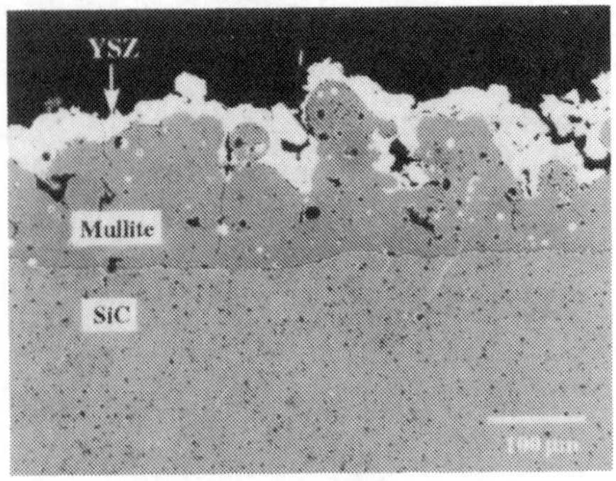

Fig. 2. Mullite/YSZ-coated SiC after 100h with 2h cycles at 1300°C in 90% H_2O/O_2, showing excellent oxidation resistance.

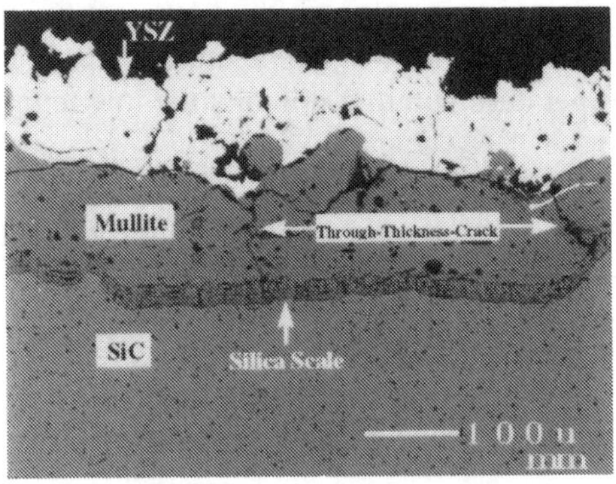

Fig. 3. Mullite/YSZ-coated SiC after 200h with 2h cycles at 1300°C in 90% H_2O/O_2, showing water vapor-enhanced oxidation of SiC along the interface.

As-sprayed mullite coatings (Figure 4) are free of macrocracks (crack width >1 μm). However, in thermal exposure, they develop through-thickness-cracks. Some cracks are as wide as 5-10 μm. Water vapor appears to transport through these cracks and attack the SiC. Figure 5 shows an area where the water vapor-enhanced oxidation has just begun at the mullite/SiC interface near through-thickness-cracks. The silica scale formed in water vapor is porous, allowing the oxidation to propagate readily along the interface. This is in contrast to the scale formed in dry air where the silica scale at the bottom of cracks is dense, preventing the rapid propagation of oxidation.

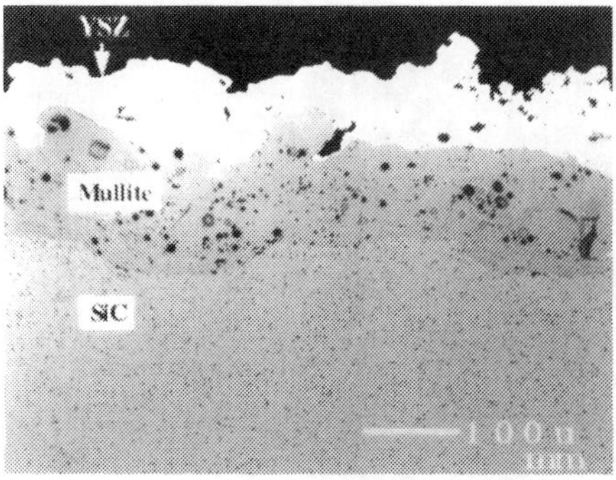

Fig. 4. As-sprayed mullite/YSZ coating on SiC.

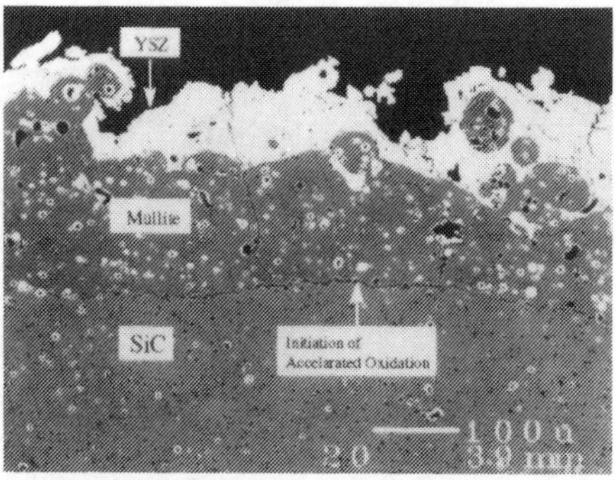

Fig. 5. Mullite/YSZ-coated SiC after 100h with 2h cycles at 1300°C in 90% H_2O/O_2, showing an area where the water vapor-enhanced oxidation has just begun.

It is believed that the development of through-thickness-cracks is due to stresses in the coating. The most likely source for the stresses is the precipitation of various second phases in the mullite coating. Amorphous phase mullite precipitates in the coating due to the rapid cooling of molten mullite [7]. We were able to eliminate most of the amorphous phase mullite by slowing down the cooling rate. This was accomplished by heating the SiC substrate above the crystallization temperature (~1000°C) of the amorphous phase mullite during the coating deposition [7]. However, it is likely that some residual amorphous phase mullite still remains in the coating. Volumetric shrinkage results during the crystallization of the residual amorphous phase mullite in subsequent thermal exposures. A significant amount of alumina also precipitates in the plasma-sprayed mullite coating (Fig. 6) [7]. The plasma-sprayed alumina typically contains substantial amount of metastable alumina phases such as γ-alumina [13]. Volumetric shrinkage results when the metastable alumina phases transform to stable α alumina in subsequent thermal exposures. The precipitation of the alumina phase is accompanied by the precipitation of silica-rich phases to maintain the chemical balance [7]. Both the alumina and silica-rich phases cause CTE mismatch stresses. Thus the precipitation of second phases in the mullite coating, the resultant volumetric shrinkage and the CTE mismatch are suggested to be the major sources for the stresses in the coating.

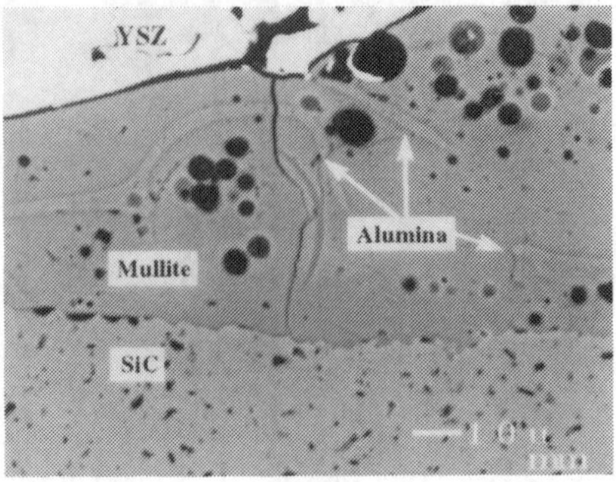

Fig. 6. As-sprayed mullite/YSZ coating on SiC, showing alumina second phase.

Modification of the Mullite/SiC Interface

Based on the degradation mode of the mullite-based coating system on SiC in water vapor, one promising approach to enhancing the durability of mullite coating in water vapor is to reduce the coating stresses. Mullite does not form a strong chemical bond with SiC according to our diffusion couple study [14]. Thus the mullite/SiC bond of as-sprayed coatings is mainly due to mechanical interlocking. The lack of chemical bond may be another reason why the oxidation propagates readily along the interface in water vapor. Therefore, another approach to enhancing the durability of mullite coating in water vapor is the modification of mullite/SiC interface to improve the bond.

In our effort to improve the mullite/SiC bond through the interface modification, a thin layer (1-5 μm) of Ni, Mo, Y, Mg, or Zr was sputter coated onto SiC and annealed in Ar-5% H_2, prior to the application of the mullite coating. All systems except for the Mo layer readily spalled during the annealing. Figures 7a and 7b show the as-coated and annealed Mo coating on SiC. It appears that silicon diffuses into the molybdenum forming molybdenum silicide with varying composition through the thickness. Similar results were reported in a Mo/SiC diffusion couple study [15,16]. The annealed coating is thicker and partially delaminated, presumably due to the volume expansion as the silicon diffuses into the molybdenum. Figure 8 shows the mullite/molybdenum silicide/SiC system after 500h at 1300°C with 2h cycles in 90% H_2O/O_2. Some areas of the coating, where the molybdenum silicide layer remained intact, showed excellent durability. Even the silica scale at the interface attacked by water vapor was thinner than the scale at the unmodified interface, indicating that the interfacial modification delayed the water vapor attack. This study suggests the potential of interface modification for improving the coating durability in water vapor. Future work will be focused on understanding the role of molybdenum silicide layer, in improving the coating durability, and on further exploring this approach.

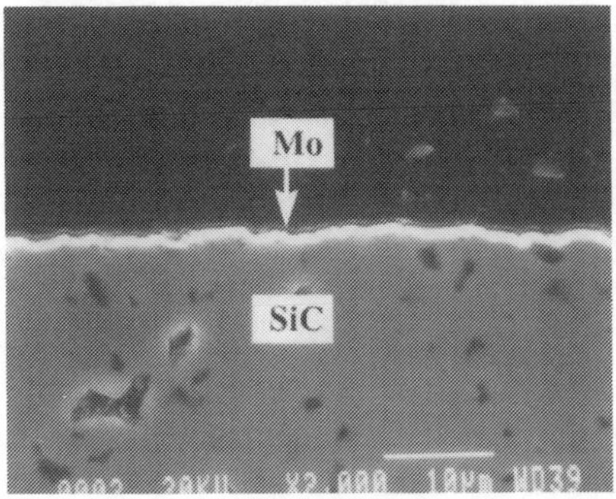

Fig. 7a. As-deposited Mo coating on SiC.

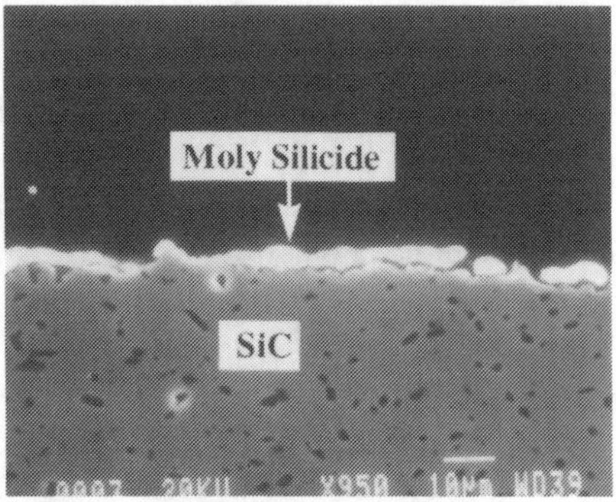

Fig. 7b. Mo coating converted to molybdenum silicide after 20h at 1300°C in Ar-5% H_2.

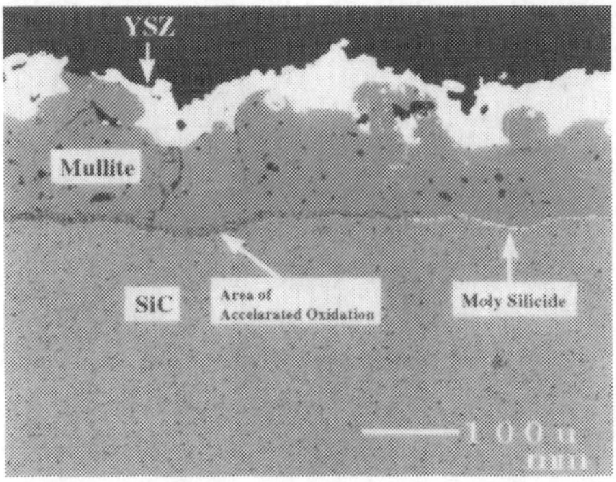

Fig. 8. Mullite/Moly Silicide/YSZ coating on SiC after 500h with 2h cycles in 90% H_2O/O_2.

CONCLUSIONS

Mullite-based coating systems on SiC suffer from the water-vapor enhanced oxidation of SiC in high water vapor pressure environments. Water vapor appears to transport through the cracks in the mullite-based coatings and attack the SiC along the interface. Interfaces modified with a thin molybdenum silicide layer converted from sputter coated Mo provided significantly improved water vapor resistance. This suggests the potential of interface modification for improving the water vapor resistance of mullite-based coating systems.

REFERENCES

1. N.S. Jacobson, J.L. Smialek, and D.S. Fox, "Molten Salt Corrosion of SiC and Si_3N_4,"; pp. 99-135 in *Handbook of Ceramics and Composites*, Vol. 1. Edited by N.S. Cheremisinoff. Marcel Dekker, New York, 1990.

2. V. Pareek and D.A. Shores, "Oxidation of Silicon Carbide in Environments Containing Potassium Salt Vapor," *J. Am. Ceram. Soc.*, **74**[3], 556-563 (1993).

3. A. Hashimoto, "The Effect of H_2O Gas on Volatilities of Planet-Forming Major Elements: I. Experimental Determination of Thermodynamic Properties of Ca-, Al-, and Si-hydroxide Gas Molecules and Its Application to the Solar Nebula," *Geochim. et. Cosmochim. Acta,* **56**, 511-32 (1992).

4. E.J. Opila and R. Hann, "Paralinear Oxidation of CVD SiC in Water Vapor," *J. Am. Ceram. Soc.*, **80**[1], 197-205 (1997).

5. F.S. Pettit and C.S. Giggins, *Superalloy II* (C.T. Sims, N.S. Stoloff and W.C. Hagel eds.), Wiley, New York, 327 (1987).

6. N.S. Jacobson, "Corrosion of Silicon-Based Ceramics in Combustion Environments," *J. Am. Ceram. Soc.*, **76**[1], 3-28 (1993).

7. K.N. Lee, R.A. Miller, and N.S. Jacobson, "New Generation of Plasma-Sprayed Mullite Coatings on Silicon-Carbide," *J. Am. Ceram. Soc.*, **78**[3], 705-10 (1995).

8. K.N. Lee, N.S. Jacobson, and R.A. Miller, "Refractory Oxide Coatings on SiC Ceramics," *MRS Bulletin*, **XIX**[10], 35-38 (1994).

9. K.N. Lee and R.A. Miller, "Oxidation Behavior of Mullite-Coated SiC and SiC/SiC Composites under Thermal Cycling between Room Temperature and 1200-1400°C," *J. Am. Ceram. Soc.*, **79**[3], 620-26 (1996).

10. K.N. Lee and R.A. Miller, "Development and Environmental Durability of Mullite and Mullite/YSZ Dual Layer Coatings for SiC and Si_3N_4 Ceramics,"

Surface and Coatings Technology, **86-87**, 142-48 (1996).

11. N.S. Jacobson, K.N. Lee, and T.Y. Yoshio, "Corrosion of Mullite By Molten Salts," *J. Am. Ceram. Soc.*, **79**(8), 2161-67 (1996).

12. K.N. Lee, R.A. Miller, N.S. Jacobson, and E.J. Opila, "Environmental Durability of Mullite Coating/SiC and Mullite-YSZ Coating/SiC Systems," *Ceram. Eng. & Sci. Proc.*, September-October, 1037-1044 (1995).

13. G.N. Heintz and U. Uematsu, "Preparation and Structures of plasma-Sprayed γ and α–alumina Coatings," *Surface and Coatings Technology,* **50** 213-22 (1992).

14. K.N.Lee, NASA Lewis Research Center, unpublished research.

15. A.E. Martinelli, R.A.L. Drew, "Microstructure Development during Diffusion Bonding of α-Silicon Carbide to Molybdenum," *Mat. Sci. And Eng. A*, **191**, 239-247 (1995).

16. A.E. Martinelli, R.A.L. Drew and R. Berriche, "Correlation Between the Strength of SiC-Mo Diffusion Couples and the Mechanical Properties of the Interfacial Reaction Products," *J. Mat. Sci. Letters*, **15**, 307-310 (1996).

OXIDATION RESISTANT COATINGS FOR CERAMIC MATRIX COMPOSITE COMPONENTS

Virginie M. Vaubert and David P. Stinton
Oak Ridge National Laboratory
P.O. Box 2008, Bldg. 4515, MS-6063
Oak Ridge, TN 37831-6063

Deidre A. Hirschfeld
Department of Materials & Metallurgical Engineering
159 Jones Hall
New Mexico Institute of Mining and Technology
Socorro, NM 87801

ABSTRACT

Corrosion resistant $Ca_{0.6}Mg_{0.4}Zr_4(PO_4)_6$ (CMZP) and $Ca_{0.5}Sr_{0.5}Zr_4(PO_4)_6$ (CS-50) coatings for fiber-reinforced SiC-matrix composite heat exchanger tubes have been developed. Aqueous slurries of both oxides were prepared with high solids loading. One coating process consisted of dipping the samples in a slip. A tape casting process has also been created that produced relatively thin and dense coatings covering a large area. A processing technique was developed, utilizing a pre-sintering step, which produced coatings with minimal cracking.

INTRODUCTION

The necessity of operating in a high temperature and highly corrosive environment will require the use of ceramic materials in many fossil energy applications. Unfortunately, monolithic ceramics are brittle and exhibit very low toughness. These undesirable features limit the reliability, damage tolerance, and durability of monolithics and make them unsuitable for many fossil energy applications. Emphasis has thus been placed on the development of ceramic matrices reinforced with strong ceramic fibers. Such ceramic matrix composites (CMC's) combine the temperature stability of monolithic ceramics with the "graceful" failure of fiber reinforced composites.

The corrosion of silicon carbide based fibers and matrix materials in the presence of alkali species[1,2] is a major problem in certain fossil energy applications, such as hot-gas filters and heat exchangers. The alkali species, mainly sodium compounds, are by-products from the combustion of fossil fuels. Therefore, corrosion/oxidation resistant coatings for the exterior of CMC components are needed for durability and reliability in such environments.

To the extent authorized under the laws of the United States of America, all copyright interests in this publication are the property of The American Ceramic Society. Any duplication, reproduction, or republication of this publication or any part thereof, without the express written consent of The American Ceramic Society or fee paid to the Copyright Clearance Center, is prohibited.

Thirty years ago, a new family of ceramic materials was developed based on the crystal structure of sodium zirconium phosphate $NaZr_2(PO_4)_3$ (NZP).[3,4] $Ca_{0.6}Mg_{0.4}Zr_4(PO_4)_6$ (CMZP)[5] and $Ca_{0.5}Sr_{0.5}Zr_4(PO_4)_6$ (CS-50)[6] have been derived from this family and possess a near-zero thermal expansion coefficient, low thermal conductivity and good thermal stability up to 1500°C which make them excellent candidates for high temperature applications. Additionally, these oxides have shown great promise for operating in an alkali environment. CMZP coatings produced by sol-gel techniques on silicon carbide and silicon nitride have provided protection to alkali exposure at 1000°C.[7,8,9] The interaction between plasma-sprayed CS-50 coatings, as well as bulk CS-50, and sodium sulfite at high temperature has been studied, and CS-50 remained unaffected after alkali exposure.[10,11] The goal of this study is to develop a protective coating of these two NZP formulations. Processing techniques focused on two cost-effective, water-based slurry-coating methods: A slurry dipping process and a tape-casting process.

EXPERIMENTAL PROCEDURE

Substrate Preparation

Ceramic grade Nicalon™ fiber-reinforced SiC matrix composites prepared by chemical vapor infiltration by the Ceramic Surface Systems Group at Oak Ridge National Laboratory (ORNL) were used in this study. Flat samples as well as ring samples cut from prototypical heat exchanger tubes were machined to the appropriate size.

The fibers and woven pattern of the composite offer a very textured surface that provides good interlocking sites for the coating. To further improve the infiltration of the slurry into crevices in the composite, the effort focused on improving the wettability of the substrate by the slurry. The impact of an intermediate oxide layer on substrate wettability was evaluated. A first oxidation process was assessed which consisted of heating the specimens to 1000° C for four hours. The second process was developed by Li et al.[8] and involves a chemical route. The substrates were washed in acetone, dried, washed in HF, rinsed, dried, calcined at 1200°C, washed in HCl and finally rinsed and dried. The wettability of untreated and oxidized monolithic SiC was then measured with water in an environmental chamber. Surface oxidation of monolithic SiC improved its wettability since contact measurements were ~55° for untreated samples and ~40° for oxidized specimen. In order to reduce the use of harmful chemicals in the coating process, the thermal oxidation process was chosen over the chemical oxidation process. Additionally, some samples presenting large smooth areas where most of the fibers were completely embedded in the matrix exhibited poor coating adhesion. Roughening of the surface with 60-grit sandpaper before oxidation improved coating adherence of these samples.

Powder and Slip Preparation

Powder Preparation

CMZP powder was prepared at Virginia Polytechnic Institute and State University and New Mexico Institute of Mining and Technology by a proprietary method and was screened to 100-mesh. The powder was then fully crystallized by heating it to 1200°C for 4 h. CS-50 powder was obtained from LoTEC Inc. (Salt Lake City, UT). The powder was fully calcined with an initial average particle size of 10 μm.

Particle size distribution is important and must be optimized to obtain a high quality coating. A small and narrow particle size distribution allows higher solids loading, thus milling trials were performed to optimize powder size and distribution. The desired particle size and distribution was reached after 48 h of wet milling with a pearled zirconia media. Particle size analysis was completed using a Horiba LA-700 particle size analyzer. A mean particle size of ~0.9 μm was obtained for CS-50 and CMZP. The particle size distribution was very narrow with a standard deviation of 0.25 μm for both powders.

Slip Composition

A proprietary CMZP slip composition was adjusted to obtain the best rheology for dip coating. Dolapix PC 21 (Zschimmer and Schwartz, Lahnstein Germany) replaced the original dispersant. A CS-50 slurry was also prepared according to a proprietary composition. The powder was gradually added to the slurry to obtain the maximum powder-to-water ratio. Maximizing the solids loading will increase contact between particles and reduce the amount of water which must be removed during drying. A solids loading of 33 vol% was initially used for coatings with a goal of reaching 55 to 60 vol% solids loading. Currently, a solids loading of 60 vol% was obtained for CMZP and 63 vol% for CS-50. The binder content of the slurries was then modified according to the coating technology. Dipping required a low viscosity slurry to obtain thin coatings and no binder was added to the dispersant. Tape casting required a thicker, more plastic slurry obtained with the addition of binders, Methocel (Dow Corning, Midland, MI) or polyvinyl alcohol (PVA, grade 75-15, Dupont, Wilmington, DE), and plasticizers such as polyethylene glycol (PEG, ME Science, Gibbstown, NJ). The gel-like slurry was easily sprayed on the substrate and the casting process was improved.

ZnO (ME Science) was added to both slurries as a sintering aid. The effect of ZnO on the sintering behavior of NZP materials has been found to create a liquid phase which promotes the sintering reaction[12,5,13]. In CMZP, Ca^{2+} and Mg^{2+} dissolve in a liquid phase of $Zn_3(PO_4)_6$ that forms at 1000°C. In order to reduce coating cracking during densification, it was believed that the coating should behave more like a glaze. Therefore, the amount of ZnO additions, as well as particle size and sintering temperature have been optimized to obtain the maximum of liquid phase. Aging of the slurry by milling proved detrimental to the coating quality. As the viscosity increased, the pH decreased causing the slurry to become less stable and to flocculate. Milling had to be limited to a maximum of 48 h.

Thermal Expansion Adjustment

Matching the coefficient of thermal expansion (CTE) of the coating to that of the substrate to prevent cracking and debonding, without detrimentally affecting the properties of the materials was investigated. Mixtures of CS-50 with a CTE of ~3.0x10-6/°C and 4% Y_2O_3 stabilized ZrO_2 (Tosoh Soda, Tokyo Japan) with a CTE of ~10x10-6/°C have been prepared to match the CTE of SiC ceramic matrix composites of ~4.5x10-6/°C[14]. Cold pressed samples of mixtures were prepared, sintered at 1300 °C for 8 h and analyzed by X-ray diffraction. No reaction between CMZP and 4 wt % Y_2O_3-stabilized ZrO_2 additive was found.

Coating application

Dip coating

Coatings were applied to composite rings by manually dipping the samples in the CMZP or CS-50 slurry. The samples were only dipped once since multiple dipping formed thick and non-uniform coatings which cracked in the thickest regions. It is anticipated that more controlled dipping of the samples will yield thinner, more uniform, higher quality coatings. Increasing withdrawal speed of the sample from the solution might also limit the formation of drops at the bottom of the specimen. Further improvements in the process will include mechanical dipping at varying rates.

Tape casting

The second slurry coating was obtained with a doctor-blade coating process also known as tape casting (Fig. 1). Tape casting is mostly used in the electronics industry.[15,16,17] This process was chosen for evaluation since it provides thin, large area and uniformly dense coatings. The use of a blade to level the coating allows an increase in the viscosity and the density of the slurry, which reduces shrinkage stresses. Additionally, coating thickness is precisely controlled (± 0.01 mm). The coating process is very simple and consists of covering the surface of the specimen with the slurry, leveling the coating with a doctor-blade, drying the specimen in a controlled environment and finally sintering the samples at a closely monitored rate. The quality of the coating is influenced by many different factors which were optimized as part of this program. Slurries were initially prepared as if for slip-casting then adjusted for tape casting. Low viscosity slurry flows off the substrate while a thick slurry cracks during processing.

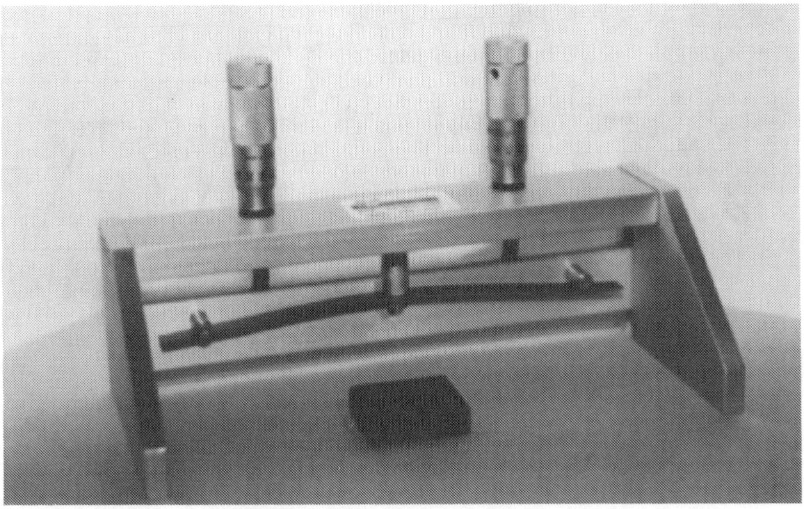

Figure 1: Doctor blade apparatus

Drying and sintering

After coating, the samples were kept at 80 % humidity for 48 h and then at ambient humidity for the same amount of time. The samples were then slowly heated at 0.5 °C/min up to 600 °C for 4 h to enable binder burnout. The samples were crack-free after this treatment. The specimens were then ramped at 1°C/min up to 1200°C, soaked for 1 min and cooled at the same rate down to room temperature. The same heating schedule was performed a second time, with a soaking time of 24 h at 1200°C. Stresses and deformations occurring during heating and cooling were minimized with the pre-sintering stage. A summary of the coating process is presented in Fig. 2.

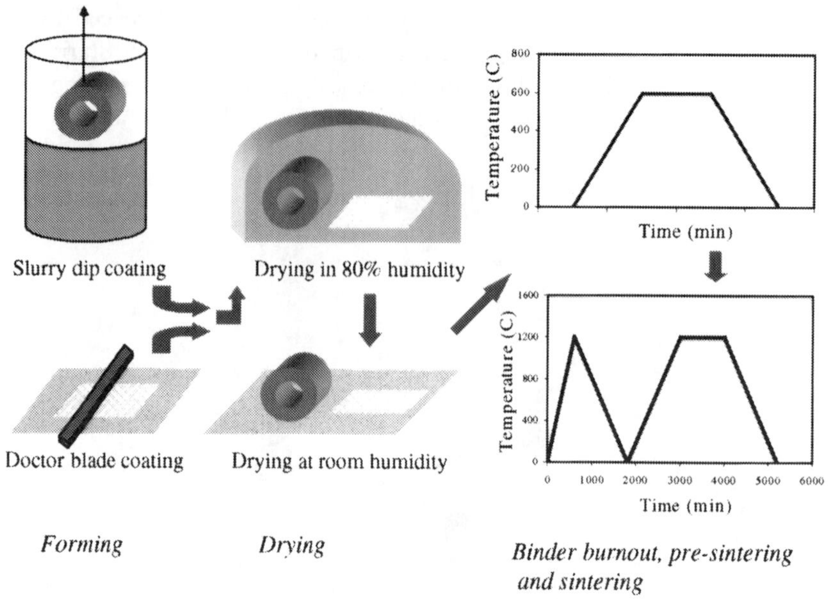

Figure 2: Coating deposition, drying and sintering process

Microscopy

Samples were embedded in epoxy resin, cut with a diamond saw and polished. Characterization of the coating/composite interface was first performed with an optical microscope, Nikon 4X, and then with a scanning electron microscope (SEM) Hitachi S800.

Corrosion Resistance of the Coatings

The surface of SiC oxidizes at high temperature to form a silica layer, which is responsible for the oxidation resistance of the material. There have been excellent reviews of the oxidation of SiC and Si_3N_4 which have shown that the protective silica layer, when in contact with molten sodium salts, reacts to form a sodium silicate liquid, resulting in an increase in the oxidation rate[1,2]. Several types of corrosion studies have been evaluated; (a) burner rig corrosion test,[18] (b) application of a thin film of sodium sulfite or sodium carbonate and heating the coated specimen to the desired temperature,[19] (c) partial or complete immersion of the specimen in a crucible containing the alkali melt. A thin film procedure was performed for this evaluation. A total of five specimens were used; two coated with CS-50, one coated with CMZP and two reference composites. The dimensions of the samples were carefully measured and the specimens were weighed. Then the specimens were washed in an ultrasonic cleaner in acetone followed by distilled water. The samples were dried overnight in an oven at 60°C. The sodium sulfate loading was performed by heating the specimen on a hot plate to 90°C and dispensing drops of saturated Na_2SO_4 solution onto the surface. The samples were left to dry for an hour, weighed and more sodium sulfate was added until the loading reached 8 mg/cm^2. The samples were then heated in a furnace to 1000°C for 100 h in flowing O_2 (200 cm^3/min). The specimens were then weighed, washed in hot distilled water for two hours to dissolve any residual sodium sulfate and then weighed again. Phases present on the sample surface were determined before and after corrosion by standard XRD. Some samples were then cut, polished using kerosene as a lubricant and viewed with the SEM to determine any microstructural changes. Energy dispersive X-ray analysis (EDX) was also used to evaluate the penetration depth of sodium compounds.

RESULTS

Visual Evaluation

Good quality coatings exhibiting few cracks and a smooth surface were obtained and are illustrated in Fig. 3 and 4.

ZnO additions were beneficial to the coating quality as seen in Fig. 5. Because of the liquid phase present, the coating flowed around the substrate and was less likely to crack during densification.

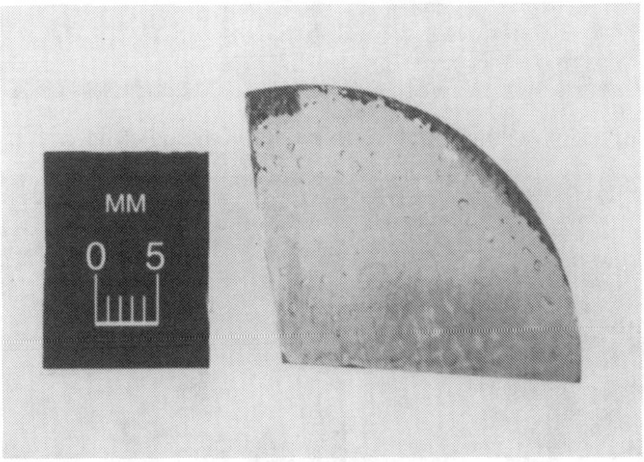

Figure 3: Nicalon fiber-reinforced SiC matrix composite coated with sintered CMZP (doctor-blade process)

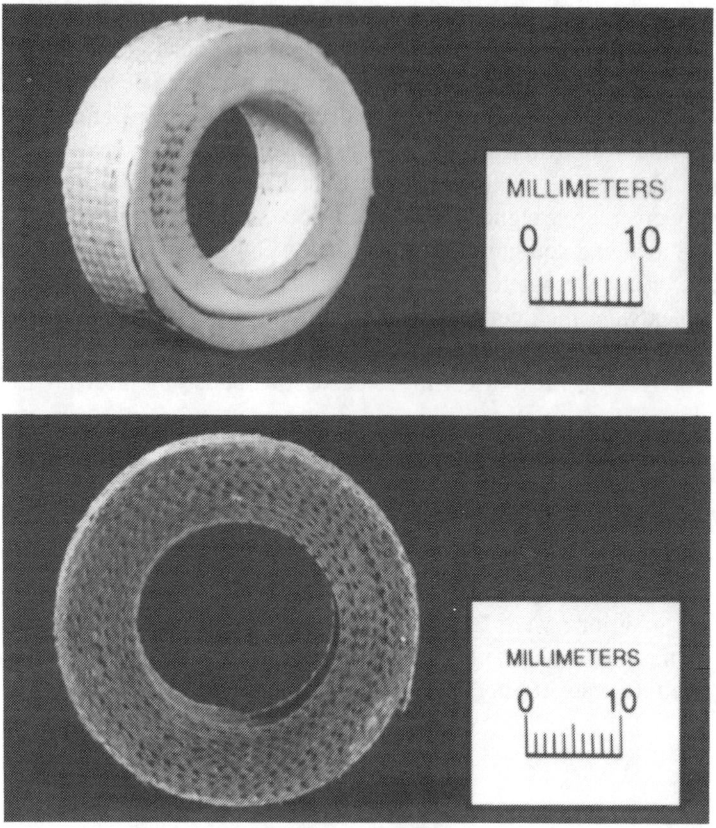

Figure 4: Examples of (a) thick and (b) thin CS50 coatings (dip-coating process)

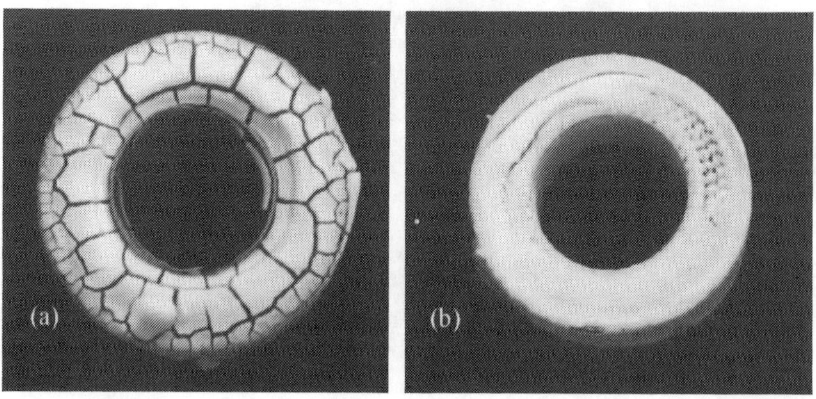

Figure 5: Influence of ZnO on coating quality, (a) without ZnO, (b) with ZnO

Analysis

X-ray Diffraction

The diffraction patterns indicated that CMZP and CS-50 were present. No other phases were detected.

Microscopy

Pictures of different CS50 and CMZP coatings obtained by optical microscopy are seen in Fig. 6. The coatings are very uniform, and no gap or debonding can be seen at the coating matrix interface. The CMZP coating, obtained by the tape casting method, is much thinner than the CS-50 coating obtained by dip-coating.

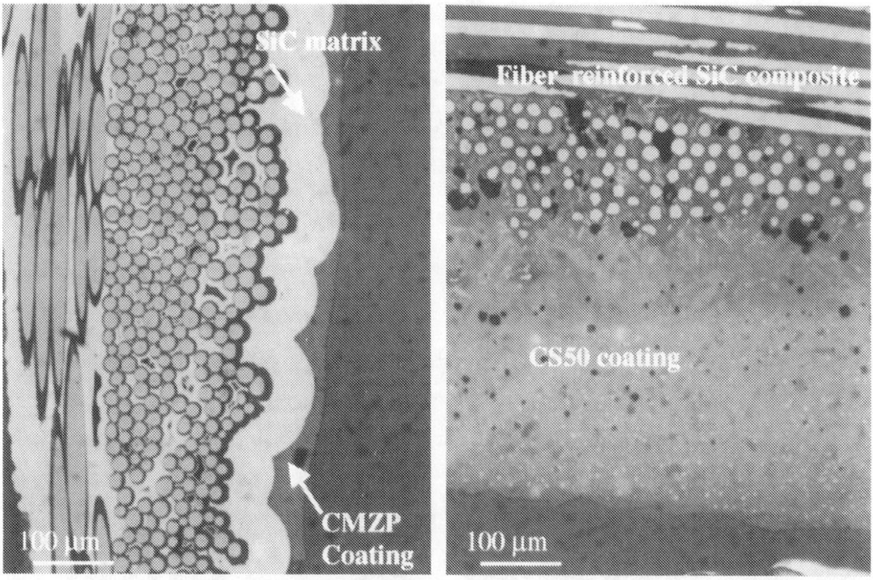

Figure 6: Optical micrographs of CS-50 and CMZP coatings

Corrosion Resistance of the Coatings

No major weight loss was observed for both coated and reference samples after exposure. The samples used in the study were very porous, consequently, the sodium sulfate flowed into the pores and lowered the surface loading to less than 8 mg/cm^2. No sign of degradation was found on the coated samples, but a white deposit was visible on both non-coated composites. No sign of microstructure degradation was observed during microscopic evaluation of the polished samples. Samples were then broken and observed with a SEM. Fig. 7 and 8 are SEM micrographs of non-coated Nextel fiber-reinforced SiC matrix composites before and after corrosion. The corroded, non-coated samples contain a very glassy phase indicating degradation of the SiC matrix. In contrast, coated (CS50) samples show very little degradation as illustrated in Fig. 9 and 10 which are different samples (A,B). EDX results indicate that sodium was present in the CS50 coating, but not in the composite. Thus, most of the sodium was trapped in the coating, which prevented the degradation of the substrate.

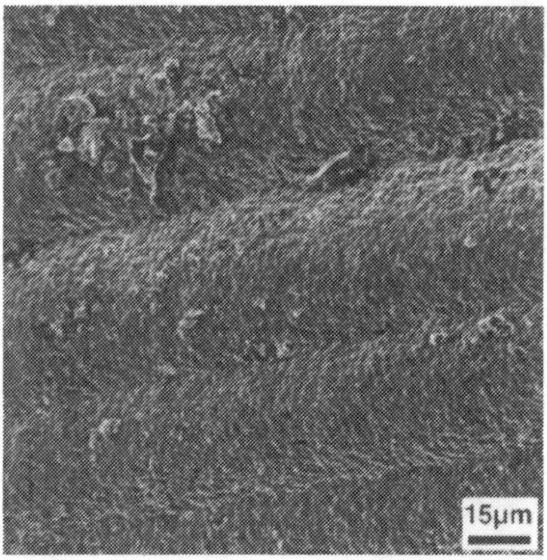

Figure 7: Non-corroded, non-coated composite

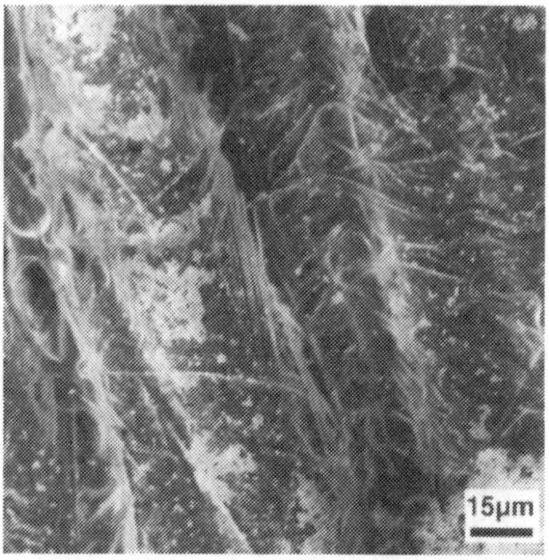

Figure 8: Corroded, non-coated composite

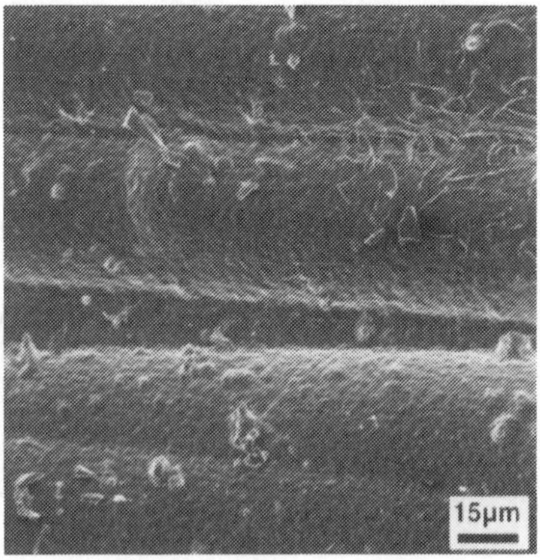

Figure 9: Corroded, coated with CS-50, A

Figure 10: Corroded, coated with CS-50, B

CONCLUSIONS

A good quality, relatively inexpensive coating process for applying protective oxide coatings on SiC matrix materials has been developed, and an understanding of the most critical factors controlling the process has been obtained. The glaze-like behavior of both NZP materials obtained with ZnO additions was beneficial to the quality of the coating. The corrosion resistance behavior of the coatings is very encouraging, and further improvements in coating quality should make NZP coatings even more efficient in terms of corrosion protection.

ACKNOWLEDGMENTS

This work was supported by the Advanced Research and Technology Development Materials Program of the Office of Fossil Energy, U.S. Department of Energy under contract DE-AC05-84OR21400 with Lockheed Martin Energy Corporation. We would like to thank Carolyn Angel for the photography work, Tom Geer for the metallography, John Stuecker for the processing of CMZP powder at New Mexico Tech and LoTEC, Inc. for providing us with CS-50 powder. We would also like to thank Claudia Walls for her help and advice on coating processing.

REFERENCES

[1] N. S. Jacobson, "Corrosion of Silicon-Based Ceramics in Combustion Environments", *Journal of the American Ceramic Society*, **76**, [1], 3-28, 1993

[2] G. R. Pickrell, T. Sun, J. J. Brown, "High temperature Corrosion of SiC and Si_3N_4", *Fuel processing technology*, **44**, 213-236, 1995

[3] L. O. Hagman and P. Kierkegaard, "The Crystal Structure of $NaMe_2^{IV}(PO_4)_3$; Me^{IV}=Ge, Ti, Zr", *Acta Chemica Scandinavice*, **22**, 1822-1832, 1968

[4] R. Roy, D. K. Agrawal, J. Alamo and R.A. Roy, "[CTP]: A New Structural Family of Near Zero Expansion Ceramics," *Materials Research Bulletin*, **19**, 471-77, 1984

[5] T. K. Li, D. A. Hirschfeld, S. VanAken, Y. P. Yang and J. J. Brown, "The Synthesis, Sintering, and Thermal Properties of $Ca_{0.6}Mg_{0.4}Zr_4(PO_4)_6$ (CMZP) Ceramics, *Journal of the Materials Research Society*, **8**, 2954-67, 1993

[6] S. Y. Limaye, D. K. Agrawal, and H. A. McKinstry, "Synthesis and Thermal Expansion of $MZr_4(PO_4)_6$, (M=Mg, Ca, Sr, Ba)", *Journal of the American Ceramic Society*, **70**, C232-36, 1987

[7] M. Kang, "Alkali/Steam Corrosion Resistance of Commercial SiC Products Coated with Sol-Gel Deposited Mg-Doped Al_2TiO_5 and $Ca_{0.6}Mg_{0.4}Zr_4(PO_4)_6$" Masters Thesis, Virginia Polytechnic Institute and State University, 1994

[8] T. K. Li, D. A. Hirschfeld, and J. J. Brown, "Thin Film Coatings of $Ca_{0.6}Mg_{0.4}Zr_4(PO_4)_6$ (CMZP) Ceramics" *Journal of the Materials Research Society*, **9**, 2004-18, 1994

[9] T. K. Li, D. A. Hirschfeld, and J. J. Brown, "Alkali Corrosion Resistant Coatings and Ceramic Foams Having Superfine Open Cell Structure and Method of Processing", U.S. Patent No. 5,268,199, Granted December 1993.

[10] W. Y. Lee, K. M. Cooley, C. C. Berndt, D. L. Joslin and D. P. Stinton, "High-Temperature Chemical Stability of Plasma-sprayed $Ca_{0.5}Sr_{0.5}Zr_4P_6O_{24}$ Coatings on Nicalon/SiC Ceramic Matrix Composite and Ni-based Superalloy Subtrates", *Journal of the American Ceramic Society*, **79**, [10], 2759-62, 1996

[11] W. Y. Lee, D. P. Stinton, and Debra L. Joslin, "Interaction of low-expansion NZP Ceramics with Na_2SO_4 at 1000°C," *Journal of the American Ceramic Society*, **79**, [2], 484-486, 1996

[12] D. K. Agrawal and V.S. Stubican, "Synthesis and sintering of $Ca_{0.5}Zr_2P_3O_{12}$- A Low Thermal Expansion Material", *Materials Research Bulletin*, **20**, 99-106, 1985

[13] J. R. Clarcke, "The Sintering Effects of Time, Temperature and ZnO Additions on $Ca_{0.6}Mg_{0.4}Zr_4(PO_4)_6$ [CMZP] Ceramics", M.S. Thesis, Materials Science and Engineering, Virginia Polytechnic Institute and State University, Blacksburg, VA, 1996

[14] D. P. Stinton, D. M. Kupp, and R. L. Martin, "Oxide Coating Development", *Fossil Energy Program Annual, Progress Report for April 1996 Through March 1997*, ORNL –6924, 17-27, July 1997

[15] R. E. Mistler, "Tape/Casting", Ceramics and Glasses, Engineered Materials Handbook Volume 4, ASM international, 161-165, 1991

[16] R. E. Mistler, "Tape Casting: The basic Process for Meeting the Needs of the Electronics Industry", *Ceramic Bulletin*, **69**, [6], 1990

[17] R. E. Mistler, R. B. Runk, and D. J. Shanefield, "Ceramic Fabrication Before Firing", G. Y. Onoda and L. L. Henc. Wiley, N.Y., 411-418, 1978

[18] D. S. Fox and J. L. Smialek, "Burner Rig Hot Corrosion of Silicon Carbide and Silicon Nitride." *Journal of the American Ceramic Society*, **73**, [2], 303-311, 1990

[19] N. S. Jacobson, J. L. Smialek, "Hot Corrosion of Sintered α-SiC at 1000°C.", *Journal of the American Ceramic Society*, **68**, [8], 432-39, 1985

MICROSTRUCTURE OF Ti_3SiC_2 COATINGS SYNTHESIZED BY CVD

E. Pickering*, W. J. Lackey‡, and S. Crain*
* School of Materials Science and Engineering,
‡ G.W. Woodruff School of Mechanical Engineering
Georgia Institute of Technology, Atlanta, Georgia, 30332

ABSTRACT

The search for oxidation resistant, high temperature fiber/matrix interface coatings for ceramic matrix composites has lead to the study of new materials systems. One such material, Ti_3SiC_2 has recently received attention as a possible fiber/matrix interface candidate because of its excellent oxidation resistance and layered crystal structure. In this paper, the microstructural characteristics of chemically vapor deposited Ti_3SiC_2 was investigated with x-ray diffraction, energy dispersive spectrometry, and electron microscopy. A CVD study was devised using the $TiCl_4$-$SiCl_4$-CCl_4-H_2 reagent system. Deposits show a complicated microstructure consisting of various combinations of Ti_3SiC_2, TiC, and $TiSi_2$. The orientation of Ti_3SiC_2 basal planes was shown to be perpendicular to the substrate, and the microstructural characteristics of films depended strongly on the deposition temperature.

INTRODUCTION AND REVIEW

Toughening of composites for applications at high temperatures has been an elusive goal of the ceramics community for over a decade. The insertion of a thin interface coating separating the fiber and matrix has been shown to improve composite fracture toughness by empowering fiber pullout and crack deflection along weakly bonded planes inherent to the interface material.[1-10] Interface materials such as carbon, B-doped carbon, and BN have been quite successful in improving fracture toughness and lifetime to failure of composite parts at modest temperatures (300-500°C for carbon and 900°C for BN).[1-5] The hexagonal, layered crystal structure of carbon (graphite) and BN is characterized by poorly bonded basal planes, which when oriented parallel to the fiber reinforcement, protect fiber integrity against cracks propagating through the matrix. However, the poor

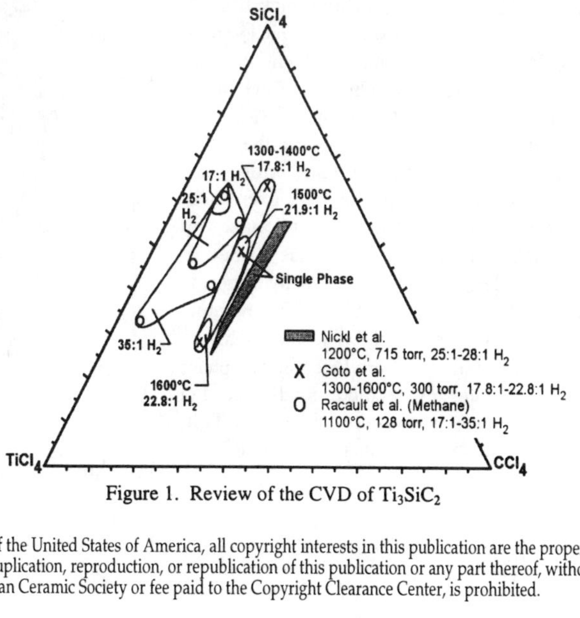

Figure 1. Review of the CVD of Ti_3SiC_2

To the extent authorized under the laws of the United States of America, all copyright interests in this publication are the property of The American Ceramic Society. Any duplication, reproduction, or republication of this publication or any part thereof, without the express written consent of The American Ceramic Society or fee paid to the Copyright Clearance Center, is prohibited.

oxidation resistance of these materials has fueled a massive effort to find high temperature, oxidation resistant alternatives. Layered oxide compounds in the monazite, and magnetoplumbite families have received extensive attention in recent years.[5-10]

The goal of the present work was to develop a sufficient understanding of the CVD of Ti_3SiC_2 to permit its evaluation as an interface coating or matrix material. Once the CVD of Ti_3SiC_2 is understood, it should be possible to utilize conventional CVI or the forced-flow-thermal gradient variant (FCVI) to fabricate composites containing this material.[11-15]

Ti_3SiC_2 was first synthesized in 1967 by Nowotny and Jeitschko[16] by hot pressing graphite, silicon, and TiH_2 powders in excess of 2000°C. Recently, Barsoum, El-Raghy and coworkers[17-20] have done extensive work with solid state synthesis of Ti_3SiC_2 and characterization of its properties. They report excellent oxidation resistance due to the formation of two oxide layers on exposed surfaces. These are composed of 1) rutile (TiO_2) over 2) titania and silica (TiO_2+SiO_2).[20] In addition to its oxidation behavior, the crystal structure of Ti_3SiC_2 is appropriate for interface applications. The unit cell is described as planes of TiC octahedra stacked between planes of silicon atoms.[17] Observed ductile mechanical properties and machinability, especially at high temperature, have been credited to this crystal structure. The combination of layered crystal structure and high temperature oxidation resistance suggest that Ti_3SiC_2 should be investigated for interface and matrix applications in ceramic composites.

CVD synthesis of Ti_3SiC_2 has previously been addressed by three authors.[21-23] Figure 1 is a summary of their work, showing reagent compositions on a ternary deposition diagram, along with the deposition temperature, pressure and hydrogen dilution used during their respective CVD studies. Nickl et al.[21] completed the most comprehensive study of the $TiCl_4$-$SiCl_4$-CCl_4-H_2 deposition system in 1972, experimentally determining the entire range of the ternary deposition diagram with more than 100 experiments. Nickl et al.[21] report polycrystalline layers of Ti_3SiC_2 below 1200° C and single crystal, hexagonal plates above 1300° C. Goto and Hirari[22] reported the next successful attempt at single phase Ti_3SiC_2 deposition in 1987. Like Nickl et al., they prepared thick films of monolithic Ti_3SiC_2 using the $TiCl_4$-$SiCl_4$-CCl_4-H_2 reagent system. Deposition rates were reported at 200 µm/hr. The final author to publish on the CVD of Ti_3SiC_2 was Racault et al.[23] in 1994. They used the $TiCl_4$-$SiCl_4$-CH_4-H_2 reagent system, citing the reduced thermal stability of CH_4 over CCl_4. Their work demonstrated the importance of hydrogen dilution, a parameter they called α. This was defined as the ratio of hydrogen to the total amount of Si-Ti-C containing reagents or $\alpha = X_{H2} / (X_{TiCl4} + X_{SiCl4} + X_{CH4})$. This work failed to produce single phase Ti_3SiC_2 under any studied conditions, but instead Ti_3SiC_2 was co-deposited with $TiSi_2$ or Ti_5Si_3 and small amounts SiC and TiC.

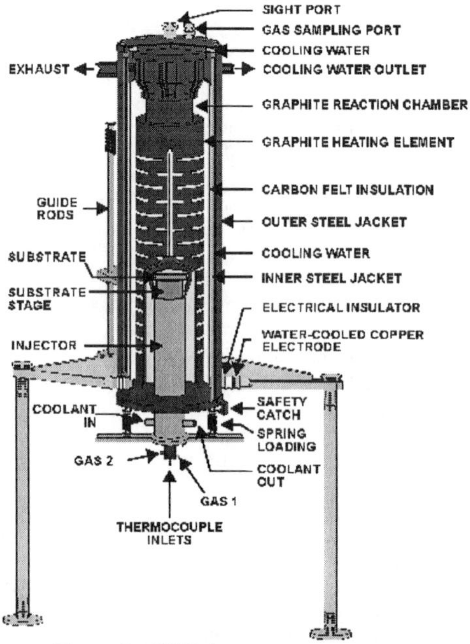

Figure 2. CVD Apparatus

DESIGN OF EXPERIMENT

In previous work by our group, the $TiCl_4$-$SiCl_4$-CCl_4-H_2 system was thermodynamically modeled in order to suggest the optimum conditions for deposition of Ti_3SiC_2.[24] Furthermore, a series of over 40 preliminary experiments were performed in an effort to verify the results of previous authors. Conditions for these preliminary experiments covered a wide range of conditions. Compositions varied from 5-30% CCl_4, 25-55% $TiCl_4$, and 25-60% $SiCl_4$; temperature and hydrogen dilutions varied from 1000-1400°C and 10-40, respectively. Several months of experimentation resulted in deposits containing only TiC and SiC. Only after increasing the hydrogen dilution above 25 was Ti_3SiC_2 finally deposited. Results from these preliminary experiments, and thermodynamic modeling, along with the previous work by other authors mentioned earlier, were all used to plan a CVD study.

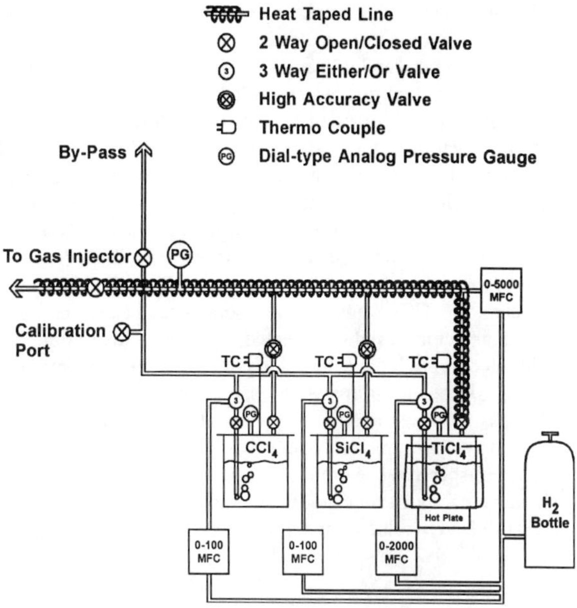

Figure 3. Reagent Supply to Gas Injector

The $TiCl_4$-$SiCl_4$-CCl_4-H_2 system was chosen because of previous success in obtaining single phase Ti_3SiC_2. Four reagent compositions were chosen which fell within the region of the deposition diagram in Figure 1 in which Ti_3SiC_2 was deposited as a single phase. The hydrogen flow rate was kept constant at 2000 sccm. Because the hydrogen remained fixed while the total flow rate of reagent varied, hydrogen dilution, α, varied from 42.1 to 32.0. The four reagent compositions were repeated at three different temperatures (1100, 1200, 1300° C) for a total of 12 experiments. Pressure and run time were kept constant at 300 torr and 150 minutes, respectfully. Table 1 shows the run conditions listing reagent flow rates, α, and temperature for each experiment.

Table I: Run Conditions for CVD of Ti_3SiC_2

Run Number	Target Flow Rates (sccm) $TiCl_4$	$SiCl_4$	CCl_4	H_2	α	Temp (°C)	Press (torr)	Time (min)
1,9,5	17.5	17.5	12.5	2000	42.1	1100,1200,1300	300	150
2,10,6	25.0	17.5	12.5	2000	36.4	1100,1200,1300	300	150
3,11,7	17.5	25.0	12.5	2000	36.4	1100,1200,1300	300	150
4,12,8	25.0	25.0	12.5	2000	32.0	1100,1200,1300	300	150

The CVD apparatus was a vertical, hot-walled reactor resistively heated with a graphite heating element. The apparatus is shown in Figure 2. The heating element, reaction chamber, substrate stage, and substrate were graphite. The furnace was evacuated, leak checked, and back filled with argon before each experiment. The substrates were ATJ graphite rectangles cut to 20.5 x 5.1 x 58.5 mm, and suspended ~ 90 mm above the gas injector. The outer skin and gas injector was stainless steel and water-cooled. Reactant gases were introduced through the gas injector, which in turn was supplied reagent gases through a series of quarter inch diameter stainless steel tubes and various 2 and 3-way valves depicted in Figure 3. The tetrachloride reagents are liquids at room temperature and were vaporized using hydrogen as the carrier gas. Each vaporizer was equipped with a thermocouple and pressure gauge. Hydrogen flow to the vaporizers was adjusted according to the partial pressures of the reagents and controlled with MKS mass flow controllers. Excess hydrogen was routed through a separate line. Actual reagent flow rates were determined by weight loss in the vaporizer from beginning to end of an experiment. Due to its very low vapor pressure, the $TiCl_4$ vaporizer was sometimes heated, and the stainless steel tubing between it and the gas injector was always heated to avoid condensation in the lines. Mass flow controllers were calibrated through the calibration port before each experiment.

Deposited films were characterized using x-ray diffraction and electron microscopy. A Phillips 1800 automated diffractometer identified deposited phases, while an Hitachi 4100 scanning electron microscope was used to explore surface morphology and cross section features. Energy dispersive spectrometry was used to help characterize the microstructure by determining chemical composition of individual phases. Coating cross sections were polished and etched with $HF:H_2O:HNO_4$ solution at 1:1:1 concentration for 2-15 seconds, depending on the composition of the coatings.

RESULTS

Table II gives the results from the study, listing run numbers with actual flow rates and temperatures, deposited phases as determined with XRD and EDS, and coating thickness. Ti_3SiC_2 was never found as a single phase, even though XRD patterns for experiments at 1300°C suggest otherwise. SEM showed the microstructure of these films was quite complicated, involving co-deposition of Ti_3SiC_2 with TiC and $TiSi_2$. Ti_3SiC_2 deposits were highly textured, with the basal planes oriented perpendicular to the substrate. Although XRD shows significant amounts of Ti_3SiC_2 at 1300 and 1100°C, very little was deposited at 1200°C. Figure 4 is a reagent composition triangle showing the composition of the reagent stream for each experiment in the study.

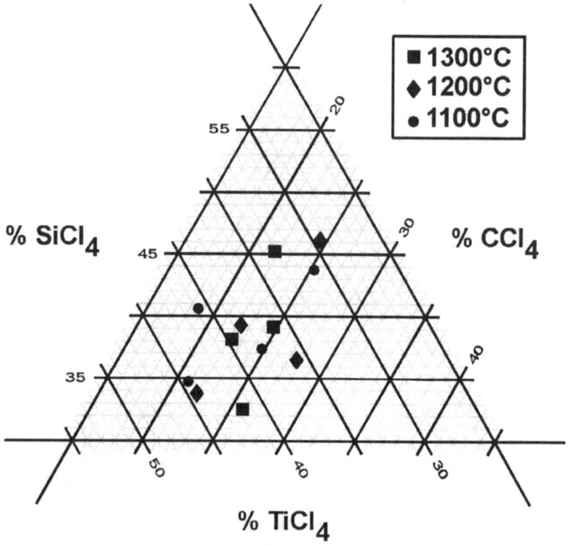

Figure 4. Reagent Composition Triangle

Table II. Results from Ti_3SiC_2 CVD Study

Run Number	Run Designation	Actual Flow Rate (sccm) TiCl4	SiCl4	CCl4	H2	α	Temp (°C)	Phases Identified With XRD, EDS	Thickness (μm)
1	EP29	16.46	16.27	10.88	2000	41.86	1100	Ti_3SiC_2, TiC	184
2	EP23	21.10	16.60	10.00	2000	41.89	1100	Ti_3SiC_2, TiC	99
3	EP26	17.50	24.30	14.20	2000	35.69	1100	Ti_3SiC_2, TiC, $TiSi_2$	160
4	EP32	23.32	23.31	10.78	2000	34.84	1100	Ti_3SiC_2, TiC	161
5	EP24	16.70	17.80	11.08	2000	43.83	1300	Ti_3SiC_2	143
6	EP31	24.73	19.26	15.06	2000	33.87	1300	Ti_3SiC_2 $TiSi_2$	209
7	EP27	17.50	24.40	11.30	2000	37.47	1300	Ti_3SiC_2	197
8	EP22	22.92	22.25	12.73	2000	34.54	1300	Ti_3SiC_2, TiC	175
9	EP25	17.50	17.70	13.40	2000	41.10	1200	Ti_3SiC_2, TiC	129
10	EP19	24.10	18.73	11.76	2000	36.64	1200	Ti_3SiC_2, TiC, $TiSi_2$	215
11	EP15	16.70	26.03	13.89	2000	35.32	1200	Ti_3SiC_2, TiC	144
12	EP14	24.26	24.98	14.09	2000	31.58	1200	Ti_3SiC_2, TiC	171

X-Ray Diffraction

Several XRD patterns from the study are shown in Figure 5. The peaks for each of the three phases are labeled along with the corresponding (hkl) values. The top scan, EP31, is from a 1300°C experiment, in which only Ti_3SiC_2 peaks are evident. It is representative of two other high temperature experiments, EP24 and EP27. All show a great deal of preferred orientation. Notice the (008) peak is totally missing in this scan, but according to the PDF card (40-1132) it should be the 100% peak. As further evidence, the (110) reflection is quite strong, which would be expected as the intensity from (00l) planes disappear because these planes lie at right angles to each other. Other peaks are missing between 50-60° 2θ such as (107), (108), and (109). The middle and bottom patterns, EP23 and EP32, were taken from 1100°C experiments. They show peaks from both Ti_3SiC_2 and TiC and two $TiSi_2$ peaks are evident in EP32. Here, the same preferred orientation is apparent in Ti_3SiC_2, but to a lesser degree. The (008) is present, but clearly a weaker peak than (101), (104), (105) and others. The TiC peaks in these patterns do not show preferred orientation. Their relative intensities roughly match those of the TiC PDF card (32-1383). Notice the TiC (220) peak directly overlaps the Ti_3SiC_2 (110) peak. Without knowing the relative amounts of each phase, it is difficult to quantify how much peak intensity is due to which phase. The pattern for EP32 looks quite similar to that of EP23 accept for the presence of $TiSi_2$ peaks. XRD patterns for the four 1200°C runs, EP14, EP15, EP19, and EP25 show almost all TiC, with very weak peak intensity from Ti_3SiC_2. Only EP14 has Ti_3SiC_2 peaks of significant intensity.

Several discrepancies were noted between the XRD patterns collected for this CVD study and the pattern for Ti_3SiC_2 recorded on the PDF card. Three peaks were consistently obtained during the course of this research that were not previously accounted for on the card. During the process of identifying these peaks several key factors were considered. First, the PDF card was taken from material synthesized via CVD by Goto and Hirai,[22] who used the same reagents used in this work and report the same preferred orientation seen here. However, the (008) peak is the 100% peak on the PDF card, which clearly is not consistent with the texture they reported and we confirmed. To generate the pattern on the PDF card, Goto and Hirai must have removed the coating from the substrate and crushed it into a powder. Unfortunately, neither their paper nor the PDF card is explicit on this. Second, the locations of these three peaks do not correspond to TiC, $TiSi_2$, Ti_5Si_3, SiC, Si, C, Ti or any other solid in the Ti-Si-C system. Because Cl and H are used in this reagent system, $TiCl_3$ and TiH_2 were also considered but eliminated. Third, these peaks appear in XRD patterns published by Racault et al.,[25] who prepared Ti_3SiC_2 through solid state means, and on patterns we prepared from Ti_3SiC_2 powder obtained from Barsoum and El-Raghy. Because these peaks are found in XRD patterns taken from samples prepared through different means, synthesized by different authors, and because they do not belong to other solids in the Ti-Si-C system, we concluded that the peaks belong to Ti_3SiC_2.

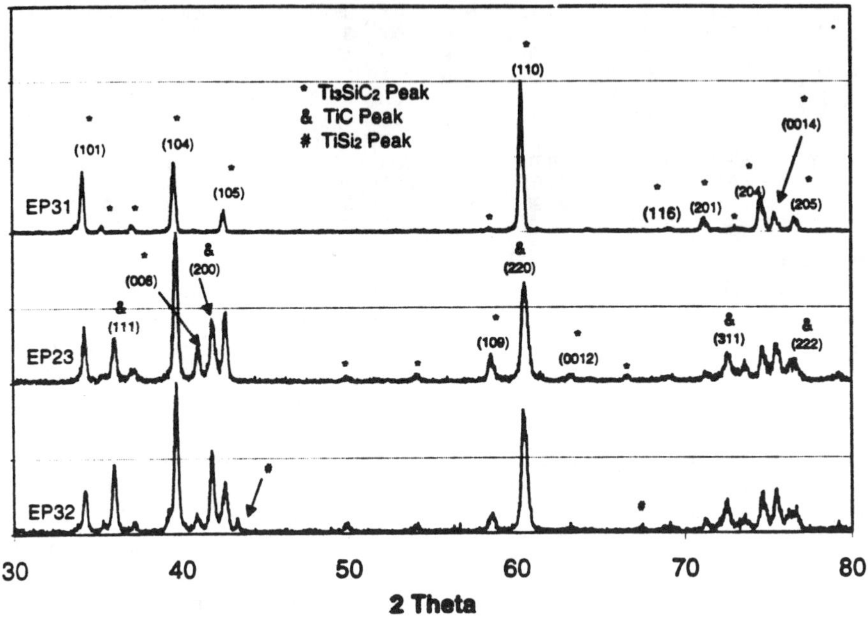

Figure 5: XRD Patterns showing Ti_3SiC_2, TiC, and $TiSi_2$ peaks. Ti_3SiC_2 is highly textured.

If these peaks are indeed reflections from the Ti_3SiC_2 matrix, then they should have corresponding (hkl) values. The plane-spacing equation for hexagonal crystals was taken from Culity[26] and is shown below:

$$d^{-2} = (4/3)*(h^2 + hk + k^2)*a^{-2} + (l^2/c^2)$$

where d is the plane spacing, a and c are lattice parameters of the unit cell, and h,k,l are the indicies of the plane. The lattice parameters for this unit cell are 1.767 and 0.3067 nm, and the d-spacing for each of the three peaks was taken directly from XRD patterns. Plugging in various (hkl) values and solving for d yielded the following matches.

Table III: Indexing Ti_3SiC_2 Peaks

(hkl)	Calculated d-spacing	Actual d-spacing	2θ
(116)	1.357	1.358	69.100
(201)	1.322	1.322	71.225
(204)	1.270	1.270	74.625

The intensity of each peak is a function of texture. The (116) and (201) planes are minor reflections, yielding relative intensities less than 5%. The (204) reflection, however, shows considerable relative intensity, greater than 20% for 1300°C experiments in this study.

X-ray penetration depth was determined to be less than 20 μm using calculated mass absorption coefficients for Ti_3SiC_2 and TiC, and operating conditions for the diffractometer. The thinnest coating, however, was just under 100 μm, meaning the XRD data was not inclusive for the entire thickness of the coating. Variation in the composition of the coating, or substrate/coating interaction would not be sufficiently characterized by XRD alone. As SEM characterization will show, small variations in process conditions lead to variation in coating microstructure through the course of a deposition experiment.

Electron Microscopy

Various microstructural characteristics are evident at different temperatures. Figure 6 shows a) the surface morphology, and b) the fractured cross section of Ti_3SiC_2 grown at 1300°C. The coating is highly oriented and characterized by large, plate-like grains oriented perpendicular to the substrate. This corroborates the preferred orientation seen with x-ray diffraction. Regions of the coating appear to have the flat side of the plate parallel to the fracture surface, while other regions appear to show the plate edge on. The surface morphology is similar in appearance to Ti_3SiC_2 coatings reported by Goto and Hirai.[22] The surface is an intricate pattern of grains, which are oriented in random directions. Taken together, the micrographs suggest that though the basal planes stand perpendicular to the substrate, there are no additional restrictions on the orientation of Ti_3SiC_2. Figure 7 is the a) surface morphology and b) fractured cross section of a coating grown at 1100°C. XRD suggests the sample is a mixture of Ti_3SiC_2 and TiC but individual phases and grains are not discernable at this magnification. Instead the coating

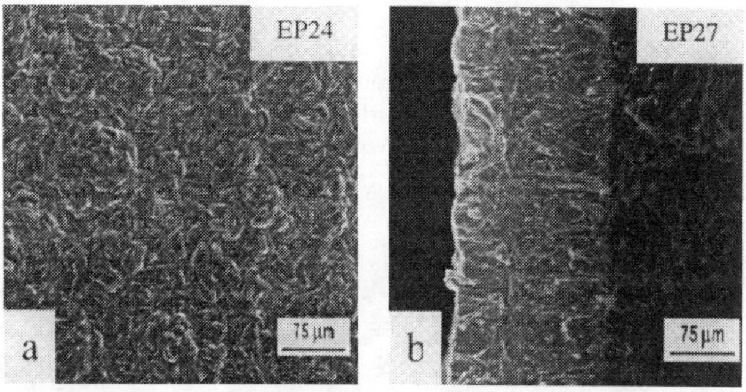

Figure 6. a) Surface morphology and b) cross section of coatings grown at 1300°C.

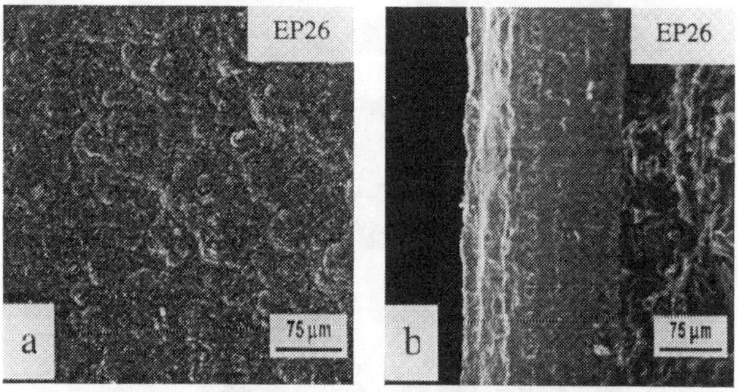

Figure 7. a) Surface morphology and b) cross section of coatings grown at 1100°C.

cross section appears rather featureless. The surface appears to consist of small, nodule-like features growing on larger nodules. Compared with Figure 6, the micro-structure is much finer. There is clearly a marked difference in deposition of highly oriented large grained Ti_3SiC_2 at 1300°C and the fine grained co-deposition of TiC and Ti_3SiC_2 at 1100°C. Strangely, the 1200°C experiments did not yield significant amounts of Ti_3SiC_2, probably due to deposition at inappropriate reagent compositions. Whether coatings deposited at 1200°C with the proper reagent composition mimic the microstructure of higher temp or lower temp coatings can not be determined from this study.

Polished and etched cross sections reveal more details about the microstructure of these coatings. Figure 8 is a series of micrographs taken from experiments performed at 1100°C. Figure 8a shows the cross section of an over etched sample. The vertical bands running parallel to the substrate are the result of different etch rates on the different phases being deposited. EDS has shown these bands to be composed of Ti_3SiC_2, TiC and $TiSi_2$, again confirming data from XRD. Figure 8b was taken at higher magnification. Continuous layers of $TiSi_2$ appear early in the growth process, and are believed to be cause by fluctuations in reagent flow rates. As the carbon content of the reagent mix becomes too low, $TiSi_2$ begins to form at the leading edge of the growing surface. The remainder of the coating consists of Ti_3SiC_2 and TiC growing simultaneously, as shown in Figure 8c. The layered crystal structure is apparent in the Ti_3SiC_2 grains, identified with

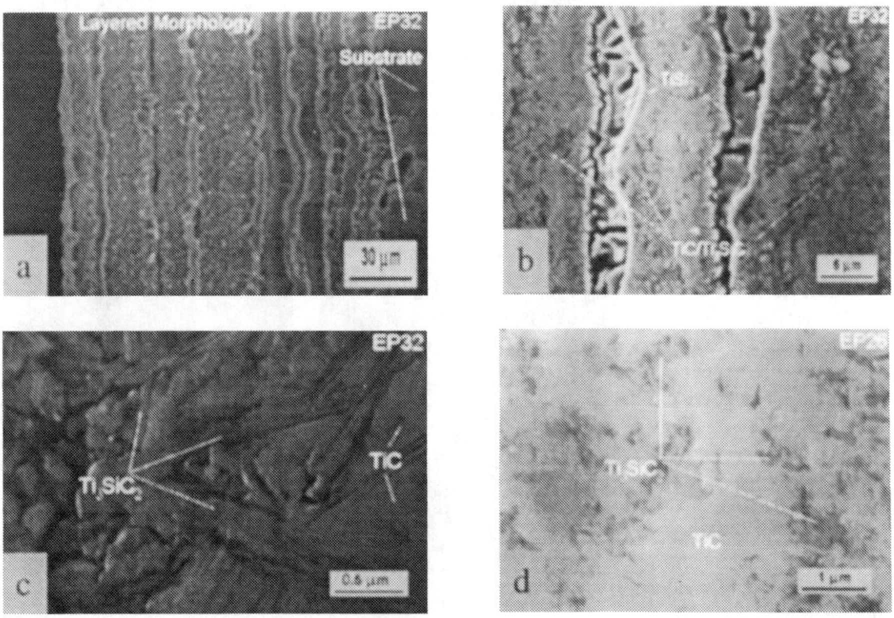

Figure 8. SEM micrographs of polished and etched cross sections of films grown at 1100°C showing a) layers of TiC, Ti_3SiC_2, and $TiSi_2$ formed by differing etch rates, b) continuous $TiSi_2$ layers and TiC/Ti_3SiC_2 co-deposition, c) high resolution of Ti_3SiC_2 grains in TiC matrix, and d) lower resolution Ti_3SiC_2 grains in TiC matrix.

EDS. Surrounding these grains are regions of TiC. The EDS spectrum from the TiC shows some silicon, but not enough to suggest Ti_3SiC_2. It is not clear whether the silicon signal is due to solid solubility of silicon in TiC or due to Ti_3SiC_2 grains hidden beneath the polished surface. Finally, Figure 8d also shows the microstructure of this two phase region at a lower magnification. The dark phase was identified with EDS as Ti_3SiC_2, while the lighter, matrix phase was identified as TiC. Figure 9 shows a series of micrographs for 1300°C experiments. Figure 9a is the polished and etched cross section of one of these coatings. The layered structure growing perpendicular to the substrate in fan-like grains is readily apparent. A boundary layer appears between these grains and the substrate. EDS identified this boundary layer as TiC with a small amount of silicon. A similar layer appears midway through the coating, parallel to the substrate, but much thinner. It is labeled the intermittent layer. EDS shows this to be TiC as well. Figure 9b shows the interface

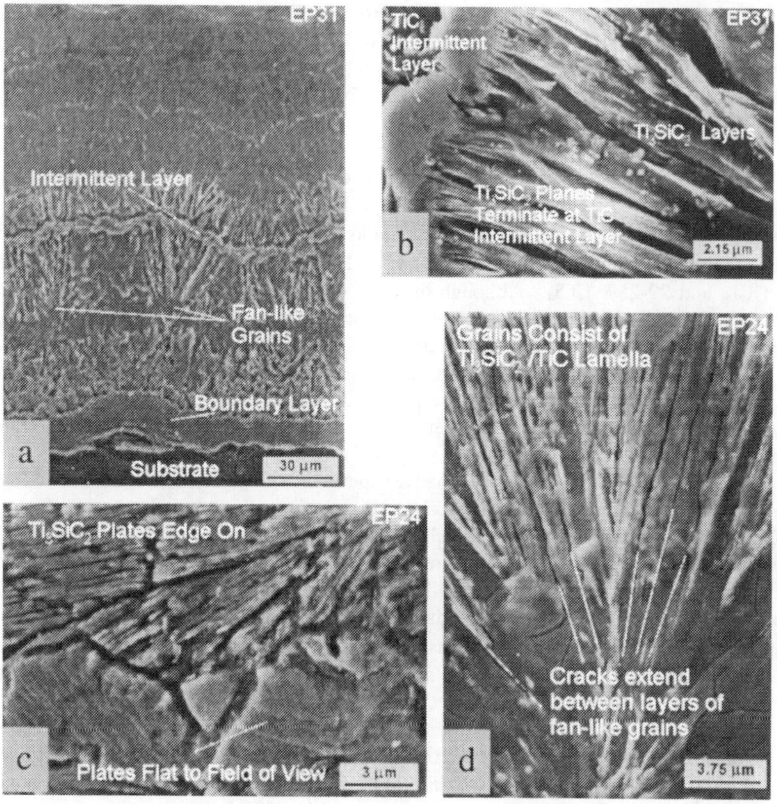

Figure 9. SEM micrographs of polished and etched cross sections of films grown at 1300°C showing a) TiC boundary and intermittent layers between fan-like grains of Ti_3SiC_2, b) interface between Ti_3SiC_2 and intermittent TiC layer, c) Ti_3SiC_2 grains seen with various orientation, and d) crack deflection in Ti_3SiC_2 and lamella structure of fan-like grains.

between this intermittent TiC layer and Ti_3SiC_2 grains at high magnification. Notice how individual layers of the Ti_3SiC_2 grains grow directly into the TiC layer. Figure 9c shows Ti_3SiC_2 grains near the surface of the coating, away from the substrate. The layered nature of the plate-like crystals is unmistakable, with some grains oriented edge on and some oriented with their broad face more parallel the field of view. EDS from these regions confirmed the presence of Ti_3SiC_2. Figure 9d shows the fan-like structure of Ti_3SiC_2 grains. Notice the cracks stretching from the base of the fan to the outer edges, guided by the basal planes of the crystal. This demonstrates the crack deflecting properties of layered Ti_3SiC_2. These grains appear to consist of a light and a dark phase, even though the layered appearance of the structure suggests Ti_3SiC_2, alone. EDS on individual layers showed the lighter phase was too silicon deficient to be Ti_3SiC_2. Instead, the grain may consist of sandwiched layers of TiC and Ti_3SiC_2. This may correspond to the lamella TiC/Ti_3SiC_2 structure first described by Nickl et al.[21] Referring again to Figure 9a, the presence of TiC is further evidenced by a distinct line approximately half way through the coating, below which a great deal of the lighter phase, believed to be TiC, exists. Above this line, there is much less of the light phase. It is from this upper region that XRD data was collected, and perhaps one reason TiC was not seen with XRD.

DISCUSSION

Results from the analysis of this complicated microstructure suggest several things about the growth of Ti_3SiC_2 films. First, Ti_3SiC_2 grows large, oriented crystals at 1300°C in a specific range of reagent compositions. These compositions, found in Figure 4, can be summarized in the following manner. In this study, Ti_3SiC_2 grown at 1300°C was found between 32-45% $SiCl_4$, 33-42% $TiCl_4$, and 22-25% CCl_4. Although these 12 experiments are not sufficient to describe the entire deposition diagram, they do corroborate the sensitivity of Ti_3SiC_2 deposition to reagent composition, and especially CCl_4 concentration.

Second, Ti_3SiC_2 probably requires the presence of TiC to grow. This is evidenced by the appearance of a TiC boundary layer separating the substrate from the remainder of the coating, and by the occasional appearance of intermittent TiC layers between the fan-like Ti_3SiC_2 grains. Both Racault et al,[23] and Barsoum et al.[17] have proposed a two step growth mechanism for Ti_3SiC_2, for solid state and CVD synthesis, which involves the intermediate formation of TiC. There is no evidence to suggest otherwise in our work. As seen from the orientation of Ti_3SiC_2, the fast growing planes are the (110) planes, which have the same d-spacing as (220) planes in TiC. We know this from overlapping peaks in X-ray diffraction patterns for each phase. Parallel to the basal plane, the absence of planes of silicon atoms would resolve directly into a TiC crystal structure. In fact, the only thing differentiating the TiC and Ti_3SiC_2 unit cells is the incorporation of planes of silicon atoms in Ti_3SiC_2. This may also explain the structure of the 1100°C coatings. A TiC matrix growing in excess of $SiCl_4$, may incorporate enough silicon to initiate the growth of hexagonal Ti_3SiC_2 grains, thus creating pockets of the secondary phase.

Third, the basal plane of chemically vapor deposited Ti_3SiC_2 is clearly oriented perpendicular to the substrate for the experimental conditions used here. Irrespective of deposition temperature, the (00l) peaks from XRD had consistently smaller relative intensities as compared with the PDF card. This is not conducive to a good fiber/matrix interface material as the orientation of the basal planes must direct propagating cracks away from the fiber. The use of this material in interface applications via CVD synthesis is limited until this orientation problem can be solved. However, the extent and type of preferred orientation for CVD materials often depends on processing variables, particularly deposition temperature. Figure 5, in fact, shows that going from a deposition temperature of 1300 to 1100°C, the basal plane is beginning to orient parallel to the substrate for some grains, as evidenced by the appearance of the (00l) peaks. We hope in future work to identify these variables in order to control the orientation of Ti_3SiC_2.

SUMMARY AND CONCLUSIONS

Chemical vapor deposition of Ti_3SiC_2 was investigated using XRD, SEM, and EDS. The complicated microstructure of these coatings involved the incorporation of TiC and $TiSi_2$. It is believed the appearance of $TiSi_2$ was the result of fluctuations in reagent composition to lower carbon content. On the other hand, TiC is believed to be necessary for the formation of Ti_3SiC_2, either as an intermediate phase or deposited concurrently. At 1300°C, films consisted of large grained structures in which Ti_3SiC_2 basal planes were oriented perpendicular to the substrate. A TiC boundary layer separated these grains from the substrate, and often these grains terminated in intermittent TiC layers. Some of these fan-like grains were shown to consist of layers of TiC sandwiched between layers of Ti_3SiC_2. At 1100°C, the microstructure consisted of a TiC matrix with pockets of Ti_3SiC_2. These pockets still demonstrate the same preferred orientation of the high temperature experiments, but to a lesser degree. More work must be done to understand the kinetics of deposition before this material can be effectively utilized in CVI processes.

ACKNOWLEDGEMENTS

We appreciate the financial support and technical guidance provided by Dr. Alexander Pechenik of the Air Force Office of Scientific Research, Dr. Liselotte Schioler of the National Science Foundation, and First Lieutenant Kenneth A. Self of Wright Laboratories.

REFERENCES

1) S. F. Schuler, J. W. Holmes, X. Wu, and D. Roach, "Influence of Loading Frequency on the Room Temperature Fatigue of a Carbon-Fiber/SiC-Matrix Composite," *J. Am. Cer. Soc* .**76** [9] 2327-36 (1993).
2) B. F. Sorenson and J. W. Holmes, "Improvement in the Fatigue Life of Fiber-Reinforced Ceramics by Use of Interfacial Lubrication," *Scr. Met. et Mat.* **32** [9] 1393-98 (1995).
3) S. Kumar and R. N. Singh, "Effects of Fiber Coating Properties on the Crack Deflection and Penetration in Fiber-Reinforced Ceramic Composites," *Acta Met. et Mat.* **45** [1] 4721-31 (1997).
4) L. R. Hwang, J. W. Fergus, H. P. Chen, and B. Z. Jang, "Interface Compatibility in Ceramic-Matrix Composites," *Comp. Sci. Tech.*, **56** 1341-48 (1996).
5) N. Chawla, J. W. Holmes, and R. A. Lowden, "The Role of Interfacial Coatings on the High Frequency Fatigue Behavior of Nicalon/C/SiC Composites," *Scr. Met. et Mat.*, **35** [12] 1411-16 (1996).
6) N. Iyi, S. Takekawa, and S. Kimura, "Crystal Chemistry of Hexaaluminates: β-Alumina and Magnetoplumbite Structures," *J. Sol. St. Chem.*, **83** 8-19 (1989).
7) M. K. Cinibulk and R. S. Hay, "Textured Magnetoplumbite Fiber-Materix Interphase Derived from Sol-Gel Fiber Coatings," *J. Am. Cer. Soc.,* **79** [5] 1233-46 (1996).
8) D. H. Kuo, W. M. Kriven, and T. J. Mackin. "Control of Interfacial Properties through Fiber Coatings: Monazite Coatings in Oxide-Oxide Composites," *J. Am. Cer. Soc.*, **80** [12] 2987-96 (1997).
9) P. Morgan and D. Marshal, "Functional Interfaces for Oxide/Oxide Composites," *Mat. Sci. Eng.*, **A162** 15-25 (1993).

10) K. K. Chawla, M. K. Ferber, Z. R. Xu, and R. Venkatesh, "Interface Engineering in Alumina/Glass Composites," *Mat. Sci. Eng.*, **A162** 35-44 (1993).

11) W. J. Lackey, S. Vaidyaraman, and K. L. More, "Laminated C-SiC Matrix Composites Produced by CVI," *J. Am. Cer. Soc.*, **80** [1] 113-16 (1997)

12) W. J. Lackey and A. J. Caputo, "Process for the Preparation of Fiber-Reinforced Composites by Chemical Vapor Deposition," U.S Pat. No. 4,580,524 (1986)

13) A. J. Caputo and W. J. Lackey, "Fabrication of Fiber-Reinforced Ceramic Composites by Chemical Vapor Infiltration," *Cer. Eng. Sci. Proc.*, **5**, 654 (1984).

14) A. J. Caputo, W. J. Lackey, and D. P. Stinton, "Development of a New, Faster Process for the Fabrication of Ceramic Fiber-Reinforced Ceramic Composites by Chemical Vapor Infiltration," *Cer. Eng. Sci. Proc.*, **6**, 694 (1985).

15) S. Vaidyaraman, W. J. Lackey, G. B. Freeman, P. K. Agrawal, and M. D. Langman, "Fabrication of Carbon-Carbon Composites by Forced Flow-Thermal Gradient Chemical Vapor Infiltration," *J. Mat. Res.*, **10** [6], 1469-77 (1995).

16) W. Jeitschko and H. Nowotny, "Die Kristallstructur von Ti_3SiC_2-Ein Neuer Komplexcarbid-Type," *Monatash Chemical*, **98** 329-337 (1967)

17) M. W. Barsoum and T. El-Raghy, "Synthesis and Characterization of a Remarkable Ceramic: Ti_3SiC_2," *J. Am. Cer. Soc.*, **79** [7] 1953-56 (1996).

18) T. El-Raghy, A. Zavaliangos, M. W. Barsoum, and S. R. Kalidindi, "Damage Mechanisms Around Hardness Indentations in Ti_3SiC_2," *J. Am. Cer. Soc.*, **80** [2] 513-16 (1997).

19) M. W. Barsoum, T. El-Raghy, and D. Brodkin, "Layered Machinable Ceramics for High Temperature Applications," submitted to: *Scripta Materials.*

20) M. W. Barsoum, T. El-Raghy, "Oxidation of Ti_3SiC_2 in Air," submitted to : *J. Elec. Soc.*

21) J. Nickl, K. K. Schweitzer, and P. Luxenberg, "Chemical Vapor Deposition of the System Ti-Si-C and Ti-Ge-C," Proc. 3rd Conf. on CVD, Salt Lake City, Utah (1972).

22) T. Goto and T. Hirari, "Chemical Vapor Deposition of Ti_3SiC_2," *Mat. Res. Bul.*, **22**, 1195-1201 (1987).

23) C. Racault, F. Langlais, R. Naslain, and Y. Kihn, "On the Chemical Vapor Deposition of Ti_3SiC_2 from $TiCl_4$-$SiCl_4$-CH_4-H_2 Gas Mixture: Part II, An Experimental Approach," *J. Mat. Sci.*, **29** [15] 3941-48 (1994).

24) E. Pickering, W. J. Lackey, and S. Crain, "Thermodynamic Modeling of the Ti-Si-C-H-Cl-Ar System to Determine Optimum Conditions for Chemical Vapor Deposition of Ti_3SiC_2," Proceedings of the Am. Cer. Soc. Conference, CoCoa Beach Fl. January (1998).

25) C. Racault, F. Langlais, and R. Naslain, "Solid State Synthesis and Characterization of the Ternary Phase Ti_3SiC_2," *J. of Mat. Sci.*, **29** [13] 3384-3391 (1994).

26) B. D. Culity, "Elements of X-Ray Diffraction," 2nd Ed. Addison-Wesley Publishing Company, (1978).

THE MECHANICAL RESPONSE OF SINTERED ALUMINA BUNDLES AND POROUS ALUMINA MATRIX COMPOSITES

Paul E. Cantonwine and Haydn N.G. Wadley
Materials Science and Engineering Department
University of Virginia, Charlottesville, VA 22903.

ABSTRACT

The effect of high temperature sintering upon the mechanical response of Nextel 610™ alumina bundles has been investigated prior to and after infiltration of an alumina loaded slurry. Room temperature single filament and dry bundle strengths were found to decrease with sintering temperature (1100 to 1450 °C). The drop in strength was correlated to a change in the defect population. Initially 50% of the fibers failed from internal defects while at the highest sintering temperature, most filaments failed from weld-line defects caused by filament-to-filament sintering. For sintering temperatures less than or equal to 1300 °C, an ideal bundle model predicted the dry bundle response well. However, the model overpredicted the failure stress after sintering above 1400 °C. The discrepancy between theory and experiment was linked to the collective failure of filament clusters. The ambient strength of single tow porous alumina matrix composites were also found to decrease with sintering temperature and deviated from ideal bundle behavior. The temperature of deviation decreased as the matrix density increased. At a low matrix density (~ 55% porosity), the behavior paralleled that of the dry bundles. At a higher matrix density (~ 40% porosity), the lower temperature of deviation from ideal bundle behavior was linked to a change in the fracture mode. This change appeared to be associated with an increase in the interfacial strength and/or sintering induced surface defects.

INTRODUCTION

Our objective here was to gain some fundamental insight into how high temperature sintering affected the structure and properties of oxide/oxide ceramic matrix composites (CMCs). With that in mind we decided to initially eliminate the complexities of coatings (e.g. Monazite [1]) and investigate a simple porous alumina CMC. Significant interest in porous oxide CMCs has recently developed [1-5] because the porous matrix provides a weak interface without incurring the expense of chemical vapor deposited coatings necessary for dense matrix composites [6]. Single tow porous oxide CMCs have also been proposed as a hybrid composite fiber for reinforcing metals [7].

The essential microstructural features of a porous matrix CMC are shown in figure 1. When heated the powder-to-powder, powder-to-filament, and filament-to-filament contacts can all grow. Here we will focus on describing how the mechanical properties of heat treated alumina bundles are affected by filament-to-filament sintering. We will then briefly evaluate the effects of the powder-to-powder and powder-to-filament sintering phenomena.

To the extent authorized under the laws of the United States of America, all copyright interests in this publication are the property of The American Ceramic Society. Any duplication, reproduction, or republication of this publication or any part thereof, without the express written consent of The American Ceramic Society or fee paid to the Copyright Clearance Center, is prohibited.

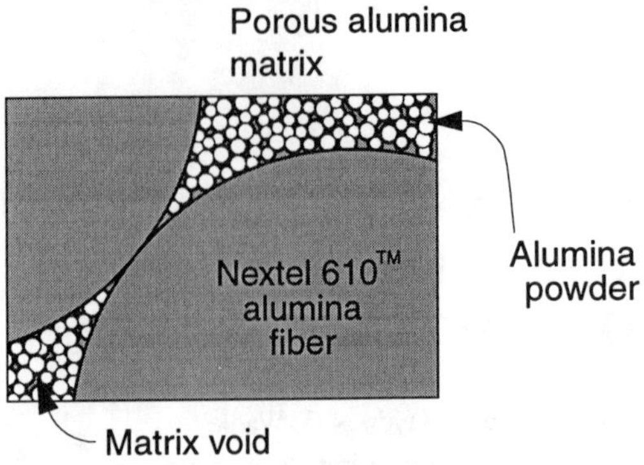

Figure 1. Schematic microstructure of a porous alumina matrix composite.

EXPERIMENTAL PROCEDURES

Materials and Processing

The Nextel 610™ alumina multifilament tow made by 3M was used for this research. Nextel 610™ is 99.5% Al_2O_3 with approximately 400 filaments per tow (fiber diameter ≅ 12 μm [8]). This fiber was designed as a high strength fiber for use below 1000 °C [9]. The high strength was achieved by keeping the grain size small (less than 100 nm) and limiting the processing defects [9]. The tow was sized into a flat shape previous to winding onto a spool to allow for easy removal.

Before heat treating and testing, the tows were resized and reshaped into "cylinders". The flat tows were first desized. A methanol/polyvinyl butyral (PVB) solution was then infiltrated into the tow, and the tow was reshaped by drawing it through a glass tube with a circular orifice (320 μm). The solvent was then driven off in a drying furnace. Single tow CMC fibers were similarly manufactured, but with the PVB solution replaced with one of two slurries containing either 15 or 30 vol% of 1 μm alumina powder. A silicone resin was also added to the 30 vol% slurry which during sintering can oxidize and act as a sintering aide.

The effects of sintering were investigated via an in-line sintering (ILS) process. Filament bundles and single tow CMC fibers were continuously drawn at 30 cm/min through two furnaces: The line tension was around 0.7 N to avoid significant sagging. To burn off the residual organics from the slurries, the temperature of the first furnace was set at 750 °C. The second or sintering furnace was set at temperatures ranging from 1100 to 1450 °C. Since, the hot zone was about 30 cm in length, the ILS fibers were at temperature for about 1 minute. An experiment to represent the as-received condition was done with the sintering furnace at room temperature. Before winding onto a spool, the filament bundles were again re-sized using a horizontal glass tube (orifice diameter, 320 μm) and a PVB/methanol solution.

Testing

Single filaments tests were conducted at room temperature following each sintering experiment. The mounting procedure was as follows: Two pieces of tape were placed on the underside of a glass slide with a strip of paper connecting them (2.5 cm gauge length). A single filament was removed from a bundle and placed onto the adhesive side of the tape. Another small piece of tape was then put over the filament effectively sandwiching the filament between the tape. The mounted filament, now ready for testing, was then removed from the glass slide.

In four out of seven heat treatments (as-received, ILS 1300 °C, 1400 °C, and 1450 °C), the failure load of at least 45 single filaments was measured and the fracture surfaces saved. The fracture surfaces were saved by embedding the mounted filaments in petroleum jelly during the testing. In the other heat treatments, about 25 single filaments were tested and no attempt was made to save the fracture surfaces. The fracture initiation sites of the saved samples were observed and the filament diameters measured via scanning electron microscopy (SEM).

To prepare the filament bundles and single tow CMC fibers for testing, they were first cut into 4 to 5 inch specimens. The bundle specimens were then put into a 1" long furnace to burn-off the organic sizing within the approximate 2.5 cm gauge length. The bundles and CMC fibers were then mounted into paper tabs. The CMC fibers were not laterally constrained which allowed the filaments to spread out upon failure, similar to the bundles. The failure loads of the single filaments and the stress-strain response of the bundles and CMC fibers were measured at room temperature using an ATS series 1100 tensile testing machine with a 50 lb. load cell, 0.25 cm/min crosshead speed, rubber face grips and an OPTRA 3000 laser extensometer.

The grain size was measured from SEM photomicrographs of etched filament surfaces.

RESULTS AND DISCUSSION

Alumina Single Filaments

Filament strengths were calculated from the measured failure loads and filament diameters. The strengths were then sorted in ascending order and assigned an integer value, *i*, representing their position. The probability of failure was then calculated for each strength using the relationship:

$$P_f = \frac{i - 0.5}{N} \tag{1}$$

where N is the total number of tests.

The strength of ceramic materials can be represented by a Weibull distribution. Ignoring volume effects from slight variations in filament diameters (d = 11.9 ± 1.5 μm), the probability of failure, P_f, can be written as:

$$P_f = 1 - exp\left\{\frac{-L}{L_o}\left(\frac{\sigma}{\sigma_o}\right)^m\right\} \tag{2}$$

Where σ is the filament stress, σ_o and m are the Weibull reference stress and modulus, respectively, L is the test gauge length of interest, and L_o is the reference gauge length where σ_o and m were calculated.

Plotting $Ln(Ln(1/(1-P_f))$ versus $Ln(\sigma)$ gives m as the slope. σ_o is the stress at a probability of failure of 0.63. The Weibull parameters are given in Table I. There is a slight decrease in reference stress after heat treating at 1100 and 1200 °C followed by a 0.2 to 0.3 GPa decrease at 1300, 1350 and 1400 °C. At 1450 °C the reference stress decreased almost 1 GPa. Also note that the Weibull modulus appears to decrease and then increase as the sintering temperature increases. The reason for this will be discussed later.

Wilson [10] recently reported on the strength distribution of Nextel 610™. Using the single gauge length technique described above, Wilson found m = 11.5. He also reported the average strength varied between 3.0 and 3.5 GPa which compares favorably to our results for the as-received filaments.

Table I: The Effect of In-Line Sintering on Nextel 610™: Filament Strength Parameters and Grain Size

Heat Treatment	Weibull Reference Stress (GPa)	Weibull Modulus	Average Grain Size (nm)
As-Received	3.37	11.0	65 ± 6
ILS 1100 °C	3.30	7.1	69 ± 7
ILS 1200 °C	3.30	7.0	69 ± 8
ILS 1300 °C	3.09	7.0	75 ± 8
ILS 1350 °C	3.10	8.1	78 ± 9
ILS 1400 °C	3.00	9.0	91 ± 11
ILS 1450 °C	2.51	11.0	100 ± 15

The average grain size and standard deviations are also shown in Table I. Grain growth is a possible mechanism for the decrease in filament strength [11]. The mean filament strength is plotted versus the inverse square root of grain size in figure 2 and compared to previous experiments on Nextel 610™ by Xu *et al.* [12]. The large increase in strength compared with Xu *et al.* was a result of improved processing procedures after 1993 [10]. The possible role of grain size on filament strength will be discussed in a subsequent section.

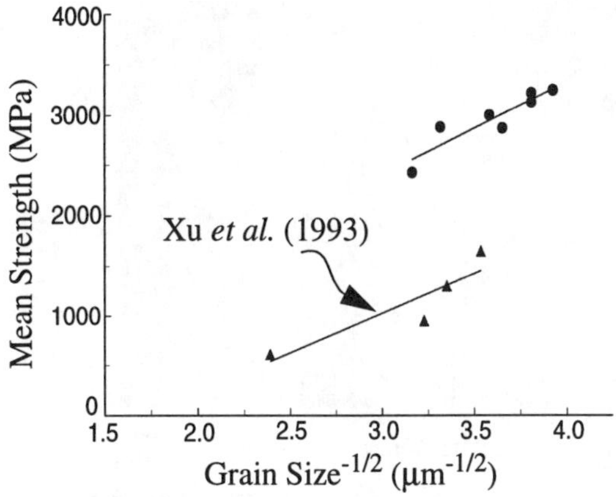

Figure 2. Mean filament strength versus inverse square root of grain size for the in-line sintered samples and the results of Xu *et al.* [12]

SEM observations were used to identify the defects which caused crack initiation and catastrophic failure. Defects were initially classified as either internal or surface. Three types of internal defects are shown in figure 3: a single spherical pore, non-spherical pores and an unidentified defect (an area of intergranular fracture or an inclusion). Five different types of surface defects were also observed (see figure 4). One defect was non-geometric and associated with a small area of intergranular fracture. The four other defects are geometric in nature. The weld-line defect shown in figure 4b was caused by filament-to-filament sintering. Figure 4c shows an uneven surface area that appears similar to a body sore or blister, hence referred to as a blister defect. A surface crack is shown in figure 4d, and a surface nodule associated with the crack initiation site is shown in figures 4e and 4f.

The statistical distribution of the defects are presented in Table II. It can be seen that the flaw population changed with the sintering temperature. In the as-received condition, 50% of the filaments failed from a surface defect and 50%, from an internal defect. At 1400 °C no internal defects initiated fracture while weld-line defects accounted for over 30%. At 1450 °C the percentage of weld-line defects increased to more than 80%.

The variation of the Weibull modulus with sintering temperature can be qualitatively explained by looking at the average defect strengths and their standard deviations in Table III. The average strengths and standard deviations at ILS 1300 °C vary more than the as-received case which would translate into a decrease in Weibull modulus. At the higher heat treatments, the variations in strength and standard deviations become less corresponding to the observed increase in the Weibull modulus.

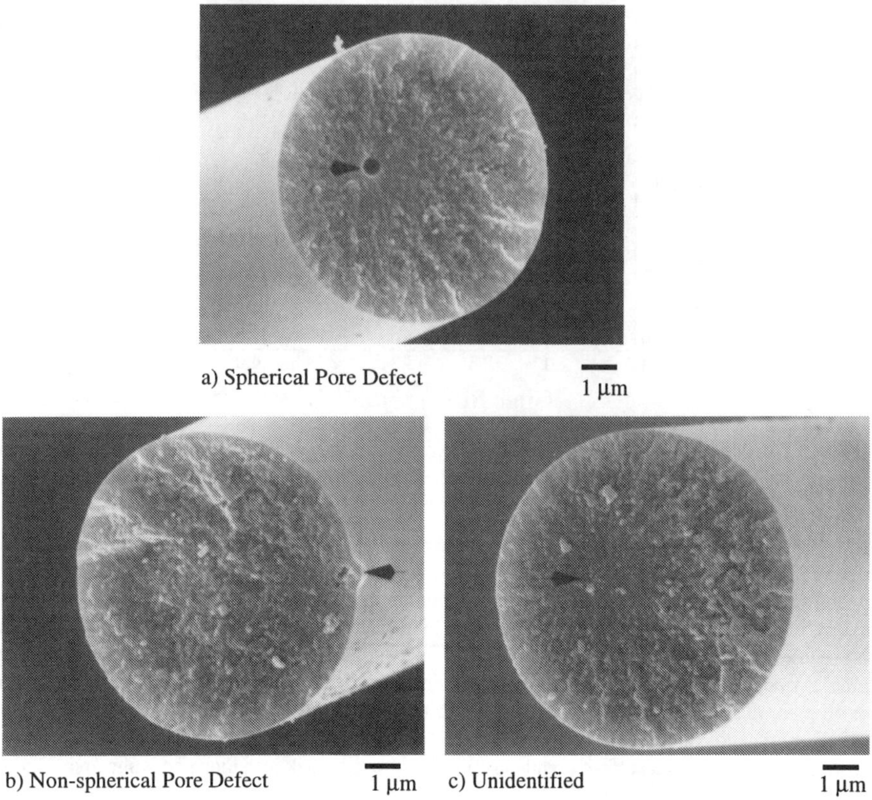

a) Spherical Pore Defect 1 μm

b) Non-spherical Pore Defect 1 μm c) Unidentified 1 μm

Figure 3. Fracture surfaces of single alumina filament tests; internal defects: a) spherical pore, b) non-spherical pores, c) unidentified

In 1993, Xu *et al.* [12] concluded that the primary reason for the filament strength loss was grain growth. Their rational was that since there was a linear relationship between strength and the inverse square root of the grain size, the most severe microflaw had a length equal to the grain size. However, calculated defect sizes are much larger then the average grain size. The failure stress can be related to crack depth using $\sigma = K_{1c}Y/(\pi b)^{1/2}$; where Y is the geometric factor for a semi-elliptical surface crack in a long shaft under tension [13], K_{1c} is the fracture toughness (assumed to be 2.5 MPa $m^{1/2}$ for alumina [14]), and b is the crack depth. The predicted crack depth for a fiber with a strength of 3 GPa would be approximately 400 nm which is well above the average grain size of even the fiber heat treated at 1450 ^{o}C. The measured defect sizes also varied from 300 to 800 nm. This follows the usually explanation of the Petch branch in the Orowan-Petch plots where the flaws controlling the strength are extrinsic [11,15]. Therefor, it is proposed that the strength loss with sintering temperature is primarily caused by the dramatic change in the defect population. However, this does not preclude grain growth from playing a role in the strength loss especially if there is any sub-critical crack growth along the grain boundaries.

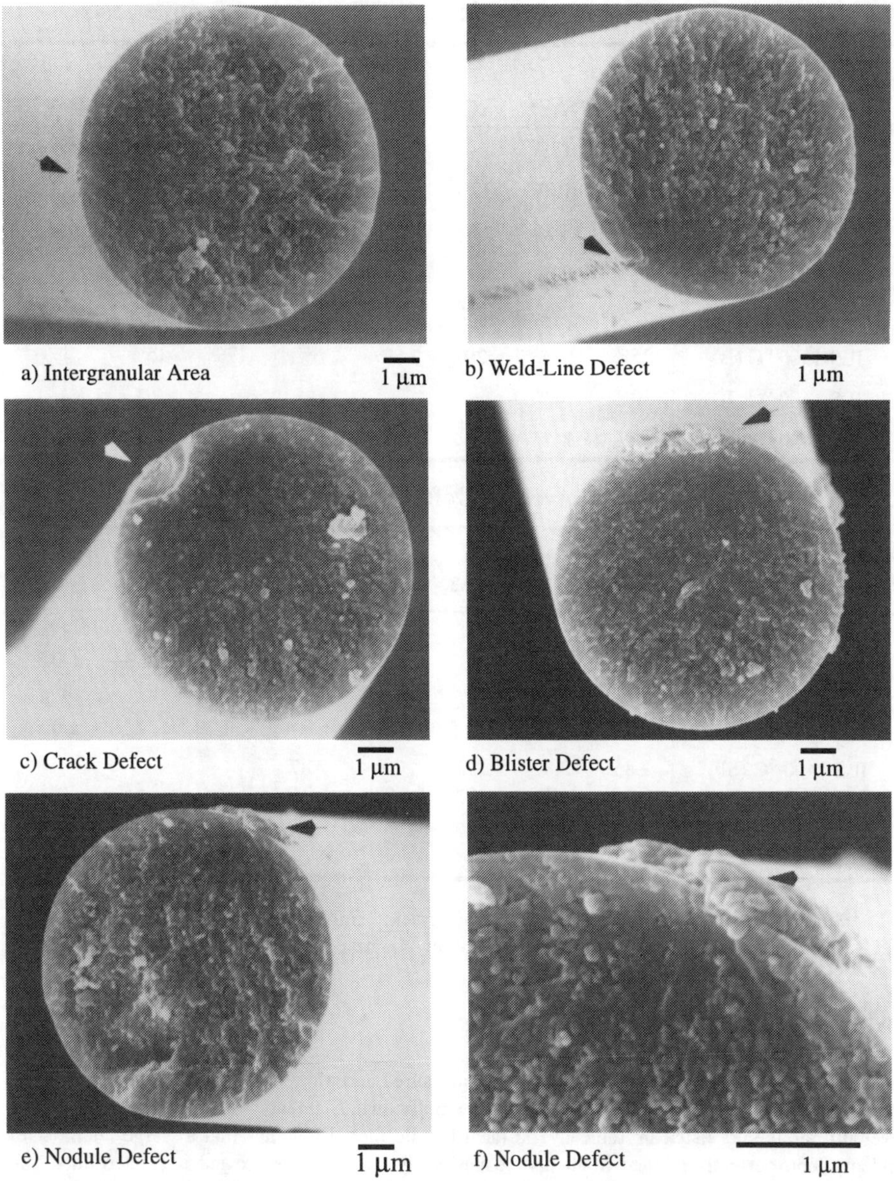

Figure 4. Fracture surfaces of single alumina filament tests; surface defects: a) area of intergranular fracture, b) weld, c) crack, d) blister, e) nodule, f) nodule

Table II: Defect Distribution of Single Filament and Bundle Tests

	Defect Type							
Heat Treatment	SG[1]	SW[2]	SC[3]	SN[4]	SB[5]	IU[6]	IP[7]	IPY[8]
As-Received (SF[9])	24%	17%	-	5%	2%	14%	26%	12%
ILS 1300 °C (SF)	24%	15%	12%	27%	12%	-	-	10%
ILS 1400 °C (SF)	23%	33%	14%	23%	7%	-	-	-
ILS 1450 °C (SF)	12%	82%	-	-	4%	-	2%	-
As-Received (B[10])	32%	4%	4%	7%	-	21%	23%	9%
ILS 1300 °C (B)	25%	25%	2%	6%	6%	17%	15%	4%
ILS 1400 °C (B)	30%	68%	-	-	-	-	2%	-
ILS 1450 °C (B)	13%	81%	-	2%	-	-	4%	-

Table III: Average Defect Strength (GPa)

	Defect Type							
Heat Treatment	SG[1]	SW[2]	SC[3]	SN[4]	SB[5]	IU[6]	IP[7]	IPY[8]
As-Received (SF[9])	3.2 ±0.4	3.3 ±0.3	-	-	-	3.3 ±0.3	3.5 ±0.3	3.0 ±0.3
ILS 1300 °C (SF)	3.16 ±0.4	3.15 ±0.4	2.7 ±0.5	3.2 ±0.4	2.6 ±0.4	-	-	2.6 ±0.5
ILS 1400 °C (SF)	2.85 ±0.3	2.87 ±0.5	2.7 ±0.5	2.8 ±0.2	2.87 ±0.2	-	-	-
ILS 1450 °C (SF)	2.38 ±0.3	2.44 ±0.4	-	-	-	-	-	-

[1] Surface grain; [2] Surface weld-line; [3] Surface crack; [4] Surface nodule; [5] Surface blister; [6] Internal unidentified; [7] Internal spherical pore; [8] Internal non-spherical pores; [9] Single filament tests; [10] Bundle tests;

Alumina Filament Bundles

The stress-strain response of dry bundles were measured and plotted in figure 5 together with the predicted response of an "ideal" bundle whose individual filaments were characterized by the Weibull parameters listed in Table I. The ideal bundle model assumes that a "large" number of independent, perfectly aligned filaments are uniformly loaded; Phoenix and Raj determined that bundles should contain 200 or more filaments to exhibit ideal behavior [17]. Global load sharing is also assumed (i.e. when a filament fails, its load is evenly distributed among the remaining unbroken filaments). The ideal bundle stress, σ_b, is given by [16]:

$$\sigma_b = \sigma(1 - P_f) \tag{3}$$

Where σ is the filament stress obtained from the bundle strain, ε_b, and filament modulus, E_f, using Hooke's law ($\sigma = E_f\varepsilon_b$).

Substituting equation (2) into equation (3) gives an expression for the ideal bundle stress-strain response:

$$\sigma_b = E_f\varepsilon_b \exp\left(-\frac{L}{L_o}\left(\frac{E_f\varepsilon_b}{\sigma_o}\right)^m\right) \tag{4}$$

This result is compared with data collected up to the ultimate stress, since the testing was performed with crosshead control which did not allow for the capture of the response during unstable failure.

The average modulus for all the tests was measured to be 355 ± 25 GPa. This was slightly less than the reported modulus of the Nextel 610™ filaments, 380 GPa [9]. The less than reported modulus was expected since there were typically a few broken filaments in the as-received bundles. After in-line sintering at or below 1300 °C, the ideal bundle model was found to predict the measured stress-strain response well. Note that the measurements also capture the predicted non-linear behavior due to individual filament fracture. At 1400 and 1450 °C, the ideal bundle model overpredicts the stress and strain at failure.

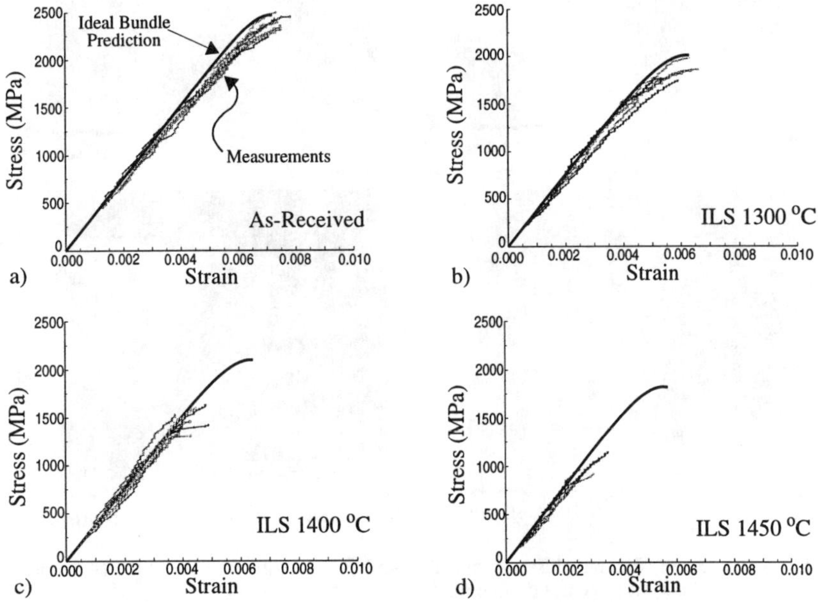

Figure 5. Stress-Strain measurements and ideal bundle predictions for as-received alumina bundles after various in-line sintering treatments.

SEM micrographs of filament failures in the bundles indicated a similar change in the defect population compared with the single filament results (see Table II). Initially over 50% of the failures are due to internal defects. After in-line sintering at 1450 °C, over 80% of the failures are due to a weld-line defect. In the as-received and 1450 °C specimens, the defect population sampled in the single filament tests were essentially identical to that observed in the bundle tests. However, there were some slight differences in the defect distributions when sintering at 1300 and 1400 °C. At 1300 °C, there were more internal pore and unidentified defects observed in the bundle tests versus the single filament tests. Since the internal pore and unidentified defects were generally the strongest, the measure strengths from the single filament tests may be slightly lower than the actual filament strength. At 1400 °C there were significantly more weld defects observed in the bundle tests, but there appears to be little difference in the strengths of the surface defects after heat treating at this temperature. Since there were no dramatic differences in the defect populations of the single filament and bundle tests, the strengths determined in the single filament test should accurately describe the strengths of the filaments in the bundles.

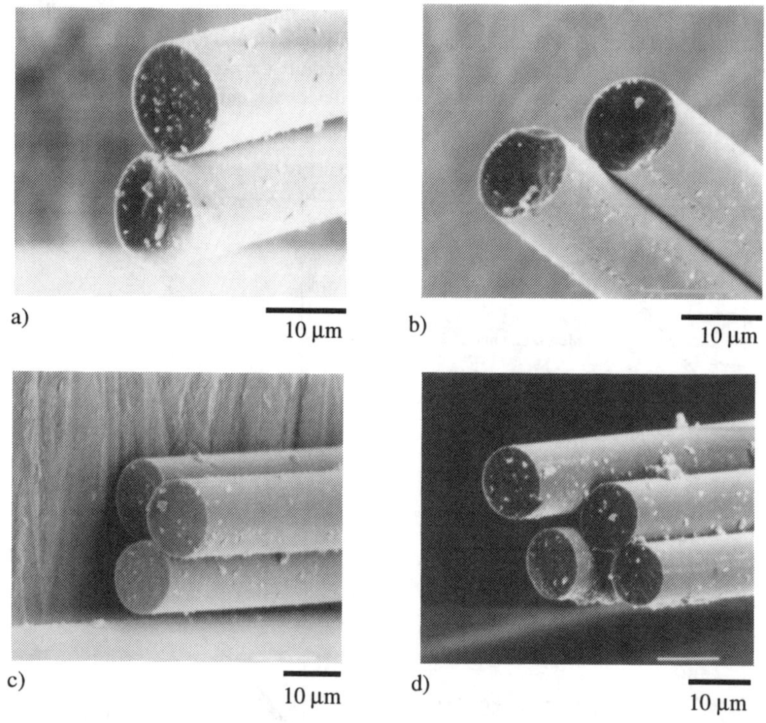

Figure 6. Failure of filament clusters from ILS 1450 °C. Note the fracture of the filament clusters occur on or near the same plane.

The observations of the weld-line defects implies the existence of bonded filament clusters which brings into question the assumption that the filaments fail independent of their neighbors. With this assumption the probability that a filament would fail within a diameter of its neighbor is equal to the filament diameter (12 μm) divided by the gauge length (2.54 cm): $5x10^{-4}$. Similarly, the probability of a three filament cluster would be $2x10^{-7}$, and for a four filament cluster, 10^{-10}. However filament clusters with as many as four filaments have been observed (see figure 6). Therefor, the assumption of independence from neighbors is invalid when filament clustering occurs.

Single Tow CMC Fibers

The interfacial shear strength in a porous matrix CMC is expected to increase as the matrix density increases (i.e. powder-to-powder and powder-to-filament contacts increase). Two single tow CMC fibers with different matrix densities were made by varying the powder content in the slurry that infiltrated the tow prior to heat treatment. The first slurry contained 15 vol% alumina powder and the second, 30 vol%. These single tow CMC fibers were then in-line sintered at various temperatures and tested in tension at room temperature. The percent porosity in the matrix was determined from denier and cross sectional area measurements. The matrix porosity in the 15 vol% sample was approximately 55% while the 30 vol% sample was approximately 40%. To avoid the difficulties of defining cross sectional area when comparing bundles to CMCs, the failure loads are plotted versus in-line sintering temperature in figure 7.

The single tow CMC fibers made with 15 vol% alumina powder paralleled the results of the dry bundles. This indicated that the mechanism causing the under-performance compared to the ideal bundle prediction was the same as in the dry bundles (i.e. filament-to-filament sintering). However, the failure load of the CMC fibers made with 30 vol% alumina powder departed from ideal bundle behavior more rapidly with sintering temperature. Fractography indicated that the decrease in the failure load correlated to a transition from a "tough" bundle failure to a more brittle, localized failure. This would indicate a transition from global load sharing to local load sharing or matrix crack penetration of the fibers. The sintering of the porous matrix to the filaments may also be creating surface nodule type defects.

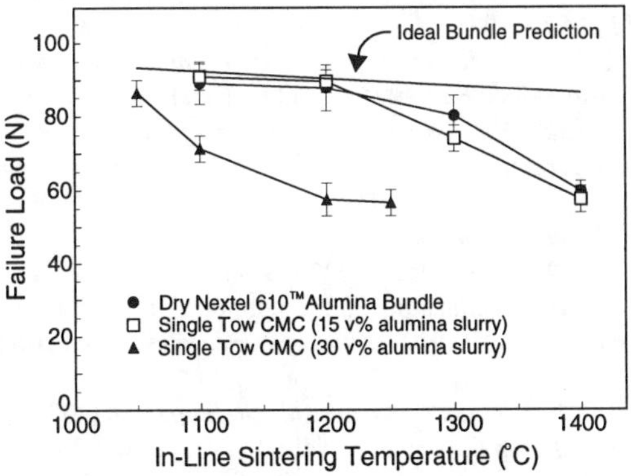

Figure 7. Failure load versus in-line sintering temperature for the alumina bundles and single tow CMC fibers containing different volume fractions of alumina. The time at temperature was about 1 minute.

CONCLUSIONS

The room temperature strength of Nextel 610^{TM} single filaments decreased as the sintering temperature increased which correlated to a change in the defect population. In the as-received fiber, 50% failed from internal defects and 50% failed from surface defects. At the highest sintering temperature, weld-line defects resulting from filament-to-filament sintering controlled the strength. For sintering temperatures less than or equal to 1300 ^{o}C, the ideal bundle model predicted the Nextel 610^{TM} bundle behavior well. However, for temperatures at or above 1400 ^{o}C, the ideal bundle model overpredicted the strength. This was caused by the formation of filament clusters via filament-to-filament sintering. The failure loads of single tow porous matrix CMCs with 55 or 40% matrix porosity were compared to the measured and predicted bundle failure loads. The strength of the 55% porous matrix composites were controlled by the filament clustering mechanism similar to the dry bundles. Decreasing the matrix porosity to 40% caused a premature drop in the strength. This appears to result from a combination of local load sharing and powder-to-filament sintering induced defects.

ACKNOWLEDGEMENTS

We are grateful to Dana Elzey, Carl Herakovich, Dave Wilson and Herve Deve for their helpful discussions, and to Tommy Eanes and Willy Shoupe for their practical assistance with the experimental work. This work has been supported by an AFOSR-URI grant #F49620-93-1-0359.

REFERENCES

1. P.E.D. Morgan and D.B. Marshall, *J. Am. Cer. Soc.,* 78 [6], pp. 1553-63 (1995).
2. W.Tu, F. Lange, and A. Evans, *J. Am. Ceram. Soc.*, 79 [2], pp. 417-424 (1996).
3. C.G. Levi, J.Y. Yang, B.J. Dalgeish, F.W. Zok and A.G. Evans, *J. Am. Ceram. Soc.*, in press.
4. J.A. Heathcote, X.Y. Gong, J. Yang, U. Ramamurty and F.W. Zok, *J. Am. Ceram. Soc.*, submitted.
5. E. Mouchon and Ph. Colomban, *Composites*, 26 [3], pp. 175-182 (1995).
6. J. B. Davis, J. P.A. Lofvander, A. G. Evans, E. Bischoff and M.L. Emiliani, *J. Am. Ceram. Soc.,* 76 [5], pp. 1249-57 (1993).
7. P. Cantonwine and H. Deve, *Recent Advances in Titanium Metal Matrix Composites; Proceedings from Materials Week Fall 1994,* edited by F.H. Froes and J. Storer, TMS (1995).
8. M.M. Vogel-Martin and D.M. Wilson, *Proceedings from the 16th Conference on Metal Matrix, Carbon and Ceramic Matrix Composites;* NASA conference publication 3175 part 2, pp. 519-533 (1992)
9. D. Wilson, *personal communication*
10. D. Wilson, *J Mat Sci*, 32 [10], pp. 2535-2542 (1997)
11. B. Lawn, *Fracture of Brittle Solids,* pp. 332, Cambridge University Press (1993)
12. Z.R. Xu, K.K. Chawla and X. Li, *Materials Science and Engineering,* A171, pp. 249-256 (1993).
13. H. Nisitani and D.H. Chen, *Trans Japan Soc. Mech. Engrs.*, 50 [453], pp. 1077-1082 (1984).
14. J. Seidel and J. Rodel, *J. Am. Ceram. Soc.,* vol 80, no 2, pp. 433-38 (1997).
15. R.E. Tressler, R.A. Langensiepen, and R.C. Bradt, *J. Am. Ceram. Soc.*, 57 [5], pp. 226-7 (1974)
16. H.E. Daniels, *Proc. Royal Soc. (London),* 183A pp. 405, (1945).
17. S. Phoenix and R.Raj, *Acta Metall. Mater.,* vol 40, no. 10, pp. 2813-2828 (1992)

SILICON CARBIDE LAMINATES WITH POROUS INTERLAYERS MADE BY EPD

Luc J. Vandeperre & Omer O. Van Der Biest
Department of Metallurgy and Materials Engineering
Catholic University Leuven
De Croylaan 2
3001 Heverlee
Belgium

ABSTRACT

The production of Silicon Carbide (SiC) laminates with porous interlayers via electrophoretic deposition was performed by depositing layers of SiC and layers of SiC with incorporation of starch, which after pyrolisis leaves pores in the sample. Attention is given to the difference in deposition rate of the two types of powders during co-deposition; the obtained microstructures are characterized and the mechanical behaviour in bending of the laminates is discussed.

INTRODUCTION

Silicon carbide/graphite laminates have been introduced as an economical way to obtain tough ceramics[1]. However, the poor oxidation resistance of the graphite interlayers at temperatures above 600°C, does not allow to use these laminates at elevated temperatures without protective coatings. Fortunately, graphite interlayers are not the only type of interlayer that can cause cracks to deflect. In alumina based laminates for instance porous interlayers have been used

To the extent authorized under the laws of the United States of America, all copyright interests in this publication are the property of The American Ceramic Society. Any duplication, reproduction, or republication of this publication or any part thereof, without the express written consent of The American Ceramic Society or fee paid to the Copyright Clearance Center, is prohibited.

successfully to obtain tough laminates[2]. Since SiC has a relatively good oxidation resistance due to the formation of a protective silica coating, it can be expected that a silicon carbide laminate with porous interlayers would be applicable at high temperatures, i.e. above 1500°C.

Further, for production of laminated components, versatile forming techniques are required. Ceramic laminates have been made using a variety of ceramic forming techniques such as powder pressing[3], slip casting[4], tape casting[5], solid freeform techniques[6] and electrophoretic deposition[7]. The latter technique is especially attractive as it allows to perform the lamination of the object and the shaping simultaneously, whilst allowing to form complex shapes. Previous work on electrophoretic shaping in our laboratory included the development of a suspension from which SiC green shapes can obtained with incorporation of the sintering additives so that pressureless sintering leads to densities greater than 95%[8], the electrophoretic deposition of silicon carbide laminates with graphite[9] and with SiC/graphite composite interlayers[10]. As an example of the shaping capabilities of the process, laminated SiC/graphite tubes were produced[11]. In this paper the electrophoretic deposition of silicon carbide laminates with porous interlayers will be described together with the mechanical behaviour of these laminates as a function of the porosity level of the interlayers. Furthermore, using the results on electrophoretic co-deposition of two powders with different deposition rates, the effect of these different deposition rates on the composition of the deposited layers will be discussed.

EXPERIMENTAL

Suspensions containing either a boron doped silicon carbide powder (SiC, Superior Graphite, HSC-059s), a starch powder (unmodified rice or corn starch or an analytical starch from Merckx) or a mixture of these powders were prepared by vigourously stirring and ultrasonicating. The suspension medium was technical acetone with addition of 20 vol% technical iso-propyl alcohol and 5 vol% n-butylamine. More details on the effect of iso-propyl alcohol and n-butylamine on the deposition rate and stability of the suspension can be found elsewhere[12]. The solids' loading of the suspensions was 100 g/l.

After suspension preparation, deposits were made by applying 310 V for a given time to the suspensions using two stainless steel electrodes, which are 3.5 cm apart. The volume of suspension contained within the deposition cell was 50 ml.

The deposition rate of the respective powders was determined first by using suspensions containing only one type of powder. The electrophoretic mobility was estimated from the yield, which for the conditions used follows the following simple equation[13]:

$$Y = M(1 - e^{-k.t})$$

with : Y : yield of deposition (g)
M : initial amount of powder in suspension (g)
k : a kinetic constant (s^{-1})
t : deposition time (s)

The kinetic constant, k, can be related to the experimental conditions using :

$$k = \frac{\mu.E.S}{H}$$

with : μ : the electrophoretic mobility ($m^2/V.s$)
E : the electric field strength (V/m)
S : the deposition electrode surface area (m^2)
H : the volume of the suspension (m^3)

Hence, the electrophoretic mobility can be estimated from the yield.

For production of laminates, the deposit was moved from one suspension to another repetitively so that a laminate consisting of 4 SiC and 3 SiC+starch layers was obtained.

TGA measurements were used to determine the optimal burn-out schedule for the starch particles in the deposit prior to sintering. Sintering was performed in a graphite furnace under vacuum at 2050°C for 30 minutes.

Preparation for microstructural characterisation with SEM (JEOL 733) consisted of grinding and polishing with a final surface finish of 0.5 μm.

Three point bend tests were performed on samples with typical dimensions of 25 mm x 3 mm x 2 mm. The span length was 20 mm and the load was measured using a load-cell of 500 N. A cross-head spead of 50 μm/min. was used.

RESULTS

ELECTROPHORETIC DEPOSITION

Deposition occurred on the positive electrode for starch as well as for SiC, indicating a negative charge on the powder. The iso-electric point in water of the SiC powder used in this investigation was determined previously and found to be pH 3[11] The iso-electric point of the starch powders was also situated at pH 3. Taking into account these acid iso-electric points, the negative powder charge on the powder in a basic medium is in accordance with what can be expected. The electrophoretic mobilities as estimated from the yield are reported in table I, and show that the deposition rate is higher for starch than for SiC.

Table I : particle size and electrophoretic mobility (μ)of SiC and starch. The electrophoretic mobility was estimated from the yield of deposition. (a large part of the deposit slides of the electrode upon removal from the suspension)*

solid	size (μm)	μ (cm^2/Vs)
SiC	0.7	1.9×10^{-4}
rice starch	5	3×10^{-4}
corn starch	15	3×10^{-4}
analytical starch	40	$>1 \times 10^{-4}$ *

In order to determine whether during co-deposition, preferential deposition of SiC or starch occurred, a series of consecutive deposits was made from the same suspension containing 50 wt% of corn or rice starch and 50 wt% SiC. For the first two deposits the deposition time was 2.5 minutes (A&B), for the last two 5 minutes (C&D). Hence this series of deposits can be regarded as a number of slices in the thickness direction of a deposit and if both types of particles deposit at an equal rate, the amount of starch and SiC incorporated in the deposits A through D should be the same. The actual amount of starch incorporated in the deposits was determined by burning out the starch by heating in air in a thermogravimetric apparatus. The obtained TGA-curves are shown in figure 1 for the case of corn starch. It is clear that the amount of starch in the deposits is higher than 50 wt% for the first deposit, and decreases with each new deposit. Therefore the starch powder, which also has a higher electrophoretic mobility, deposits faster than the SiC powder. Figure 2 compares the measured composition of the deposits with the calculated composition of the deposits. For the calculation it was assumed that there is no interaction between the depositing powder particles, i.e.

each type of powder deposits at a rate imposed by its own electrophoretic mobility given in table 1. As can be seen there is a good agreement between these calculations and the measured composition.

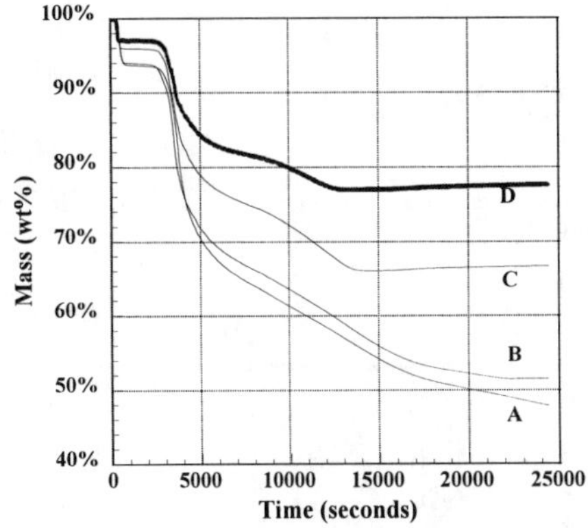

Figure 1: weight loss during heating in air to 600 °C with a heating rate of 1.66 °C/min of consecutive deposits (A-D) from a suspension containing 50wt% of corn starch and 50 wt% of SiC.

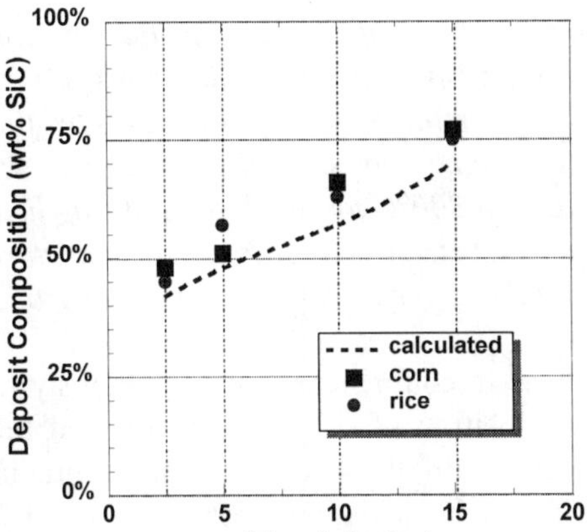

Figure 2 : measured and calculated composition of consecutive deposits from a suspension containing 50 wt% corn or rice starch and 50 wt% SiC

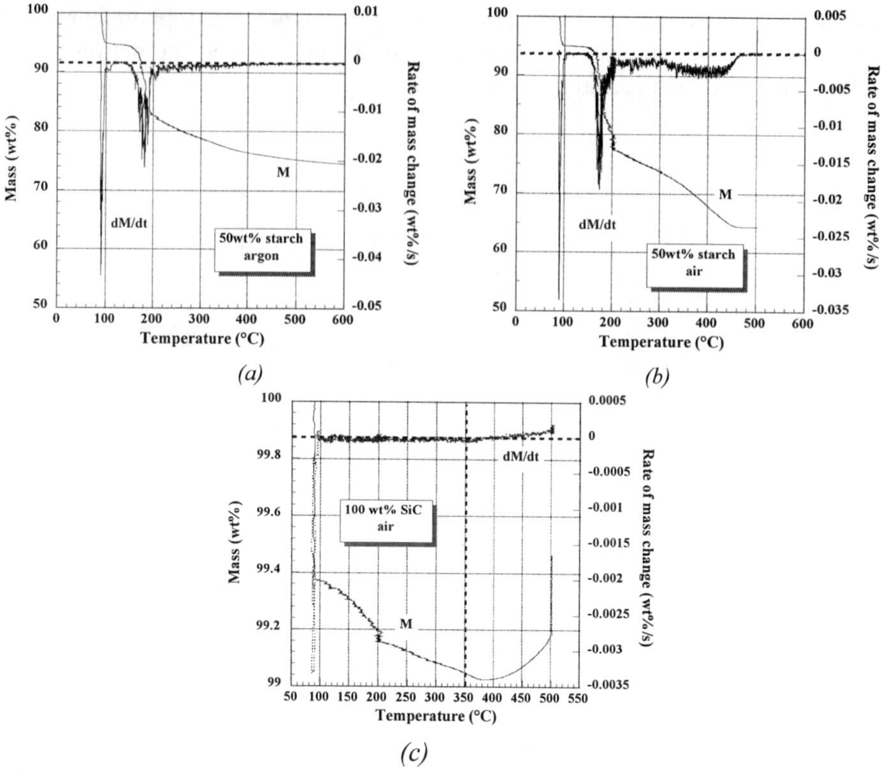

Figure 3: mass change and rate of mass change during (a) heating in argon of a sample deposited from a suspension containing 50wt% starch, (b) heating in air of a sample deposited from the same suspension, (c) heating in air of a deposit from a suspension containing only SiC. The heating rate was 1.67°C/min for all but the sample was held at 200°C for 30 minutes during heating in air. Note that during pyrolisis in air an extra weight loss step occurs starting at about 300°C, explaining the lower carbon yield during pyrolisis in air.

These results seem to be in contradiction with previously reported ones where it was said that the composition of the deposit followed the composition of the suspension[14]. However, as in those experiments a gradient was intentionally formed in the deposits, slight deviations from the suspension composition went probably unnoted. Moreover as the deposition rates for the individual components were not reported, it is impossible to conclude whether a spontaneous gradient could be expected in their work. Our results, however, leave no doubt that for co-deposition of SiC and starch from acetone, a gradient forms spontaneously within a deposit, caused by a faster deposition of the starch compared to the SiC.

PYROLYSIS OF THE STARCH FROM THE DEPOSITS

In order to introduce porosity in the interlayers, the starch particles have to be removed from the SiC+starch layers, e.g. by pyrolysis. In contrast to oxide based laminates where this starch removal step can be performed in air, pyrolysis in air can lead to oxidation of the silicon carbide, which would limit the densification during sintering.

TGA measurements on a deposit containing only SiC, indicate that in order to avoid the oxidation of the SiC powder, the temperature should remain below 400°C (Figure 3c). Alternatively, one could envisage to remove the starch by heating in an inert atmosphere (e.g. argon), but even if the temperature is limited to 350 °C in air, the carbon residue of the starch is lower than when the pyrolysis is performed in argon up to 600°C (Figure 3a&b). Therefore the starch particles were removed as well as possible by heating the laminates slowly (30°C/h) in air to 350 °C.

LAMINATES WITH POROUS INTERLAYERS

Microstructures

Figure 4, 5 and 6 give an overview of the microstructures of the porous interlayers after starch removal and sintering. The size of the pores is clearly related to the size of the original starch particles (see table I) : smaller starch particles lead to smaller pores. Figure 7 shows the gradient that forms spontaneously during deposition from a suspension containing SiC and starch due to the faster deposition rate of the starch. As starch deposits more rapidly, the porosity is initially higher than the nominal composition of the suspension. Near depletion of the suspension most starch has already deposited and therefore the outer layer consists mainly of SiC.

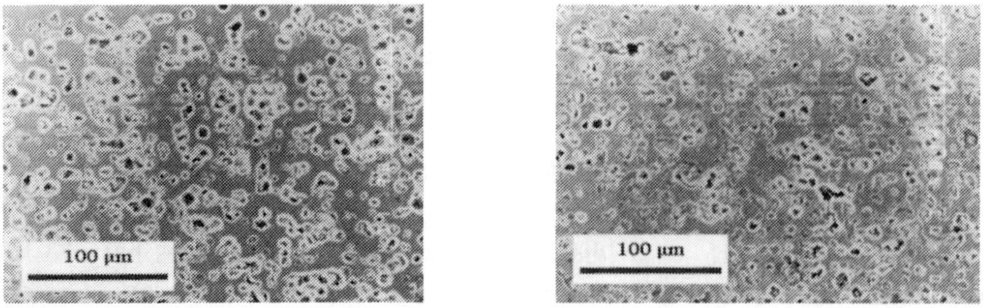

(a) *(b)*

Figure 4 : microstructure of the porous interlayers deposited from a suspension containing nominally (a) 40 vol% and (b) 50 vol% of corn starch

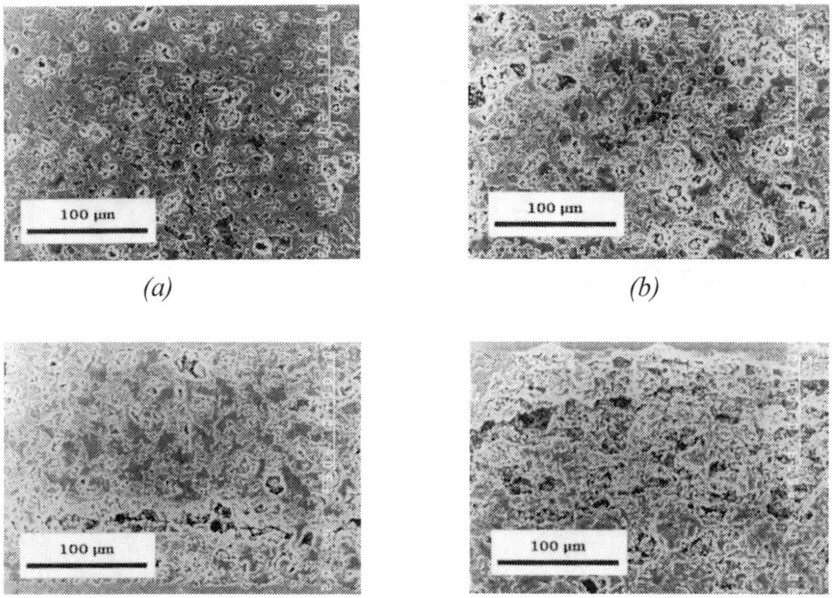

(a) *(b)*

(c) *(d)*

Figure 5 : microstructure of the porous interlayers deposited from a suspension containing nominally (a) 40 vol%, (b) 50 vol%, (c) 55 vol% and (d) 60 vol% of rice starch

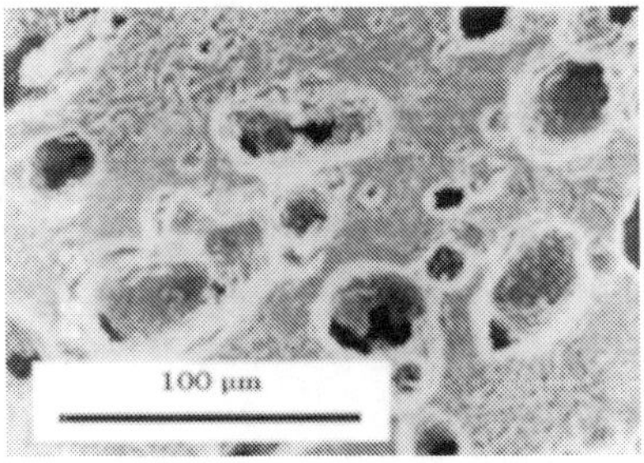

(a)

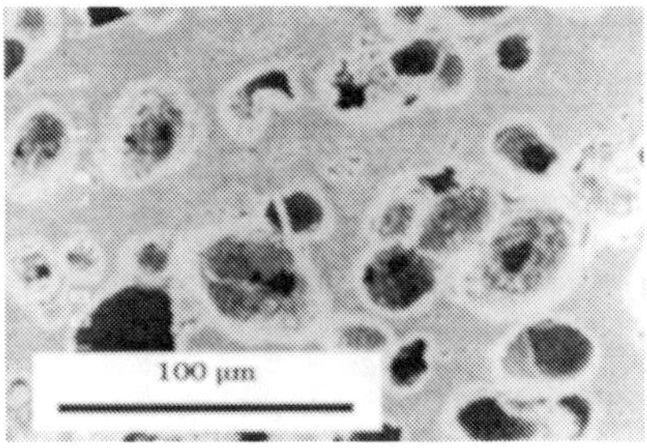

(b)

Figure 6 : microstructure of the porous interlayers deposited from a suspension containing nominally (a) 40 vol% and (b) 50 vol% of analytic starch

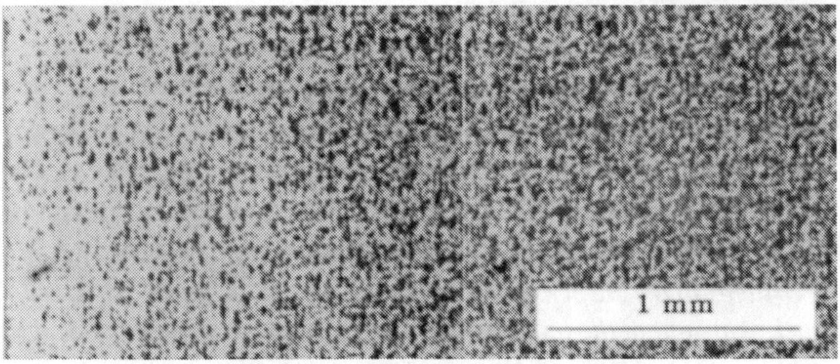

Figure 7 : overview of the microstructure (nominal suspension composition: 58 vol% SiC and 42 vol% corn starch). Deposition occurred from right to left. The gradient in porosity is due to a higher deposition rate of the starch powder, which results in an earlier depletion of starch than of SiC.

Mechanical behaviour

Laminates for bending tests were only made using rice and corn starch in order to limit the pore size. A smaller pore size offers the advantage of obtaining an interlayer structure which is more homogeneous on a microscopic level. In laminates with interlayers deposited from suspensions containing nominally only 40 vol% of starch, no crack deflection was observed. For all interlayers with a higher porosity cracks were deflected (Figure 8a). However, only in the case of the first interlayer deposited from a suspension containing nominally 60 vol% of starch, the deflected crack did not kink out of the interlayer again (Figure 8b). As usefull toughening is only obtained if indeed the crack does not kink out of the interlayer again, and taking into account the different deposition rates for SiC and starch, the required volume fraction of starch particles in the interlayer for a tough behaviour of the laminate is estimated to be at least 66 vol%.

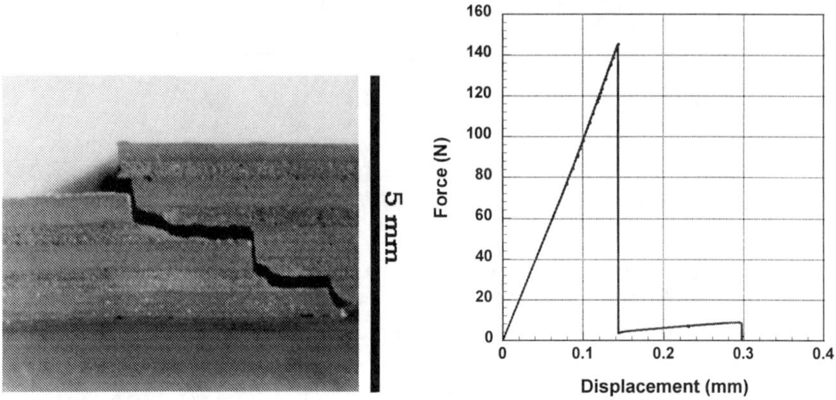

(a) *(b)*

Figure 8: (a) cracks are deflected in porous interlayers when 50 vol% or more starch particles in the suspension are used to deposit the interlayers, but in most cases cracks kink out of the interlayer again, (b) useful toughening, i.e. a deflected crack which does not kink out of the interlayer, was only observed in the porous interlayer in which the porosity was estimated to be 66 vol%.

CONCLUSIONS

The electrophoretic forming of SiC laminates with porous interlayers was achieved by alternatingly depositing SiC layers from a suspension containing SiC and a suspension containing a mixture of SiC and starch particles. A range of pore sizes can be introduced by selecting the appropriate type of starch particles.

During co-deposition of SiC and starch particles, each type of particle deposits at a rate proportional to its electrophoretic mobility. As a result consecutive deposits from the same suspension differ in composition.
Crack deflection is observed in interlayers deposited from suspensions containing nominally 50 vol% or more starch particles. However, the cracks kink out of the interlayers again, resulting in brittle failure. Only when at least 66 vol% starch

particles are incorporated in the interlayers, is useful toughening obtained: the deflected cracks remain in the interlayer.

ACKNOWLEDGEMENTS

This work was supported by the Brite-Euram programme of the Commission of the European Communities under project BE-7414/BRE2-0564. The authors would like to thank F. Boyer and A. Foissy from the Université de Franche Comte for the mobility measurements on the starch powders.

REFERENCES

[1] W.J. Clegg, K. Kendall, N. McN. Alford, T.W. Button, J.D. Birchall, A simple way to make tough ceramics, Nature, Vol. 347, p455-57, 1990

[2] J.B. Davies, W.J. Clegg, Ceramic Laminates for High Temperature Structural Applications, , 97th Meeting of the American Ceramic Society, Cincinatti, May, 3, 1995, Joint Engineering Ceramics and Basic Science Divisions, (paper no. C-37-95)

[3] H. Wang, X. Hu, Surface Properties of Ceramic Laminates Produced by Die Pressing, J. Am. Ceram. Soc., 79, (12), 553-556, (1996)

[4] J. Requena, R. Moreno, J.S. Moya, Alumina and Alumina/Zirconia Multilayer Composites obtained by Slip Casting, J. Am. Ceram. Soc., 72, (8), 1511-13, (1989)

[5] R. Ham-Su, D.S. Wilkinson, Strength of Tape Cast and Laminated Ceramics, J. Am. Ceram. Soc., 78, (6), 1580-84, (1995)

[6] E.A. Griffin, D.R. Mumm, D.B. Marshall, Rapid Prototyping of Functional Ceramic Composites, Ceramic Bulletin, Vol. 75, No. 7, 1996, 65-68

[7] P.Sarkar, X. Huang, P.S. Nicholson, Microlaminate ceramic/ceramic composites (YSZ/Alumina) by electrophoretic deposition, J. Am. Ceram. Soc., 75, (10), 2907-909, (1992)

[8] L. Vandeperre, O. Van Der Biest, F. Bouyer, J. Persello, A. Foissy, Electrophoretic Deposition of Silicon Carbide Ceramics, J. Eur. Ceram. Soc., 17, No.2, (1997), 373-376

[9] L. Vandeperre, O. Van Der Biest, W.J. Clegg, Silicon Carbide Laminates with Carbon Interlayers by Electrophoretic Deposition, Key Engineering Materials, Vols. 127-131 (1997), pp567-574, Trans Tech. Publications (Switzerland)

[10] L. Vandeperre, O. Van Der Biest, Composite SiC-Graphite Interlayers for Crack Deflection in Ceramic Laminates, Silicates Industriels, *in press*

[11] L. Vandeperre, F. Bouyer, A. Foissy, O. Van Der Biest, SiC-Graphite Laminates Shaped by EPD, Ceramic Bulletin, Vol. 77, No. 1, p53-58

[12] L. Vandeperre, O. Van Der Biest, Influence of Iso-Propyl Alcohol on Electrophoretic Deposition of SiC from a mixture of acetone and n-butylamine, Proceedings of the Symposium on Innovative Processing and Synthesis of Ceramics, Glasses and Composites, American Ceramic Society, in press

[13] Z. Zhang, Y. Huang, Z. Jiang, Electrophoretic Deposition Forming of SiC-TZP Composites in a Nonaqueous Sol Media, J. Am. Ceram. Soc. , 77, 1946-1950, (1994)

[14] P. Sarkar, S. Datta, P.S. Nicholson, Functionally Graded Ceramic/Ceramic and Metal/Ceramic Composites by Electrophoretic Deposition, Composites Part B, 28B, 1997, p49-56

Continuous Fiber Ceramic Composites: Processing and Microstructure/Property Relationships

HERMETIC SIC-SIC COMPOSITE TUBES

Witold Kowbel,* Yin Liu, Calvin Bruce and James C. Withers, MER Corporation, Tucson, AZ 85706
Lynne E. Kolaya and Nathan Lewis, Lockheed Martin, Schenectady, NY 12301

Abstract:

SiC-SiC composites have good potential for structural applications but are limited by expensive forming techniques. A high purity β-SiC fiber produced by MER, and a polymer derived SiC matrix were used to fabricate small diameter hermetic SiC-SiC tubes. The process was optimized to prevent the formation of a brittle structure while rapidly forming a dense matrix. The tube was made hermetic by first coating the surface of the tube with a silicon carbide particle filled polymer slurry, followed by a Chemical Vapor Infiltration/ Deposition (CVI/CVD) SiC deposition which was performed to close any residual porosity on the composite tube surface. X-ray diffraction and Transmission Electron Microscopy (TEM) examination was performed to determine the fiber and matrix structures. These tubes were found to be impermeable to helium with leak rates below 10^{-9} cc/sec as determined by testing similar to MIL-STD-883D, method 1014.10. This high level of impermeability was sustained following thermal cycling between room temperature and 1550°C.

1. INTRODUCTION

High temperature hermetic materials are critical to many engineering applications, e.g., thermocouple wells, coal plants, and fusion power plants. In addition to hermeticity, these high temperature applications usually require reliable materials with high strength, toughness, and thermal-chemical stability. Monolithic ceramic materials have high strength, good hermeticity and thermal-chemical stability; but suffer from poor toughness and reliability. [1] In contrast, ceramic matrix composites (CMCs) reinforced with ceramic fibers offer high strength and toughness. However, fabricating hermetic CMC materials is technically difficult. For instance, making a CMC with the polymer infiltration and pyrolysis (PIP) method, can take nine or more PIP cycles to reduce the void proportion to < 10%.

To the extent authorized under the laws of the United States of America, all copyright interests in this publication are the property of The American Ceramic Society. Any duplication, reproduction, or republication of this publication or any part thereof, without the express written consent of The American Ceramic Society or fee paid to the Copyright Clearance Center, is prohibited.

In addition to voids, shrinkage cracks due to pyrolysis are also potential leakage paths. CMCs made from the chemical vapor infiltration (CVI) method also have voids trapped into the matrix. The architecture of the fiber reinforcements influences the hermeticity of a CMC component. Component shape also affects hermeticity, e.g., a closed end tube where fibers bend close to 90° is difficult to make hermetic.

There are only a few reports of hermetic CMCs. Streckert et al. reported the fabrication of hermetic SiC composite tubes using the CVI method.[2] Nicalon SiC fibers were filament wound on a graphite-rod mandrel at a low bias angle (the fiber tow was almost normal to the mandrel) to minimize large pores due to fiber over lap. Then CVI SiC was applied on the SiC fibers to densify the composites producing hermetic SiC CMC tubes. Several limitations with these hermetic SiC CMC tubes are: (1) no fiber reinforcement along the tube axis direction to increase the strength and toughness; (2) the proportion of fiber reinforcement to the overcoated chemical vapor infiltrated/deported (CVI/CVD) monolithic SiC was ~1:1, so the mechanical properties are not dominated by composite behavior; and (3) the method does not provide for the fabrication of tubes with one end closed. This paper describes a process for fabrication of a hermetic CMC tube which solves these problems.

2. EXPERIMENTAL PROCEDURE

2.1 Materials

Plain weave SiC fabrics fabricated by the MER chemical vapor reaction (CVR) process were utilized as the reinforcement.[3-6] Ceraset™ inorganic polymer from Du Pont Lanxide Inc. (Newark, DE), was used as a matrix precursor. Toluene (Mallinckrodt Co) and either submicron SiC powders (Superior Graphite Co., Chicago, IL) or MER produced SiC nanoparticles were mixed with the Ceraset™ polymer to prepare a slurry for coating the composite tube surface.

2.2 Closed End CMC Tube Molding

The as manufactured tube dimensions were 7-8 mm in diameter by 125 mm long with a 0.75-1.5 mm thick wall and one end closed. To fabricate this tube, carbon coated (0.1 μm) SiC fabrics were first impregnated with Ceraset polymer, then the fabrics were wrapped on an aluminum mandrel. One side of the pre-preg was cut into tabs, which were folded over one end of the mandrel to form a closed end as illustrated in Figure 1. Three fabric layers were needed for hermeticity, however, this resulted in walls ~1 mm thick.

The SiC pre-pregg wrapped on an Al mandrel was placed into a flexible rubber mold, and heated to 185°C for 1 hour (Figure 2).

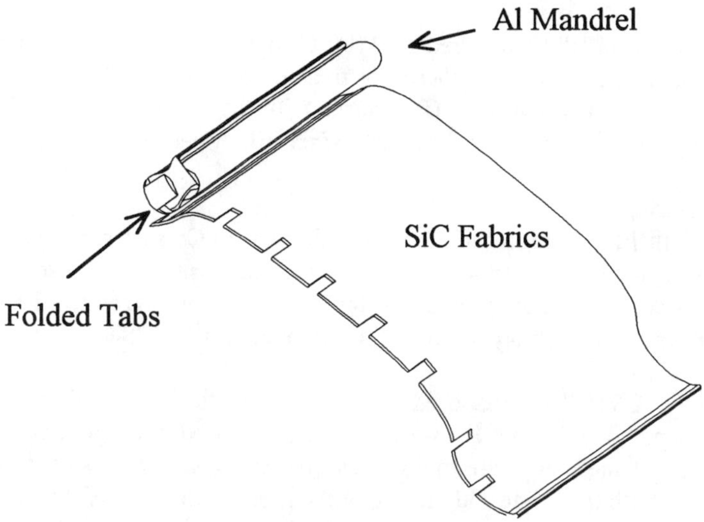

Figure 1. SiC fabric wrapping for one end closed CMC tube

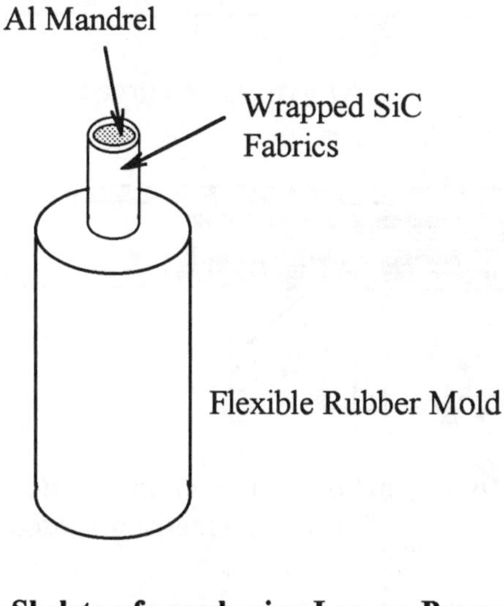

Figure 2. CMC Tube Skeleton formed using Lay-up-Pressure Molding Technique

2.3 CMC Densifications

The tubes were initially pyrolyzed to 1100°C in argon and then were re-infiltrated with Ceraset polymer and thermally treated to 1100°C. There was a third polymer infiltration and pyrolysis (PIP) cycle with the temperature raised to 1400°C. This sequence was repeated for a total of six PIP cycles.

2.4 CMC Tube Sealing

After the sixth PIP cycle, a SiC powder (sub-micron)/Ceraset slurry (2 : 1 weight ratio) was coated on the outside surface of the composite tubes to seal the large pores. After curing at 150°C/1 hour/air, the tubes were pyrolyzed to 1100°C in Ar to convert the Ceraset polymer into SiC producing a SiC_p/SiC composite layer.

Subsequently, CVI/CVD was used to provide the final seal (Figure 3). Methyltrichlorosilane (MTS) and H_2 were separately passed through capillary tubes to the inside of the composite tube. This processing scheme resulted in a CVI/CVD layer on both the inside and outside of the tube. The CVI/CVD process was carried out at a deposition temperature of 1120°C, and a pressure of 25 Torr.

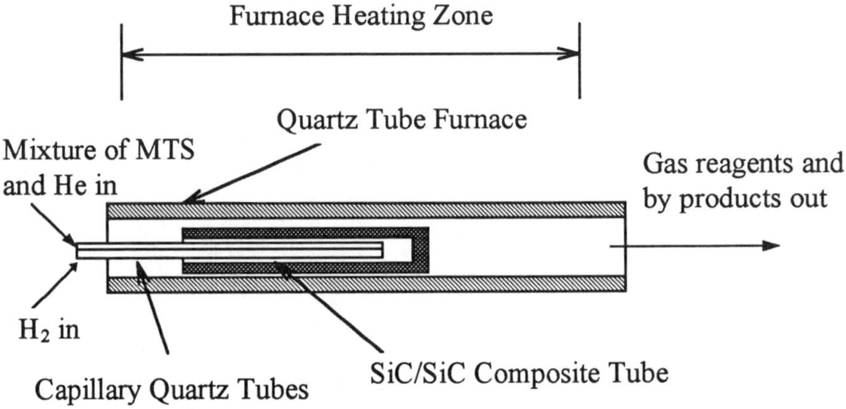

Figure 3. CVI/CVD SiC Deposition on the Inside Surface of a Composite Tube with One-End-Closed

2.5 CMC Tube Testing

The hermeticity testing of the CVI/CVD sealed CMC tubes was conducted using a Varian Porta-Test Leak Detector (Model 938-41), which has a sensitivity limit of 1×10^{-9} cc/sec helium leak rate. A transmission electronic microscope

(TEM) was used to examine the CMC and coating microstructure, and a scanning electronic microscope (SEM) was used to examine the structural integrity. The thermal shock test consisted of cycling from 25°C to 1520°C with a 10°C/min heating/cooling rate in argon (Ar) three times.

3. RESULTS AND DISCUSSION

3.1 Hermeticity

During development, the CMC tubes were frequently checked for hermeticity between fabrication steps. Tubes without the CVI/CVD sealing were not hermetic. Studies were performed to determine how much PIP densification was needed before CVI/CVD. It was found that only a few PIP cycles were needed yielding tubes which were still quite porous. These tubes only held a vacuum of 700 millitorr, which is too high a pressure for the leak detector to operate at. After CVI/CVD the hermeticity was measurable, yielding leak rates $< 1 \times 10^{-9}$ cc/sec (the lower limit of the detector). More than five CMC tubes have been fabricated with leak rates $< 1 \times 10^{-9}$ cc/sec. Studies to optimize the thickness of the CVI/CVD layer indicated that a minimum of ~30 μm is needed for helium impermeability. The polymer infiltrations and slurry application provide a uniform surface for the CVI/CVD.

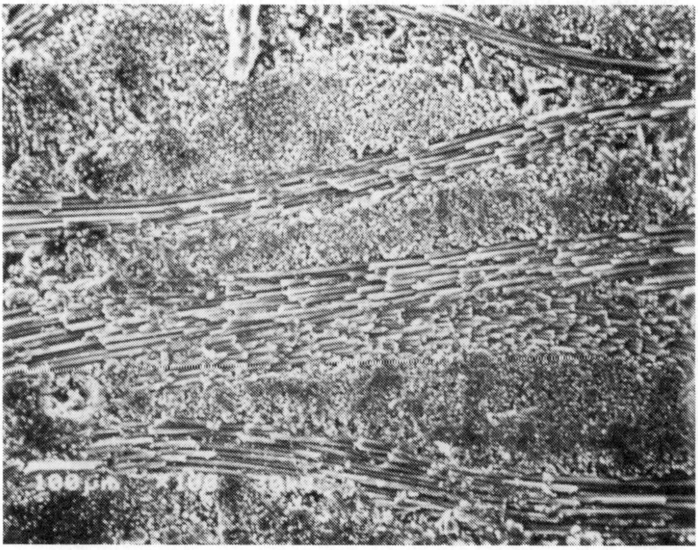

Figure 4. CMC after thermal shock for three cycles

3.2 Thermal shock

Two hermetic CMC tubes with leak rates < 1×10^{-9} cc/sec were thermally shocked three times and retained the same leak rates. SEM inspection of the shocked tubes (Figure 4) revealed no cracks on the outer surface, thus indicating that residual cracks, either from the polymer conversion or the thermal processing during fabrication, did not propagate a leakage path through the slurry and CVD layers. Also, the lack of surface cracks or spalling indicates good adhesion and matching of thermal properties between the polymer derived CMC and the CVD seal layers.

3.3 CMC Tube Microstructure

The overall hermetic behavior of the CMCs depends on several factors: (1) type of SiC fiber, (2) type of SiC matrix, (3) type and quality of the seal. In order to elucidate the influence of these factors on the overall CMC tube hermeticity, a thorough microstructural analysis of all these elements was conducted. Figure 5 shows a cross section of a CMC tube. The coefficient of thermal expansion (CTE) mismatch between the CVR-SiC fibers and the PIP matrix appears to be minimized; thus resulting in no transverse cracking. In addition, no gross defects such as large pores or voids are observed. The adherence between the outer CVD layer and the CMC tube appears to be good.

Figure 6 shows TEM images of the CVR-SiC fiber. The outer shell of the fiber consists of fine grain β-SiC (10-20 nm), while the interior has regions of turbostratic carbon. This indicates that the CVR conversion proceeds from the fiber surface to the fiber center. The fine β-SiC grains may have resulted from the fine grains of the initial T-300 PAN based carbon fibers which acted as nuclei for the CVR process.

Figure 7 shows TEM images of the Ceraset derived β-SiC microstructure. Very fine (2-5 nm) and fine (10-20 nm) β-SiC grains were observed suggesting that the 1400°C heat treatment employed during the CMC processing resulted in the crystallization of the Ceraset derived matrix.

The PIP processing studies for this material combination show that a higher heat treatment temperature has a stronger influence on the final product toughness. The 1100°C pyrolysis produces a brittle composite, which has no fiber pull out during crack propagation, as shown in Figure 8a. The ≥ 1400°C pyrolysis offers a tough fracture structure, which has good fiber pull out, as shown in Figure 8b.

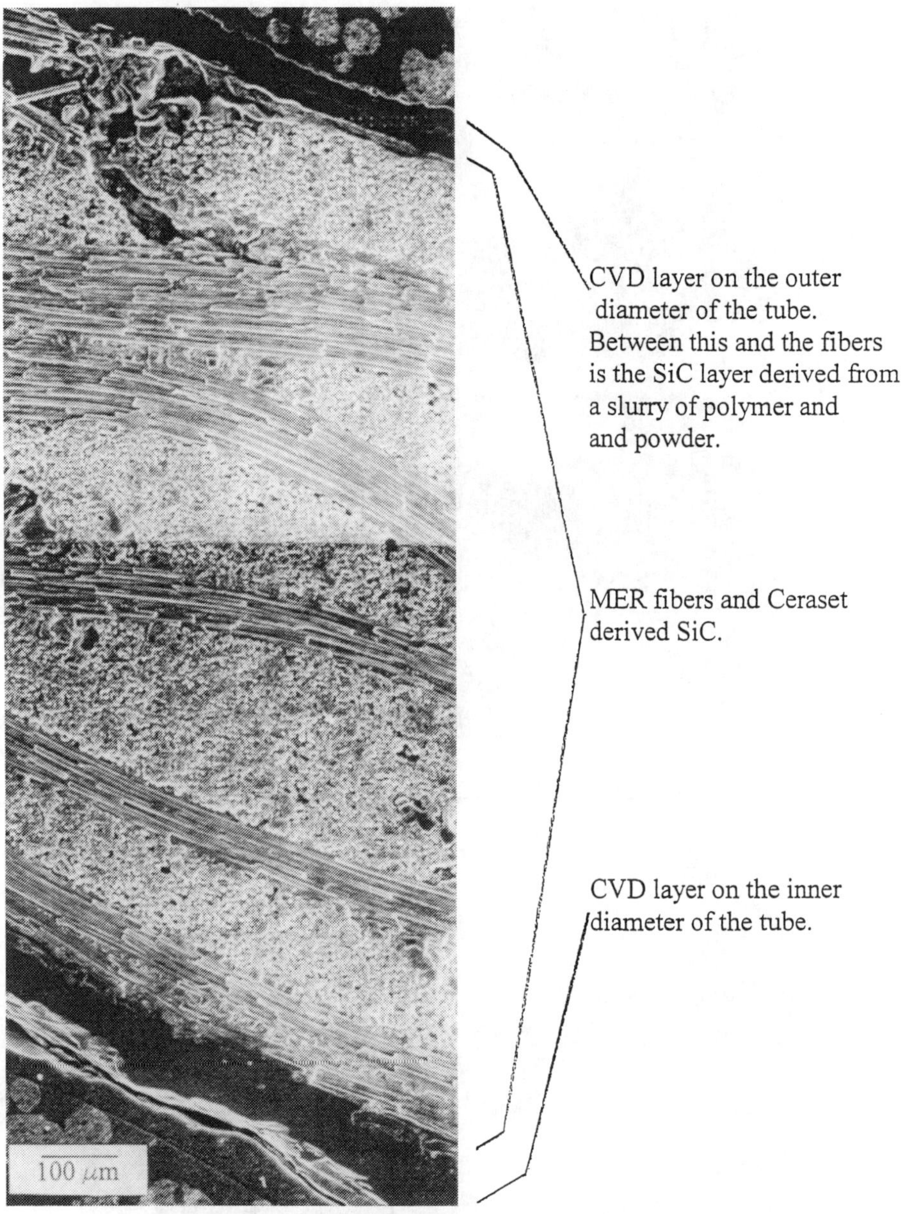

Figure 5. Cross section of a CMC tube

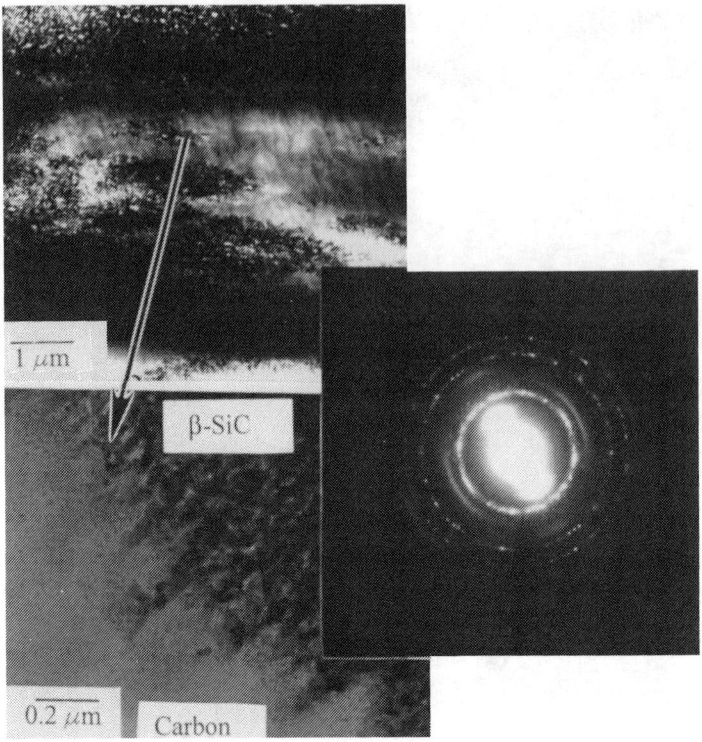

Figure 6. TEM of CVR-SiC fibers (longitudinal section)

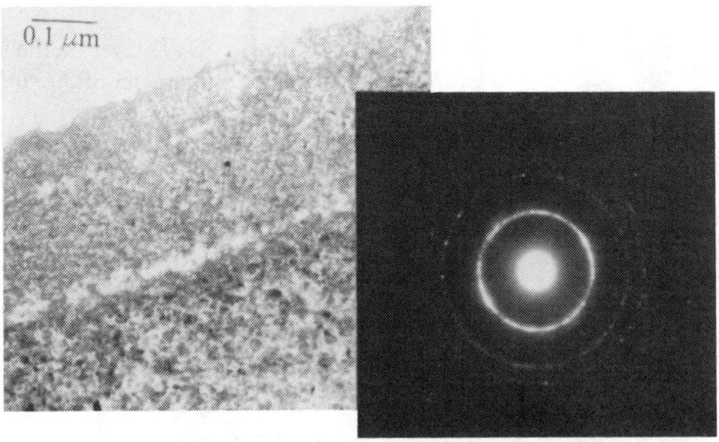

Figure 7. TEM of Ceraset-derived SiC matrix

(a)

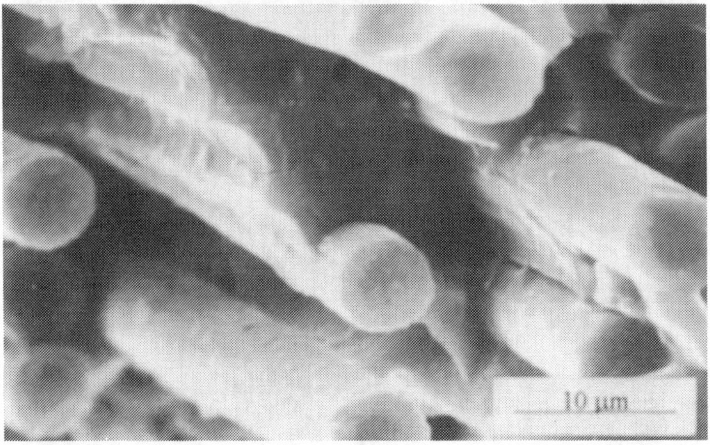

(b)

Figure 8. a) An SEM micrograph of the brittle structure resulting from the 1100°C pyrolysis. b) An SEM image of the tough fracture structure (fiber-pull-out) after pyrolysis at 1400°C/Ar/1h

3.2 Microstructure of the PIP/CVD Sealant

Figure 9a and 9b are SEM micrographs of the SiC slurry derived coating as applied to the SiC CMC tube. As a result of the volume change during the Ceraset pyrolysis, voids (about 100 μm) are formed within the slurry coating. However, no cracks due to thermal processing are observed on the slurry coating surface. The absence of the thermal processing cracks is the key factor enabling the subsequent pore infiltration with CVI to form a hermetic seal because thermal processing cracks open and close with thermal exposure, so they usually cannot be sealed.

Figure 10 shows TEM micrographs corresponding to the slurry coating. It reveals large β-SiC grains embedded in a Ceraset derived fine β-SiC crystallites. Subsequently to the slurry coating application, a CVI/CVD overcoating was applied to yield a hermetic seal.

Figure 11 shows SEM micrographs corresponding to the CVD SiC overcoat after thermal shock testing. No surface cracks or pores are observed. The CVI/CVD conditions employed provided for the full infiltration of the pores in the slurry coating.

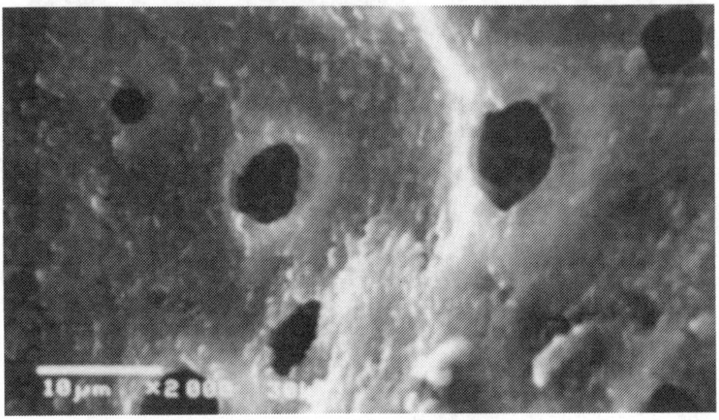

(a)

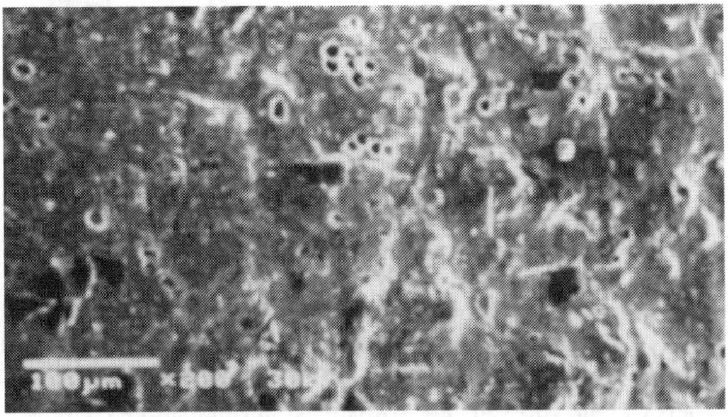

(b)

Figure 9 a and b. SEM of slurry coating

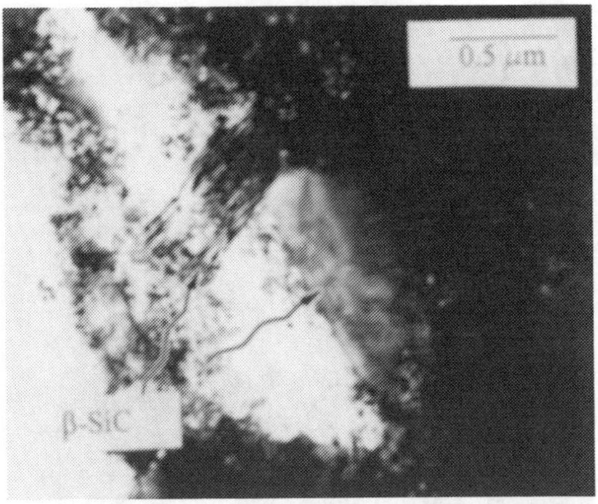

Figure 10. TEM of slurry coating

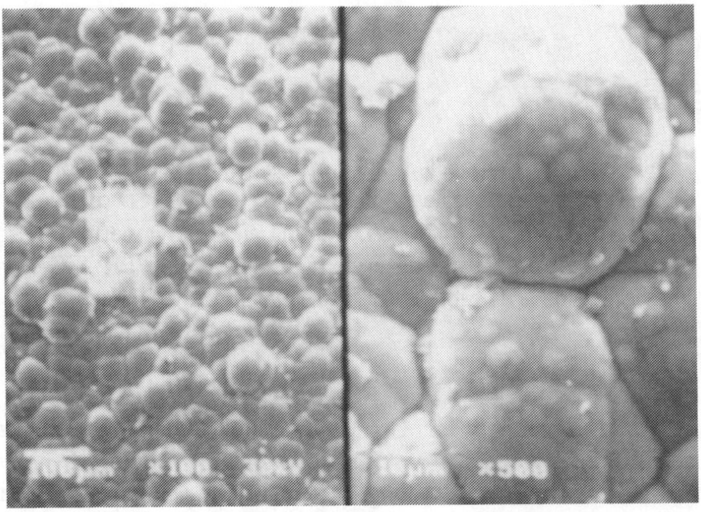

Figure 11. SEM of CVD overcoat

Figure 12 shows TEM micrographs of the CVD SiC overcoat. A columnar growth is observed.

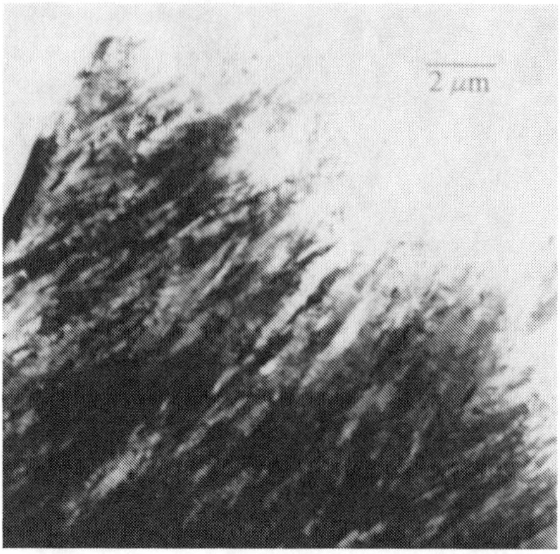

Figure 12. TEM of CVD overcoat

4. CONCLUSIONS

1. Small diameter, thin walled SiC composite tubes can be fabricated using a PIP matrix, a polymer slurry coating, and a CVI/CVD overcoat.
2. About a 30 μm thick CVI/CVD coating yields an impermeable seal.
3. No transverse cracks are observed within CVR-SiC/PIP SiC composites.
4. The hermetic CMC tubes exhibit thermal shock resistance.

REFERENCES:

[1] Y. Liu, "Fiber Processing for Composite Applications," Ph.D dissertation, Department of Material Science and Engineering, University of Michigan, 1997.

[2] H. Streckert, K. Norton and C. Wong, "Hermetic SiC composite Tubes," *ACers Bulletin*, **75** [12] 61-64 (1996).

[3] W. Kowbel, J.C. Withers, R.O. Loutfy, C. Bruce, and C. Kyriacou, "Silicon Carbide Fibers and Composites from Graphite Precursors for Fusion Energy Applications", *J. Nuc. Mat.*, Vol. 219, 15 (1995).

[4] W. Kowbel et al., MRS 365 (1995), p197

[5] W. Kowbel et al., MRS 410 (1996) p417

[6] W. Kowbel, et al., 2 IEA, Workshop on SiC-SiC, Sepdai, Japan, to be published.

PROCESSING AND PROPERTIES OF NICALON-REINFORCED ZIRCONIUM PHOSPHATE COMPOSITES

Barry A. Bender, Todd L. Jessen and Scott Browning
U.S. Naval Research Laboratory
Washington, DC 20375-5343

Hot-pressed and phosphate-bonded Nicalon SiC fiber-reinforced ZrP_2O_7 matrix composites were made. The effects of variations in processing and fiber reinforcement on the composites' mechanical properties (strength, toughness, and thermal up shock resistance) were measured. ZrP_2O_7 reinforced with carbon-coated Hi-Nicalon SiC fibers has the best combination of strength, toughness, thermal up shock resistance, and low dielectric constant for use as an inexpensive composite faceplate in a laminated radome.

INTRODUCTION

To reduce the threat of future missile attacks to our Naval fleet at sea, the next generation of advanced air defense missile interceptors will have to travel at hypersonic speeds of Mach 4 or greater. These higher velocities create a harsh aerothermal environment of increased thermal shock and mechanical stress that must be withstood by future radomes. To meet these new performance requirements the next generation of radomes must be: (1) stable up to 1400°C (both physically and electro- magnetically), (2) strong and tough enough to survive environmental erosion from rain and dust particles, and (3) capable to resist 1200°C thermal up shock. Last of all, in this era of reduced military spending, the new radomes have to be cost effective.

To meet these demands there has been a resurgence in radome development to improve existing radome materials, which are either slipped cast fused silica (SCFS) or Pyroceram 9606 (Corning Glass Works, Corning, NY). SCFS possesses excellent dielectric and thermal shock properties but it is very susceptible to rain erosion. Pyroceram is a relatively stronger and tougher material but its dielectric properties vary too much with increasing temperature. One newly developed radome material is Nitroxyceram (Loral Aeronautics,

To the extent authorized under the laws of the United States of America, all copyright interests in this publication are the property of The American Ceramic Society. Any duplication, reproduction, or republication of this publication or any part thereof, without the express written consent of The American Ceramic Society or fee paid to the Copyright Clearance Center, is prohibited.

Dearborn, MI). This material meets most of the requirements for today's advanced radomes.[1] Unfortunately, the material must be hot isostatically pressed, which is a very expensive process.

One relatively new class of materials that can meet these neoteric requirements is failure-resistant ceramic laminate structures. Jessen *et al.*[2] have shown that the failure resistance of a weak, brittle inner core can be enhanced by the presence of a ceramic matrix composite (CMC) faceplate. Also CMCs are highly resistant to thermal shock.[3-6]

Low-temperature synthesis phosphate bonding offers the potential for fabrication of inexpensive radome CMC faceplates. This technology can be used for net shape forming.[7] Phosphates can be refractory and possess low dielectric constants. ZrP_2O_7 was chosen as a potential material candidate for use as a matrix for a radome CMC faceplate because it is refractory to 1500°C and has a low dielectric constant.[8] It can be synthesized from inexpensive starting materials- ZrO_2 powder and concentrated H_3PO_4. SiC fibers were chosen to be the reinforcement for CMC faceplate because SiC is more compatible with H_3PO_4 than oxide fibers. Nicalon (Nippon Carbon, Tokyo, Japan) SiC fiber was chosen due to its availability in different grades (Ceramic grade, low dielectric HVR, and Hi-Nicalon), styles (tow versus 2D-weave), and excellent strength.

This paper reports the development of a SiC fiber-reinforced ZrP_2O_7 composite for use as a faceplate in a sandwich structure radome wall. The first section of the paper discusses the processing and properties of model SiC/ZrP_2O_7 composites fabricated by hot pressing. The second section discusses the fabrication of Nicalon-reinforced ZrP_2O_7 CMCs derived from phosphate bonding. The strength, thermal up shock resistance, and electromagnetic properties are reported and correlated with their microstructure as characterized by scanning and transmission electron microscopy (SEM and TEM).

EXPERIMENTAL PROCEDURE

Processing

Three different types of hot-pressed Nicalon-reinforced ZrP_2O_7 composites were fabricated using hot pressing. The composites were reinforced using either BN-coated ceramic grade (CG) Nicalon SiC fibers in tow form or plain 2D-weaves or with tows of carbon-coated Hi-Nicalon SiC fibers. The fabrication of the CMCs parallels that of a SiC/$ZrTiO_4$ composite which has been described elsewhere.[9] In this case the alkoxide is zirconium n-propoxide (Morton Thiokol, Chicago, IL) and triisopropyl phosphite (Eastman Kodak, Rochester, NY). The hydrolysis process was carried out in a humidity chamber (95% humidity) at room temperature instead of a steam bath. Calcination of the green composite was done at 540°C in flowing air and hot pressing was carried out in an overpressure of CO at a temperature of 1270°C.

The initial phosphate-bonded composites were made using HVR Nicalon SiC fibers (plain 2D-weave) in a ZrP_2O_7 matrix derived from the reaction of concentrated H_3PO_4 (85%) and fine ZrO_2 powder (Zircar Products, Florida, NY):

$$ZrO_2 + 2H_3PO_4 \rightarrow ZrP_2O_7 + 3H_2O \qquad (1)$$

To summarize the processing steps (see Bender *et al.*[10] for further details) the weaves were soaked in a stoichiometric ZrO_2-H_3PO_4 mixture (zrpo), stacked, dead-loaded to prevent bloating, and slowly fired to 700°C in a humid atmosphere. This entire process was repeated two more times to reduce the porosity of the composite.

The Hi-Nicalon phosphate-bonded composite consists of tows of C-coated Hi-Nicalon SiC fibers in a ZrP_2O_7 matrix. The CMC was made by adding tows of fibers into a Teflon boat containing zrpo until 50 vol% fiber was obtained. The boat was then heated slowly to 140°C and held there for 10 h to harden the matrix. Next, the composite was placed in a carbon boat and packed in carbon (to prevent oxidation of the C-coated fibers) and slowly fired to 700°C. The composite was immersed in zrpo a second time, fired slowly to 140°C on Teflon supports, and fired again in carbon to 700°C.

Thermal Up Shocking

The composites were placed on mullite foam supports on a square of mullite foam. The foam was then placed in a furnace at 250°C for 3 hours to remove moisture. The foam was then placed into the thermal shock furnace where the power was increased to raise the furnace back to its designated temperature of 1200°C within 60 s. Thermal up shocking was done to 1200°C because this is the maximum thermal transient expected during hypersonic flight.[11] The composite was held at the 1200°C for 300 s. To avoid oxygen embrittlement problems, the up shocked sample was cooled to 600°C at a rate of 1.67°C/s. The specimen was then furnace cooled to room temperature and cut into bend bars.

Mechanical, Microstructural, and Electromagnetic Characterization

Mechanical characterization was done using test bars that varied in size but with a fixed width to thickness ratio of 2 on a testing machine (Model 1122, Instron Corp., Canton, MA). Strength was measured using the four-point flexure test. Four bars were broken for each composite using an outer span of 27 mm, an inner span of 9, and a strain rate of 0.003/s. Toughness was measured using the single-edge notch-beam (SENB) test. Four bars were broken for each composite using the same loading configuration. The notch depth was a constant percentage (28%) of the specimen's thickness.

Fracture surfaces and polished cross sections were examined in the SEM. TEM (Model 200 CX, JEOL Ltd., Tokyo, Japan) was done on samples whose cross section was normal to the long axis of the fiber. The samples were mechanically polished to 100 μm in thickness, dimpled to 25 μm in thickness, and then ion milled until perforation occurred, using argon at 5 keV.

Permittivity was measured using a vector network analyzer (VNA-Hewlett Packard 8510C, Santa Clara, CA) and a S-Parameter test set (45 MHz to 40 GHz- Hewlett Packard 8516A, Santa Clara, CA). The CMCs were measured at 35 GHz (K_a band). The VNA was calibrated to remove sources of error in the measurement. The samples were placed in the waveguide and the complex reflection (S11) and transmission (S21) coefficients measured from which the dielectric constant was calculated.

RESULTS AND DISCUSSION

Hot-Pressed Composites

The development of a hot-pressed BN-coated Nicalon-reinforced zirconium phosphate composite was pursued for several reasons. An optimized hot-pressed CMC would determine if the Nicalon/ZrP_2O_7 system had the potential to be a strong and tough composite material to serve as an effective composite laminate radome faceplate. Also an optimized composite would give us an indication of the potential of the system to survive thermal up shocking.

The first hot pressing effort led to a CMC with relatively poor mechanical properties (see Table I). However, the test bars did break in a fibrous manner indicating that the system had potential to yield a better CMC. Further analysis showed that the sample had 25% porosity which was the contributing factor to the sample's poor mechanical properties.

Table I. Mechanical Properties of Hot-Pressed CMCs

Composite	Strength* (MPa)	Toughness* (MPa√m)
BN-coated CG (undoped)	200 (30)	4.2 (0.5)
BN-coated CG (2% MgO)	423 (57)	8.2 (0.5)
BN-coated CG 2D-weave	238 (13)	4.5 (1.1)
C-coated Hi-Nicalon	237 (45)	3.8 (0.9)

*Numbers in parentheses represent one standard deviation.

Yamai and Oota[12] showed that ZrP_2O_7 could be sintered to theoretical density using sintering aids. After some experimentation, 2 w/o MgO was added to the starting precursors as a sintering aid. The resultant composite had dramatically improved mechanical properties (see Table II) and the porosity was reduced to 10%.

SEM indicated that the matrix infiltration and fiber distribution was good. The fibers show little degradation as only small pockets of silica can be observed around the edges of the fibers. SEM also indicated that the BN coating was still intact and had not reacted with the zirconium phosphate. This was confirmed by TEM as shown in Fig. 1A. This figure also shows that the BN is acting as an effective debond coating as the crack runs through the matrix and the pocket of silica, but at the BN layer it is deflected so that it runs around the fiber at this interface, which leads to the improved properties (see Table I).

The last hot-pressed composite fabricated was a BN-coated CG Nicalon 2D-weave SiC fiber-reinforced ZrP_2O_7. This composite was fabricated to determine the effect of fiber architecture on mechanical properties by switching to a 2D-cloth reinforcement. The net result was that the composite was tough with the fracture surface being very brushy in appearance due to the presence of fiber pullout. The measured strength and toughness (see Table I) of the composite was about 50% of the tow-reinforced composite. This was expected because the 2D-weave cloth has about 50% less fiber support in the tensile testing direction. Last of all, there were some problems with infiltration between the tows of the cloth which was also observed with the composites synthesized from ZrO_2 powder and H_3PO_4.

Having shown that the SiC/ZrP_2O_7 composite system has excellent room temperature mechanical properties, thermal shock tests were run to see if the system had the potential to survive thermal up shocking to 1200°C. The CG hot-pressed composite survives quite well losing only 15% of its room temperature strength. However, the up shocked composite fails in a bimodal fashion. The outer shell (about 90 μm thick) fractures in a woody manner, while the interior fails in a tough manner with lots of fiber pullout (see Fig. 1B). As observed in other BN-coated CG Nicalon-reinforced CMCs[13], the BN coating oxidizes to boria during exposure to oxygen at high temperatures. The boria reacts with the silica present to form a less viscous interphase. This leads to the formation of elongated pores (see Fig. 1C) and strong bonding between the fiber and the matrix in the outer shell of the composite. Also, the fibers in the outer region had weakened considerably. However, the core of the composite is not affected because the short high temperature exposure limits the ingress of oxygen into the interior of the CMC. The net result is a strong (355 MPa), tough (8.7 $MPa\cdot\sqrt{m}$) composite even after thermal shocking up to 1200°C.

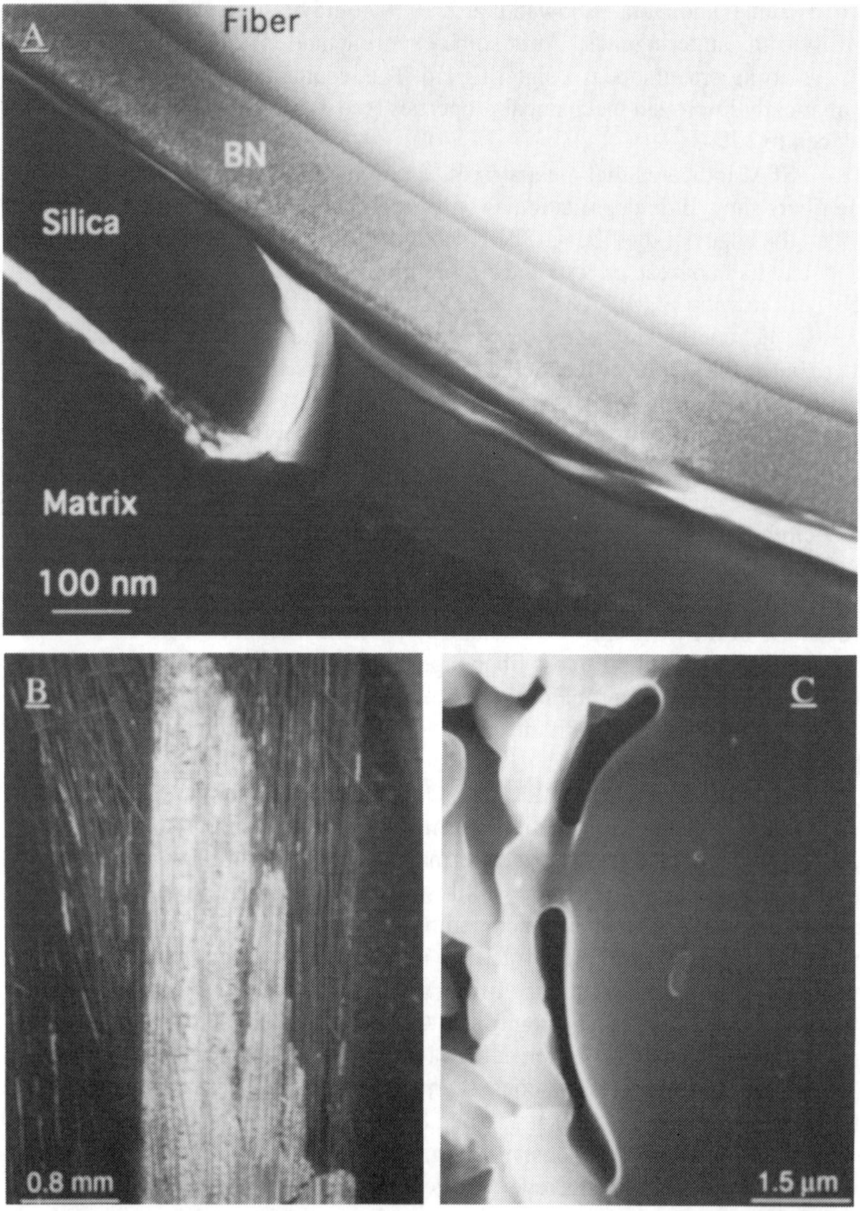

Fig. 1 Hot-pressed CG Nicalon-reinforced CMCs- (A) TEM image of the fiber/matrix interface, and (B & C) SEM fractographs of the CMC after up shocking.

In an effort to further increase the thermal shock resistance of the phosphate composite system carbon-coated Hi-Nicalon SiC fibers were used. This is because Hi-Nicalon was designed to be a more refractory fiber than CG Nicalon. Unexpectantly, the composite was weaker than the CG CMC (see Table I) and broke in a semi-brittle manner. TEM analysis (see Fig. 2) of the fiber/matrix interface shows that there is no reaction between the phosphate matrix and the SiC fiber, but that the carbon debond layer has been lost during processing (calcination in flowing air at 540°C for 1 h may have oxidized the thin layer of carbon). This explains why the composite showed very little fiber pullout.

HVR Nicalon/ZrP_2O_7 Phosphate Bonded Composites

The mechanical properties of the hot-pressed composites demonstrated that the Nicalon/ZrP_2O_7 system is a viable radome composite material. Additional research was initiated to develop a composite radome faceplate incorporating both a lower dielectric reinforcement (HVR Nicalon) and less expensive processing (phosphate bonding). Due to the ease of handling and fabrication, initial research was done using uncoated HVR cloth as the reinforcement for the development of phosphate bonded CMCs. The net result was that a relatively tough and strong composite could be fabricated without the use of a fiber debond coating. At room temperature the cloth-reinforced composite had decent strength (see Table II) and broke in a tough brushy manner. TEM showed (see Fig. 3A) that the debonding was occurring at the fiber/matrix interface. Due to the reaction of the bare HVR fiber with H_3PO_4, a complex interface is formed during processing. Thin layers of SiP_2O_7 form next to the matrix. Separating these layers from the fiber is a thin layer of amorphous silica about 90 nm in thickness. A discontinuous layer of thin elliptical pores bisects this layer. Debonding occurs through tearing of this thin porous interphase.

Table II. Measured Properties of the Phosphate-Bonded ZrP_2O_7 Composites

CMC System	Strength* (MPa)	Up Shock Strength*	Up Shock Failure	Up Shock Wt. Loss	Dielectric Constant#
HVR Cloth	94 (6)	21 (2)	Semi-brittle	2.9%	4.6 [0.028]
HVR 'Ironed'	86 (5)	19 (4)	Semi-brittle	4.6%	not measured
HVR Tow	137 (11)	20 (6)	Semi-brittle	1.5%	not measured
Hi-Nicalon	239 (61)	90 (6)	Brushy	0.2%	6.2 [0.004]

*Numbers in parentheses represent one standard deviation
#Numbers in brackets represent loss tangent values

Fig. 2 TEM bright field of the fiber/matrix interface of a zirconium phosphate composite reinforced with carbon-coated Hi-Nicalon SiC fibers.

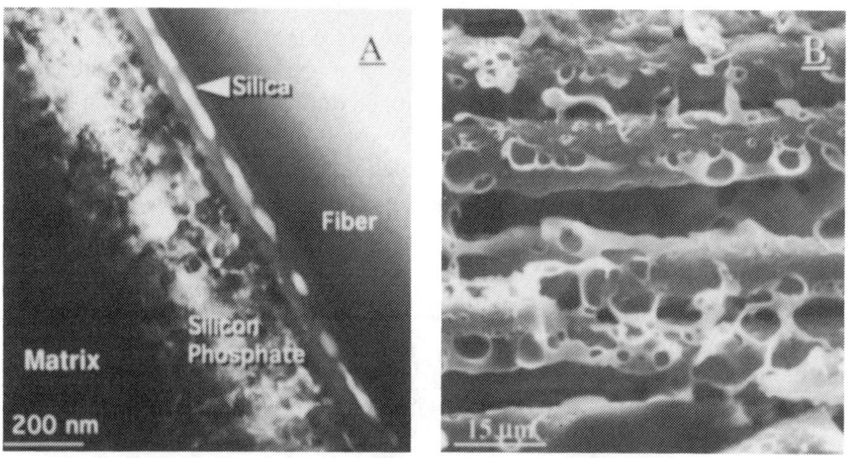

Fig. 3 (A) TEM bright field of the fiber/matrix interface of a phosphate-bonded uncoated HVR Nicalon fiber-reinforced zirconium phosphate CMC. (B) SEM fractograph of the same composite after thermal up shocking to 1200°C.

However, to form this in-situ debond layer and reduce the porosity present in the phosphate-bonded composites firing times of 101 hours are needed. To reduce this firing time 'ironed' composites were developed. These composites were made by placing the stacked zrpo-infiltrated cloth mats in between sheets of Teflon on a 180°C hot plate and dead loading with a heated weight for 1 h. Then the stack was placed on heated carbon planchets on the 180°C hot plate and 'basted' with the zrpo precursor. This basting reduced the number of infiltration cycles from three to one with the net result of shortening firing times to 32 hours. The net result was that the 'ironed' CMC had a room temperature strength of 86 MPa, which was similar to the strength of the 'standard' cloth-reinforced CMC.

Unfortunately, design requirements for the radome composite faceplate call for a strength in the range of 125 to 150 MPa. The research on the hot-pressed ZrP_2O_7 CMCs had shown that the strength of the CMC almost doubled when the reinforcement was switched from cloth to tow. Processing was then developed to incorporate uncoated HVR tow into a phosphate-bonded ZrP_2O_7 CMC. Green CMCs were fabricated by immersing HVR tow into a Teflon boat filled with zrpo precursor until 50 vol% fiber was obtained. The Teflon boat was heated slowly up to 140°C and held there for 10 h to form a green CMC. Standard processing was used after that. The net result after three reinfiltration cycles was a relatively tough composite whose strength improved by 45% to 137 MPa (see Table II).

Having developed an uncoated HVR Nicalon phosphate-bonded ZrP_2O_7 CMC with mechanical properties that met design requirements, the composite's resistance to thermal up shock had to be determined. Research showed that the HVR CMC survive up shocking to 1100°C[10] but start to fail when up shocked to 1200°C (see Table II). All three composites broke in a semi-brittle manner and showed significant weight loss (see Table II). This is because the SiP_2O_7 interphase starts to decompose leading to a porous interphase (see Fig. 3B) that bonds strongly the matrix to the fiber. This coupled along with loss of fiber strength (the average fiber fracture mirror diameter has quadrupled in size) results in a CMC which loses 75% of its starting strength after 1200°C up shocking.

Hi-Nicalon/ZrP_2O_7 Phosphate Bonded Composites

To develop a more thermal shock resistant phosphate-bonded composite, carbon-coated Hi-Nicalon SiC fibers were substituted for the low dielectric HVR Nicalon. This was done because Hi-Nicalon has improved thermal stability even though its EM properties are higher than HVR Nicalon. Since the Hi-Nicalon fibers have an applied debond coat, slow firing times above 210°C, which are necessary to form the in-situ debond interface with uncoated HVR Nicalon, were eliminated. The net result was a tough strong composite with an average strength of 181 ± 40 MPa. TEM revealed that the composites had lots of fiber pullout with

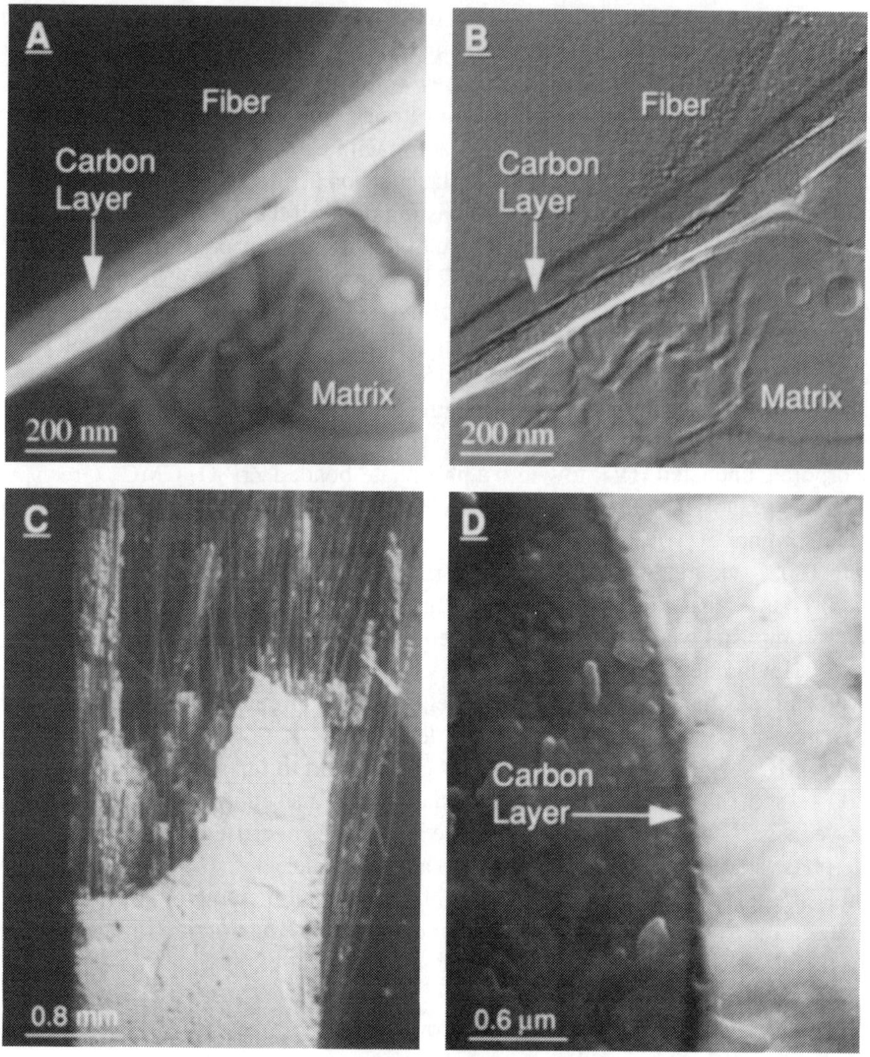

Fig. 4 (A) TEM bright field and (B) embossed image of (A) of a fiber/matrix interface in a phosphate-bonded carbon-coated Hi-Nicalon-reinforced composite. (C) and (D) are fractographs of the composite after 1200°C thermal up shocking.

debonding occurring at the coating/matrix interface (see Fig. 4A). As shown in Fig. 4B debonding occurs by tearing of the matrix away from the carbon coating.

To further improve the mechanical properties of the Hi-Nicalon SiC fiber-reinforced phosphate CMC, research was initiated on the effect of firing temperature on composite strength. Work by Talmy *et al.*[7] showed that in the Si_3N_4-ZrP_2O_7 system the maximum strength was achieved with firing temperatures of 900°C. This same trend was observed for the Hi-Nicalon/ZrP_2O_7 system as the strength of the composites after firing at 800°C dropped to 147 ± 6 MPa, but increased to 239 MPa (see Table II) when fired at 900°C. Microstructural observations of the three differently fired CMCs (700°C,800°C, and 900°C) offer no clues for this improvement in strength.

The 900°C fired Hi-Nicalon CMC was then up shocked to 1200°C. Fig. 4C shows that the CMC fails in a graceful manner with lots of fiber pullout. The retained strength of the CMC has improved from the average 20 MPa strength for the HVR composites to 90 MPa (see Table II). SEM (Fig. 4D) shows that the carbon layer is still intact with hints of the beginning of oxidation of the carbon coating. Unlike the up-shocked, hot-pressed CG composites (see Fig. 1C), no degradation of the fiber/matrix interface has occurred. The primary reason for loss in CMC strength after up shocking is that the fibers have weakened.

Last of all, the Hi-Nicalon CMCs have a relatively low loss tangent (see Table II) with a dielectric constant that is slightly greater than the HVR fiber-reinforced CMCs. This was unexpected because Hi-Nicalon was not designed to be a low dielectric polymer-derived fiber like HVR Nicalon. The fiber was designed to minimize the amount of free O within its structure. As a result, it contains a lot more free C than the HVR fiber. Also the fiber is coated with a thin 100 nm layer of C which should also increase its dielectric constant. Despite this, the Hi-Nicalon CMCs have EM properties that are suitable for a radome material.

CONCLUSION

Processing was developed to fabricate a dense BN-coated CG Nicalon SiC fiber-reinforced ZrP_2O_7 matrix composite. The resultant CMC, which was strong (423 MPa), tough (8.2 MPa$\sqrt{m}$), and resistant to 1200°C thermal up shock, showed that the Nicalon/ZrP_2O_7 CMC system had the potential to be a viable faceplate material in a laminate radome wall. Relatively tough and strong phosphate-bonded uncoated HVR Nicalon-reinforced ZrP_2O_7 CMCs were fabricated. Firing times were reduced by 67% without loss of strength. Thermal up shocking of these composites to 1200°C degrades the mechanical properties of these CMCs as they become semi-brittle and lose 75% of their room temperature strength. However, C-coated Hi-Nicalon phosphate-bonded ZrP_2O_7 CMCs have improved resistance to thermal up shocking to 1200°C as they remain tough and retain a strength of 90 MPa. Firing these CMCs at different temperatures showed that the 900°C-fired Hi-Nicalon CMCs were the strongest with a strength of 239 MPa. The combination of good mechanical properties and decent electromagnetic properties demonstrates that an inexpensive phosphate-bonded CMC can be made that has the capacity to be a suitable faceplate for a laminate sandwich radome.

REFERENCES

1. J. B. Kouroupis, "Flight Capabilities of High-Speed-Missile Radome Materials," *John Hopkins APL Tech. Dig.*, **13** [3] 386-393 (1992).

2. T. L. Jessen, B. A. Bender, and D. Lewis, III, "Failure Resistant Ceramic Laminate Structures," *Ceram. Eng. Sci. Proc.*, **14** [9-10] 914-921 (1993).

3. H. Wang and R. N. Singh, "Thermal Shock Behavior of Two-Dimensional Woven-Fiber-Reinforced Ceramic Composites," *J. Am. Ceram. Soc.,* **79** [7] 1783-92 (1996).

4. Y. Kagawa, N. Kurosawa, and T. Kishi, "Thermal Shock Resistance of SiC-Fibre-Reinforced Borosilicate Glass and Lithium Aluminosilicate Matrix Composites," *J. Mater. Sci.*, **28**, 735-41 (1993).

5. A. Eckel, T. Herbell, and E. Generazio, "Thermal Shock Fiber-Reinforced Ceramic Matrix Composites," *Ceram. Eng. Sci. Proc.*, **12** [7-8] 1500-08 (1991).

6. H. Wang, R. N. Singh, and R. A. Lowden, "Thermal Shock Behaviour of Unidirectional, 0°/90°, and 2-D Woven Fibre-Reinforced CVI SiC Matrix Composites," *J. Mat. Sci.*, **32**, 3305-13 (1997).

7. I. G. Talmy, C. A. Martin, A. H. Le, and A. Janovsky, "Phosphate Bonding in Processing of Non-Oxide Ceramics," pp. 291-297 in *Metal Matrix, Carbon, and Ceramic Matrix Composites*. Edited by J. Buckley. NASA Conference Publication 3133, 1991.

8. M. Omori, K. Hsu, S. Tsunekawa, and T. Hirai, "Thermal and Dielectric Properties of Zirconyl Phosphate Compact," *J. Mater. Sci.*, **27** [2] 408-12 (1992).

9. B. A. Bender, T. L. Jessen, and D. Lewis III, "Optimization of the Mechanical Properties of Silicon Carbide-Fiber-Reinforced Zirconia Titanate Matrix Composites Through Controlled Processing," *J. Am. Ceram. Soc.*, **75** [6] 1628-35 (1992).

10. B. A. Bender, R. J. Rayne, T. L. Jessen, and S. Browning, "The Effect of High Temperatures on the Microstructure and Properties of Nicalon-Reinforced Zirconium Phosphate Composites," *Ceram. Eng. Sci. Proc.,* **18** [3] 689-96 (1997).

11. R. K. Frazer, "Evaluation of Silicon Nitride as an Advanced Radome Material," *John Hopkins APL Tech. Dig.*, **13** [3] 393-99 (1992).

12. I. Yamai and T. Oota, "Low-Thermal-Expansion Polycrystalline Zirconyl Phosphate Ceramic," *J. Am. Ceram. Soc.*, **68** [5] 273-78 (1985).

13. B. A. Bender, T. L. Jessen, R. P. Ingel, and D. Lewis, "High Temperature Exposure of BN-Coated Polymer-Derived SiC Fiber/$ZrTiO_4$ Composites," *Ceram. Eng. Sci. Proc.*, **13** [9-10] 889-96 (1992).

14. R. Bodet, X. Bourrat, J. Lamon, and R. Naslain, "Tensile Creep Behavior of a Silicon Carbide-Based Fiber with a Low Oxygen Content," *J. Mater. Sci.,* **30** [3] 661-677 (1995).

THE EFFECT OF MICROSTRUCTURE ON THE MECHANICAL BEHAVIOR OF CFCMCS UNDER DYNAMIC AND STATIC LOADING CONDITIONS

T.L. Jessen and B.A. Bender
U.S. Naval Research Laboratory, Washington, D.C. 20375-5343

The effect of dynamic, static, and combined static/dynamic loading on the flexure strength and fracture behavior of a unidirectional fiber composite was investigated. The 50 volume percent Nicalon™ fiber/zirconium titanate matrix composite was similar to that used in an earlier study which demonstrated microstructure-sensitive mechanical behavior. In the present study, statistically significant differences in time-dependent strength were observed as a function of composite microstructure across five decades of constant crosshead displacement (dynamic loading). Composite strength was independent of time-to-failure (TTF) when the composite contained distributed matrix-rich channels aligned perpendicular to the applied load. The strength was observed to decrease with increasing TTF when the matrix-rich channels were aligned parallel to the applied load. Under both static and combined static/dynamic loads similar mechanical/microstructure behavior was observed. The strength decreased when the channels were aligned parallel to the applied load and was constant when the channels were perpendicular.

INTRODUCTION

For more than 25 years, enhancement of the mechanical properties of ceramics through the use of continuous fibers has been an area of active research. Increasing attention has been focused on the fatigue and loading rate behavior of this class of materials [e.g. 1-9], since the observations by Marshall and Evans [1] in 1985 of hysteresis during cyclic loading of fiber reinforced composites. Fatigue behavior is an important aspect of continuous fiber ceramic matrix composite (CFCMC) performance because of their anticipated use as structural materials in such applications as aircraft engine components and heat exchangers. A common observation among the fatigue studies [5-9] is that the mechanical behavior of CFCMCs (i.e. matrix-cracking strength and fiber pullout length) is a function of loading rate. In general, the higher the load rate, the higher the observed mechanical properties.

The primary method for improving the mechanical properties of CFCMCs has been through the incorporation of a weakly bonded interface between the fiber and matrix. These interfacial layers have typically been either C or BN-based, although

To the extent authorized under the laws of the United States of America, all copyright interests in this publication are the property of The American Ceramic Society. Any duplication, reproduction, or republication of this publication or any part thereof, without the express written consent of The American Ceramic Society or fee paid to the Copyright Clearance Center, is prohibited.

$LaPO_4$ [10] has been increasingly investigated. Reliance on debond layers for property improvements has led to the general consensus that CFCMCs are microstructure insensitive. Recently, we [11] observed significant (99.9% confidence level) differences in both ultimate strength and fracture resistance in a unidirectionally-aligned CFCMC as a function of fiber distribution. The presence of thin (<90 µm in width), periodic matrix-rich channels gave the CFCMC laminate-like mechanical behavior. The CFCMC exhibited high toughness (36% improvement) when loaded normal to the channel orientation and high strength (25% improvement) when loaded parallel to the channel orientation.

The effect of CFCMC microstructure on the fatigue behavior was not addressed in the earlier studies. The present work uses similar composites to that of ref. 11 to investigate the relationship between microstructure and time dependent failure. Three distinct conditions were considered: (1) dynamic loading, (2) static loading, and (3) combined static/dynamic loading.

EXPERIMENTAL PROCEDURE

The composite billets which contained 50 volume percent BN-coated CG-Nicalon (Nippon Carbon Co., Tokyo, Japan) were prepared with as close to identical processing as possible. The procedure detailed in ref. 12 is summarized as follows. The Nicalon fiber tows were cut in 7.5 cm lengths and stacked lengthwise in a 1.27 cm wide by 7.7 cm long wire-mesh cage to develop a fiber preform. Typically, 4.5 or 6.0 g of fibers were stacked in each preform. The fiber preform was then immersed in a solution of zirconium n-propoxide (Morton Thiokol, Chicago, IL USA) and titanium isopropoxide (Alfa Products, Danvers, MA USA) mixed in proportions to yield 50 mol% each ZrO_2 and TiO_2, further diluted by 50% with absolute ethanol. The fiber preform was then dried for 4 h over a beaker of water in a drying oven at 110°C, which hydrolyzed the alkoxides into a mixture of Zr and Ti hydroxides. After three infiltrations, the preform was calcined in flowing air at 540°C, further converting the matrix phase into oxides. Typically, four iterations of the infiltration/calcination procedure were required to achieve a matrix buildup of 50 v/o. The composite was then hot pressed at 1270°C for 30 minutes in an over-pressure (1.1 atm) of CO, with an applied pressure of 17.25 MPa. Typically, these billets were densified to 93-95% of theoretical via this processing procedure.

Following hot pressing, the only intentional difference in billet processing was performed during the specimen machining operation. Billets were machined in such a manner that two sets of flexural specimens oriented 90° to each other were produced. This difference, which affects the matrix-rich channel orientation, is shown in Fig. 1. The different orientations were identified as N and P; N denoting channel orientation normal to the main consolidation direction, and P representing a parallel orientation with respect to consolidation.

The specimens, with nominal dimensions of 3.0 mm in width, 1.4 mm in thickness, and length ranging from 32-34 mm, were ground to a 20 µm finish. The specimens were tested on a universal testing machine (Model TT-C-l, Instron Engineering Corp., Canton MA) in four-point flexure, with a 27 mm outer span and inner supports spaced 9 mm apart.

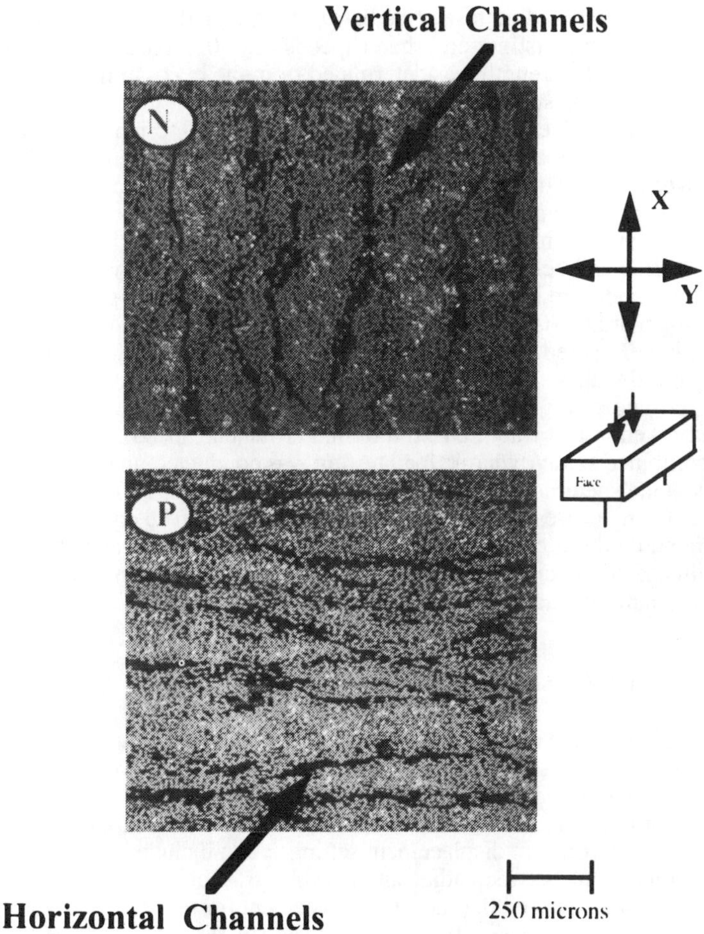

Figure 1. Inverted backscattered electron SEM micrographs of typical N and P oriented microstructures, with the long axes of the fibers perpendicular to the plane of the page. In these images the fibers are light and the matrix phase dark. The arrows highlight the orientation of the distributed matrix-rich channels (nominally <90 micrometer in width). The diagram on the right indicates channel orientation during 4 pt. flexure testing. The X-Y reference grid indicates the orientations for the flexural modulus measurements.

For the dynamic loading tests, five bars of both N and P orientation were broken at each of five distinct crosshead speeds (2.0, 0.2, 0.02, 0.002, and 0.0002 in/min). The ultimate strength was determined from the maximum deflection on the load/time trace and the specimen's dimensions.

For the static load tests, one specimen of each N and P orientation were used for each of three deadload durations: (1) 1000 s, (2) 10,000 s, and (3) 100,000 s. The loads were applied using the universal testing machine in the four-point configuration described above. The stress levels were kept constant for each specimen by adjusting the load based on the specimen cross-sectional area. Prior to the initial loading and before each successive load, the flexural modulus of the specimens was measured in both the X and Y orientation (see Fig. 1) using a commercial apparatus (GrindoSonic, J.W. Lemmens, Inc., St. Louis, MO). The modulus value measured in each orientation were averaged and then normalized versus the initial values. The static loads were increased until a drop in modulus was observed- indicating a change in the specimen's compliance. The specimens were then broken in fast fracture (2.0 mm/min crosshead speed).

The combined static/dynamic load testing was conducted on only the P specimens. One specimen for each duration (1000, 10,000, 100,000 s) was deadloaded at single stress level. This level was selected to be equal to the highest level where no modulus change had been detected in the static loaded specimens. The modulus of the specimens was measured pre and post-loading to insure there was no compliance change. The specimens were then broken in fast fracture.

RESULTS and DISCUSSION

Dynamic Loading

Table I displays the composite flexure strength data as a function of crosshead displacement. Three sets of data are listed in the Table. The 'Average Flexure Strength' and standard error for the 10 specimens tested at each crosshead speed. The P and N-oriented data sets include the five samples with similar microstructure (see Fig. 1) at each constant displacement setting. Also included in the Table are statistical comparisons between adjacent data subsets. For example: the comparison between 2.0 and 0.2 in/min displacement for the 'Average Flexure Strength' data show that these population means differ at a 75% confidence level. Other statistical comparators in Table I include: OVLP- population means which overlap at a one sigma confidence level, 90 and 99%- population means differ at these levels of statistical confidence. The 'Average Flexure Strength' data in Table I give the impression that these CFCMCs are relatively time-to-failure (TTF) insensitive. However, when the data are examined using the microstructure as a discriminator (P vs. N-oriented matrix channels), a different interpretation is obtained. Figure 2 compares plots of the N and P strength data versus the time to achieve maximum load.

The P-oriented specimens display a TTF insensitive flexure strength (approximately 670 MPa) across the five decades of constant crosshead displacement investigated. The time to maximum load ranged from less than two seconds at the 2.0 in/min rate to an average of 12,094 s at the 0.0002 in/min rate for the P samples. The star on the strength vs. log TTF plot indicates the relative

Table I. CFCMC Strength as a Function of Crosshead Speed

Crosshead Speed (in/min)	Average Flexure Strength (MPa)	P-oriented Strength (MPa)	N-oriented Strength (MPa)
2.0	774 ± 48*	663 ± 34	914 ± 14
	> 75.0%	> OVLP	> 99.0%
0.2	684 ± 27	670 ± 29	698 ± 27
	> OVLP	> OVLP	> OVLP
0.02	691 ± 33	673 ± 45	705 ± 35
	> OVLP	> OVLP	> OVLP
0.002	691 ± 35	679 ± 52	700 ± 40
	> OVLP	> OVLP	> 90.0%
0.0002	653 ± 11	665 ± 19	640 ± 5

* represents the standard error

positioning of the data from ref. 11 with similar microstructure. The previous data [11] which represents 46 specimens are in good agreement with the current results.

The N-oriented specimens have a very different TTF strength response compared with the P samples. A decrease in average strength of more than 270 MPa across the five decade testing span is observed with the N specimens. At the highest rate (2.0 in/min) the N strength exceeds that of the P samples at a 99.9% confidence level but quickly drops to comparable values by the second decade. The average time to maximum load while comparable between the specimen types for the first four decades (N vs. P- 2.0 in/min: <2 s vs. <2 s, 0.2: 9 s vs. 11 s, 0.02: 80 s vs. 99 s, and 0.002: 821 s vs. 836 s) differ at a 90% confidence level at the slowest crosshead speed (8071 s vs. 12094 s). The data (star symbol) from the earlier study [11] (72 similar specimens) plot along the trendline of the current N specimen results.

In addition to the different strength vs. TTF response between the N and P microstructures, there was a difference in the failure processes. Figure 3 displays the two types of failure behavior observed for all specimens. Figure 3A shows the behavior seen for all N-oriented samples and the P samples tested at the highest loading rate. Figure 3B is typical of the remaining P-oriented specimens. This failure behavior is similar to that observed in ref. 11 for these two microstructures. The different behavior is a result of the differing matrix channel orientation between the specimens. In both cases, failure proceeds along the matrix channels. For the N specimens this results in the traditional fibrous fracture observed by numerous researchers. In the P samples, the failure occurs via bulk delamination along a matrix channel perpendicular to the applied load, with a corresponding significant increase in fracture resistance. The exception to this behavior was the P-oriented samples tested at 2.0 in/min- it is likely that the energy imparted to the specimens at this rate was sufficient to overwhelm any microstructural considerations and they failed via the most direct path.

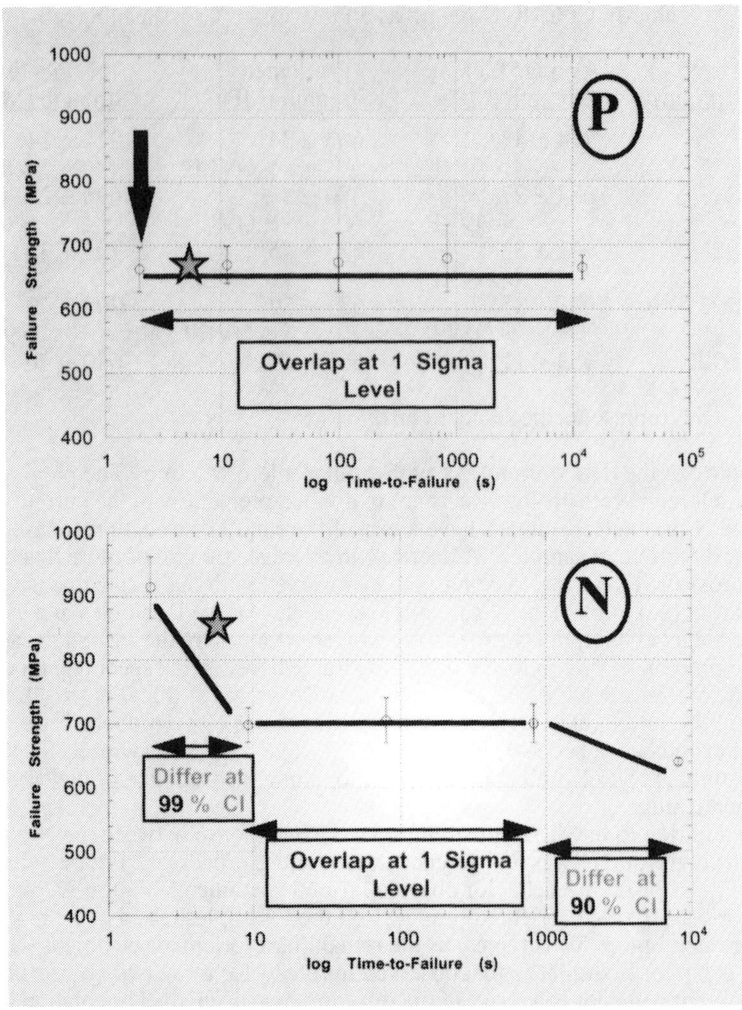

Figure 2. Log/normal plots of the flexure strength vs. time to maximum load for the N and P-oriented CFCMCs. The lines overlaid on the data are trendlines and do not represent 'best fit' approximations. The error bars are the standard error. The stars indicate the relative positioning of the N and P data collected for ref. 11. The double-ended arrows represent the range over which the indicated confidence predictions apply. The bold arrow on the P-oriented plot is an aid for Fig. 3.

The difference in TTF behavior between the N and P specimens can be directly related to their differences in microstructure. The P specimens exhibit time independent flexure strength because as matrix microcracks coalesce in the channels, their propagation to critical dimension in the direction of the applied load is limited by the fiber-rich regions. Linked micro-to-macrocracks which penetrate one fiber-rich layer in the P specimens encounters another fiber-rich region. This self-similar layering throughout the microstructure limits crack coalescence and results in no macroproperty (strength or fracture resistance) degradation. In contrast, the N specimen strength is highly sensitive to time under load because the microcracks which form in the matrix can link in the matrix channels into a critical dimension with only limited interaction with the reinforcing fibers.

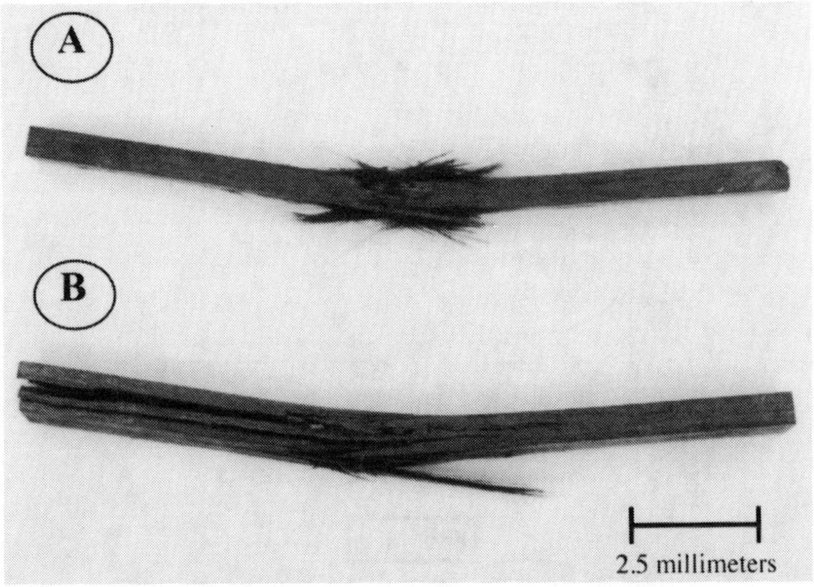

Figure 3. Two distinct failure behaviors were observed for the data displayed in Table I. Fig. 3A is representative of all the N-type specimens and the P specimens indicated by the bold arrow in Fig. 2. Fig. 3B represents the failure behavior observed among the remaining P-oriented samples.

Static Loading

Figures 4 and 5 display the normalized composite modulus as a function of both applied stress and time-under-load for the N and P-oriented specimens. A general trend observed for all specimens was that the composite modulus decreased with increasing time-under-load at increasingly lower stress levels. This change in modulus is an indication that the composite compliance is increasing due to the formation and growth of microcracks in the matrix. To study the effect of

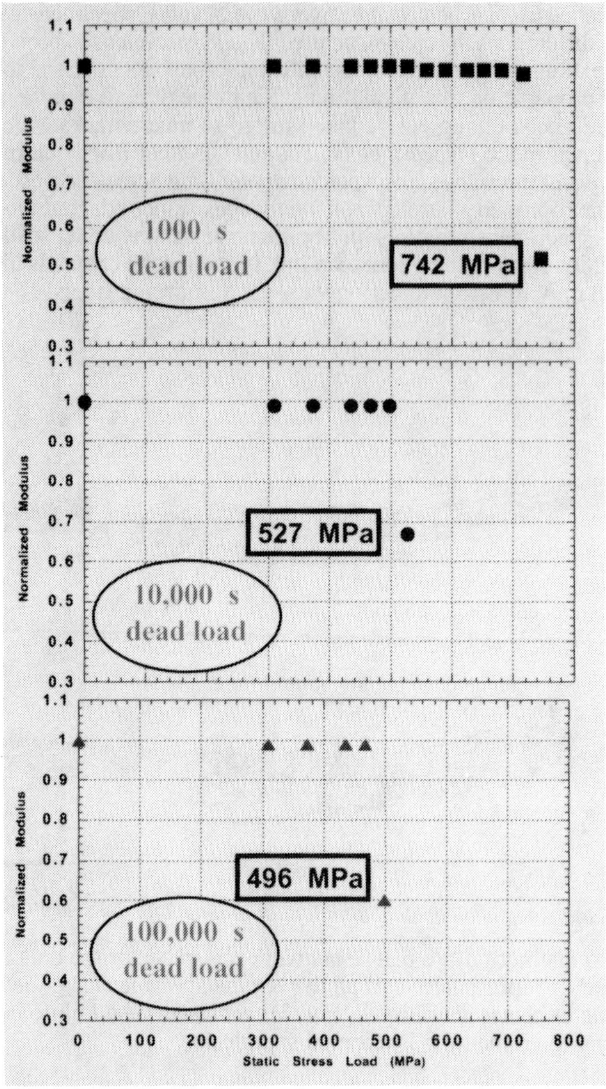

Figure 4. Plots of the normalized moduli values for the N-oriented specimens as a function of both applied stress and time-under-load. The stress highlighted on each plot was the level at which the compliance of the specimen changed significantly.

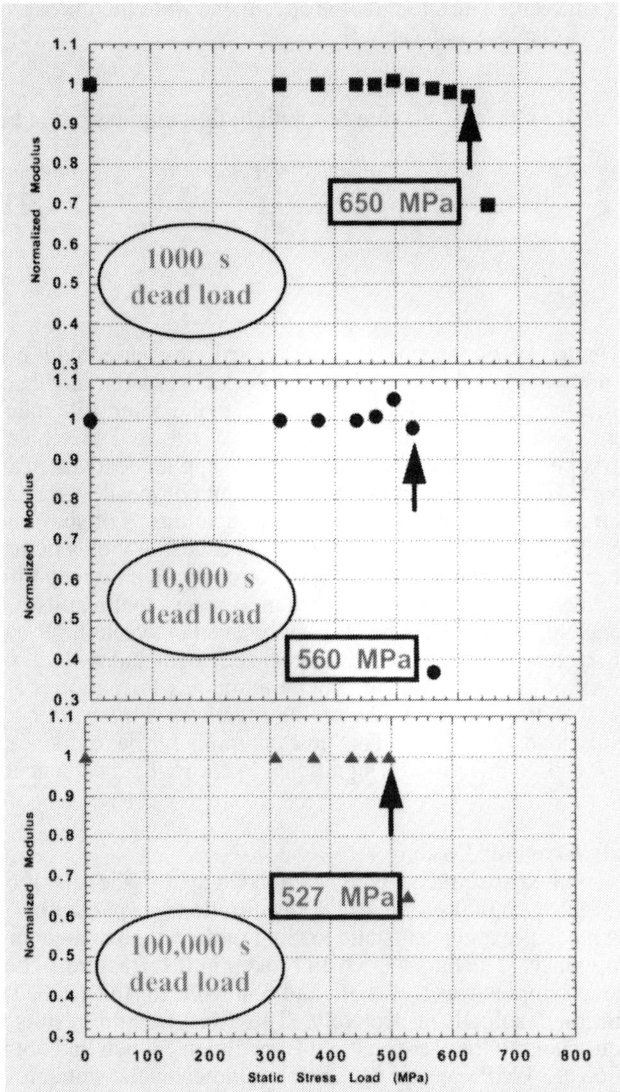

Figure 5. Plots of the normalized moduli values for the P-oriented specimens as a function of both applied stress and time-under-load. The stress highlighted on each plot was the level at which the compliance of the specimen changed significantly. The arrows indicate the deadload stress used for the combined static/dynamic testing.

microcracking on composite strength, the specimens were then broken in fast fracture, which is displayed in Table II.

Table II. CFCMC Flexure Strength Following Static Loading

Composite Type	Time-Under-Load 1000 s	10,000 s	100,000 s
N-Oriented	520 MPa	450	210
P-Oriented	390	370	380

The N-oriented specimen strength continuously decreased with increasing time-under-load, whereas, the strength of the P specimens remained constant. This difference in mechanical properties is attributable to the composite microstructure in a similar manner to that observed for the dynamically-loaded specimens. The change in matrix channel orientation relative to the applied load alters the crack propagation path. This result suggests that different composite microstructural arrangements may be beneficial for different applications. For short durations under load, the N orientation might be preferred because it yields higher strength for a given fiber loading. However, for applications with extended time-under-load, the P-oriented microstructure provides more consistent results.

A difference in failure mechanism was observed between the P specimens broken in fast fracture (2.0 in/min crosshead speed) with and without the static deadloading. All of the specimens in which microcrack formation is assumed (based on modulus drop) showed macrofailure features similar to Fig. 3B unlike the non-statically loaded samples. This observation is further evidence that the microstructure/crack interaction is a significant factor in the mechanical properties of CFCMCs.

Combined Static/Dynamic Loading

Single P-oriented specimens were deadloaded at stress levels of 620, 527, and 496 MPa for 1000 s, 10,000 s, and 100,000 s durations respectively. Moduli measurements made pre and post- static loading indicated no change in the specimen compliance. The bars were then broken in fast fracture. The flexure strength of these samples was 681, 636, and 654 MPa for the 1000, 10,000, and 100,000 s duration deadloads respectively. These fast fracture results were statistically equivalent to the fast-fractured P specimens which had not been deadloaded (663 ± 34 MPa, where the error bar indicates the standard error). However, the macro-failure features were quite different. The deadloaded/fast-fractured specimens resembled Fig. 3B. This observation suggests that some level of microcracking occurred during the deadloading but was not sufficient to either be detected by the modulus technique or reduce the fast fracture strength. These results combined with the dynamic and static loading data reinforce the observations of ref. 11 which found that the P-oriented microstructure has a higher fracture or damage resistance than the N microstructure.

SUMMARY

The flexure strength of a unidirectional CFCMC was measured across five decades of crosshead displacement speed ranging from 2.0 to 0.0002 in/min. Overall the strength was found to generally be independent of time under load. However, when the composites were subdivided into groupings based on microstructural similarities, distinct differences in mechanical behavior were evident. When the microstructure contained continuous, distributed matrix-rich channels (<90 μm in width) aligned parallel to the applied load, composite strength decreased with increasing time-to-failure (TTF). In contrast, when the matrix channels were aligned perpendicular to the applied load, the strength was TTF independent. Differences in failure behavior corresponding to differences in microstructure were also observed.

Static and combined static/dynamic load tests also showed a mechanical property difference based on microstructure for CFCMCs. These experiments showed that the P microstructure had higher damage-resistance than the N-oriented structures.

These observations are structurally-dependent rather than compositionally-dependent thus similar behavior can be expected in all brittle matrix/fiber-reinforced composites. Further, they suggest that composite microstructures can be optimized depending on the anticipated stress level and load duration of specific applications.

REFERENCES

1. D.B. Marshall and A.G. Evans, "Failure Mechanisms in Ceramic-Fiber/ Ceramic Matrix Composites," J. Am. Ceram. Soc., 68[5], pp. 225-31 (1985).

2. E. Minford and K.M. Prewo, "Fatigue of Silicon Carbide-Reinforced Lithium-Alumino-Silicate Glass Ceramics," pp. 561-70 in Tailoring Multiphase and Composite Ceramics, edited by C.G. Pantano and G.L. Messing, Plenum Press, New York (1986).

3. D.B. Marshall and W.C. Oliver, "Measurement of Interfacial Mechanical Properties in Fiber-Reinforced Ceramic Composites," J. Am. Ceram. Soc., 70[8], pp. 542-8 (1987).

4. T. Kotil, J.W. Holmes, and M. Comninou, "Origin of Hysteresis Observed During Fatigue of Ceramic-Matrix Composites," J. Am. Ceram. Soc., 73[7], pp. 1879-83 (1990).

5. P.G. Karandikar and T.W. Chou, "Damage Development and Moduli Reductions in Nicalon-Calcium Aluminosilicate Composites under Static Fatigue and Cyclic Fatigue," J. Am. Ceram. Soc., 76[7], pp. 1720-8 (1993).

6. S.M. Spearing, F.W. Zok, and A.G. Evans, "Stress Corrosion Cracking in a Unidirectional Ceramic-Matrix Composite," J. Am. Ceram. Soc., 77[2], pp. 562-70 (1994).

7. B.F. Sorensen and J.W. Holmes, "Effect of Loading Rate on the Monotonic Tensile Behavior of a Continuous-Fiber-Reinforced Glass-Ceramic Matrix Composite," J. Am. Ceram. Soc., 79[2[, pp. 313-20 (1996).

8. H. Tsuda, M. Enoki, and T. Kishi, "Influence of Loading Rate and Processing Atmosphere on Pullout of SiC Fibers from Pyrex Glass," J. Am. Ceram. Soc., 79[5], pp. 1319-23 (1996).

9. M.E. Walter and G. Ravichandran, "Effect of Quasi-Static Loading Rate on Damage Formation in Silicon Carbide Reinforced Calcium Aluminosilicate Unidirectional Composites," J. Am. Ceram. Soc., 80[5], pp. 1289-92 (1997).

10. P.E.D. Morgan and D.B. Marshall, "Ceramic Composites of Monazite and Alumina," J. Am. Ceram. Soc., 78[6], pp. 1553-63 (1995).

11. T.L. Jessen, V.A. Greenhut, D. Lewis, III, and J.J. Friel, "Effect of Microstructure on the Mechanical Behavior of Continuous Fiber Reinforced Ceramic Matrix Composites," submitted to J. Am. Ceram. Soc., (1997).

12. B.A. Bender, T.L. Jessen, and D. Lewis, III, "Optimization of the Mechanical Properties of Silicon Carbide-Fiber-Reinforced Zirconia Titanate Matrix Composites Through Controlled Processing," J. Am. Ceram. Soc., 75[6] pp. 1628-35 (1992).

SYLRAMIC™ CERAMIC-MATRIX COMPOSITES PROCESSING AND PROPERTIES

T.E. Easler, D.R. Petrak, and A. Szweda
Advanced Ceramics Program
Dow Corning Corporation
Midland MI 48686

ABSTRACT

The current processing and properties of Dow Corning SYLRAMIC™ S200 CMC materials are reviewed. These CMCs are fabricated using CG Nicalon™ fibers, and a preceramic polymer precursor which yields an SiNC matrix. The versatile PIP (polymer impregnation and pyrolysis) process method allows a variety of different fiber architectures to be used to fabricate components with simple or complex shapes. Examples of various fabrication methods and components are shown. The thermal and mechanical properties and microstructure are described.

INTRODUCTION

Dow Corning Corporation has been actively developing chemical routes to ceramics for over thirty years. During the past ten years, the Advanced Ceramics Program has been developing continuous-fiber reinforced ceramic-matrix composites (CMCs). These materials are non-oxide silicon-based ceramics with a matrix derived from polymeric precursors. Ceramic composites fabricated under the tradename SYLRAMIC™ S200 CMC are the focus of this article.

Dow Corning fabricates ceramic-matrix composites using a Polymer Impregnation and Pyrolysis (PIP) process. This process is widely recognized as a versatile method to fabricate large, complex shaped structures. It consists of the impregnation of ceramic fiber or fabric with the matrix precursor, pyrolysis to convert the polymeric precursor to a ceramic phase, and then repeated reimpregnation of the component until the desired level of density is attained. In comparison to other ceramic composite fabrication processes, the PIP process offers significant flexibility. By utilizing low- temperature forming and molding steps typically used in organic matrix composites, the PIP approach allows the use of existing equipment and processing technology developed for plastic matrix composites (PMCs).

The primary constituents of a continuous fiber reinforced CMC are shown schematically in Figure 1. The ceramic fibers are coated with a thin interface layer to control the degree of bonding of the fibers to the matrix. This is required

To the extent authorized under the laws of the United States of America, all copyright interests in this publication are the property of The American Ceramic Society. Any duplication, reproduction, or republication of this publication or any part thereof, without the express written consent of The American Ceramic Society or fee paid to the Copyright Clearance Center, is prohibited.

in order to obtain the desired crack deflection behavior which translates to a graceful failure mode. A ceramic filler material is typically added to the matrix phase to modify properties and processing behavior. A small amount of residual porosity and matrix cracks are also present, the latter being formed during conversion of the matrix precursor to a ceramic phase.

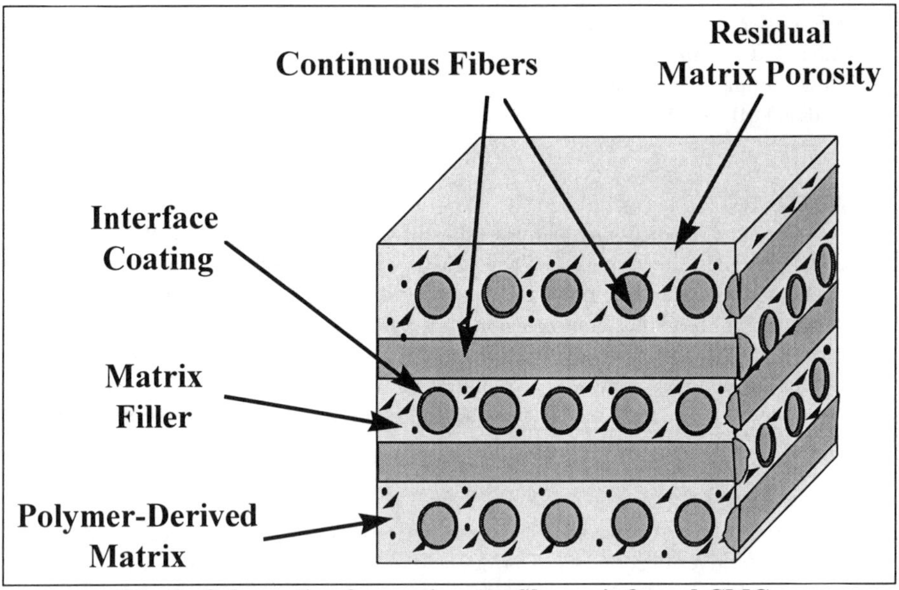

Fig. 1. Schematic of a continuous-fiber reinforced CMC.

The ability to fabricate preforms having the desired shape and architecture is of primary importance for ceramic composites. Real-world applications require versatile approaches to produce the wide range of structural shapes, wall thicknesses and tolerances required. Techniques such as filament winding, resin transfer molding, and ply lay-up followed by autoclaving can be used to fabricate a variety of shapes and sizes of components. When the PIP process is completed, net shape parts result that require a minimum of final machining. When final machining is required, conventional diamond tooling as well as laser and ultrasonic machining have been used.

MATERIALS AND PROCESSING METHODS

Materials

The ceramic composites fabricated by Dow Corning have a versatile chemistry and can consist of various fibers, interface coatings, and matrix chemistries. The currently available systems are divided into 3 classes of materials denoted:

- SYLRAMIC™ S100 CMCs* : carbon-coated CG NICALON™ fiber** in an amorphous SiOC matrix. The maximum use temperature is < 450°C in oxidizing environments since the carbon coating will be lost, but temperatures up to 1100°C can be tolerated in inert environments. High-temperature oxidative exposure can be acceptable in limited life applications.
- SYLRAMIC™ S200 CMCs: various coatings (typically boron-containing) on CG NICALON™ fiber in an amorphous SiNC matrix. These materials can be used at temperatures up to 1200 - 1250°C in an oxidizing or inert environment, and can tolerate temperatures up to 1400°C for limited lifetime applications.
- SYLRAMIC™ S300 CMCs: various coatings on SYLRAMIC™ SiC fiber in an amorphous SiNC, or crystalline Si_3N_4/SiC or SiC matrix for higher temperature applications.

Fabrication Methods

In the PIP process shown schematically in Figure 2, one of three methods is typically used to fabricate the preforms for the components, depending on the geometry of the component. These methods are described individually below.

Complex shapes may require resin transfer molding. The ceramic fibers, in the form of a woven fabric, are assembled inside a mold to create a specific shape. The mold is heated and then injected with a solventless liquid resin. When enough resin has been injected to completely impregnate the fabric, the temperature of the mold is raised to cure the resin. After cooling the mold, the part is removed and pyrolyzed. This method is useful for components with flanges or other complex features.

For cylindrical shapes, the method of choice is frequently filament winding, in which the component is created by winding a continuous length of fiber onto a spool or mandrel until the desired component length and thickness are attained. Resin may be applied to the fibers before they are wound onto the mandrel by running the fibers through a slurry of the matrix precursor and filler, or the resin

---

* Trademark of Dow Corning Corporation.

** Trademark of Nippon Carbon Company.

may be applied after winding by pouring or painting the slurry onto the wound fibers. The entire spool is then heated to cure the resin, and then the component is cooled and removed from the mandrel for pyrolysis.

A third fabrication method is typically used for making flat panels. This method consists of laying-up flat plies of fabric which have been prepregged with the resin and filler slurry. The stacked plies are then vacuum-bagged and autoclaved under pressure to consolidate the layers and cure the resin.

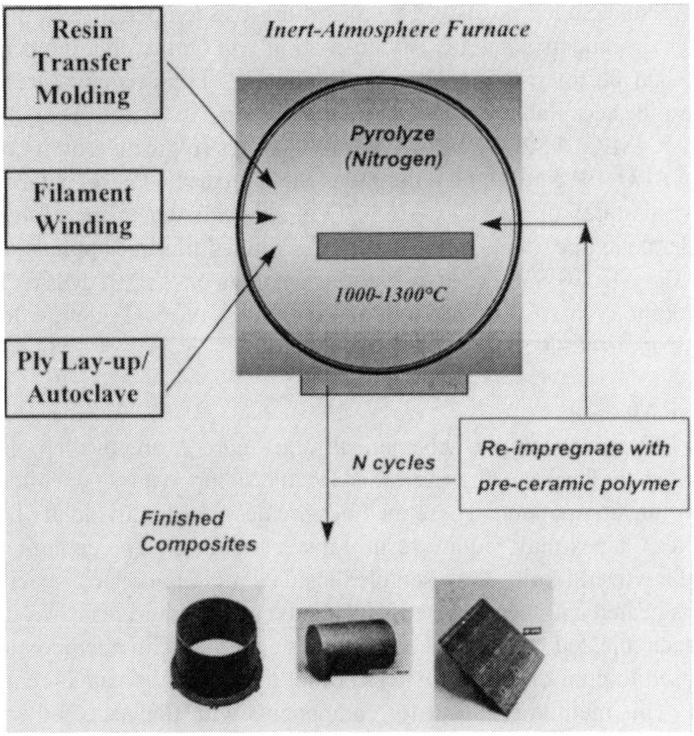

Fig. 2. The PIP Process for Fabricating CMCs.

Regardless of the method chosen for fabricating the preform, all components are processed the same way after the initial steps. As shown in Figure 2, an inert-atmosphere furnace is used to pyrolyze the matrix precursor at temperatures of 1000-1300°C, which converts the polymer to a ceramic phase. During this process, there is a reduction in the volume of matrix material as it is converted from resin to ceramic and increases in density. The resultant void space is filled by reimpregnating the composites with additional resin, followed by another pyrolysis run. This process loop is repeated until the desired levels of density and porosity are achieved, typically about 10-15 cycles. A bulk density of 2.1-2.2 g/cc and an open porosity of <5% are typical.

RESULTS

Microstructure

The microstructure of a typical CMC is shown in Fig. 3. The alternating layers of fibers oriented perpendicular and parallel to the page can be seen. In the case of this symmetric layup, there are two layers of fibers parallel to the page in the center of the CMC. Individual fibers are approximately 15 microns in diameter. The fibers are surrounded by the matrix phase which is continuous except for a small number of voids and cracks which form during processing. Filler particles are less than one micrometer in size and therefore not visible at this magnification. Some fiber tips are lost during sample preparation for microscopy.

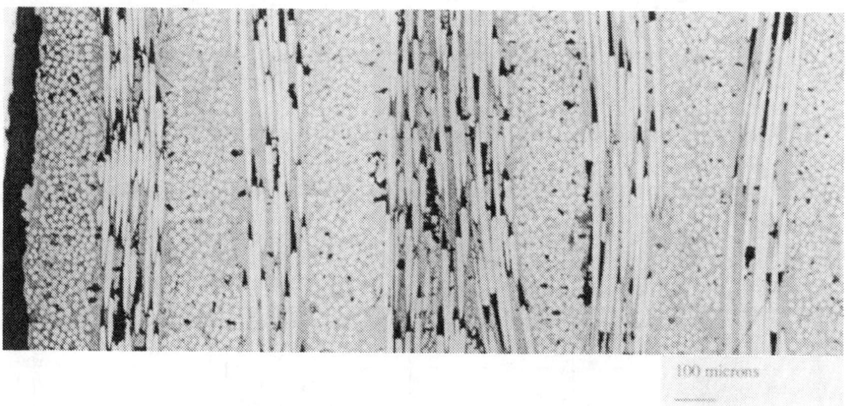

Fig. 3. Microstructure of a CMC.

Density and Porosity

The increase in bulk density and decrease in open porosity as a function of PIP cycle number are shown in Figure 4. These parameters and the weight gain are tracked to monitor progress during PIP processing. Both the density and porosity tend to level-out as processing is completed, and a point of diminishing returns is reached with further PIP cycles.

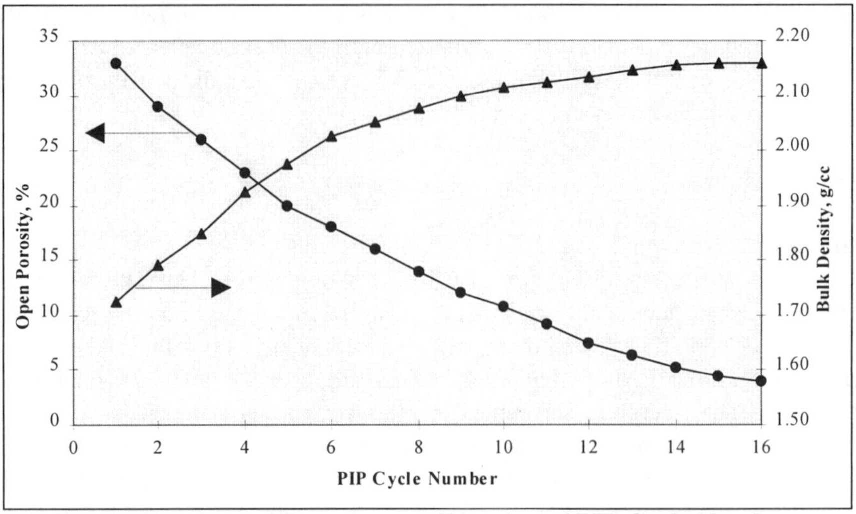

Fig. 4. Density and Porosity as a Function of PIP Cycle Number.

Mechanical Properties

The mechanical properties at 20°C and 1000°C are given in Table 1. There is about a 15% increase in tensile strength at the elevated temperature which is attributed to the redistribution of residual stresses. Concurrently, there is a decrease in shear strength with temperature. The latter behavior is a matrix-dominated property. Effectively, the higher temperature condition results in a decrease in stiffness but an increase in tensile strength. This stress-strain behavior is shown graphically in Figure 5 for tensile data at three temperatures.

Table 1. Mechanical Properties of SYLRAMIC™ S200 CMCs.

	Tensile Strength, MPa (ksi)	Shear Strength, MPa (ksi)	Flexure Strength, MPa (ksi)	Compressive Strength, MPa (ksi)
20°C	262 (38)	38 (5.5)	379 (55)	434 (63)
1000°C	300 (43)	24 (3.5)	414 (60)	NA

	Tensile Modulus, MPa (ksi)	Proportional Limit, MPa (ksi)	Proportional Strain, %	Strain-to-Failure, %
20°C	96 (14)	96 (14)	0.09	0.5
1000°C	90 (13)	90 (13)	0.08	0.7

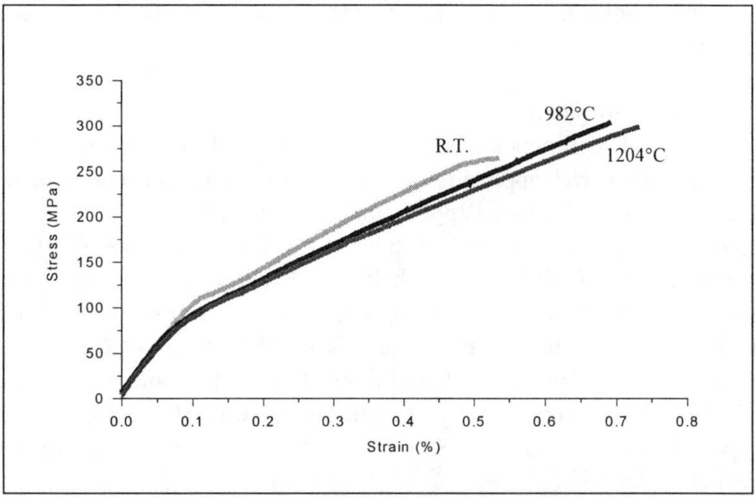

Fig. 5. Tensile Stress-Strain Behavior vs. Temperature.

Thermal Properties

The thermal conductivity as a function of temperature is shown in Fig. 6. This data shows that thermal conductivity is a relatively isotropic property and is only slightly affected by temperature over the range from room temperature to 1200°C. The constant nature of this property in the x, y, and z planes in the CMC

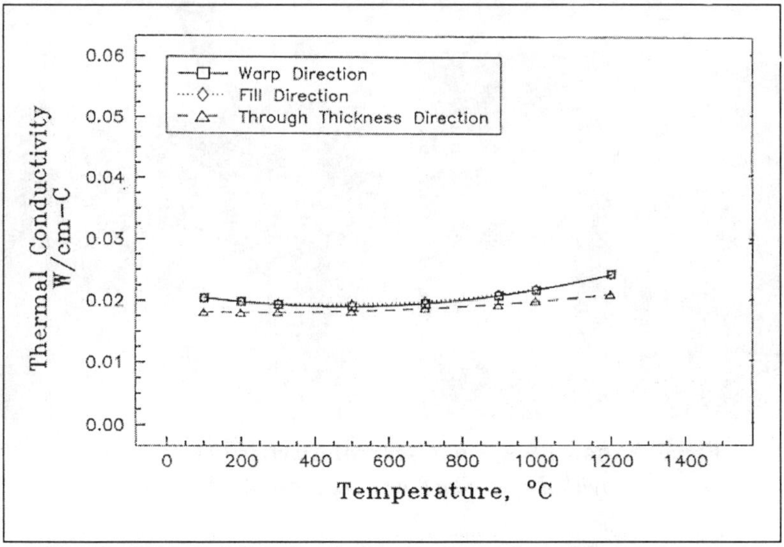

Fig. 6. Thermal Conductivity as a Function of Temperature.

occurs because both the fiber and matrix are amorphous and have similar conductivities.

Representative Components

A number of components have been fabricated for evaluation in both aerospace and industrial applications. After fabrication, the components were subjected to testing in simulated application environments.

Figure 7 illustrates a number of exhaust flaps for a General Electric F-110 aircraft engine. The CMC flaps are being evaluated as replacements for metal flaps which have limited life capability under the severe thermal and oxidative environment of the engine exhaust. The SYLRAMIC™ S200 CMC flap as well as flaps fabricated by GE, HITCO, and Dupont Lanxide Composites were tested in a GE F110-129 engine for up to 120 hours. The HITCO and Dow Corning flaps survived well, whereas the GE and DLC flaps exhibited some cracking. More testing is planned.

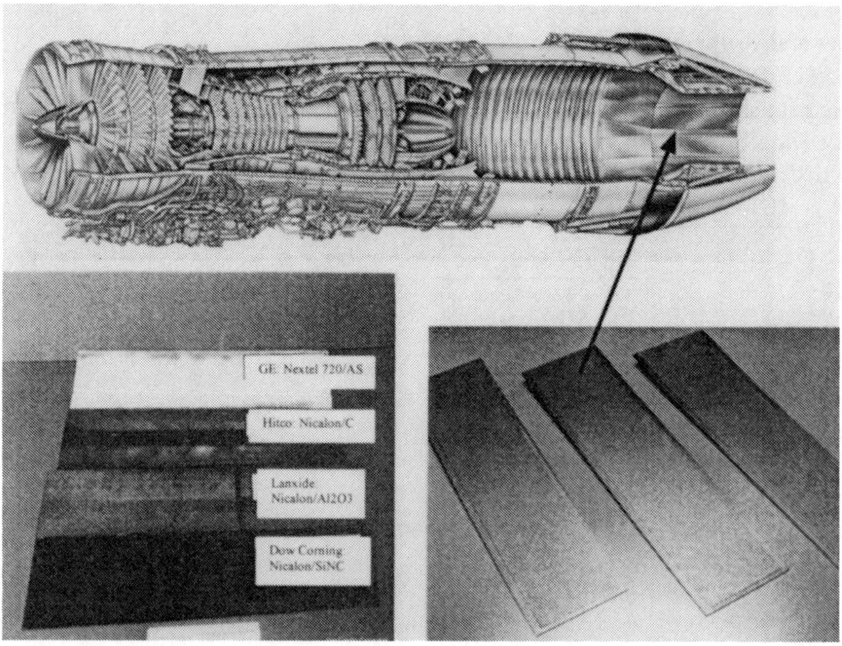

Fig. 7. Exhaust Flaps for General Electric F-110 Engine.
(provided by J. Staehler & L. Zawada, AFML)

Figure 8 shows an exhaust panel for the proposed NASA High-Speed Civil Transport aircraft. The CMC tiles would line the exhaust to provide structural stability to the acoustic damping system. The tile shown in Figure 8 has six fastener locations as well as a large number of perforations which affect the contribution of the CMC to the overall characteristics of the acoustic damping system.

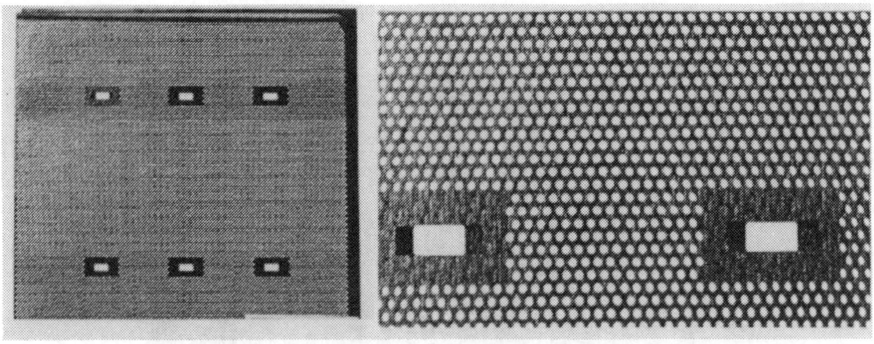

Fig. 8. NASA-HSCT (High-Speed Civil Transport) Exhaust Panels

Figure 9 illustrates a turbine exhaust nozzle, which is comprised of an inner ring and an outer ring connected by three struts. These components were fabricated separately and then joined to form the final assembly.

Fig. 9. PrototypeTurbine Exhaust Assembly.

Figure 10 shows several prototype components that have been fabricated for a number of industrial applications where high-temperature and corrosive environments limit the use of metals. They include containment housings for sealless pumps, combustor liners for gas turbines, and pipe hangers for moving crude oil through refinery furnaces.

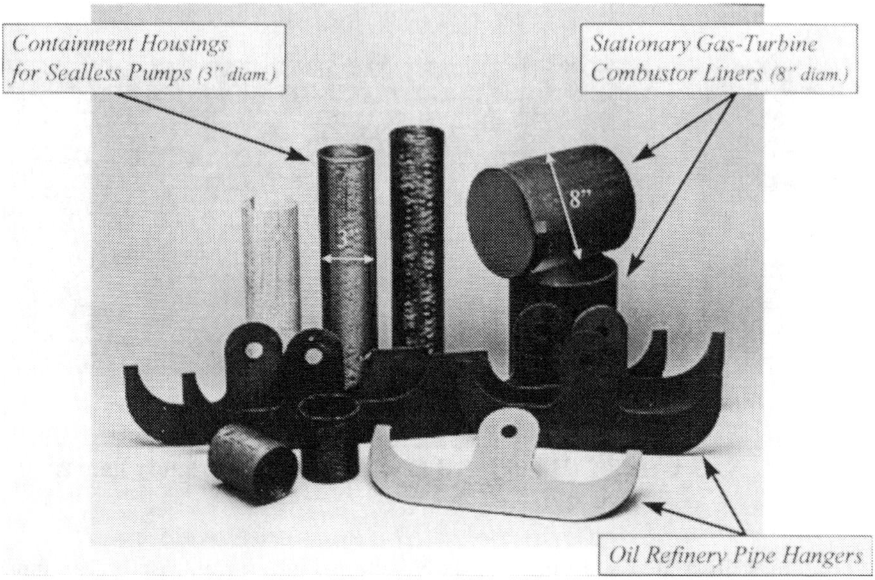

Fig. 10. Prototype Components for Industrial Applications.

CONCLUSIONS

The polymer impregnation and pyrolysis process used by Dow Corning Corporation for the fabrication of CMCs is a versatile method for producing both simple and complex net-shaped components. These CMCs have excellent mechanical properties at both room and elevated temperatures. A number of prototype components fabricated by Dow Corning are being evaluated in aerospace and industrial applications.

ACKNOWLEDGMENTS

The contributions of V. Black and B. Mutsuddy of Dow Corning towards sample fabrication are acknowledged. This work was funded in part by the U.S. Air Force, the U.S. Department of Energy, and NASA.

HIGH THERMAL CONDUCTIVITY SiC-SiC COMPOSITES

W. Kowbel*, C. Bruce, K.T. Tsou and J.C. Withers, MER Corp., Tucson AZ 85706 and G.E. Youngblood, PNNL, Richland, WA 99352

ABSTRACT

SiC-SiC composite offer excellent potential for numerous commercial applications including thermal management. Very high manufacturing cost ($3,000/lb) and very low through-the-thickness thermal conductivity, greatly hinders their commercial applications. A new chemical vapor reaction (CVR) process was applied to produce pure β-SiC/β1-SiC composites. The projected cost is greatly reduced as compared to the state-of-the-art CVI-SiC composites ($200/lb vs $3,000/lb). In addition, a breakthrough is reported in the through-the-thickness, thermal conductivity (80 W/mK vs 12 W/mK for the state-of-the-art SiC-SiC composite). In addition, a PIP-derived SiC-SiC composite exhibited through the thickness thermal conductivity in excess of 80 W/mK (2 W/mK for commercial PIP-derived SiC-SiC composites). CVR and PIP derived SiC-SiC composites exhibited 30 W/mK through the thickness thermal conductivity at 1000°C. Processing structure property relationship for these SiC-SiC composites will be presented.

INTRODUCTION

Silicon carbide (SiC) has been considered as a structural material because of its outstanding mechanical properties [1-3]. To provide improved strength and toughness, continuous fiber reinforced SiC composites (SiC_f-SiC), are being developed primarily for advanced aerospace applications, and also are now being considered for fusion reactor applications [4]. Such composites typically are fabricated with about 40% by volume fiber reinforcement, while the matrix can be fabricated by either chemical vapor infiltration (CVI) or polymer infiltration and pyrolysis (PIP). The higher strengths as well as the improvement in toughness and strain-to-failure originate from an engineered interphase.

Unfortunately, while the thermomechanical properties of SiC/SiC can be improved over that of monolithic SiC, the thermal conductivity is significantly reduced. For instance, a high purity SiC of theoretically full density made by chemical vapor deposition (CVD) generally will have a room temperature thermal conductivity value >300 W/mK [5], while through-the-thickness conductivities of SiC/SiC CVI-composites have been observed to be <12 W/mK at RT [1]. Although porosity in the composite can account for some of this reduction in thermal conductivity, reasonable porosity correction factors cannot account for the observed reduction.

Recent studies have shown that in the direction perpendicular to the fiber planes in 2-D plain weave SiC_f-SiC, the SiC fibers themselves contribute very little to the thermal conductivity of the composites [6]. The weakly bonded fiber structures, necessary

To the extent authorized under the laws of the United States of America, all copyright interests in this publication are the property of The American Ceramic Society. Any duplication, reproduction, or republication of this publication or any part thereof, without the express written consent of The American Ceramic Society or fee paid to the Copyright Clearance Center, is prohibited.

for optimized mechanical properties in SiC_f-SiC, apparently act as an insulating phase, while heat transport takes place primarily through the matrix phase. Thus, differences in thermal conductivity between different types of SiC_f-SiC will depend upon the morphology and substructure of the matrix as well as the degree of crystallinity and purity of the matrix.

To achieve high-purity, crystalline β-SiC fiber, an economical process was developed at MER Corporation to directly convert graphite fiber to SiC by a chemical vapor reaction (CVR) [3]. The CVR process can also be used to convert a pitch-derived graphite matrix directly to a crystalline β-SiC matrix. By combining these two processes into a simultaneous fiber/matrix CVR-conversion, a CVR-CVR conversion process potentially offers major time and operational cost savings over the conventional CVI process. The crystalline β-SiC microstructures which result from CVR-processing exhibit exceedingly clean grain boundaries which contribute to maintaining overall high phonon thermal transport in a CVR-converted SiC fiber or matrix.

This paper examines SiC-SiC composites made with CVR-SiC fibers and two types of SiC matrices: CVR-SiC and PIP/CVI-SiC. Effect of processing conditions on thermomechanical properties is reported.

EXPERIMENTAL

A T-300 1K fabric was used as the graphite precursor. The CVR process was carried out using a SiO gas generator. The conversion of the carbon precursor to SiC was controlled by the generator's chemistry and the operating temperature. The conversion level was measured by direct oxidation of the CVR converted fibers and by SEM examination of cross section of the oxidized SiC converted fibers. Figure 1 shows the two processing routes employed in this work to fabricate SiC-SiC composites without any engineered interface or C-coating as an interface: i) CVR-SiC conversion of carbon matrix, and ii) Polysilazene (PZS) consolidation with B-doping or undoped as a variable and CVI densification. Density of the densified SiC-SiC composites was measured using the Archimedes principle. Mechanical testing was performed using a four point bending fixture with a 15:1 span to depth ratio. SEM was used to examine fracture surfaces, as well as the topography of SiC converted carbon fibers. Thermal diffusivity was measured using the laser flash method. Thermal conductivity was calculated as a product of density, thermal diffusivity and heat capacity. XRD was used to perform the phase identification of the SiC matrix.

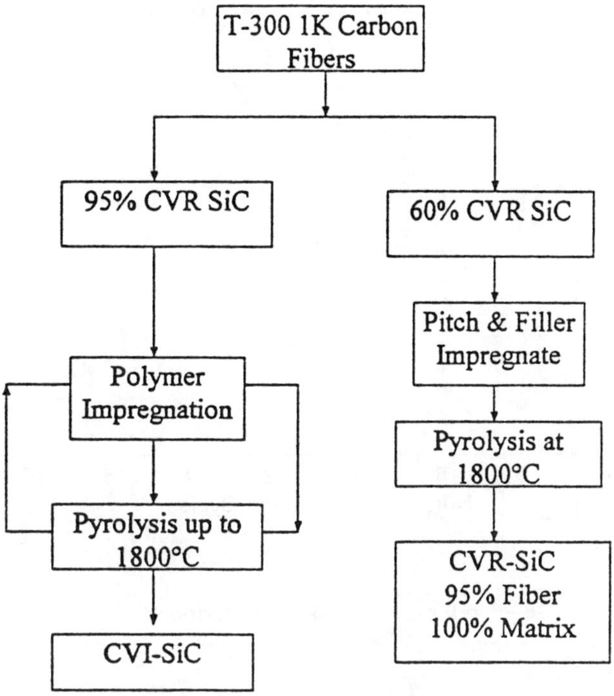

Figure 1. Composite processing schemes.

RESULTS AND DISCUSSION

Figure 2 shows the thermal diffusivity for the undoped and the 0.25 wt% boron doped SiC-SiC composite (0.05 μm carbon coating) as a function of PIP infiltration cycles. The thermal diffusivity generally increased up to a value of $45 \pm 1 \times 10^{-2}$ cm^2/s, and $35 \pm 1 \times 10^{-2}$ cm^2/s after 18 PIP cycles, in the case of the undoped and the 0.25 wt% boron-doped SiC-SiC composites, respectively. It is worth noting, that the un-doped SiC-SiC composite exhibited higher thermal diffusivity after 18 PIP cycles. Secondly, the CVI densification (undoped SiC-SiC composite) increased the RT thermal diffusivity to 0.53 cm^2/s. Boron doping was found to increase the composite thermal diffusivity up to the 9th PIP cycle (p_9).

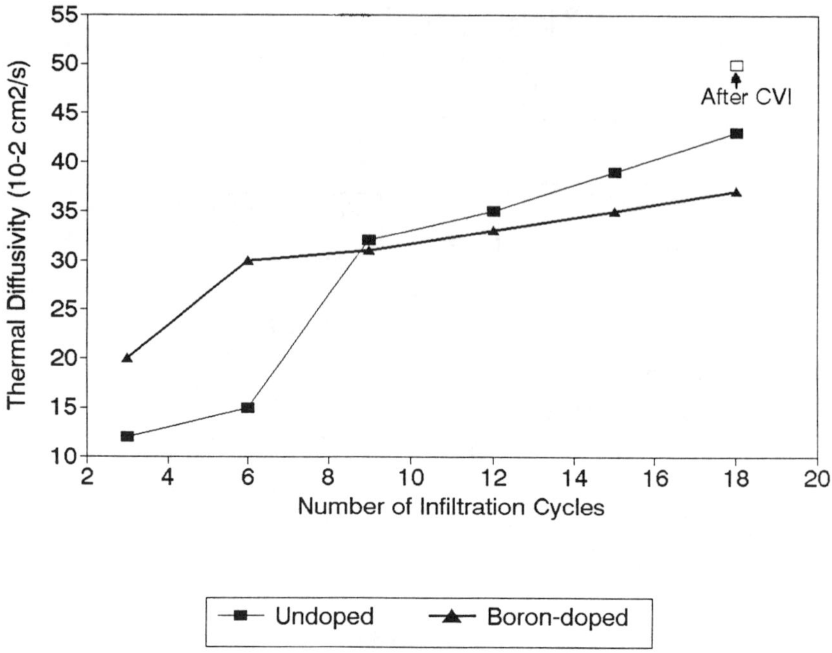

Figure 2. **Effect of Densification on Thermal Diffusivity of SiC-SiC composites (0.05 μm carbon).**

Figure 3 shows the effect of densification on thermal diffusivity for SiC-SiC (doped and undoped) composites made with a 0.2 μm interfacial carbon coating. The undoped SiC-SiC composite exhibits higher thermal diffusivity than the doped SiC-SiC composites at all processing levels. However, the thicker interfacial coating lowers the final thermal diffusivity. Thus, the interfacial carbon acts as a thermal diffusion barrier. The 1800°C pyrolysis is insufficient to graphitize the turbostratic interfacial carbon layer.

Figure 4 shows the temperature dependence of thermal conductivity for an undoped SiC-SiC composite with 0.05μm carbon coating. The 86 W/mK RT value drops to about 28 W/mK at 1000°C, which is 5 times higher than the state-of-the-art CVI SiC value and about 30 times higher than the commercial PIP-SiC composite. The high temperature (up to 1800°C) heat treatment was found to be critical to achieving such a dramatic improvement in the thermal conductivity of PIP-derived SiC-SiC composites. Thus, a high thermal stability CVR-SiC fiber [3] is critical to achieving high thermal conductivity with a PIP-derived SiC matrix composite.

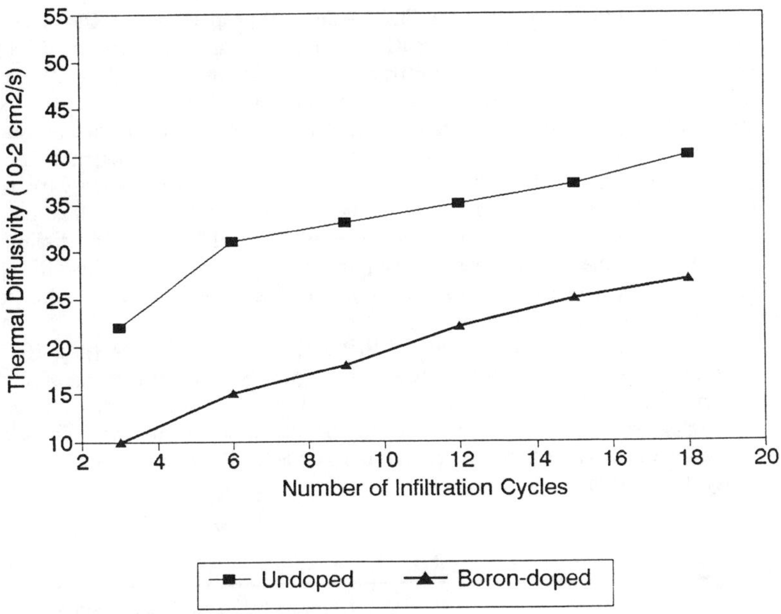

Figure 3. **Effect of Densification on Thermal Diffusivity of SiC-SiC composites (0.2 μm carbon).**

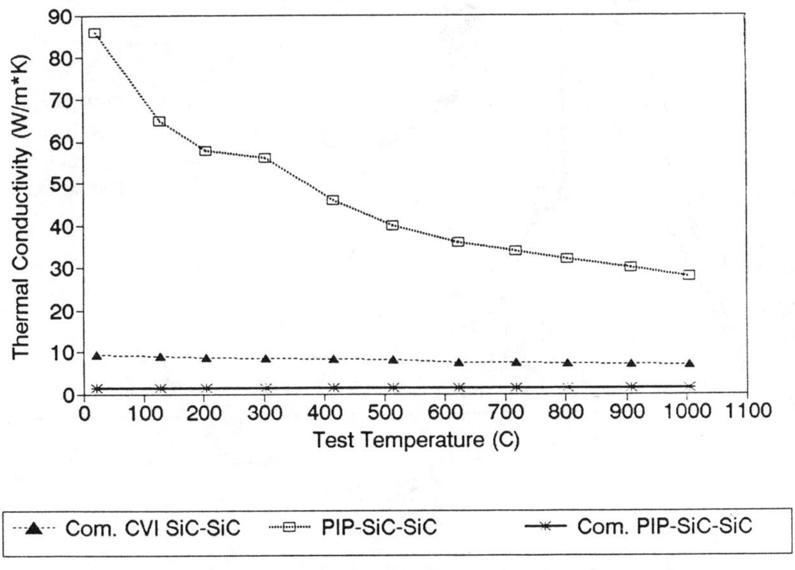

Figure 4. **Temperature dependence of thermal conductivity.**

Figure 5 shows the corresponding mechanical properties. In the case of T-300 (95% SiC fibers, 0.05 μm interfacial coating), a high flexural strength of 350 MPa was observed, combined with a semi-brittle fracture. In the case of T-300 (95% SiC) fibers with a 0.2 μm carbon interfacial coating, a flexural strength of 320 MPa was exhibited, combined with a non-brittle behavior. Thus, the thicker interfacial carbon coating promotes better fiber pull-out, but reduces the through-the-thickness thermal conductivity.

Figure 6b shows the XRD spectrum of the SiC-SiC composite with the PZS-derived matrix after a 1400 °C H.T. SiC peaks are observed with no clear presence of carbon. The XRD pattern of this composite after a 1000 °C H.T. yielded no distinguishable peaks indicating an amorphous structure (Figure 6a). Figure 6c shows the XRD spectrum of the SiC-SiC composite with the PZS derived-matrix after a 1800°C HT. SiC peaks are observed with some carbon residue.

Figure 7 shows a cross section of PIP-derived SiC composite (p_{18}, 1800°C). A small amount of residual porosity is observed. Figure 8 shows the corresponding TEM examination of the PIP matrix (undoped composite). Very small β-SiC crystallites are observed. Similar observations are true for the boron-doped matrix. Thus, the final differences in the thermal behavior of PIP-derived SiC-SiC composites are possibly due to porosity distributions.

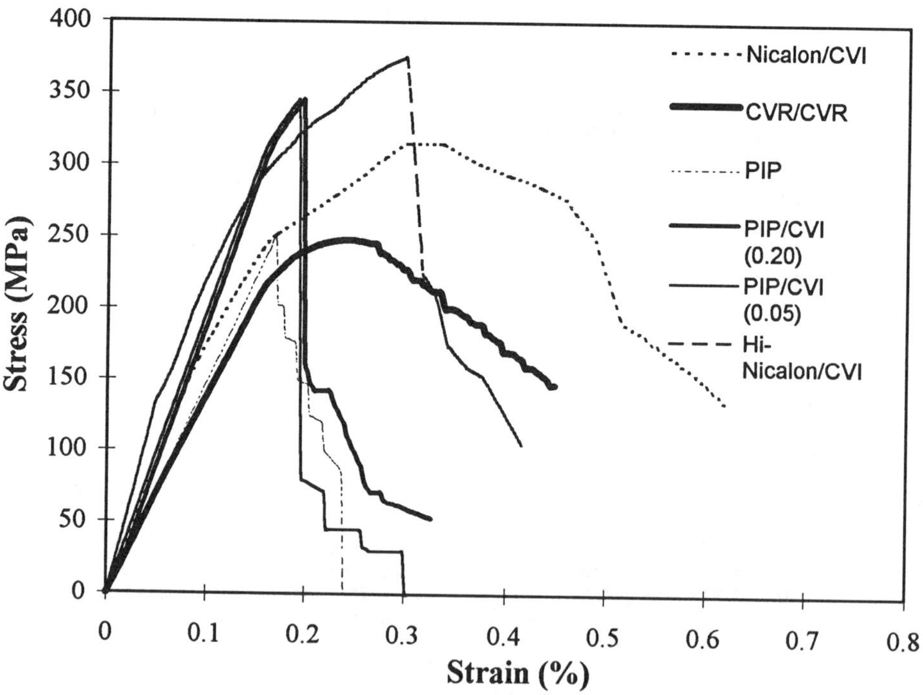

Figure 5. Stress-strain curve for SiC-SiC composite.

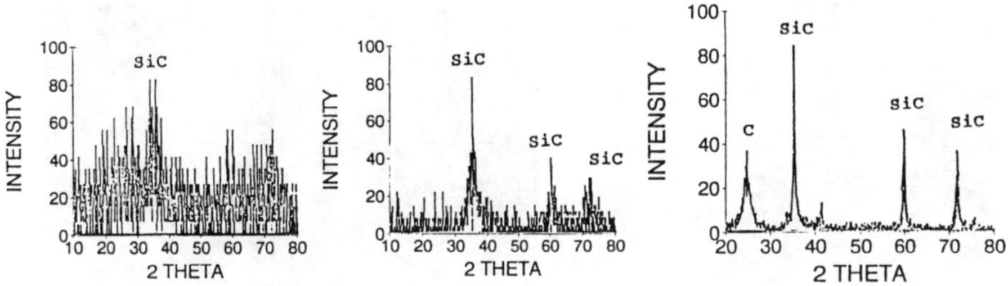

1000°C HT 1400°C HT 1800°C HT

a) b) c)

Figure 6. XRD Pattern of SiC-SiC Composite With the PZS Derived Matrix: a) 1000°C HT, b) 1400°C HT, c) 1800°C HT.

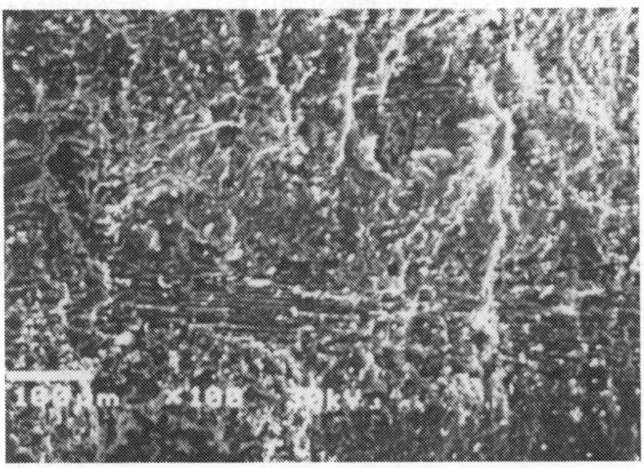

Figure 7. Cross section of PIP-derived SiC composite (p_{18}).

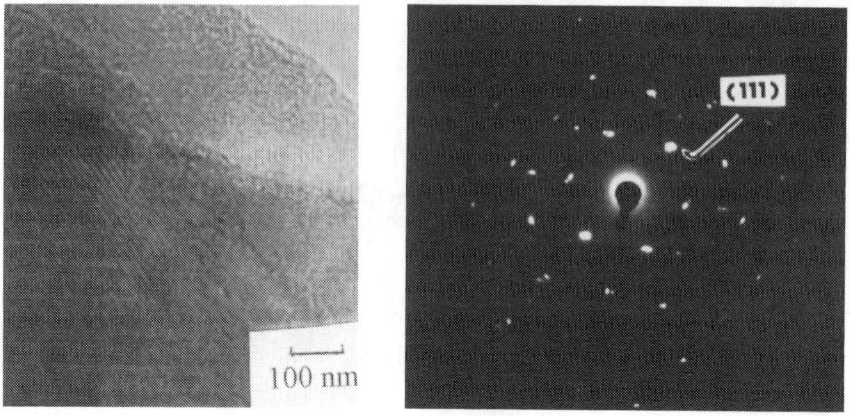

a) b)

Figure 8. TEM of PIP-SiC matrix: a) bright field, b) microdiffraction.

CVR-derived matrix

The CVR-SiC/CVR-SiC composite had a RT through-the-thickness thermal conductivity of 75 W/mK, Figure 9. This represents almost an 8-fold increase over the state-of-the-art SiC-SiC composite thermal conductivity. Figure 9 also shows the thermal conductivity of the CVR-SiC/CVR-SiC composite as a function of temperature. The thermal conductivity experiences a parabolic behavior reaching 35 W/mK at 1000°C. The importance of this result can be further emphasized by comparing it to the 6 W/mK, 1000°C thermal conductivity exhibited by the state-of-the-art CVI-matrix SiC-SiC composite.

Figure 10 shows an XRD pattern corresponding to the CVR-SiC derived composite. Some residual carbon is observed combined with β-SiC diffraction lines. The residual carbon is the unreacted core of the CVR SiC fiber. The overall conversion of T-300 fiber was about 95%. In addition, ultra-thin window EDS was used to assess the chemistry of the CVR-SiC matrix. Pure ß-SiC matrix was confirmed. Thus, the CVR process can produce virtually pure ß-SiC matrix with highly re-converted (95% SiC fibers).

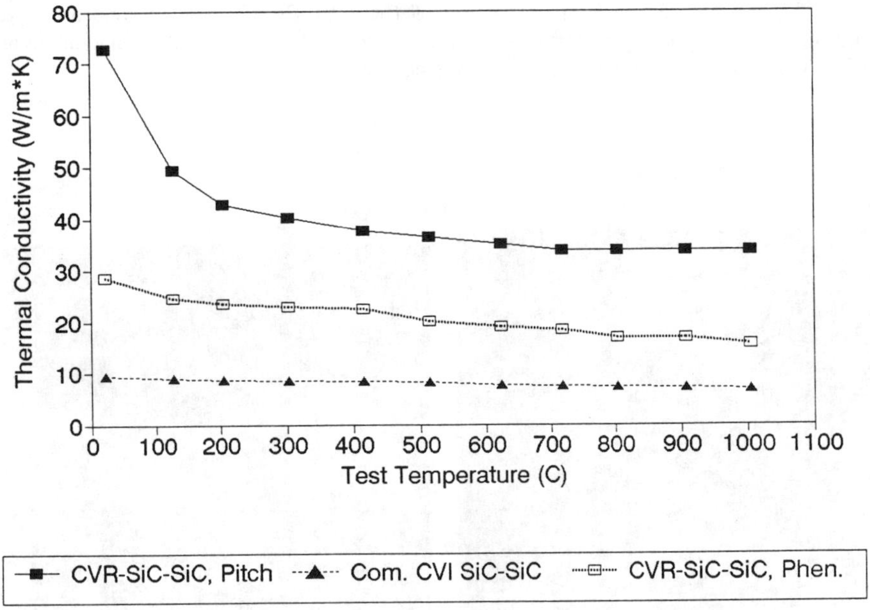

Figure 9. **Thermal conductivity of CVR-SiC/CVR-SiC composite.**

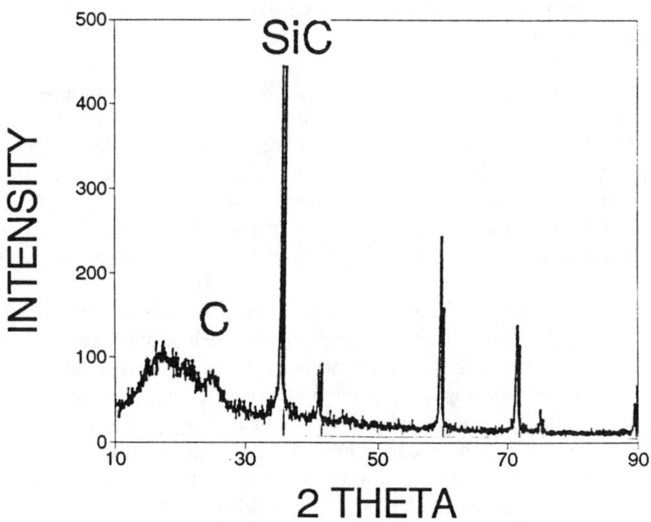

Figure 10. **XRD of CVR-SiC/CVR-SiC composite.**

Figure 11a shows a cross section of the CVR-SiC/CVR-SiC composite. No transverse cracking or delamination is observed. Figure 11b shows a higher magnification SEM. Very good fiber bundle penetration is achieved.

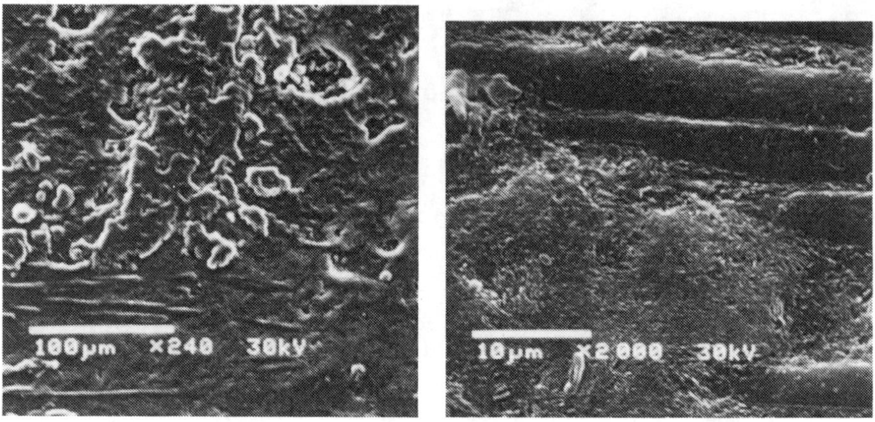

a) b)

Figure 11. Cross section of CVR-SiC/CVR-SiC composite made with pitch-derived matrix: a) low magnification, b) higher magnification.

Figure 12 shows a stress-strain curve of CVR-SiC/CVR-SiC composite. The proportional limit occurs at about 200 MPa (similar to the state-of-the-art SiC-SiC composites made with CVI-SiC matrix and CG Nicalon and Hi-Nicalon fibers, respectively). The flexural strength of 250 MPa is observed combined with a non-brittle fracture. The flexural strength of the CVR-SiC/CVR-SiC composite is lower, however it is not yet fully optimized, than that of the CVI-matrix composites (300 MPa and 350 MPa for CG Nicalon and Hi-Nicalon, respectively). However, the CVR-SiC/CVR-SiC composite was made with no interfacial coating.

The non-brittle behavior of CVR-SiC/CVR-SiC composites is controlled by fiber-matrix interactions during the composite processing. At the carbon matrix stage, the SiC-C composite can be made very ductile via the control of the processing conditions.

At the CVR-SiC/CVR-SiC stage, the fiber matrix interactions are also controlled by mechanical bonding and interfacial porosity created during the CVR conversion. Thus, a porous SiC interfacial layer can be created in-situ. In addition, there should be no strong bonding between the SiC porous interfacial layer and the CVR fiber (only mechanical interaction is envisioned). Thus, the CVR-SiC process is envisioned to generate a porous SiC interfacial layer poorly bonded to the CVR-SiC fibers. The combination of these two factors can possibly explain the non-brittle behavior of the CVR-SiC/CVR-SiC composite.

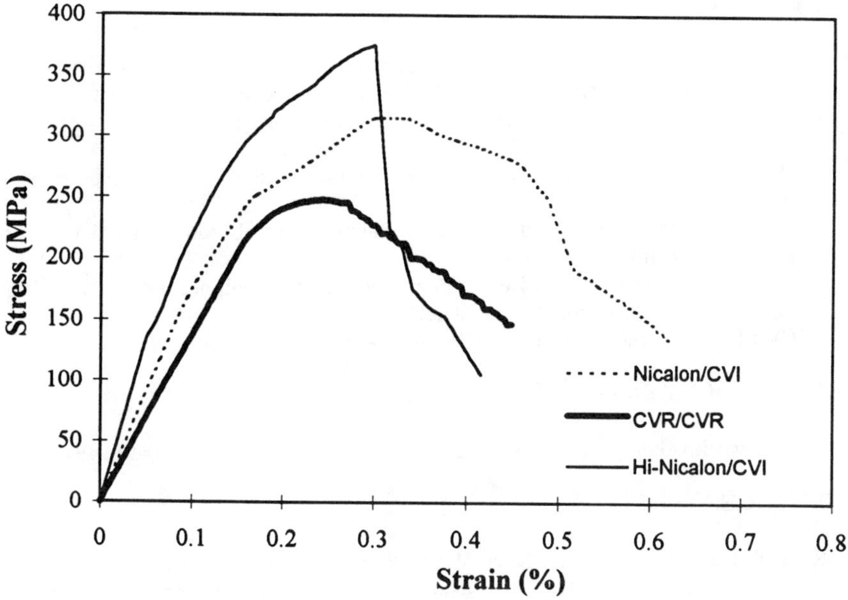

Figure 12. Stress-strain curve for CVR-SiC/CVR-SiC composite.

Table I shows the thermomechanical properties of different SiC-SiC composites. It illustrates a clear improvement in through-the-thickness thermal conductivity of CVR-SiC fiber based composites over Hi-Nicalon based composites. In addition, CVR-SiC derived fiber offers greatly reduced cost over Hi-Nicalon SiC fiber, and CVR-based composite processing greatly reduces the composite processing cost (it is a one-step gas-based processing).

TABLE I. Properties of SiC-SiC Composites

Type	Density (g/cc)	Ther. Cond. RT (W/mK)	Ther. Cond. 1000°C (W/mK)	Flex. Strength RT (MPa)	Open Porosity (%)
Hi Nicalon/CVI	2.55	13	10	350	15
(PIP/CVI hybrid)	2.55	86	28	350	8
(CVR-CVR)	2.65	75	35	250	15

CONCLUSIONS

1. PIP-derived SiC matrix SiC-SiC composites made with CVR-SiC fibers exhibit excellent thermal properties (86 W/mK at RT and 28 W/mK at 1000°C).
2. PIP-derived SiC-SiC composite made with CVR-SiC fibers exhibit very good mechanical properties (350 MPa flexural strength).
3. Thicker interfacial carbon coating improves fiber pull-out, but reduces thermal conductivity.
4. CVR-SiC/CVR-SiC composites exhibit excellent thermal properties (75 W/mK at RT and 35 W/mK at 1000°C).
5. CVR-SiC/CVR-SiC composites exhibit good mechanical properties (250 MPa flexural strength, non-brittle behavior).
6. CVR-SiC/CVR-SiC composites offer excellent cost reduction potential.

ACKNOWLEDGEMENTS

This work is being supported by DOE Contract No. DE-FG0294ER81487.

REFERENCES

1. G.W. Hollenberg et al., J. Nucl. Mater. **219**, 70-86 (1995).
2. J.C. Withers et al., p. 87 in the Fusion Materials Semiannual Progress Report for Period Ending December 31, 1996, DOE/ER-0313/21.
3. W. Kowbel et al., J. Nucl. Mater, **219**, 15-25 (1995).
4. J. C. Withers et al., p. 69 in the Proceedings of the IEA International Workshop on SiC/SiC Ceramic Composites for Fusion Structural Applications, 28-29 October 1996, Ispra, Italy.
5. M.A. Pickering, Mat. Res. Soc. Symp. Proc. **250**, 145 (1993).
6. D.P. Hasselman, Thermal Conductivity **20**, pp. 141-152, ed. D.P. Hasselman, Plenum Press, New York, NY (1989).

TEMPERATURE DEPENDENCY OF STRENGTH OF A UNIDIRECTIONAL SiC FIBER-REINFORCED $(BaSr)Al_2Si_2O_8$ CELSIAN COMPOSITE

Ö. Ünal and N. P. Bansal*
Ames Laboratory-USDOE, Ames, IA 50010
* NASA-Lewis Research Center, Cleveland, OH 44135

ABSTRACT

Temperature dependency of strength of a high Nicalon-$(BaSr)Al_2Si_2O_8$ (BSAS) celsian composite was studied between room temperature and 1100°C by the three-point bend test. Test results showed that there is a definite transition in the failure mode with temperature. Up to about 800°C, tensile was the dominant failure mode. Above 800°C, however, the shear failure became prominent. It was shown that mode shift and rapid degradation of shear strength is due to the softening of the untransformed glassy phase in microstructure. The failure mode also exhibited a strong dependency on the L/h ratio of test specimen; large L/h ratio favored the tensile failure, while small L/h ratio promoted the shear failure.

INTRODUCTION

Fiber-reinforced ceramic composites are being projected to replace conventional metallic materials for high temperature structural applications because of their promising properties of high strength, high creep resistance, environmental stability and chemical inertness. Monoclinic celsian is an oxide based refractory material with the phase stability up to about 1600°C [1]. Thus, it has been targeted as a potential matrix material in the development of high performance composites [2]. Since fiber-reinforced composites are orthotropic, their behavior is much more complex as compared to isotropic materials. Moreover, composites are often used in loading situations where they are subjected to complex state of stresses. Thus, determination of material properties, such as strength, requires the knowledge of the stress states in the specimen and the post failure analysis.

This study utilized the three-point bend test method to measure both the tensile and shear strength values. Figure 1 shows the stress states in the three-point bend specimen. Figure 1a is the force diagram, while Fig. 1b and Fig. 1c are

To the extent authorized under the laws of the United States of America, all copyright interests in this publication are the property of The American Ceramic Society. Any duplication, reproduction, or republication of this publication or any part thereof, without the express written consent of The American Ceramic Society or fee paid to the Copyright Clearance Center, is prohibited.

the corresponding shear and moment diagrams, respectively. Elastic beam theory predicts that the shear stress has a parabolic distribution across the specimen depth, being zero at both surfaces and maximum at the neutral axis. In contrast, normal stress has a linear distribution being maximum tensile at the bottom surface and maximum compressive at the top surface. Since the specimen cross-section is symmetrical, zero normal stress is present at the neutral axis. The maximum shear and normal stress values in Fig. 1 are given,

$$\tau_{max} = \frac{3P}{4bh} \tag{1}$$

$$\sigma_{max} = \frac{3PL}{2bh^2} \tag{2}$$

Notice that the shear stress is independent of span-length, L, while the normal stress is not. Which means that regardless of the fixture dimensions the given load, P, produces the same shear stress distribution in the specimen if the cross-section is the same. The same thing cannot be said for normal stress since it has L dependency. Understanding this difference between shear and normal stresses is very important because one can selectively induce either shear or normal failures by simply changing the fixture and specimen dimensions. This unique feature was exploited in this study to measure the shear strength values at low temperatures. So, the goal of this study was to study the temperature dependency of strength of high Nicalon fiber-reinforced celsian composite up to 1100°C.

EXPERIMENTAL

High Nicalon SiC fibers with the chemical composition (wt %) of 62.4 % Si, 37.1 % C, and 0.5 % O were used as the reinforcement. These fibers had an average diameter of about 14 μm, the room temperature tensile strength of about 2.8 GPa and the elastic modulus of 270 GPa. Prior to composite processing the fibers were coated with layers of 0.4 μm-thick amorphous BN and ~ 0.2 μm-thick crystalline SiC by chemical vapor deposition (CVD). The BN coat was applied onto the fiber surface to produce a relatively compliant layer, thus to promote crack deflection away from the fibers. The SiC coat was made to form a diffusion barrier between BN and the matrix, and to protect fibers from damage during handling and composite processing. The precursor to the celsian matrix, 0.75 BaO-0.25SrO-Al_2O_3-2SiO_2, was made by a solid-state reaction method [2]. Precursor powder consisted of SiO_2, $BaAl_2O_4$ and small amounts of Ba_2SiO_4, Al_2O_3, and $Ba_2Sr_2Al_2O_7$. The composite was produced by impregnation of the fiber tows with the matrix slurry [3]. Unidirectional fiber reinforced composites were prepared by tape lay up (12 plies) and warm pressing. Fully densified composites were obtained by hot pressing under vacuum in a graphite die. The monoclinic celsian matrix phase formed from the oxide precursor in-situ during hot pressing through solid state reactions. The details of this processing is provided elsewhere [3].

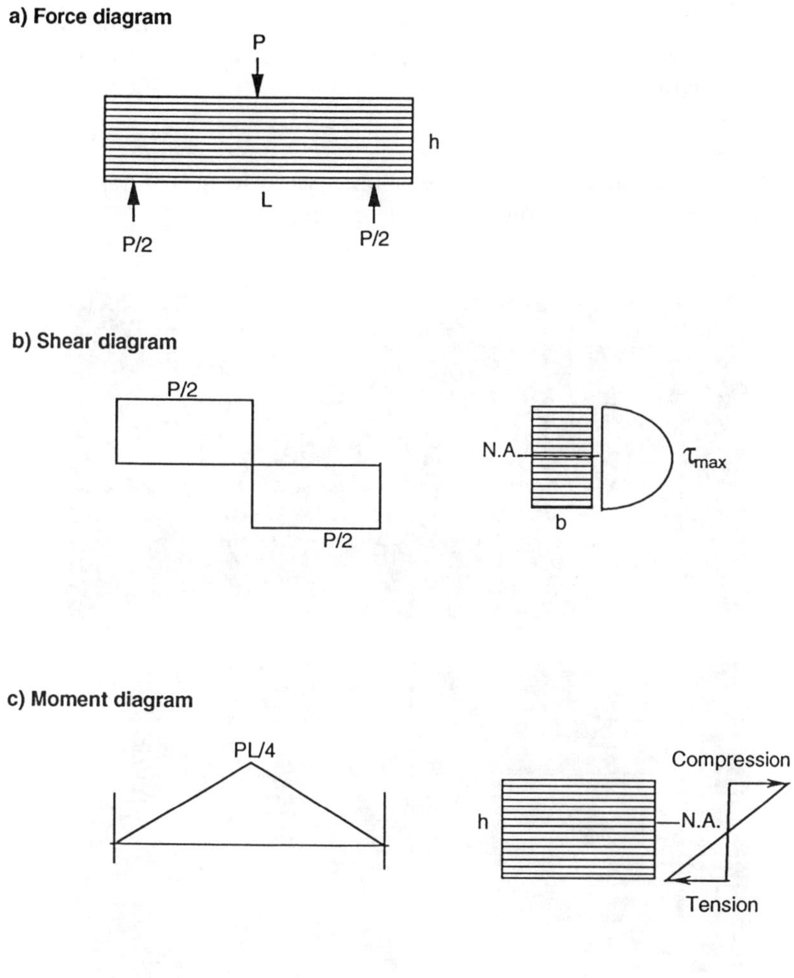

Fig. 1 Stress states in three-point bend test.

The composite panel was machined into 3x1.75x25 mm test bars. The flexure tests were conducted using two different three-point bend fixtures; the one with 20 mm span-length was used to determine low temperature tensile and high temperature shear strength values while the other with 10 mm span-length was used to measure shear strength at low temperatures. All tests were conducted in displacement control with the cross-head speed of 2 μm/s (0.12 mm/min). The high temperature tests were conducted in flowing N_2 gas up to 1100°C. Except at 850°C and 1100°C, where only one specimen was tested, at least four specimens were tested at each temperature. Microstructural evaluation was made by transmission electron microscopy (TEM), scanning electron microscopy (SEM) and optical microscopy.

RESULTS

Figures 2a and 2b show the optical microscopy images of transverse and longitudinal sections of the composite, respectively.

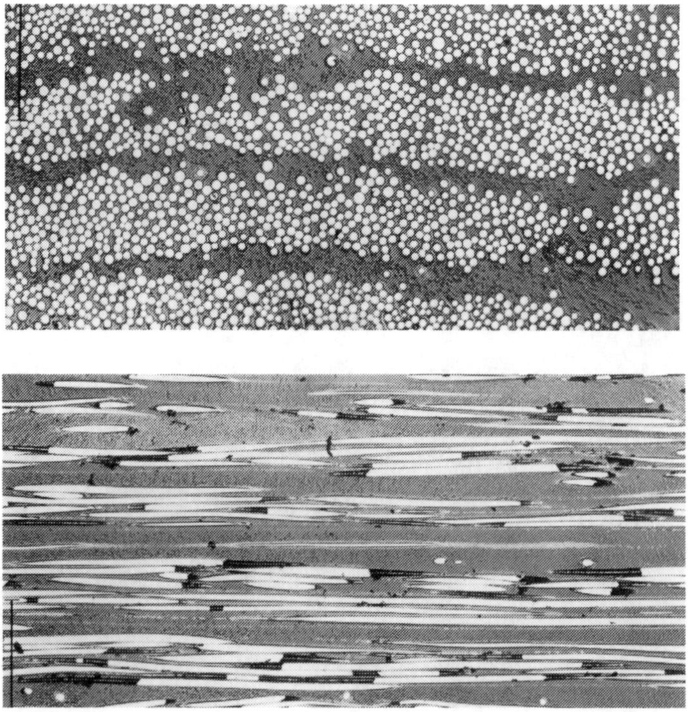

Fig. 2 Optical micrographs a) transverse and b) longitudinal sections.

The composite was free of pores and cracks. Although the material was unidirectional composite, Fig. 2a shows that it was not transversely isotropic. This means that transverse plane properties differ with direction due to its distinct microstructural pattern. Because the composite was produced through the stacking of plies, well-defined matrix channels were formed between the fiber rich regions during densification. In Fig. 2b, the fibers are seen in the horizontal direction. The discontinuities in this figure are due to the waviness in the fiber bundles.

Figure 3 shows the plots of normal stress and shear stress versus displacement for a test carried out at 600°C. The stress values were calculated based on the equations (1) and (2).

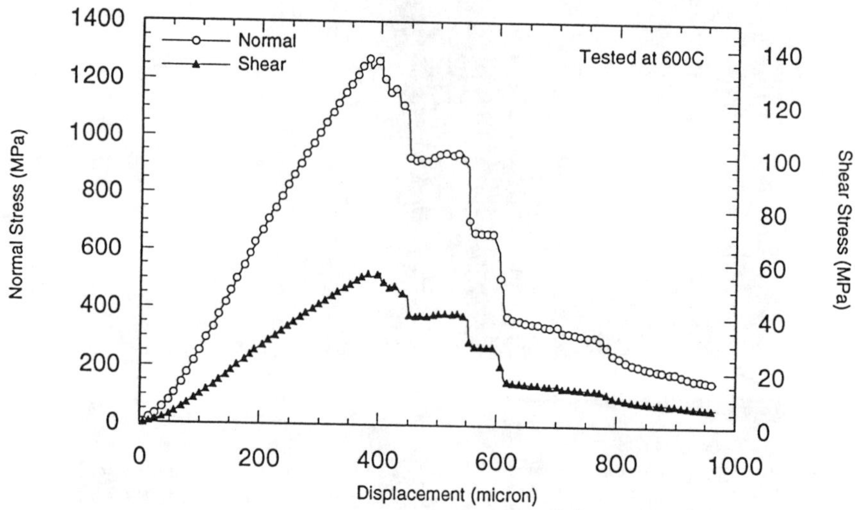

Fig. 3 Normal and shear stress plots obtained at 600°C

Note that the composite exhibited a linear region at the beginning with a small deviation at ~ 700 MPa. It had the peak normal and shear stress values of ~ 1300 MPa and ~50 MPa, respectively. Beyond the maximum stress the material developed macroscopic cracks and the crack growth took place in mixed mode (Mode-I and Mode-II). Since the composite fracture typically takes place in a complex manner beyond the maximum stress region, the identification of failure mode of specimen becomes very difficult if the specimen is tested for very large displacements. Therefore, the proper identification of the failure mode necessitated the stoppage of tests as soon as the macroscopic cracks appeared.

By using optical microscopy and SEM, the relevant strength, whether it was normal (tensile or compressive) or shear was determined. For instance, the test specimen in Fig. 3 failed in tension, thus, the data obtained from this test was the tensile strength rather than shear strength. No compressive failure was observed in this study.

The current study shows that the failure mode is strongly dependent on the test temperature. Figure 4 shows typical examples.

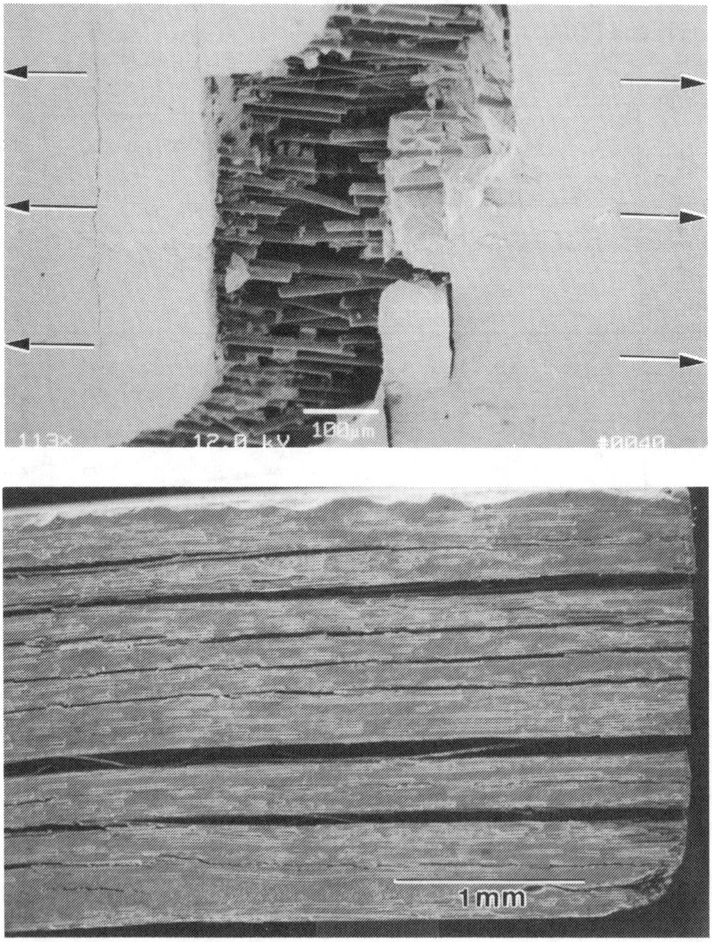

Fig. 4 Failure modes at a) room temperature and b) 1100°C

Figure 4a illustrates tensile cracks on a room temperature specimen while Fig. 4b shows interlaminar shear cracks on a specimen tested at 1100°C. The arrows in Fig. 4a indicate the direction of applied tensile stresses. Note that the cracks nucleated on the principal planes perpendicular to applied load. However, it was localized on a single plane due to the point loading in three-point test. At low temperatures, the composite behavior was fiber-dominated thus, the failure took place in tension by the fracture of load carrying fibers. At high temperatures, however, the failure was mostly interface-dominated and took place in shear, Fig. 4b. It is clear that the interlaminar cracks formed by the displacement of plies along the specimen axis. At elevated temperatures the loading curve almost always contained a plateau region beyond maximum load. This region indicated a saturation level for the shear cracks within the specimens, i.e., further loading did not increase the crack density, and instead the plies were further displaced by shear. Figure 5 shows a typical example of shear fracture surface of a specimen tested at 1000°C.

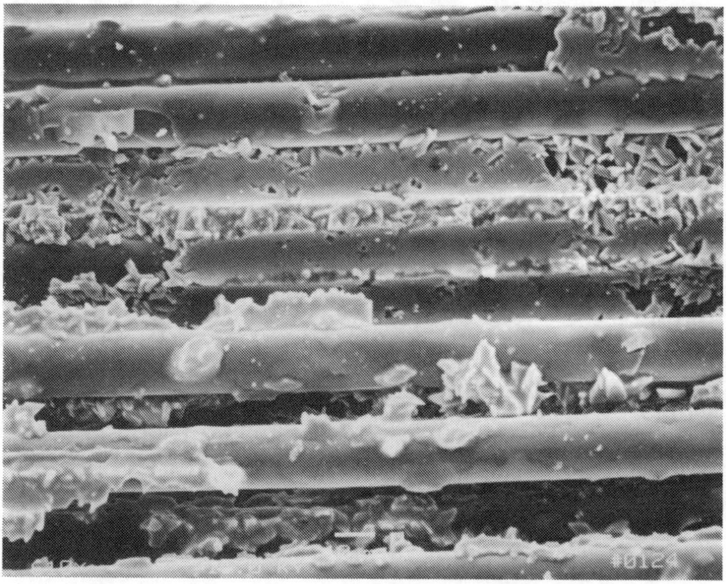

Fig. 5 Interlaminar shear fracture surface of a specimen tested at 1100°C.

It is obvious that the shear failure occurred by the slide of fibers along the specimen axis. Since the fibers/matrix interface constituted a discontinuity within the composite, it acted as a structural defect under the shear loads. The detailed study of interlaminar cracks showed that the failure occurred at the interface between the SiC coat of fiber and the matrix.

The summary of test results as a function of temperature is given in Fig. 6. It should be emphasized that tests were conducted with two different specimens. The tensile and high temperature shear strength values (above 800°C) were obtained with the same long specimens while the low temperature shear strength values (below 800°C) were obtained with short-beam specimens. The results with long specimens indicated a definite shift in the failure mode around 800°C. As indicated earlier, at low temperatures (< 800°C) tensile was the dominant failure mode. Due to the constancy of constituent properties, tensile strength was also constant, ~1250 MPa. To induce shear fracture, thus to obtain shear strength values below 800°C, the short-beam specimens were deliberately used.

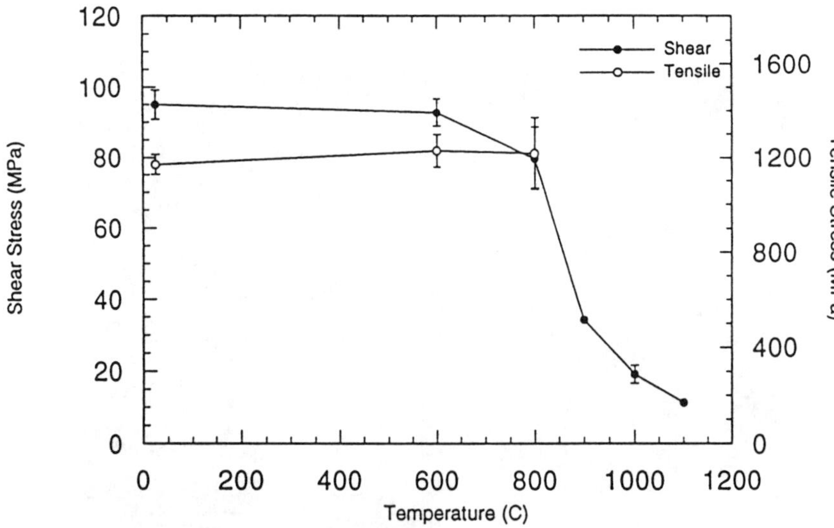

Fig. 6 Tensile and shear strength values as a function of test temperature.

Figure 6 shows that shear strength is also independent of temperature in this range, ~90 MPa. Above 800°C, however, the failure mode exclusively became shear and a rapid decrease in shear strength was also observed. For instance, the decrease in strength between 800 and 1000°C was almost 75%. The tensile strength values above 800°C could not be measured in this study due to the dominance of shear fracture.

To determine the cause of shift in failure mode and the sudden drop in shear strength above 800°C, the microstructure of composite was studied by TEM. As can be seen in Fig. 7, the microstructure contained some untransformed silicate glass within the elongated celsian grains. Although the amount of this glassy phase was relatively small, thus x-ray diffraction could not detect it, the research showed that even the small amount of glassy phase could be detrimental for high temperature properties [5]. The presence of glassy phase in microstructure was further demonstrated by an exploratory two step test and shown in Fig. 8. First, the composite material was tested at 1100°C. To avoid extensive damage to the composite, the specimen was unloaded just before plateau region. Notice that because of the cracks within the specimen, the unloading curve exhibited the residual component of deformation.

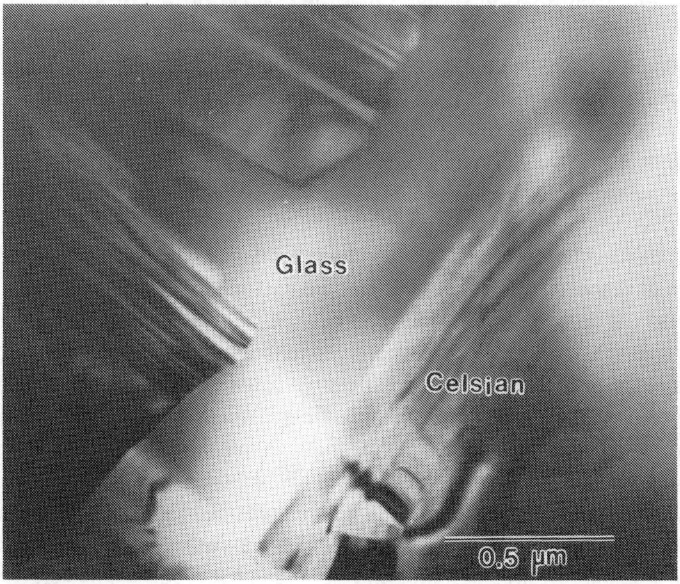

Fig. 7. TEM micrograph of residual glass between matrix grains

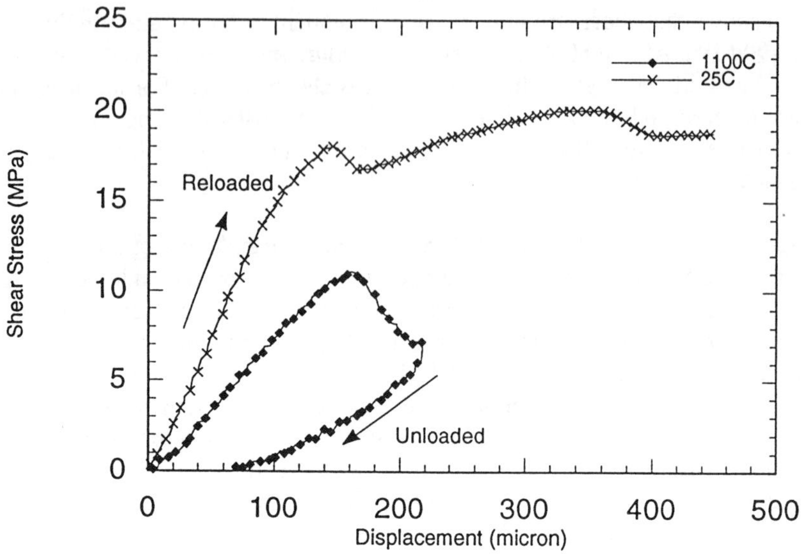

Fig. 8 Residual shear strength at 25°C after failure at 1100°C.

The microscopic examination showed that the cracks formed by shear. On the same specimen, a subsequent test was conducted at room temperature. Since the glassy phase solidified with cooling, a partial healing of interlaminar shear surfaces occurred in the process. As a result, the room temperature test showed about 100% increase in strength. Moreover, because of this limited healing of shear cracks by the glassy phase, the failure at 25°C also took place in shear.

DISCUSSION

Because of its orthotropic nature and the glassy phase content, the failure mode in this composite was dependent on two factors: (i) test temperature (intrinsic) and (ii) span-to-depth ratio of test specimen (extrinsic). Experimental results in Fig. 6 clearly showed that there was change in the failure mode from tensile to shear with increasing temperature. At low temperatures, most of the applied load was carried by the reinforcing fibers, i.e. the mechanical properties were fiber-dominated. Since fibers are stronger than matrix, the composite strength was a lot higher at low temperatures. When the fiber fracture took place, the load carrying capacity of the composite was reduced rapidly. Up to 800°C, there was no change in the thermo-mechanical behavior of constituent phases. Therefore, no change in mechanical behavior was observed. However, beyond

800°C, the glassy phase at the matrix/fiber interface softened, resulting in a major shift in fracture behavior. In three-point loading, test specimen experiences a constant shear load across the specimen in addition to maximum moment in the middle. At a low temperature, the shear force cannot lead to failure at the fiber/matrix interface since it is strongly bonded, i.e., the load transfer from matrix to fibers is sufficient. At elevated temperature, the interface was weakened due to softening and the shear stresses were localized, resulting in interlaminar fracture. So, the high temperature properties became interface dominated due to inefficient load transfer. The glass transition temperatures, T_g, for $SrO.Al_2O_3.2SiO_2$ (SAS) and $BaO.Al_2O_3.2SiO_2$ (BAS) glasses are 883°C and 905°C, respectively [6]. The glass composition in this material was not determined, thus its exact T_g is not known. However, the observed dependency of the strength on temperature and the failure mode shift around 800°C. suggest that it is somewhere between 800 and 900°C.

The span-to-depth ratio (L/h) of test specimen was also shown to be an important factor in determining the failure mode. It can be shown from equation (1) and (2) that there is a unique relationship between L/h and σ/τ developed in three-point specimens,

$$\frac{\sigma}{\tau} = 2\frac{L}{h} \qquad (3)$$

This relationship is plotted in Fig. 9. From this figure, one can see the theoretical boundaries of shear and tensile failure regions which are separated by the solid line. Equation (3) predicts that large L/h ratio favors the tensile failure, while small L/h ratio promotes the shear failure. The long specimens in this study had the L/h =11. Since the intrinsic σ/τ was about 12.6, this put the specimen in tensile failure region at low temperatures. For the same specimen geometry, with temperature the failure tendency of specimen moved along the y-axis into the shear failure region. This is because the intrinsic σ/τ ratio increased with temperature due to a drop in shear strength. Similarly, at low temperatures the shear failure was induced by reducing L/h ratio of the specimen (short-beam). The critical L/h value was 6.3. The L/h value of 5.5 was selected in this study and the shear failure was observed in all short-beam specimens. This was the specimen dimension used to determine shear strength values below 800°C. Above 800°C, the tensile strength values could not be measured with these flexure specimens. Although higher L/h values predict tensile failure, there is a practical limit for this practice due to the large deflection of composites. So, the best way to obtain tensile strength values at high temperature is to use tensile test.

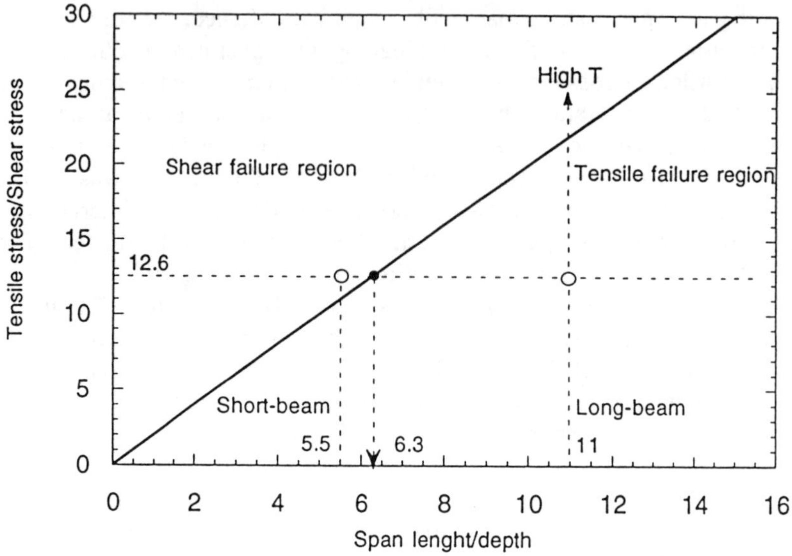

Fig. 9 Dependency of failure modes on span-to-depth ratio

CONCLUSIONS

1. Both tensile and shear strength values were independent of temperature up to about 800°C. Above 800°C, however, rapid degradation of shear strength took place due to the softening of glassy phase in microstructure.

2. At low temperatures the failure was fiber-dominated, i.e., the composite failed as the load carrying fibers fractured in tension. At high temperatures, the failure was interface-dominated since the load-transfer from the matrix to fibers was not effective in the presence of viscous glassy phase. As a result, the stresses were localized at the interface, leading to interlaminar shear fracture at the matrix/fiber interface.

3. The failure mode was also dependent on L/h ratio of the test specimens; large L/h ratio favored the tensile failure, while small L/h ratio promoted the shear failure.

ACKNOWLEDGEMENTS

The help of J.A. Setlock of NASA-Lewis in composite processing and F.C. Laabs of Ames Laboratory in SEM is greatly appreciated. The Ames Laboratory is operated for the U.S. Department of Energy (DOE) by Iowa State University under contract No. W-7405-ENG-82.

REFERENCES

1. N.P. Bansal and C.H. Drummond III, "Kinetics of Hexacelsian-to-Celsian Phase Transformation in $SrAl_2Si_2O_8$", J. Am. Ceram. Soc., 76, 1321, (1993).
2. N.P. Bansal, "Solid State Synthesis and Properties of Monoclinic Celsian", NASA TM-107355 (1996).
3. N. Bansal and J.A. Setlock, "Processing of Small Diameter Fiber-Reinforced Celsian Composites", NASA TM-07356 (1996).
4. N.P. Bansal and J.I. Eldridge, "Effects of Interface Modification on Mechanical Behavior of Hi-Nicalon Fiber-Reinforced Celsian Matrix Composites", NASA TM-107411 (1997).
5. R.L. Tsai and R. Raj, "Creep Fracture in Ceramics Containing Small Amounts of a Liquid Phase", Acta. Metall., 30, 1043, (1982).
6. N. Bansal, "$SrAl_2Si_2O_8$ Glass-Ceramic Matrix for Fiber-Reinforced Composites", NASA Confrence Publ. 10104, HITEMP Review, 62-1 (1992).

CERASET™ AND BLACKGLAS™ MINI-TOW COMPOSITES WITH CARBON-COATED NICALON™ TOWS

S. T. Gonczy, Gateway Materials Technology, Mt. Prospect, IL

K. T. Faber, Dept. of Materials Science & Engineering, Robert R. McCormick School of Engineering and Applied Science, Northwestern University, Evanston, IL

ABSTRACT

Two commercially available preceramic polymers are the polyureasilazane Ceraset™ resin from DuPont Lanxide Composites and the polysiloxane Blackglas™ resin from AlliedSignal Inc. The Ceraset™ polymer pyrolyzes to a silicon carbonitride under nitrogen and a silicon carbide under argon. The Blackglas™ polymer produces a silicon oxycarbide under nitrogen or argon pyrolysis. Two groups of ceramic composites were made with the two preceramic polymer systems and single tows of carbon-coated ceramic grade Nicalon™ fiber, using 6 infiltration steps and 1000°C argon pyrolysis. The mean tensile break loads for the two composite sets were equivalent – 75 Newtons – with coefficients of variation of 10-20% and obvious fiber pull-out. The break loads were independent of the matrix weight fraction, showing that the fibers are the dominant source of strength. There was no apparent difference in short term oxidation protection between the two types of matrix compositions, given the rapid oxidation of the pyrolytic carbon on the Nicalon™ fibers. Oxidation at 1000°C in air for one hour embrittled both sets of ceramic composites, reducing the mean break loads to 27 Newtons.

INTRODUCTION

Continuous fiber-reinforced ceramic composites (CFCCs) offer the potential of higher temperature operation, lower weight, and improved damage tolerance, compared to metal and organic materials. [1,2]. Near-term applications of interest are exhaust components and hot-gas ducting for turbine engines, where the weight savings would improve fuel efficiency and the low signature properties would have military applications. Industrial applications for CFCCs include heat exchangers, combustors, and environmental filters to improve durability, increase operating temperatures, and reduce weight.

One of the most promising fabrication approaches for CFCCs is preceramic polymer technology. With preceramic polymers, silicon-based ceramic compositions can be produced by pyrolysis of silicon-containing polymers under vacuum or inert atmosphere [3,4]. Using the preceramic polymer resin, "green-shape" composites can be fabricated by low cost polymer forming techniques, such as lamination, resin transfer molding, and

To the extent authorized under the laws of the United States of America, all copyright interests in this publication are the property of The American Ceramic Society. Any duplication, reproduction, or republication of this publication or any part thereof, without the express written consent of The American Ceramic Society or fee paid to the Copyright Clearance Center, is prohibited.

filament winding. The resulting polymer matrix component in the composite is then converted into a ceramic by pyrolysis. The porosity formed by pyrolysis of the polymer matrix is reduced by additional polymer reinfiltration/cure/pyrolysis cycles.

Two of the two commercially available preceramic polymer resins on the market are Ceraset™ resin from DuPont Lanxide and Blackglas ™ resin from AlliedSignal. The Ceraset polymer is a polyureasilazane which pyrolyzes to a silicon carbonitride under nitrogen and a silicon carbide under argon. The Blackglas polymer is a polysiloxane resin which produces a silicon oxycarbide under nitrogen/argon pyrolysis. Both systems are low viscosity resins which easily infiltrate fabric and woven preforms and pyrolyze with high char yields (75% and higher).

There has been no published experimental comparison to date of the mechanical properties and the oxidation durability of ceramic composites made with the Ceraset and Blackglas matrix compositions reinforced with equivalent ceramic fibers. The goal of this study was to compare the two resin systems by fabricating and tensile testing ceramic mini-tow composites, using ceramic grade Nicalon tows with a carbon interface coating. The specific objectives of the research were to:

- Determine the comparative fabrication results in terms of weight gains versus infiltration step for the two resin systems for the mini-tow geometry.
- Evaluate the tensile strength of mini-tow composites made with the two resins at different infiltration steps.
- Screen the oxidation resistance of the two types of composites at 600°C and 1000°C and check the retained mechanical properties of the two composite systems

The study was done with a "mini-tow" composite geometry in which a single multifilament Nicalon (C) tow was infiltrated with the preceramic polymer resin This geometry provided better utilization of fibers, reduced processing times, a uniaxial tensile stress condition, and shortened oxidation exposure times, compared to conventional flexure or tensile bars.

EXPERIMENTAL PROCEDURES

The research study used a three step experimental plan.

1. Fabricate two sets of Nicalon (carbon-coated) composites with Ceraset and Blackglas resin.
2. Measure the tensile strength of the two sets of composites in the 2X, 4 X and 6X conditions.
3. Oxidize the composites and measure the retained tensile strength of the exposed composites.

Ceramic grade Nicalon tow (1800 denier=500 filaments/tow with PVA-sizing) was purchased from Dow Corning Corp. The filaments had a pyrolytic carbon coating (nominal thickness --100-200 Angstroms).

Ceraset™ resin was provided by DuPont Lanxide Composites with dicumyl peroxide catalyst. The Ceraset is a polyureasilazane which cures by a free radical reaction. Pyrolysis is done in argon at temperatures at or above 1000°C to produce a silicon carbide composition that is amorphous in structure and may contain significant nitrogen fractions. Argon pyrolysis at higher temperatures produces a more crystalline structure and reduces the nitrogen fraction.

Blackglas™ 493A resin was provided by AlliedSignal Inc. with the 493B catalyst solution. The Blackglas is a siloxane resin which cures by additive chemistry. Pyrolysis is done in nitrogen or argon at temperatures of 1000°C, converting the cross-linked siloxane polymer into an amorphous silicon oxycarbide (45%Si, 30% O, 25%C).

Two sets of mini-tow composites were fabricated using an infiltration/curing/pyrolysis (ICP) procedure. The PVA sizing on the Nicalon (C) was removed with a hot water wash-and-dry before polymer infiltration. The tows were vacuum infiltrated with either Ceraset or Blackglas™ resin, cured at 110°C for 3-5 hours, and pyrolyzed for 1 hour at 1000°C in flowing argon. The first resin infiltration was done with the tows wrapped around an aluminum frame to hold them taut. The ICP process was repeated six times for each set to build up the matrix density and reduce porosity. Approximately one hundred (100) mini-tow composites (each five centimeter long) were fabricated for each resin type.

Oxidation exposure tests were done on 6X infiltrated ceramic mini-tow composites in a resistance heated box furnace at 600°C and 1000°C. Weighed composite samples were placed in preheated furnaces, held at temperature for one hour in stagnant air and then removed from the hot furnace for final weighing.

The as-received carbon-coated Nicalon tows were also infiltrated with acetone diluted epoxy, cured and then tensile tested. The strength of the epoxy infiltrated tows gave a "target" strength for the Nicalon, which provides a mark for comparing the ceramic matrix composites.

All tensile tests were done at 20°C on a Sintech Model # 20G screw actuation testing machine with a 2200N/500 lb calibrated load cell at a cross-head extension rate of 0.75 mm/min. The infiltrated tows were mounted with epoxy on papers tab based on the ASTM Test Method D2343. The gage length was 2.5 cm. Load and cross-head displacement data were recorded by computer data collection. There was no direct measurement of sample extension or strain. The load/extension slopes were measured by computer analysis of the linear section of the load-extension curve, discounting any initial non-linearities due to sample/fixture alignment.

The as-prepared mini-tows and the fractured composite samples were visually examined with a stereomicroscope, looking for fiber pull-out and general composite fractography.

RESULTS

Nicalon (C)™ Tow Evaluation

The as-received sized carbon-coated Nicalon tow had a unit length weight of 0.216 g/m, based on lengths and weights of the tows before first infiltration. The Nicalon (C) tows were desized and infiltrated with epoxy and then tensile tested per ASTM Test Method D2343, using a 2.5cm gage length. The mean tensile breaking load data for five as-received tows was 251 Newtons with a coefficient of variation of 14%. Based on the mean break load and the calculated tow cross-section, the experimental filament strength is 2.96 GPa, which is an excellent match to the specification strength of 2.8 GPa.

Ceraset™-Nicalon™(C) Composite Fabrication

During the six infiltration/pyrolysis cycles with the Ceraset resin, the cumulative weight changes of the infiltrated tows were recorded and are shown in Table 1 and Figure 1.

The high final weight gain (177%) for the Ceraset mini-tow composites is indicative that the Ceraset material is building up a "shell" over and around the tow. Direct infiltration into the tow interior could not account for all of the high weight gain, unless the Ceraset has pyrolyzed to fully dense silicon carbide (3.1 g/cc). This is unlikely with a 1000°C- 1 hour pyrolysis; a density of 2.5 g/cc (or less) is more likely.

The higher viscosity of the Ceraset resin (as compared to the Blackglas resin) produces more polymer pickup on the exterior of the tow, as the resin adheres to the exterior of the tow. The high surface-to-volume ratio of the mini-tow geometry accentuates this polymer pickup. This large a weight gain would not be expected in flat panels or bars with a lower surface-to-volume ratio, compared to the mini-tow geometry.

Table 1 - Fabrication Data for *Ceraset* Nicalon (C) Mini-tow Composites

Ceraset Infil/Pyro Step	1X	2X	3X	4X	5X	6X
Unit Length Wt.(g/m)	0.276	0.387	0.472	0.529	0.569	0.604
Step Wt Increase (%)	27.0%	40.1%	22.0%	12.2%	7.2%	6.1%
Cumulative Wt. Increase (%)	27.0%	77.9%	116.8%	143.2%	161.6%	177.7%

Blackglas™-Nicalon™ (C) Mini-Tow Composite Fabrication

Two sets of composites were made with the Blackglas resin. The cumulative weights and lengths of the infiltrated tows were recorded and are shown in Table 2 and Figure 1.

These weight gains are less than those observed in an earlier Blackglas-Nicalon (C) mini-tow study [5], where the cumulative weight gain was 109% at the 6th infiltration. In the 1996 study, mini-tow infiltration was done by immersing loose tows into Blackglas resin and allowing them to drain and cure with weights hanging on the tows. A likely explanation for the lower than expected weight gains in this study is that the infiltration of the taut, racked tows holds the filaments close together and "locks" in a high fiber volume fraction cross-section .

Assuming that pyrolyzed Blackglas resin has a density of 2.2 g/cc, a Blackglas weight gain of 40% for these Nicalon mini-tow composites would translate into fiber and matrix volume fractions of 68% and 32%, respectively. From previous work, the porosity is 5% or less. This assumes that there is no significant "shelling" of the Blackglas matrix around the tow.

Table 2 - Fabrication Data *Blackglas* Nicalon (C) Mini-tow Composites

SET 1 Infil/Pyro Step	**1X**	**2X**	**3X**	**4X**	**5X**	**6X**
Unit Length Wt.(g/m)	0.238	0.251	0.268	0.279	0.291	0.290
Step Wt Increase (%)	10.5%	5.5%	4.7%	3.8%	4.2%	-0.3%
Cumulative Wt. Increase (%)	10.5%	16.5%	24.4%	29.1%	35.0%	34.6%
SET 2 Infil/Pyro Step	**1X**	**2X**	**3X**	**4X**	**5X**	**6X**
Unit Length Wt.(g/m)	0.222	0.259	0.273	0.288	0.295	0.305
Step Wt Increase (%)	2.7%	16.7%	6.5%	5.5%	2.5%	3.6%
Cumulative Wt. Increase (%)	2.7%	19.9%	26.4%	33.2%	36.5%	41.4%

Figure 1 is a direct comparison of the weight gains for the Ceraset and Blackglas mini-tow composite fabrication. The much higher weight gain for the Ceraset composites is produced by the higher viscosity and greater polymer "pickup" in the infiltration steps.

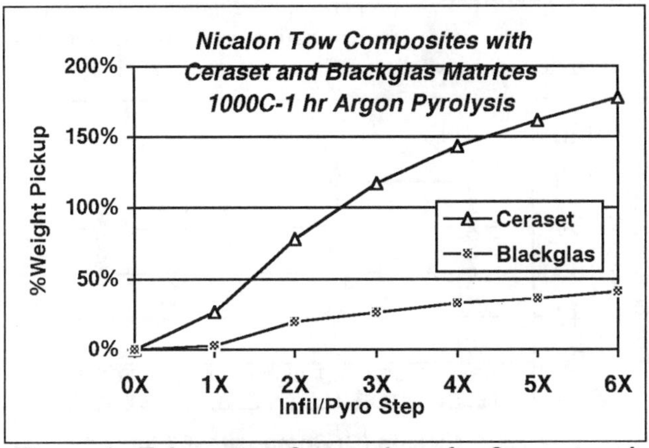

Figure 1 - Fabrication Comparison for Ceraset and Blackglas Nicalon (C) Mini-tow Composites

The 170% weight gain clearly implies that there is significant "over-coating" of the Nicalon tow. This was actually observed in the stereomicro-scope analysis. The Ceraset mini-tow composites had no visible fibers along the length; rather a surface layer of ceramic was clearly observed. On the other hand, the Blackglas mini-tow composites had ceramic filaments clearly visible along the length of the infiltrated tows.

Tensile Test of As-Prepared Nicalon™(C) Composites

The Ceraset composites were tensile tested in two test sets; the two Blackglas composite sets were each tested as a separate test set. The tensile data (Table 3) for the ceramic mini-tow composites made with the two types of matrix (Ceraset and Blackglas) show very similar values for tensile breaking loads.

Table 3 - Tensile Data for Ceraset and Blackglas Nicalon (C) Tensile Mini-tows

Nicalon (C) Mini-tow Composites 6X- Infiltrated - 2.5 cm Gage	Ceraset Test 1	Ceraset Test 2	Blackglas Set 1	Blackglas Set 2
Mean Tensile Break Load	**75.82 N**	**72.80 N**	**90.86 N**	**75.97 N**
% Coef. Variation / Count	20% / 10	14% / 6	14% / 10	11% / 10

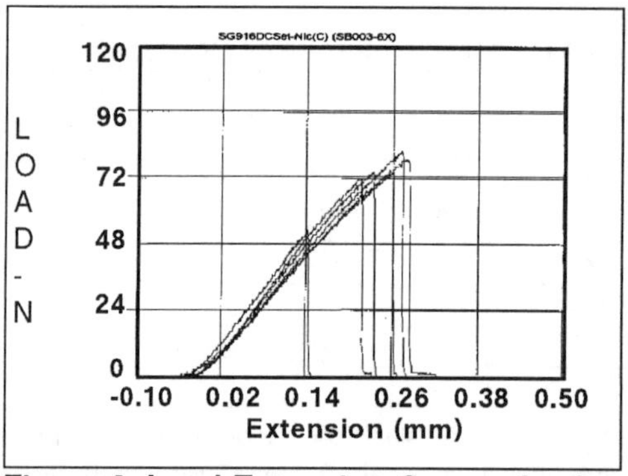

Figure 2 -Load-Extension Curves for As-Prepared Ceraset -Nicalon (C) Mini-tow Composites (6X)

Of the three measured variables, the breaking strength in Newtons is the variable that is directly and exclusively dependent on the tow properties. The values measured for the extension at break and the load-extension slopes include the compliance of the test rig and have not been corrected for the system compliance. The listed values for break extension and load/ extension slopes are therefore of limited value in characterizing the composite properties. Figure 2 shows representative load-extension curves for the tensile tests of the ceramic mini-tow composites. All the tested mini-tow composites show a rapid load drop off after fracture.

A graph of the tensile breaking loads (in Newtons) is shown in Figure 3. It details how similar the strengths of the two types of Nicalon composites are, independent of the matrix composition. However, the graph

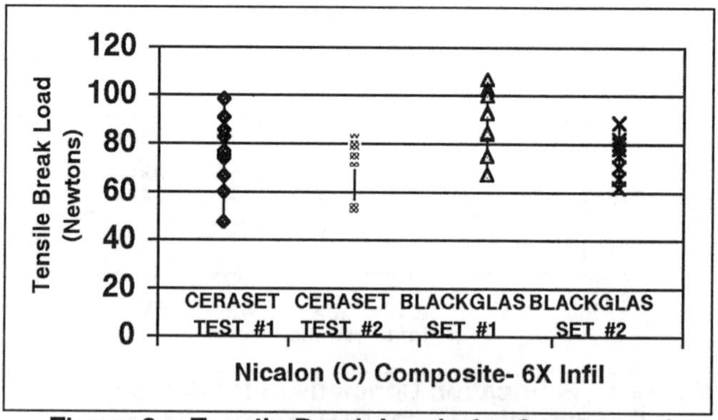

Figure 3 -- Tensile Break Loads for Ceraset and Blackglas Nicalon(C) Mini-Tow Composites

also shows the large amount of data scatter observed in these tensile tests.

The tensile breaking loads are plotted in Figure 4 against the unit length weights for the four sets of Nicalon composites. The unit weights for the Ceraset composites are markedly higher, compared to the Blackglas composites. The Ceraset composites are almost 2X heavier than the Blackglas composites. Yet their breaking loads are virtually equivalent.

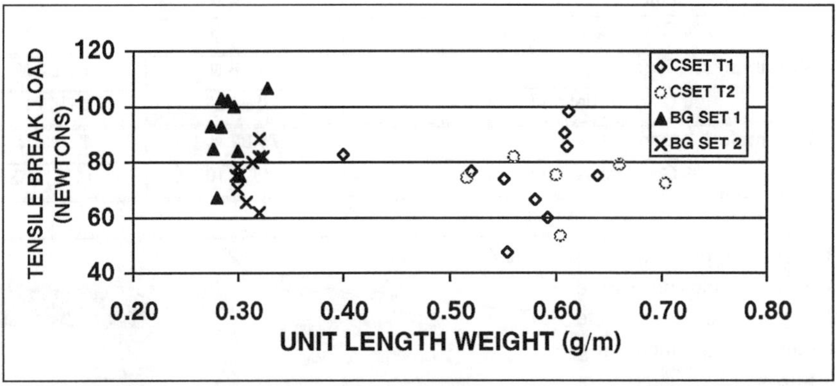

Figure 4 - Tensile Break Loads vs Unit Weights for Ceraset and Blackglas Nicalon(C) Mini-Tow Composites (6X)

Figure 4 also shows that there is no direct relationship between the unit weights and the breaking loads within a given set of composites. It is clear that, for these Nicalon mini-tow composites, the matrix carries very little of the load and the Nicalon fibers are the primary load carriers. The low modulus, microcracked matrix (whether it is Ceraset or Blackglas) only serves to transfer the load to the fibers.

Infiltration Effects

The infiltrated tows were also tested at two intermediate fabrication points -- 2X and 4X. As described in the fabrication section and as shown in Table 1 , there was a clear difference in unit length weights at the different infiltrations (2X, 4X, and 6X) for the two sets of samples. Table 4 presents the tensile test results for the 2X, 4X, and 6X infiltrations of the Ceraset and the Blackglas mini-tow composites.

As shown in Table 4, in spite of the weight differences between the two sets of samples and for the 3 levels of infiltration, there is little significant effect on the tensile break load results. This is further evidence that the fibers are the primary load carrier with little contribution from the matrix. Table 4 also shows that, for the Ceraset composites, the different infiltration levels have no significant effect on the observed breaking loads. For the Blackglas composites, there is the appearance of a trend, where higher density produces lower strength. But the data scatter among the three infiltration conditions is large enough, the apparent differences in the mean values cannot be affirmed with statistical confidence.

Table 4 - Tensile Data for Ceraset & Blackglas Nicalon (C) Mini-Tow Composites

Nicalon (C) Mini-tow Composites **2X, 4X, 6X- Infiltrated - 2.5 cm Gage**	**Ceraset Composites**	**Blackglas Composites**
	Tensile Break Load	
2XSamples Mean	**83.88 N**	**94.56 N**
% Coef. Variation / Count	12% / 5	10% / 5
4X Samples Mean	**78.16 N**	**84.27 N**
% Coef. Variation / Count	6% / 5	11% / 6
6XSamples Mean	**75.82 N**	**75.97 N**
% Coef. Variation / Count	20% / 10	11% / 10

However, there was one very important and visible change in the composite fracture with increasing infiltrations. This is shown in Figure 5. At 2X infiltrations, in both the Ceraset and the Blackglas composites, the fracture appearance was very brushy, with the fibers loose and splayed out. It looked like the matrix had powdered and flaked out during fracture.

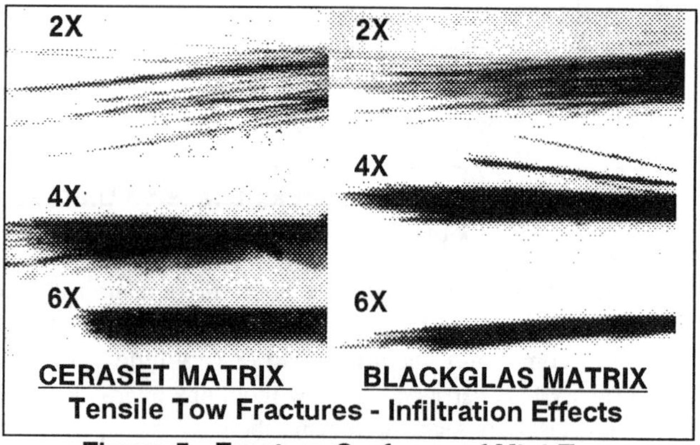

Figure 5 - Fracture Surfaces of Mini-Tow Composites

At 4X infiltrations, the fracture surface had a very fibrous appearance with extensive fiber pull-out. At 6X infiltrations the fracture surface was more brittle with signs of short fiber pull-out, on the order of a millimeter or two.

Based on the minimal effect of the unit weight lengths on the breaking load, the conclusion is that the fibers are the primary load carrier in the tensile condition with the matrix acting as a load transfer mechanism within the tow bundle. Changes in the matrix content affect the overall composite mechanical properties only when the matrix looses enough integrity that loads cannot be redistributed/shared among the filament population.

Oxidation Effects on Ceraset and Blackglas™-Nicalon (C) Composites

Samples (5 cm in length) of the as-prepared Ceraset and Blackglas-Nicalon (C) mini-tow composites were oxidized in stagnant air at 600°C and 1000°C for one (1) hour to determine if oxidation degraded the tensile properties of the mini-tow composites. Weight changes are shown in Table 5.

Table 5 - Oxid. Weight Loss for Ceraset & Blackglas Nicalon (C) Composites

	1000C- 1 Hr Oxidation		600C - 1Hr Oxidation	
Composite Set	Tow wt & count	% Change	Tow wt & count	% Change
Ceraset-Nicalon (C)	246 mg - 4	-1.9%	338.0 mg - 4	-1.7%
Blackglas-Nicalon (C) Set 1	151 mg - 5	-1.1%	120.4 mg - 4	-1.2%
Blackglas-Nicalon (C)Set 2	90.2 mg - 3	-1.4%	95.8 mg - 3	-1.3%

It appears that the Ceraset composites have a slightly higher percentage weight loss compared to the Blackglas composites. However, the Ceraset composites have a much higher weight fraction of matrix, compared to the Blackglas composites. If both matrices lose approximately the same relative amount of weight upon oxidation, the composites with the higher matrix content will have a higher overall oxidation weight loss. The only method of directly determining the oxidation susceptibility is to test the weight loss of the matrix compositions without the complication of the fiber reinforcement. Another possible difference is that the two matrices may have different pick-ups for physisorbed water. Identification of specific weight change mechanisms for these composite systems requires further study.

The oxidized Nicalon (C) mini-tow composites were tensile tested in a 2.5 cm gage length at ambient conditions, using the same methods utilized for the as-prepared composites. The tensile break load data for the Ceraset and Blackglas composites are summarized in Table 6 for the two oxidizing temperatures and the as-prepared condition.

Table 6 -Tensile Test Data on Oxidized Nicalon (C) Mini-tow Composites

Nicalon (C) Mini-tow Composites 6X-Infil 2.5 cm Gage Length	Ceraset Composites	Blackglas Composites Set 1	Blackglas Composites Set 2
	Tensile Break Load		
As-Prepared Mean	**75.82 N**	**90.86 N**	**75.97 N**
% Coef. Variation / Count	20% / 10	14% / 10	11% / 10
600°C-1 Hr Oxidized Mean	**58.18 N**	**50.73 N**	**54.04 N**
% Coef. Variation / Count	20% / 9	26% / 5	19% / 6
1000°C-1 Hr Oxidation Mean	**25.49 N**	**27.16 N**	**27.66 N**
% Coef. Variation / Count	20% / 3	27% / 6	26% / 4

The table shows how the strength and break extension are markedly reduced by short-term 600°C and 1000°C oxidation for both the Ceraset and the Blackglas mini-tow composites. Both systems are attacked by oxidation to an equal extent, as shown in Table 6, with losses in strength of 25-40% at 600°C and 70% after 1000°C oxidation.

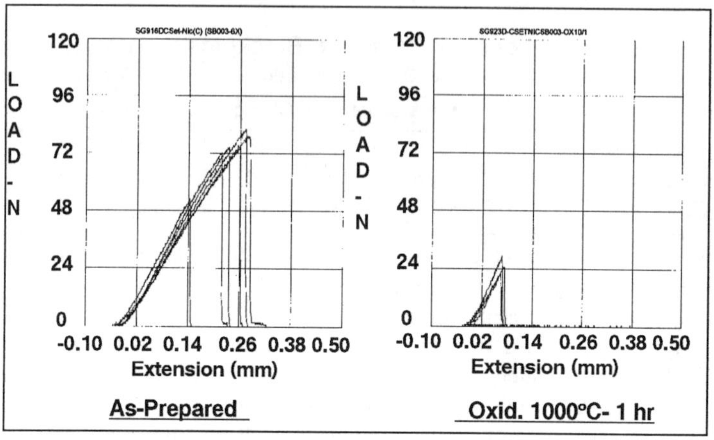

Figure 6 - Tensile Load-Extension Curves for Ceraset-Nicalon(C) Mini-Tow Composites After Oxidation

Figure 6 shows the load-extension curves for a Ceraset-Nicalon (C) mini-tow composite in the as-prepared condition and after oxidation at 1000°C for 1 hour. The loss in strength and the composite embrittlement are very apparent for the 1000°C oxidation. (The Blackglas-Nicalon curves are similar to the Ceraset-Nicalon curves.)

A similar oxidation effect has been seen in laminated 2-D Blackglas-Nicalon (C) composites with oxidation at 600°C and higher [6, 7] . The proposed embrittlement mechanism is the burn-out of the carbon interface, followed by surface oxidation of the Nicalon to form a silica layer which bonds and "locks" to the matrix. The locking produces a strong bond so that the cracks in the matrix cut through the fibers rather than diverting around them. It appears that the carbon is "burned out" at 600°C, reducing load transfer to the fibers. But insufficient silica has formed on the fibers to "lock" to the matrix. During the hotter 1000°C oxidation, the silica has formed thickly enough to embrittle the composite.

The heavier matrix loading and "overcoat" of the Ceraset did not inhibit the oxidation attack as compared to the Blackglas composites with a much lower matrix fraction. However, the Ceraset matrix may be more protective if it is pyrolyzed at a higher temperature, producing a denser and more crystalline silicon carbide type of structure.

Visual examination of the fractured 1000°C oxidized mini-tows shows a significant reduction in the amount of fiber pull-out seen on the fractured surface with a very brittle appearance.

Comparison to Other Blackglas™ Composite Studies

A direct comparison of the tensile data for this study to the data from a 1996 study on Blackglas Nicalon (C) composites (pyrolyzed in nitrogen at 1000°C), shows that there is no significant difference in any of the tensile properties (See Table 8). The mean tensile breaking loads for the three test sets are all similar, in spite of the differences in matrix compositions, unit length weights, infiltration procedures, and pyrolysis gas (argon vs nitrogen).

Table 7 - Comparison of 1996-1997 Ceramic Nicalon Composite Data

6X- Infiltrated **Nicalon (C)** - As-Prepared 2.5 cm Gage	**1996 Blackglas Composites**	**1997 Ceraset Composites**	**1997 Blackglas Composites**
	Tensile Break Load		
Mean	**69.40 N**	**75.82 N**	**75.97 N**
% Coef. Variation / Count	19% /19	20% / 10	11% / 10

CONCLUSIONS

1. The high final weight gain (170%) with 6X infiltrations of the Ceraset resin composite is confirmed visually by the build-up of matrix on the outer diameter of the tows. The most direct cause is the higher viscosity of the Ceraset resin, which produces higher polymer pick-up, as compared to the lower viscosity Blackglas resin.
2. The tensile breaking strengths of the Nicalon mini-tow composites were essentially the same for both Ceraset and Blackglas matrices. The break loads were on the order of 75 Newtons, with coefficients of variation between 10-20% for sample counts of 6-10. Fibrous fracture and fiber pull-out were visually apparent and the carbon interface functioned equally well in both Ceraset and Blackglas composites.
3. There appeared to be little effect of matrix loading or porosity on the composite strengths. The higher unit length weights of the Ceraset didn't produce higher strength, compared to the Blackglas. In both Ceraset and Blackglas there were clear differences in weight between the 2X, 4X, and 6X infiltrations, but the tensile breaking strengths for all the conditions were equivalent. The conclusion is that for tensile loading the fibers are the dominant load carrying component with little contribution from the microcracked, porous matrix. Although the tensile breaking loads didn't vary with the 2X, 4X, and 6X infiltrations, the appearance of the fracture changed markedly. With additional infiltrations, the matrix had more continuity, the fibrous appearance was markedly reduced, and the fiber pullout lengths were on the order of 1 mm or less.
4. Oxidation weight losses on the mini-tow composites at temperatures of 600°C and 1000°C for one hour produced weight losses on the order of 1-2%
5. Oxidation at 600° and 1000°C for one hour significantly reduced the strength of and embrittled the Nicalon (carbon) composites with both the Ceraset and Blackglas matrices. Strengths dropped 30-40% for the 600°C - 1 hour oxidation. After a 1000°C-1 hour oxidation, strength losses of 70% and embrittlement were observed.

6. The higher matrix fraction in the Ceraset composite did not markedly change the degree of oxidation attack on the composites at either 1000°C or 600°C.

ACKNOWLEDGMENTS

The authors would like to thank Mr. Doug Freitag and Mr. Roger Matsumoto of DuPont Lanxide Composites and Dr. Jack Sikonia of AlliedSignal for the contribution of Ceraset and Blackglas resins.

REFERENCES

1. L. M. Sheppard, " Enhancing Performance of Ceramic Composites," pp 617-631 in ACerS Bulletin, **70** [4] (1992).

2. S. Richlen, "Applications of Fiber-Reinforced Ceramic Matrix Composites," pp 495-526 in *Handbook on Continuous Fiber-Reinforced Ceramic Matrix Composites,"* Edited by R. L. Lehman, S. K. El-Rahaiby, J. B. Wachtman Jr., Purdue Research Foundation, (1995).

3. D. R.. Petrak, "Polymer -Derived Ceramics," pp. 223-226 in *Engineered Materials Handbook, Vol. 4 Ceramics and Glasses,* Edited by S. J. Schneider, Jr., J. R. Davis, G. M. Davidson, S. R. Lampman, M. S. Woods, T. B. Zorc, R. C. Uhl, ASM International, (1991).

4. J. E. French, "Ceramic Matrix Composite Fabrication and Processing: Polymer Pyrolysis," pp. 269-299 in *Handbook on Continuous Fiber-Reinforced Ceramic Matrix Composites,"* Edited by R. L. Lehman, S. K. El-Rahaiby, J. B. Wachtman Jr., Purdue Research Foundation, (1995).

5. S. T. Gonczy, R. C. Sprandel, K. T. Faber, " Tensile Test of Miniature Fiber Reinforced Ceramic Composites for Screening Process-Property-Relations," p729-736 in Ceramic Engineering & Science Proceedings, **18** [3] (1997).

6. M. G. Carswell, S. E. Schweizer, J. D. Sullivan, "Tensile Behavior of Ceramic/Ceramic Composites Before and After Exposure to a High Temperature Combustion Environment," p. 243 in Ceramic Engineering and Science Proceedings, **15**, [4]. (1994)

7. S. T. Gonczy, R. Y. Leung, & H. Dietrich, "CVD Carbide on Blackglas™-Nicalon™ (C) Composites for Oxidation Protection," p. 835 in Proceedings of the 17th Annual Conference on Composites, Materials, & Structures (Restricted Session), 10-15 January 1993, Cocoa Beach, FL., NASA Conference Publication 3235, V.2, (1993).

FLEXURE TEST METHOD INVESTIGATION FOR VARIOUS NEXTEL™ CONTINUOUS REINFORCED BLACKGLAS™ CERAMIC MATRIX COMPOSITES

R. Belardinelli and S. L. Pauletti, Northrop Grumman, Military Aircraft Systems Division, P.O. Box 655907, Dallas, TX 75265-5907

ABSTRACT

3M's family of oxide fibers offers low cost reinforcements for use as continuous fiber ceramic matrix composites (CFCMCs) in various aircraft applications. This study evaluated the mechanical properties of several 2-D Nextel fiber reinforced 312(BN), Nextel 440(C) and Nextel 550(BN-SiC) Blackglas ceramic matrix composites. The CMCs were subjected to various flexure test methods and the results evaluated. Destructive testing using 3 and 4-point established methods (ASTM D-790 and MIL-STD-1942) were used. Failure strengths, strains and moduli results were obtained. The moduli data were compared to that obtained from a non-destructive method (ASTM C-1259) utilizing an impulse excitation procedure. Also evaluated was a Nextel 312(BN)/ Blackglas mini-composite specimen having a specimen width-to-depth (b/d) ratio of 3:2 (ASTM D-1942 Configuration A) and the data compared to that of a full size specimen with a width-to-depth ratio of 5:1. Flexure testing was performed after processing and after 100 and 500 hours heat treatment at 600°C in air. Specimen span-to-depth ratios were evaluated along with the failure modes. Tests results showed 3-point testing produced higher failure strengths and lower moduli than 4-point. Good moduli correlation with the impulse excitation method was noted. Results from the mini-composite specimen showed good correlation with the large test specimens. Photomicrographs showing the CMC microstructures after processing and after heat treatment at 600°C for 100 and 500 hours are presented. Photomacrographs showing specimen fracture surfaces are also presented. The majority of specimens showed degradation of strength and moduli after heat treatment.

INTRODUCTION

Many published reports point out the advantages of ceramic matrix composites. These reports describe the strength of these materials, their outstanding ability to resist high temperatures, abrasion and harsh chemicals, and improvements in their crack resistance. There is no doubt that ideal ceramics could live up to these claims, and there are a number of applications where ceramics have the potential to outperform other materials. In jet engine applications, the ability of CMCs to withstand high temperatures would result in efficiencies that cannot be achieved by any other means. Typically ceramic components are fabricated from a material system and rigorously evaluated for their ability to perform. Such proof testing eliminates much of the uncertainty that arises from testing components independently, since the component is being subjected to exactly the loading and environment that it will experience in use. If ceramics are ever to live up to their potential, their published design data must be sufficient to characterize the material adequately for comparison with alternative materials in design use [1].

There are a number of reasons that ceramic design data is hard to come by. Material fabricators seem to be reluctant to publish much information about the material systems they

To the extent authorized under the laws of the United States of America, all copyright interests in this publication are the property of The American Ceramic Society. Any duplication, reproduction, or republication of this publication or any part thereof, without the express written consent of The American Ceramic Society or fee paid to the Copyright Clearance Center, is prohibited.

produce. Accurate data are difficult to obtain by conventional methods. Both the material manufacturers and users find that efforts to characterize ceramics by mechanical tests usually result in irreproducible data that cannot be related to the way the material will perform in use. Attempts to improve the meaningfulness of ceramic data have met with limited success, and are costly. Some standard test methods exist, but experimental error is rarely investigated. By far the most common test performed on ceramic test specimens is the flexural-strength test [2–6]. Investigations in calculating stress from these methods are (1) specimen support span to thickness (l/d) ratio, (2) support and loading bar radii, and (3) amount of material under stress (3-point versus 4-point loading).

The fiber reinforcements chosen as composite reinforcement were Nextel 312, -440 and -550 fibers manufactured domestically by 3M. These fibers are continuous polycrystalline metal oxide fibers produced by sol-gel processing and are stable in air with no chemical reactions between the fibers, matrix, fiber coatings and processing atmospheres. Minimal grain growth at 600°C occurs within the fiber [7]. All three fibers are lower-cost compared to imported silicon carbide fibers.

The matrix was a silicon oxycarbide ceramic produced from pyrolysis of a Blackglas preceramic polymer supplied by AlliedSignal Inc. Advanced Microelectronic Materials.

The focus of this paper was to perform 3 and 4-point destructive flexural tests on three different Nextel reinforced CMC materials systems after fabrication and after 100 and 500 hours of heat treatment at 600°C. Moduli measurements were also taken using a non-destructive test method and the results compared. From these analyses the specimen configuration and testing method producing the most accurate data will be identified.

EXPERIMENTAL TECHNIQUES

Experimental Design

This study incorporated the following specimen variables: (1) differing support span-to-thickness ratios (l/d), (2) differing bending conditions (3-point vs. 4-point); (3) differing heat treatment temperatures (ambient and 600°C); and (4) differing heat treatment times (0, 100 and 500 hours). A minimum of three specimens each were tested at each of the following conditions.

CMC Panels Fabrication

Six inch by nine inch panels were fabricated utilizing the following 3M reinforcements:

Nextel 312 Fiber AF-10 Fabric Weave Style: A 5 Harness Satin fabric made from 600 denier served Nextel 312 ceramic fiber roving. The fibers are composed of (by weight) 62% aluminum oxide (Al_2O_3), 24% silicon dioxide (SiO_2) and 14% boron oxide (B_2O_3). This fabric contained approximately 18 warp and 18 fill yarns per centimeter and was .254 mm thick. The fabric was exposed to AlliedSignal's process for heating Nextel fibers in an ammonia atmosphere to produce a thin BN-enriched layer at the surface of the fibers. The ammonia reacts with the boria in the fiber to produce the BN. This BN layer forms a weak interface between the Blackglas ceramic matrix and the reinforcing fibers and produces a brushy fracture characteristic of a tough ceramic composite [7].

Nextel 312 Fiber AF-30 Fabric Weave Style: A Crowfoot Satin fabric made from 1800 denier ½ twist Nextel 312 ceramic fiber yarn. The fibers are composed of (by weight) 62% aluminum oxide (Al_2O_3), 24% silicon dioxide (SiO_2) and 14% boron oxide (B_2O_3). This fabric contained approximately 19 warp and 18 fill yarns per centimeter and was .762 mm thick. The fabric was exposed to AlliedSignal's process for heating Nextel fibers in an ammonia atmosphere to produce a thin BN-enriched layer at the surface of the fibers as described above [7].

Nextel 440 Fiber BF-20 Fabric Weave Style: A 5 Harness Satin fabric made from 2000 denier Nextel 440 ceramic fiber roving. The fibers are composed of (by weight) 70% aluminum oxide (Al_2O_3), 28% silicon dioxide (SiO_2) and 2% boron oxide (B_2O_3). This fabric contained approximately 12 warp and 10 fill yarns per centimeter and was .508 mm thick. The fabric was subjected to chemical vapor deposition coating with a single layer of carbon (~ .1 μm thick) for fiber/matrix interface control. Synterials of Herndon, VA applied the coating.

***Nextel 550 Fiber XN-414-1 Fabric Weave Style**:* An 8 Harness Satin fabric made from 2000 denier Nextel 550 ceramic fiber roving. The fibers are composed of (by weight) 73% aluminum oxide (Al_2O_3), 27% silicon dioxide (SiO_2). This fabric contained approximately 9 warp and 9 fill yarns per centimeter and was .457 mm thick. Eight plies of fabric were stacked in a 0° warp aligned configuration and this preform subjected to CVD processing to provide a uniform dual layer coating of BN (~ .2 - .4 μm) followed by a SiC (~ 1 - 2 μm) overcoat. DuPont Lanxide Composites of Newark, DE applied the coating.

Impregnation of the AF-30 and BF-20 Coated Fabrics: Following fiber coating, the Nextel 312 and Nextel 440 coated fabrics were then preimpregnated with Blackglas 493 preceramic polymer solution to a resin weight of ~ 50%. AlliedSignal performed the prepreg processing.

Composite Processing of the AF-30 and BF-20 Panels: Eight ply panels were laid up, bagged and cured to transform the Blackglas preceramic polymer into a hard polymer matrix. Following cure, pyrolysis processing was performed in an inert atmosphere (gaseous argon) to a maximum temperature of 872 or 982°C. At the maximum temperature, a four-hour hold was employed. Five Blackglas resin infiltration/densifications were performed to develop the material system thermomechanical properties. The infiltrations were performed using conventional vacuum/pressure procedures developed at Northrop Grumman. Following fabrication to the CMC-5 process state (five infiltrations and six pyrolyses), the panels were submitted to non-destructive evaluation using a through-transmission ultrasonic inspection technique developed for evaluation of ceramic matrix composites. Process control data, i.e., fiber volume, apparent porosity, density and ceramic matrix weight gains were determined for the panels..

Composite Processing of the XN-414-1 Panels: The 6"x 9" eight ply panel preforms were bagged and infiltrated with Blackglas preceramic polymer solution using Northrop Grumman's resin vacuum infusion (RVI) infiltration processing techniques. The panels were autoclave cured, then pyrolyzed in an inert atmosphere to a maximum temperature of 982°C incorporating the four-hour hold. Five resin infiltrations using the RVI process were performed. Following fabrication to the CMC-5 process state, the panels were inspected and the process control data determined for the panels.

Flexure Specimen Machining

Unlike plastic test specimens, which are molded to shape, most ceramic test specimens must be cut or ground to the proper shape for testing. Machining of the various Nextel/Blackglas panels into flexure test specimens was performed using a Micro-Matic wafering machine. A diamond impregnated cutting wheel with water as coolant medium was used in cutting specimens. Extreme care was exercised in machining the specimens since defects introduced by damage during machining can weaken the material, or cause it to fracture in a location other than where the fracture is assumed to be. Specimen cutting was performed at a feed rate of 76 mm per minute with a maximum cut depth of .635 mm. After machining, the specimens were rinsed in distilled water and dried at 80°C to constant weight.

Heat Treatment

Specimens of each material system were heat treated in a laboratory muffle furnace at AlliedSignal and Northrop Grumman. The samples were weighed before and after exposure. The heat treatment procedure involved placing the specimens onto ½" square cell ceramic honeycomb core supports and placing the supports into the muffle furnace. The samples were isothermally exposed at 600°C for 100 and 500 hours in either flowing air (AlliedSignal) or stagnant air

(Northrop Grumman) atmosphere. After heat treatment, the specimens were placed into an 80°C drying oven followed by desiccation at room temperature. The specimens were then weighed and dimensionally measured.

Optical Transmission Microscopy

Photomicrographs of the three Nextel CMC systems were taken at 400X magnification following processing and heat treatment. Photomacrographs were also taken at 9.5X showing the resultant fracture surfaces and amount of fiber pull out for the various flexure test specimens.

Flexure Tests

AlliedSignal conducted the 3-point flexure tests using a support span-to-thickness ratio of approximately 6.5. The tests were performed in accordance with ASTM D-790 using a screw type universal testing machine and a 1000 lb. load cell at a crosshead rate of 1.0 mm/min. Stress-strain curves were prepared and Young's modulus values computed from the computer software.

Northrop Grumman conducted the 4-point flexure tests using a support span-to-thickness ratio of approximately 15.0 in accordance with MIL-STD-1942. Testing was conducted using a screw type universal test machine with a 1000 lb. load cell at a crosshead rate of 1.0 mm/min. A linear voltage displacement transducer (LVDT) recorded specimen deflection at the mid-point. A computer data log collection system was used to prepare load versus deflection plots for the specimens and Young's modulus values calculated from these plots.

Young's modulus was also determined by impulse excitation of vibration in accordance with ASTM C-1259. Specimen resonant frequencies were determined using a *GrindoSonic* Instrument manufactured by J.W. Lemmens, Inc.

RESULTS AND DISCUSSION

Composite Processing Results

Table I summarizes the density, apparent porosity, fiber volume and ceramic matrix weight gains for the various Nextel/Blackglas panels processed in this study. The physical properties for these panels were similar indicating composite infiltration/densification processing utilizing either the standard pressure/vacuum method or the RVI process was acceptable. The ceramic matrix weight gains for all panels ranged from 36 – 42 %. Densities ranged from 2.2 to 2.40 gm/cc and apparent porosity was 3.8 to 5.0 %. Fiber volume was highest for the Nextel 312 (BN) AF-10 UIC 50 panel (46%) followed by the Nextel 440(C) BF-20 panels (40.0-49.0%), then the AF-30 Nextel 312 (BN) panels at 29.0 – 34% and lastly the Nextel 550 (BN-SiC) XN-414-1 panels with 25% fiber content. The low fiber volume for the Nextel 550 panels was due to low preform compaction pressure used during the dual layer BN-SiC CVD coating processing.

Heat Treatment Results

Weight change results after 100 and 500 hours 600°C heat treatment are shown in Table II. These data shows that all specimens lost weight after heat treatment and the majority of the weight loss occurred in the first 100 hours of exposure. The Nextel 440(C) specimens showed the highest loss, ~2.7%, which included oxidation of the fugitive carbon coating. The Nextel 550 (BN-SiC) specimens lost ~ 2%. The Nextel 312 (BN) pyrolyzed at 872°C lost 1.9% while material pyrolyzed at 982°C lost only 1.3% indicating a more oxidatively stable matrix. Specimens exposed for 500 hours at 600°C showed additional weight loss of ~.25% over that of the 100 hours exposure; in effect the weight loss flattened out. The proposed mechanism for this is that as silica forms due to matrix oxidation, an equilibrium between matrix carbon weight loss and silica weight gain occurs, with the result that minimal weight loss is measured. Correlation to weight loss from heat treatment in flowing air vs. stagnant air could not be concluded.

Table I. Nextel/Blackglas panel properties after processing

Panel ID and CMC Material System	Fiber Volume (%)	Density (gm/cc)	Apparent Porosity (%)	Ceramic Matrix Weight Gain (%)
NG Panel 32 NX 440(C)	44.0	2.40	3.8	35.6
NG Panel 33 NX 440(C)	49.0	2.38	3.9	37.3
NG Panel 34 NX 440(C)	40.0	2.37	4.1	40.0
NG Panel 35 NX 440(C)	40.0	2.37	3.9	38.9
NG Panel 5016-03-005 NX 550(BN-SiC)	24.0	2.34	3.9	36.9
NG Panel 5016-03-006 NX 550(BN-SiC)	24.6	2.32	3.9	36.6
NG Panel 37 NX 312(BN) 982°C Pyrolysis	33.8	2.23	4.7	39.5
NG Panel 43 NX 312(BN) 982°C Pyrolysis	31.0	2.18	5.0	41.7
NG Panel 38 NX 312(BN) 872°C Pyrolysis	32.0	2.17	4.2	38.1
NG Panel 42 NX 312(BN) 872°C Pyrolysis	29.0	2.20	4.9	42.0
UIC Panel 50 NX 312(BN) 1000°C Pyrolysis	46.0	2.18	4.3	37.0

NG = Panels fabricated by Northrop Grumman.
UIC = Panel fabricated by University of Illinois at Chicago.

Table II. Nextel/Blackglas specimens weight change after heat treatment for 100 and 500 hours in flowing and stagnant air at 600°C

Panel ID and CMC Material System	Weight Change After 100 Hours @ 600°C in Flowing Air (%)	Weight Change After 100 Hours @ 600°C in Stagnant Air (%)	Weight Change After 500 Hours @ 600°C in Stagnant Air (%)
NG Panel 32 NX 440(C)	N/D	-2.62	-2.92
NG Panel 33 NX 440(C)	-2.49	-2.72	-2.89
NG Panel 35 NX 440(C)	-2.74	-2.76	-2.98
NG Panel 5016-03-005 NX 550(BN-SiC)	-1.51	-2.03	-2.25
NG Panel 5016-03-006 NX 550(BN-SiC)	-1.65	-2.17	-2.18
NG Panel 43 NX 312(BN) 982°C Pyrolysis	-1.54	-1.28	-1.32
NG Panel 38 NX 312(BN) 872°C Pyrolysis	-2.23	-1.85	-1.96

Photomicrographs After Processing and Heat Treatment at 600°C

Figure 1 presents interior photomicrographs taken at 400X magnification of the various Nextel/ Blackglas CMCs after processing and after heat treatment at 600°C for 100 and 500 hours. Figures 1a and 1b presents Nextel 312(BN)/Blackglas CMC after processing and after 500 hours at 600°C. Figures 1c and 1d show the Nextel 440(C)/Blackglas CMC after processing and after 500 hour heat treatment and Figures 1e and 1f are Nextel 550(BN-SiC)/Blackglas after processing and 500 hours 600°C heat treatment. The photomicrographs show that after processing the oval shaped Nextel fibers are encapsulated by a shiny Blackglas matrix with some porosity present. Note that in Figure 1e the Nextel 550 fibers dual layer BN-SiC fiber interphase coating can be seen as shiny rings encompassing the fibers.

Photomicrographs after the 500 hour 600°C heat treatment show that the Nextel 312 and Nextel 440 fibers became lighter in color and the amount of shiny Blackglas matrix present has diminished due to matrix oxidation. Since the fiber/matrix region contains more carbon, oxidation begins here and continues with exposure time. The primary oxidation mechanism is due to oxygen diffusion through the matrix microcrack network that results from shrinkage during pyrolysis, causing decarburization in the fiber/matrix interface region and matrix oxidation. A secondary mechanism is oxidation of the matrix surrounding the microcrack regions.

Flexure Tests of Nextel/Blackglas Composites

Figure 2 shows the specimen configurations and flexure testing methods used in this study. The support span-to-thickness (l/d) ratios for these methods were the following: for the 3-point method the l/d ratio was 6.5 and for the 4-point method 15.0. The 3-point tests were performed at AlliedSignal and the 4-point tests at Northrop Grumman. All of the 4-point specimens were tested by J.W. Lemmens on the *GrindoSonic* Instrument for Young's modulus prior to destructive testing at Northrop Grumman.

Figures 3 through 6 presents the flexure test results for the various Nextel/Blackglas CMCs after processing and after heat treatment at 600°C for 100 and 500 hours.

Analysis of these figures shows the following: Figure 3 compares the flexural strengths from 3 and 4-point testing and shows that the flexure strengths are highest for the 3-point tests. Figure 4 compares the flexure strengths after heat treatment and shows that the flexure strengths are reduced with the 3-point specimens showing higher strengths. This figure shows that the strength decreases with increased exposure time. Figure 5 shows the comparison of as-processed flexure moduli for various Nextel/Blackglas CMCs and shows that the lowest moduli values were obtained from the 3-point tests. The moduli values from the 4-point specimens were significantly higher and the moduli from impulse excitation gave the highest values. The moduli data shown for the 4-point and impulse excitation test methods was performed on the same specimens and gave good correlation with the 4-point flexure moduli data. Figure 6 compares the flexure moduli heat treated specimens and shows that the 3-point heat treated samples are less than the 4-point, while both are less than the values for impulse excitation of vibration methods. The change in moduli for the 100 hours of exposure versus the 500 hours appears negligible.

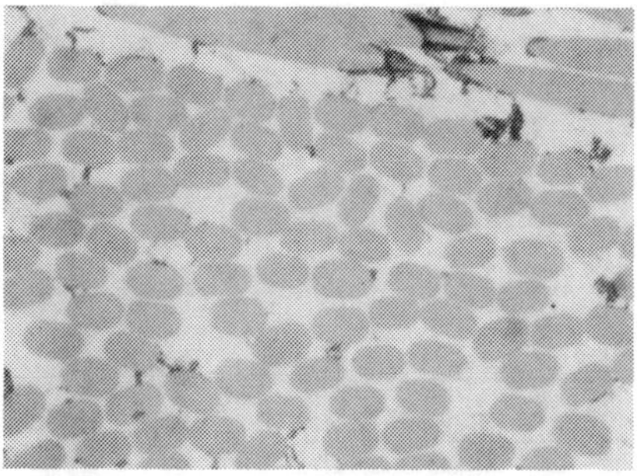

Fig. 1a. Nextel 312(BN) As-Processed.

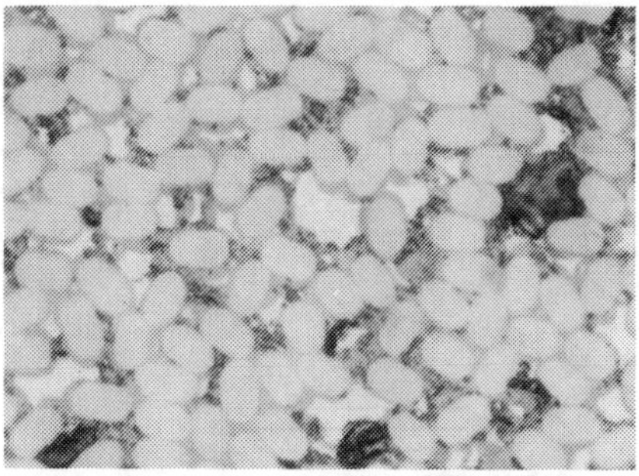

Fig. 1b. Nextel 312(BN) after 500 hour heat treatment at 600°C.

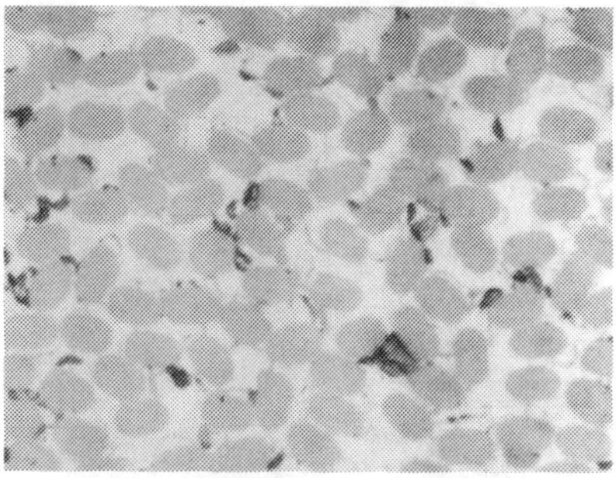

Fig. 1c. Nextel 440(C) As-Processed.

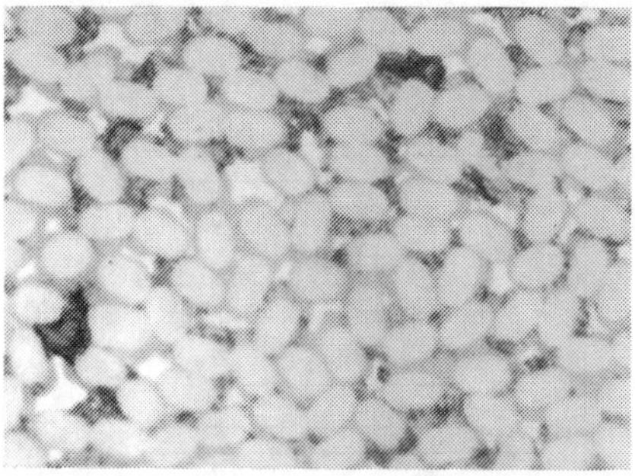

Fig.1d. Nextel 440(C) after 500 hour heat treatment at 600°C.

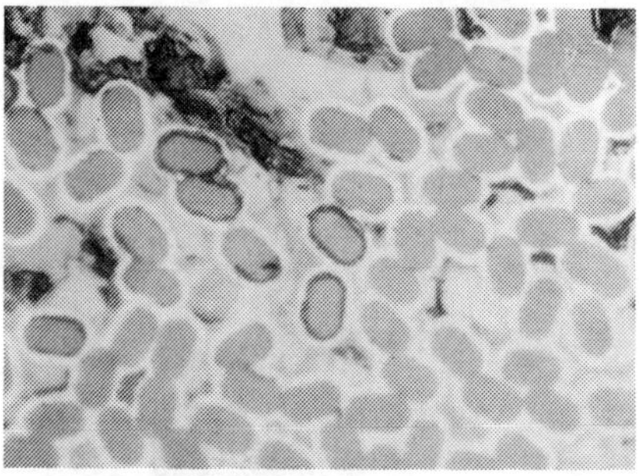

Fig. 1e. Nextel 550(BN-SiC) As-Processed.

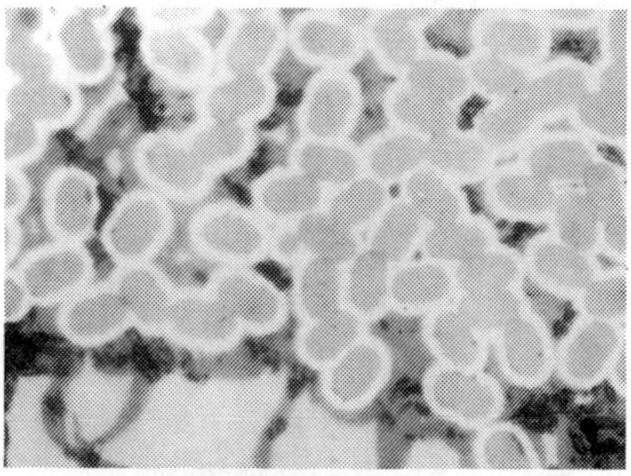

Fig. 1f. Nextel 550(BN-SiC) after 500 Hour heat treatment at 600°C.

Figure 1. Interior photomicrographs (400X magnification) showing the various Nextel/Blackglas CMCs after composite processing and after heat treatment at 600°C for 500 hours.

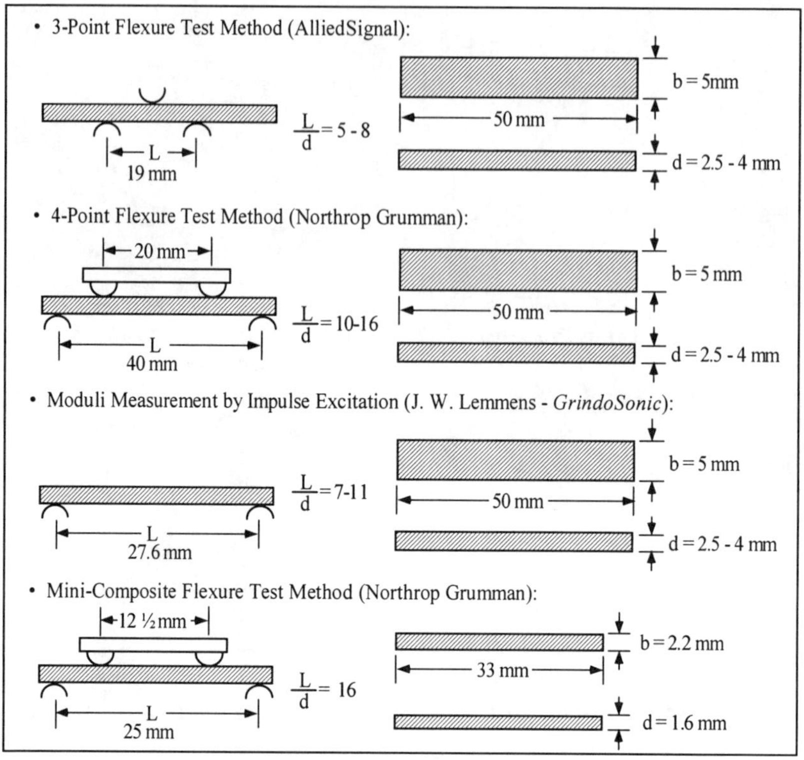

Figure 2. Sketches showing the flexure test methods and test specimen dimensions used in this investigation

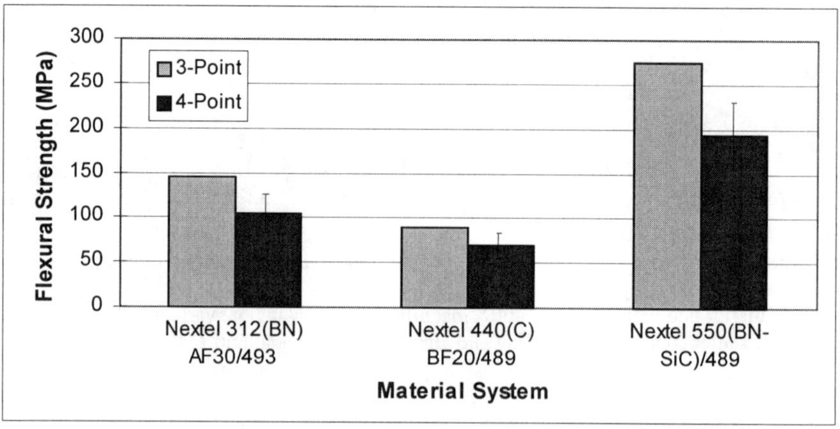

Figure 3. Comparison of as-processed flexural strengths for various Nextel/Blackglas CMCs

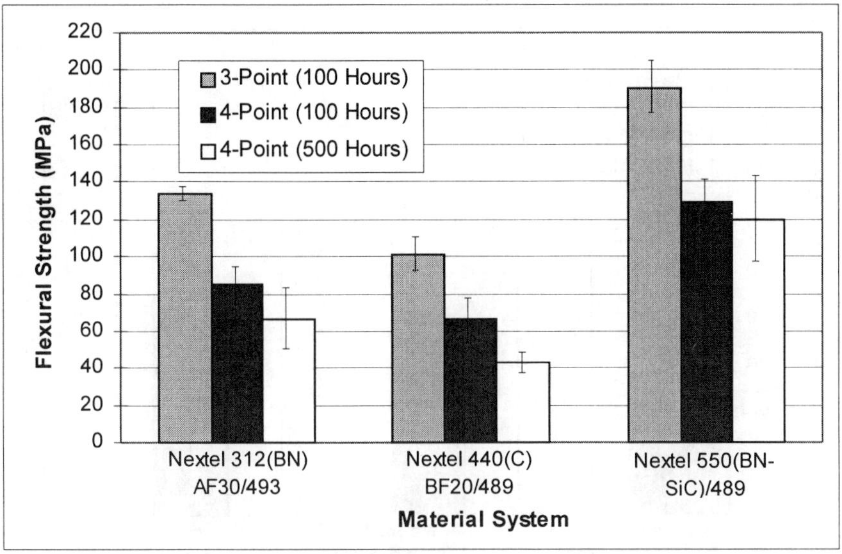

Figure 4. Comparison of flexural strength for various Nextel/Blackglas CMCs after heat treatment at 600°C for 100 and 500 hr.

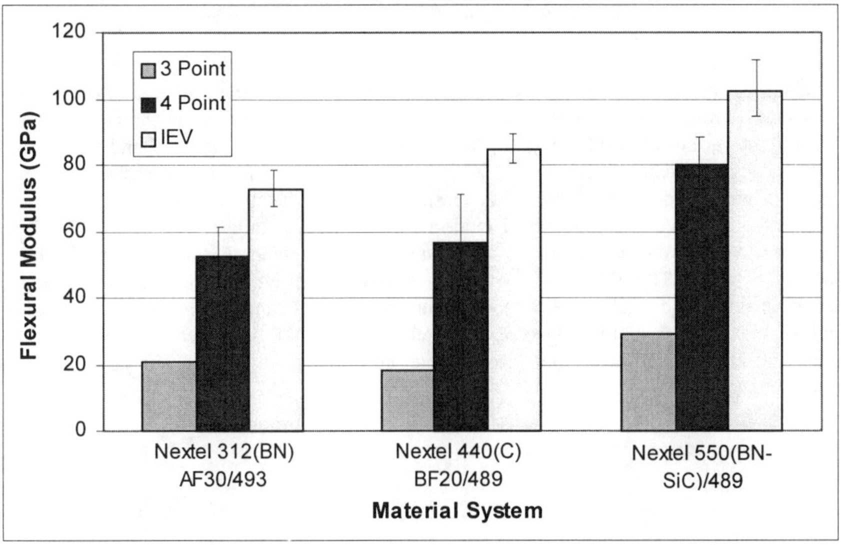

Figure 5. Comparison of as-processed flexure moduli for various Nextel/Blackglas CMCs.

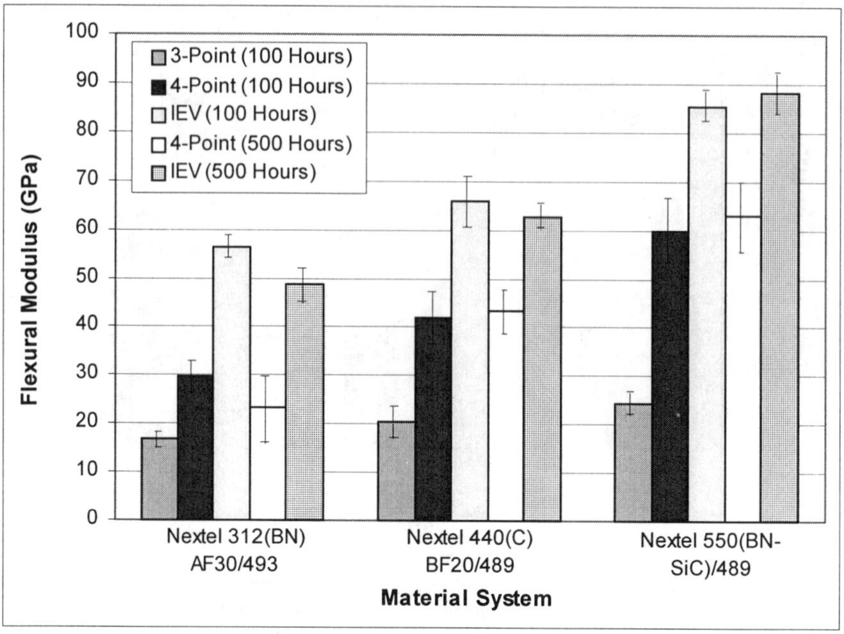

Figure 6. Comparison of moduli for various Nextel/Blackglas CMCs after heat treatment at 600°C for 100 and 500 hours

Photomacrographs of Specimen Fracture Surfaces After Processing and Heat Treatment

Figures 7 through 10 are photomacrographs taken at 9.5 X magnification showing the fracture surfaces of the various Nextel/Blackglas CMCs after flexure testing. Figures 7 and 8 compare the fracture surfaces of Nextel 550(BN-SiC) XN-414-1 specimens. The fracture surface from the 3-point test (Figure 7) specimen shows brittle tensile failure with no shear failure noted; whereas the 4-point test specimen shows the mixed mode tensile and shear failure (Figure 8). All 3-point specimens tested showed the brittle tensile failure loading fractures. Figure 9 shows the fracture surface of a Nextel 312(BN) AF-30 specimen tested in 4-point flexure after 500 hours heat treatment at 600°C; (note the mixed mode failure mechanisms and brushy fibrous appearance). Figure 10 shows the fracture surface of a Nextel 440(C) BF-20 specimen after 500 hours heat treatment at 600°C tested in 4-point flexure; note the brittle type fracture with no fiber pullout observed.

Figure 7. Fracture surface of as-processed Nextel 550(BN-SiC)/Blackglas specimen tested in 3-point flexure (9.5X).

Figure 8. Fracture surface of as-processed Nextel 550(BN-SiC)/Blackglas specimen tested in 4-point flexure (9.5X).

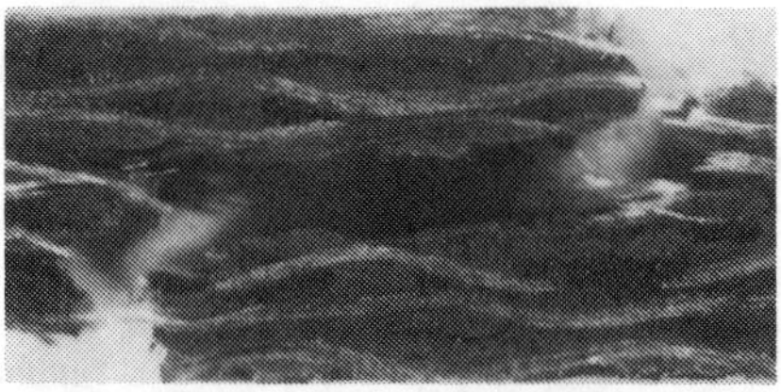

Figure 9. Fracture surface of Nextel 312(BN)/ Blackglas specimen tested in 4-point flexure following 500 hour heat treatment at 600°C (9.5X).

Figure 10. Fracture surface of Nextel 440(C)/ Blackglas specimen tested in 4-point flexure following 500 hour heat treatment at 600°C (9.5X).

Flexure Tests of Nextel/Blackglas Standard Vs. Mini-Composite Specimen Comparison

Figure 11 presents the test results from two configurations of flexure specimens prepared from a Nextel 312(BN)/Blackglas panel processed by the University of Illinois at Chicago (Panel UIC # 50). Standard 1.6 mm thick x 6.5 mm wide x 76 mm long and mini-composite 1.6 mm thick x 2.0 mm wide x 50 mm long flexure bars were prepared for 3 and 4-point testing.

The results showed that the 3-point standard flexure specimen gave higher strength and moduli than the 4-point testing of the mini-composite specimen. This is attributed to a l/d ratio of 44. As previous, AlliedSignal performed the 3-point tests, and Northrop Grumman conducted the 4-point tests.

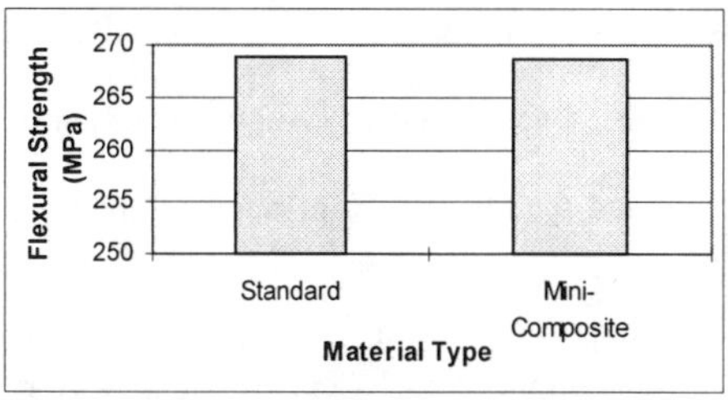

Figure 11a. Comparison of 3-point flexure strength of a large specimen (ASTM D790) and 4-point strength of a mini-composite specimen (MIL-STD-1942) for Nextel 312(BN) AF-10/Blackglas panel UIC 50.

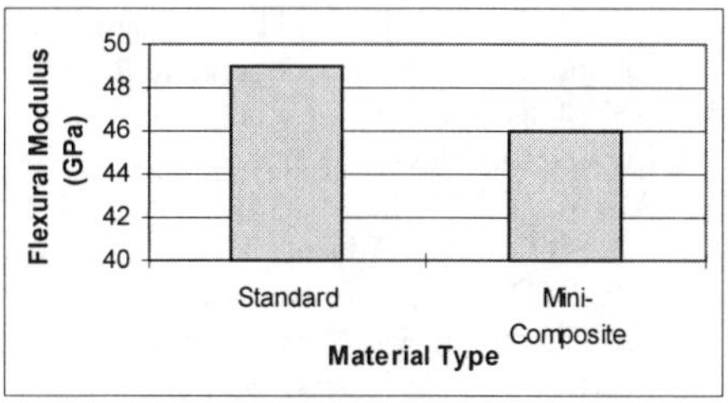

Figure 11b. Comparison of 3-point flexure moduli of a large specimen (ASTM D-790 and a 4-point strength of a mini-composite specimen (MIL-STD-1942) for Nextel 312(BN) AF-10/Blackglas panel UIC 50.

CONCLUSIONS

Three and four point flexure tests were used to investigate the fracture characteristics of various Nextel/Blackglas CMCs. Results of these tests concluded the following: fracture evaluation from 4-point flexure tests performed with span-to-depth ratios of > 10 showed the capability of the CMC to provide toughness. Nextel 312(BN) and Nextel 550(BN-SiC) CMCs showed the following fracture path: initial specimen failure caused by tension, followed by mid-plane crack defection then ultimate failure in tension. When these materials were tested using 3-point loading and span-to-depth ratios of <5, they did not demonstrate the crack tip deflection but instead an S-shaped fracture due to tension failure was observed.

Flexure tests after the 500 hour 600°C heat treatment showed oxidation of the matrix occurred, however, the fracture modes were unaffected; i.e., crack deflection was observed in the Nextel 312(BN) and the Nextel 550(BN-SiC) CMCs.

The Nextel 440(C)/Blackglas CMCs gave tension failures only, without the appearance of brushy fiber pullout. Both 3 and 4-point test methods on as-processed and after heat treatment showed pure tensile tear propagation through the CMC, with no crack deflection noted. This failure mode is indicative of an embrittled CMC without an effective fiber interface treatment. This condition may have occurred due to over-temperature processing of the Nextel 440 fabric during carbon deposition.

Flexure test results of a mini Nextel 312(BN)/Blackglas composite specimen tested in 4-point bend with a l/d ratio of 16 gave good correlation with 3-point tests incorporating and l/d of 44. With this information, the ability of providing a multitude of mini-composite samples for estimating mean and strength distribution are possible.

ACKNOWLEDGMENTS

The authors would like to thank Jim Ellis and Don Box of Northrop Grumman for their support of this effort, Dr. Jack Sikonia of AlliedSignal and André Van Leuven for performing flexural tests on the various Nextel/Blackglas CMCs, Marcus Johnson for conducting the 4-point flexure tests, Richard Duval for the microscopy effort, and Jan Davis for manuscript preparation.

REFERENCES

1. Clifford F Lewis Materials Engineering November 1988; "How Meaningful is Ceramic Test Data?", Pg. 41-44.
2. ASTM Test Method D790, "Flexure Properties of Unreinforced and Reinforced Plastics and Electrical Insulating Materials."
3. ASTM Test Method C1161, "Standard Test Method for Flexural Strength of Advanced Ceramics at Ambient Temperature R (1996)."
4. ASTM Test Method C1211, "Standard Test Method for Flexural Strength of Advanced Ceramics at Elevated Temperatures."
5. ASTM Test Method C1341, "Standard Test Method for Flexural Properties of Continuous Fiber-Reinforced Advanced Ceramic Composites."
6. MIL-STD-1942, "Flexure Test of Advanced Ceramics."
7. T.L. Tompkins 3M Ceramic Fiber Products Program, 3M Center 207-1S-23, St. Paul, MN, "Ceramic Oxide Fibers: Building Blocks for New Applications," Ceramic Industry April 1995.

NON-CONVENTIONAL, DAMAGE TOLERANT OXIDE/OXIDE FIBER-COMPOSITES

Timothy Radsick
United States Air Force Research Laboratory
Wright-Patterson AFB, OH 45433-7562

B. Saruhan
H. Schneider
Institute for Materials Research
German Aerospace Center (DLR)
Cologne, Germany D-51147

ABSTRACT

A fiber-reinforced ceramic composite was developed using matrix-infiltrated and non-infiltrated fiber layers. Fiber interface coatings were not used. Processed layers were hot-pressed together resulting in an oxide-fiber (Nextel720), oxide-matrix (mullite) composite. Matrix slurries were prepared using submicron-sized α-Al_2O_3 powder coated with amorphous SiO_2, along with additives to decrease mullite-formation temperature.

Room temperature 3-point bend tests were conducted on as-received samples and samples exposed to long-term, high-temperature oxidation. Composites demonstrate both reasonable fracture strength and damage tolerance. The favorable thermo-mechanical behavior can be explained by the composite's laminate-type structure. Matrix-infiltrated fiber layers provide strength but behave in a brittle, quasi-monolithic manner. Non-infiltrated fiber layers connect infiltrated layers, and provide damage tolerance through in-plane crack deflection and crack bridging. Varying matrix-to-fiber ratio allows tailoring of composite properties, trading off strength for damage tolerance. These composites provide damage tolerant behavior in the absence of fiber interface coatings, and when using a matrix which bonds strongly to fibers.

I. INTRODUCTION

In the most traditional sense, a fiber-reinforced ceramic-matrix composite consists of ceramic matrix material, fiber reinforcement and a tailored interface between the two. While the fiber and matrix determine for the most part the strength of the material, without a suitable interface the possibility of achieving a reasonable level of fracture toughness is limited. An interface can be tailored through in-situ reactions between fiber and matrix, but optimally an interface coating enables the necessary debonding and pullout mechanisms by deflecting cracks and dissipating crack energy. Coatings such as carbon and boron nitride are quite suitable for this purpose because of their easy-to-cleave laminate structures [1]. Non-oxide coatings work to provide damage tolerance but at the expense of one of the most valued properties of oxide-oxide composites, namely their oxidation resistance. [1]. Such oxide-oxide composites are of particular interest for use as thermal protection tiles in the combustion chambers of jet engines due to their long-term stability in oxidizing atmospheres

To the extent authorized under the laws of the United States of America, all copyright interests in this publication are the property of The American Ceramic Society. Any duplication, reproduction, or republication of this publication or any part thereof, without the express written consent of The American Ceramic Society or fee paid to the Copyright Clearance Center, is prohibited.

There are some candidate oxide-based interface coatings [2], but the selection is limited—due to high diffusion rates of oxides and the associated reaction with fibers and matrices, the synthesis of such coatings on multiple filament yarns is often complicated and costly. For these reasons, it is

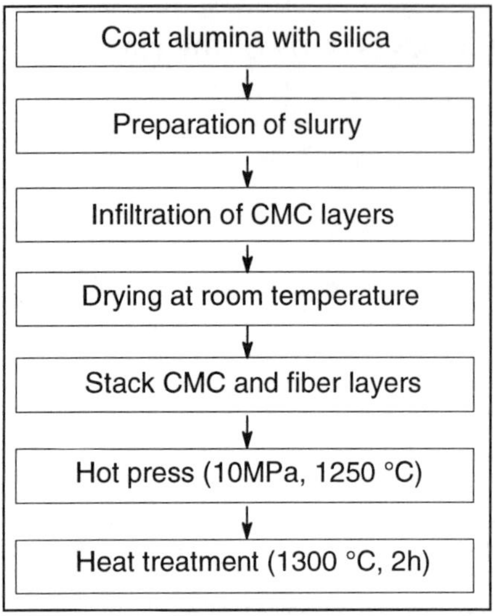

Figure 1. Flowchart of processing steps used to fabricate fiber laminate

worthwhile to explore other possible techniques to obtain damage tolerance in an oxide ceramic. Many studies have explored the use of novel construction techniques to achieve high strength and damage tolerance including several papers investigating laminate construction using monolithic and fiber-reinforced ceramic layers [3,4]. Of particular interest is the work of Tu, Lange and Evans who demonstrated the concept of "H-cracking", in-plane crack deflection resulting from inter-laminar elastic mismatches and residual stress [5].

II. FABRICATION

This paper presents initial investigations of a related concept based upon the use of non-infiltrated woven-fiber layers in a laminate to achieve crack deflection and crack bridging. In this composite, layers of matrix-infiltrated fibers (hereafter referred to as "CMC layers") are separated by non-infiltrated fiber cloth (hereafter referred to as "fiber layers") and are hot pressed together. The general method used to process fiber-laminate composites is outlined in Figure 1.

1. POWDER SYNTHESIS

These initial investigations of the Fiber Laminate Composite concept were conducted concurrently with the development of a mullite-based matrix material. This matrix material is based upon the microcomposite-powder reaction-sintering technique developed by Sacks, et. al. [6] and advanced by Bartsch, et. al [7]. Particles of alpha-alumina (AKP-50, Sumitomo Corp) were coated with amorphous silica through the hydrolization of TEOS. The ratio of TEOS to AKP-50 was selected so as to produce a final alumina/silica ratio slightly greater than the stoichiometric composition of mullite ($3Al_2O_3{\bullet}2SiO_2$). Residual alumina was favored due to the degradation of high-temperature properties caused by residual glassy silica [8].

Because the initial sintering characteristics of this coated powder are dominated by the sintering properties of the outer silica coating, a compact, high-density solid can be achieved at temperatures below 1300°C through transient liquid-phase sintering of SiO_2. In the case of reaction-sintered α-Al_2O_3/SiO_2 powder, however, mullite begins to form in significant amounts only at temperatures well above 1400°C [6]. It is known that commercial alumina and alumino-silicate fibers degrade at these temperatures, therefore other compaction techniques need to be developed [9,10]. Mechnich et. al. have shown that the addition of CeO_2 as a sintering aid can significantly reduce the mullite formation temperature down to around 1300°C [11]. This reduction in processing temperature enables the use of polycrystalline aluminosilicate fibers, albeit at a significant reduction in strength due to crystal growth and microstructural changes in the fibers at high temperatures [10].

2. COMPOSITE FABRICATION

Slurries containing microcomposite Al_2O_3-SiO_2 powder (α-Al_2O_3 particles coated with non-crystalline SiO_2) were prepared with added CeO_2 and mullite precursor powder (Siral, Condea, Brunnsbüttel, Germany.) Powders were ball milled for 30 minutes. This matrix material was cast into plaster molds, hot-pressed for 30 minutes (15MPa/1250°C) and annealed for 2h (1300°C) to obtain samples to ascertain a baseline monolithic strength value at room temperature. X-ray

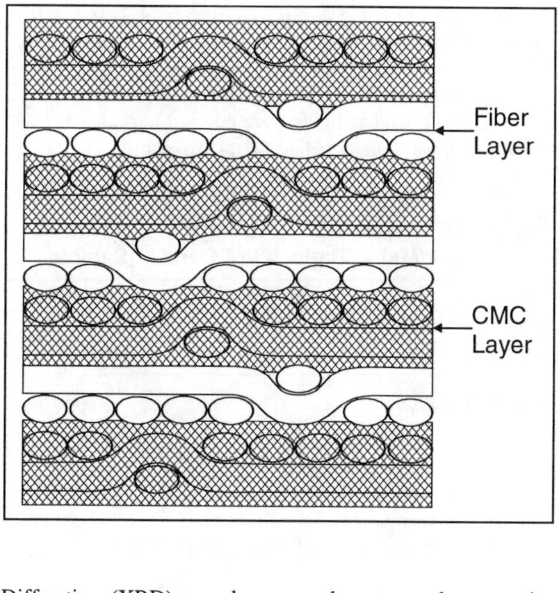

Figure 2. Conceptual Drawing of Fiber Laminate Composite. Note connection between CMC layers (matrix-infiltrated fiber layers) and fiber layers.

Diffraction (XRD) samples were also prepared to examine the crystal composition of the end matrix material. Nextel 720 (3M, Minnesota, USA) woven cloth (atlas 8-harness satin weave) was cut into squares approximately 50mm by 50mm. These squares were dipped into the slurry material, manually manipulated while immersed, allowed to absorb slurry, removed and allowed to dry. Altering the matrix solids content of the slurry along with repeated dipping allowed infiltrated fiber layers of varying matrix content and distribution to be produced. This simple dipping technique added a uniform and controlled amount of matrix to the fiber cloth—the matrix content among individual CMC layers (prepared from a given slurry mixture) varied less than 5%.

Laminate pre-forms were prepared by stacking in an alternating fashion CMC layers and fiber layers, as illustrated in Figure 2. A total of 13 layers (7 CMC and 6 fiber layers) provided an adequate specimen thickness of between 2.5 and 3.8 mm. The composite lay-up was placed in a hot-press and heated to 1250°C in air without applied pressure. At 1250°C, a pressure of 15MPa was applied and the temperature held constant for a period of 30 minutes to allow the material to consolidate and sinter. After 30 minutes, pressure was removed and the specimen was heated to a temperature of 1300°C. This temperature was held for a period of 2h to allow the components in the microcomposite powder to undergo reaction sintering to form mullite. Samples were subsequently cut into 3-point bend samples 4.0-4.5mm x 50mm in size. Some samples were exposed to high-temperatures in air for 200h.

After hot-pressing and subsequent heat treatment at 1300°C, the laminate structure of the composite was readily apparent to the eye. Though the outermost fibers in the fiber layers were bound to the matrix phase in the CMC layers, the "uninfiltrated" nature of the fiber layers was not compromised and a clear laminar discontinuity in the composite was present.

III. CHARACTERIZATION

Three series of samples, each with a different matrix content as described in Table 1, were tested in 3-point bending at Room Temperature (RT) using a UTS 10 Universal Test Machine (UTS Instruments, Inc) in order to determine the effects of matrix content. The effects of long-term exposure to high-temperature oxidizing atmospheres were also measured. For the sake of simplicity, only one series of samples was used. Fiber Laminate Composites were prepared using a 50wt% slurry and a double infiltration step. 3-point bend samples were cut and annealed for 200 hours in air at 1200, 1250 or 1300°C. Bend testing was conducted at room temperature.

Sample Description (# infil., solids wt. %)	CMC Layer Thickness	CMC Layer Fiber/Matrix Ratio	Strength (3-pt Bend)	Photo of Sample
1 Infiltration, 20% Slurry	0.28 mm	85 / 15	38 MPa	Figure 6a
1 Infiltration, 50% Slurry	0.30 mm	62 / 38	73 MPa	Figure 6b
2 Infiltrations, 50% Slurry	0.38 mm	41 / 59	88 MPa	Figure 6c

Table 1. Summary of data for three slurry preparations

Mullite formation was examined by X-ray diffraction using a Siemens D-5000 X-ray Diffractometer. Differential Scanning Calorimetry was carried out on a Netsch 404 DSC. After 3-point bend testing at RT, fracture surfaces were examined using a Philips 525M SEM.

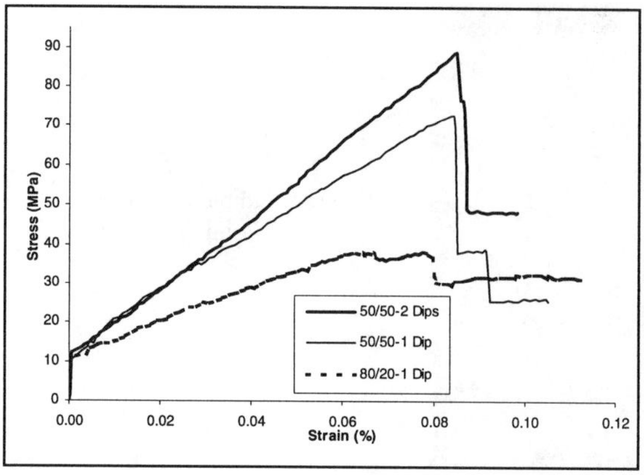

Figure 3. Stress-strain curves for as-received samples: CMC layers dipped once or twice into slurries containing 20% wt solids or 50% wt solids.

IV. RESULTS

1. Mechanical Characterization

Representative load-displacement curves for the three series of samples shown in Figure 3 clearly illustrate the differences in preparation and matrix content. Samples prepared using the slurry with 20% solids content had a high degree of damage tolerant behavior but very low strength values. The samples prepared with slurry with 50% solids content showed much higher strength values and reasonable damage tolerance. The samples prepared with two infiltration steps showed increased strength but a more brittle fracture compared with samples prepared using one infiltration step.

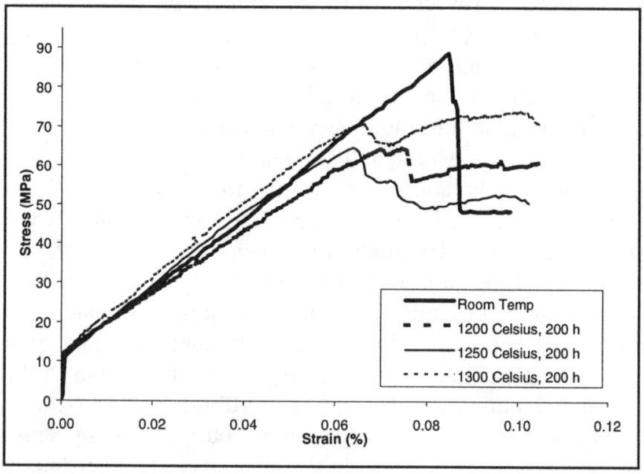

Figure 4. Stress-strain curves for heat-treated samples. All samples prepared using 50% wt solid slurry and two infiltration steps.

The effects of exposure to high-temperature oxidizing atmospheres are shown in Figure 4. The strength values of the exposed samples decreased compared with that of the as-received samples, but fail in a less catastrophic manner withstanding a higher strain before total failure.

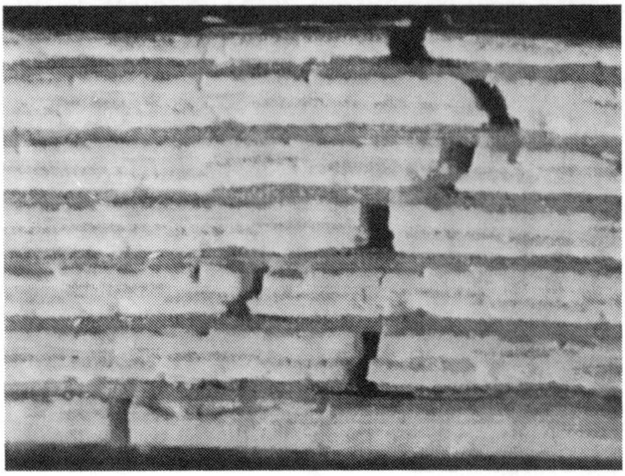

Figure 5. Photo of broken 3-Point bend sample

Figure 5 shows a representative 3-point bend sample after testing. The extent of crack deflection through the upper sample clearly shows the brittle fracture mechanism of the CMC layers and the crack bridging and deflection properties of the fiber layers. The wide crack separation shown in the upper sample did not normally occur through testing. Rather, the sample normally had to be cyclically manipulated by hand in order to cause the composite to separate. This illustrates an important advantage of fiber laminate composites, namely that after failure the composite remains intact, a desired property in aircraft engine material.

2. MICROSTRUCTURAL CHARACTERIZATION

Differential Scanning Calorimetry (DSC) and XRD results showed that mullite formation began to occur in appreciable amounts only at 1300°C. Powder samples hot-pressed for 30 minutes at 1250°C/15MPa and then annealed without pressure at 1300°C showed almost complete mullite formation. XRD analysis showed that less than 5% SiO_2 remained unreacted with Al_2O_3.

SEM photos of the three different series of samples are shown in Figures 6a-c. Figure 6a shows a sample whose CMC layer was prepared by a single infiltration step into a slurry with a 20 wt % solids content. Each CMC layer contains 14 wt % matrix and is readily discernable due to a somewhat planar fracture surface perpendicular to the laminate plane and to the absence of loose individual fibers. CMC layer thickness is approximately 0.28mm. Figure 6b shows a sample where a slurry with 50 wt % solid was used. In this case, differentiation between pure fiber layers and matrix-containing layers is easy due to a marked increase in the amount of retained matrix. Most of this matrix occupies the inter-fiber spaces, so that despite a 37% increase in matrix content, the CMC layers, with a thickness of 0.30mm, are only 0.02mm thicker than those prepared above with the thinner slurry. Figure 6c also shows a sample prepared with 50 wt % solid slurry, but in this case the individual fiber layers were infiltrated, allowed to dry and then infiltrated a second time. The photo shows not only a marked increase in CMC layer thickness, but also an increased distribution of matrix on the surface of the layers as opposed to in the inter-fiber and inter-bundle spaces. Layer thickness was approximately 0.38 mm. Table 1 summarizes the data for these three preparations.

a) (b

c)

Figures 6 a-c: Fiber laminate composites of varying matrix content and preparation technique:

(a) 20% slurry solids, 1 infiltration
(b) 50% slurry solids, 1 infiltration
(c) 50% slurry solids, 2 infiltrations

3. DISCUSSION

Because of the strong bonding between fiber and matrix and the lack of a interface coating, the CMC layers of this fiber laminate composite form a stiff material with a brittle fracture mechanism. Little or no damage tolerance results from the inclusion of fibers in CMC layers, nor would any such behavior be expected in the absence of an interface coating. Rather, in the CMC layers, fiber is used to assist in producing thin, quasi-monolithic layers, preventing excessive macro-scale cracking which can occur during the hot-pressing of thin monolithic tapes. Initial attempts to produce a fiber laminate composite using ceramic tapes instead of CMC layers failed due to this macro-scale cracking.

The characteristics of the matrix in the CMC layers affect the composite properties in several ways. The amount of matrix on the surfaces of the CMC layers influences the bonding between CMC and fiber layers arising from hot-press sintering and annealing. Insufficient matrix on the surfaces prevents bonding between CMC layers and fiber layers. In addition, the amount of matrix and its distribution within each CMC layer affect the strength and stiffness of that layer and, as a result, those of the composite. This can be seen by looking at Figure 3. Specimens with thicker CMC layers (i.e. with higher matrix content) tended to have higher strength and a sharper drop in strength after initial matrix cracking. Samples with less matrix had a fairly flat load-displacement curve after initial matrix cracking.

The fiber layers function mainly to modify the interaction between the CMC layers. Through hot-pressing, the outermost fibers in the bundles of the fiber layers are bonded to the CMC layers. The fiber layers serve as a mechanical connection between the overlying and underlying CMC layer, transferring loads and bridging inter-laminar and intra-laminar cracks. The layers provide a planar discontinuity in the matrix, adding a local anisotropic weakness which serves as an fracture path, thus improving damage tolerance.

Figure 7. Typical CMC Layer with arrows showing areas of sub-optimum matrix infiltration and fiber-to-fiber contact

Investigations to improve microcomposite-powder-derived mullite slurries need to be extended, so that slurry infiltration can be optimized and RT- and High Temperature (HT) strength increased. Figure 7 illustrates several potential causes of CMC layer weakness. Clearly present are areas of incomplete matrix infiltration into the center of the fiber bundles and areas of fiber-to-fiber contact. Not visible is the small amount of macro-scale matrix cracking in the CMC layer green body which occurs on the outermost surfaces as the slurry dries. HT-test results also show that a small amount of glass containing silicates was present, somewhat affecting the high temperature properties of the fiber laminate composites. It is likely that a better-infiltrated CMC layer will improve the properties of a Fiber Laminate Composite.

Nonetheless, in a straight comparison, fiber laminate composites prepared with the current slurry formulation and dip infiltration technique retain about 70% of the baseline monolithic strength while adding a level of damage tolerance. In applications where weight is a strong concern, Fiber Laminate Composites perform even better. When normalized according to weight, fiber laminate samples retain 83% of the baseline monolithic strength.

V. CONCLUSION

Fiber laminate composites, which involve no fiber coating but display damage tolerant fracture, were fabricated. The effect of individual components on fracture behavior and strength values were investigated. Composites show moderate strength values at RT and at HT (up to 1300°C). Properties of the fiber composites described in this context can be increased by changing the chemical composition of microcomposite particles, by improving the infiltration of the CMC layers with these slurries prepared by using microcomposite particles and by changing the type of weave of the fibers.

References

1 K. K. Chawla, Z. R. Xu, J. S. Ha, M. Schmücker and H. Schneider, "Effect of BN Coating on the Strength of a Mullite-type Fiber," *App. Comp. Mater.*, **4** (1997) 263

2 P. E. D. Morgan and D. B. Marshall, "Functional Interfaces for Oxide/Oxide Composites," *Mater. Sci. Eng.*, **162**, 15-25 (1993)

3. W. A. Cutler, F. W. Zok, F. F. Lange, "Delamination Resistance of Two Hybrid Ceramic-Composite Laminates," *J. Am. Ceram. Soc.*, **80** [12] 3029-37 (1997)

4. W. A. Cutler, F. W. Zok, F. F. Lange, "Mechanical Behavior of Several Hybrid Ceramic-Matrix-Composite Laminates," *J. Am. Ceram. Soc.*, **79** [7] 1825-33 (1996)

5. W. C. Tu, F. F. Lange, A. G. Evans, "Concept for a Damage-Tolerant Ceramic with 'Strong' Interfaces," *J. Am. Ceram. Soc.*, **79** [2] 417-24 (1996).

6. M. D. Sacks, N. Bozkurt, G. W. Scheiffele, "Fabrication of Mullite and Mullite-Matrix Composites by Transient Viscous Sintering of Composite Powders," *J. Am. Ceram. Soc*, **74** [10] 2428-37 (1991)

7 M. Bartsch, B. Saruhan, M. Schmücker, H. Schneider, "Novel Low-Temperature Processing Route of Dense Mullite Ceramics by Reaction Sintering of Amorphous SiO_2-coated γ-Al_2O_3 Particle Nanocomposites," Submitted Paper, *J. Am. Ceram. Soc.*

8. H. Schneider, K. Okada, J. A. Pask, *Mullite and Mullite Ceramics*, John Wiley and Sons, Chichester, 1994

9. 3M Company (Minneapolis MN, USA) Prospectus Data

10. C.Milz, *German Aerospace Center*, Ph.D. Research (to be published 1999)

11. P. Mechnich, H. Schneider, M. Schmücker and B. Saruhan, "Accelerated Reaction Bonding of Mullite," Accepted for publication, *J. Am. Ceram. Soc.*

Nano, Particulate, Whisker, and Self-Reinforced Composites

PRESSURELESS SINTERED ALUMINA/SILICON CARBIDE NANOCOMPOSITES

K. Niihara, Y. K. Jeong, T. Sekino, and Y. H. Choa

The Institute of Science and Industrial Research, Osaka University,
8-1 Mihogaoka, Ibaraki, Osaka 567, Japan

Abstract

Al_2O_3/SiC nanocomposites have been fabricated by pressureless sintering process using MgO and/or Y_2O_3 as sintering additives and post hot-isostatic pressing (HIP), which subsequently can break through the disadvantage of hot-pressing process. MgO sintering additive played an important role for improving densification and controlling microstructure of the nanocomposites. After HIP treatment, the Al_2O_3/SiC nanocomposites achieved the full density, the moderate increase of toughness and the high fracture strength of 1 GPa. Moreover, the measurement of high-temperature strength revealed that the high fracture strength was maintained up to 1000°C. The strength increase at high temperatures may be due to the prohibition of the grain boundary sliding and/or cavitation by intergranular SiC dispersions.

1 Introduction

Attempts to improve the strength and the strength variability of alumina ceramics have been performed by many ceramic scientists. As a result, a number of different approaches have been proposed in this area [1]. Of these, the use of second phase reinforcements such as whiskers, particulates and platelets is one of the most prominent. It attempts to create microstructures with improved fracture resistance which will also lead to an increase in fracture strength. The size of these reinforcing phases is usually several micrometers or more.

A new class of material which breaks through the microstructural control of micrometer level is the ceramic matrix nanocomposites proposed by Niihara and his coworkers [2,3,4]. They have reported that the dispersion of 5 vol% SiC nano-

To the extent authorized under the laws of the United States of America, all copyright interests in this publication are the property of The American Ceramic Society. Any duplication, reproduction, or republication of this publication or any part thereof, without the express written consent of The American Ceramic Society or fee paid to the Copyright Clearance Center, is prohibited.

particles into alumina could remarkably increase the room temperature strength from 350 MPa to over 1.0 GPa. In addition to this high fracture strength, the fracture toughness and the creep resistance were also improved by the in-grain toughening associated with the intragranular SiC particulates [5] and the grain boundary strengthening by the intergranular SiC dispersions [6,7], respectively.

Generally, the Al_2O_3/SiC nanocomposites have been fabricated to high densities by the hot-pressing process, because of the difficulty in densifying the nanocomposites. However, this process can manufacture only the ceramic articles of simple geometrical shapes, and would be expensive and unsuitable for the mass production. For many potential applications of these materials, the pressureless sintering process would be preferable if the full density could be achieved. There were some investigations for fabricating the Al_2O_3/SiC nanocomposites by pressureless sintering process [8,9,10,11,12]. Borsa *et al.* [8] fabricated Al_2O_3/5 wt% SiC nanocomposite by a pressureless sintering route to a maximum of about 95% of the theoretical density, and Zhao *et al.* [9] sintered Al_2O_3/5 vol% SiC nanocomposite to 98.3% theoretical density by the same route. Anya and Roberts [10] also reported that high density (≥99.6% theoretical) Al_2O_3/SiC nanocomposites, with up to 15 vol% SiC, have been fabricated by the pressureless sintering. However, the improved strength has not been reported for pressureless sintered Al_2O_3/SiC nanocomposites.

The objective of this work, therefore, was to fabricate Al_2O_3/SiC nanocomposites with high fracture strength by pressureless sintering with sintering additives and subsequent hot-isostatic pressing (HIP) technique, which subsequently can break through the disadvantage of hot-pressing process. Additionally, the effects of MgO and/or Y_2O_3, which were selected as sintering additives for the pressureless sintering, on densification behaviors, microstructure and mechanical properties is described for Al_2O_3/SiC nanocomposites.

We selected the sintering additives that would permit reduction of the time or temperature needed to cause sintering to a desired extent for Al_2O_3/SiC nanocomposites. It is well known that a small addition of MgO can inhibit the discontinuous grain growth and to promote the sintering of monolithic Al_2O_3, which leads to full density [13,14,15]. It has also been reported that a small addition of Y_2O_3 can enhance the densification rate [16,17], and improve the tensile and compressive creep resistance of monolithic Al_2O_3 [18,19]. The pressureless sintering and post HIP process employed in the present work consists of two steps: 1) a normal presintering procedure to make specimens without open pores, and 2) a second sintering under a high isostatic gas pressure.

2 Experimental Procedure

The α-Al_2O_3 powder used in this study (TM-DAR, Taimei Chemicals Co., Nagano, Japan) had a purity of 99.99%, mean particle size of 0.21 μm and specific surface area of 14.7 m^2/g (all quoted by manufacturer). β-SiC powder (Ibiden Co., Gifu, Japan) had a mean particle size of 0.27 μm (average grain size by TEM: 70 nm) and specific surface area of 23.5 m^2/g. High purity MgO (Ube Co., 0.1 μm) and Y_2O_3 (Shin-Etsu Chem., 1 μm) were used as sintering additives. Al_2O_3 and SiC starting powders were weighed with the fraction of 1, 5 and 10 vol% SiC. Al_2O_3/SiC powders were also doped with 0.1wt% MgO and/or 0.1 wt% Y_2O_3. The combined powders were ball-milled in ethanol for 24 h using high-purity Al_2O_3 balls of diameter 5 mm in a polyethylene pot. Then, the soft agglomerates of the dried powders were crushed by dry ball-milling for 24 h using Al_2O_3 balls of diameter 10 mm.

The mixed powders and as-received Al_2O_3 powder were uniaxially pressed at 30 MPa into bars of 5.5 mm × 6.4 mm × 53 mm, and isostatically pressed at 200 MPa. The powder compacts, which were laid on a graphite foil without any powder bed in a covered graphite crucible, were sintered at 1600-1800°C for 2 h under flowing argon. The subsequent HIP treatment was carried out at 1600°C for 1 h under argon atmosphere of 150 MPa. The specimens before and after HIP treatment were ground with a diamond wheel for mechanical testing. The tensile surfaces of the bars for the bending test were polished using diamond pastes to 1 μm finish and the edges on tensile surface were bevelled at 45°. Dimensions of the machined specimens were approximately 3 mm × 4 mm × 42 mm.

Density measurements were made using the Archimedes method by immersing in water. Fracture strength was measured by three-point bending test with a span of 30 mm at a cross-head speed of 0.5 mm/min, and high-temperature strength was measured in air. Vickers indentations were carried out for the measurement of hardness and fracture toughness. For microstructural observations by scanning electron microscopy (SEM), the polished specimens were thermally etched at 1450°C for 15 min in flowing argon. Distribution of SiC particles and structure of grain boundaries were identified by transmission electron microscopy (TEM). Specimens for TEM were thinned to electron transparency by mechanical grinding followed by ion beam milling.

3 Results and Discussion

3.1 Densification behavior

Table 1 shows relative densities before and after HIP for Al_2O_3/SiC nanocomposites sintered at 1800°C for 2h without and with sintering additives. The sintered densities of composites without additive decreased rapidly with increasing SiC content. For 10 vol% SiC nanocomposite without additive, the sintered density was found to only 89.5% of theoretical density, and the full densification was not achieved after HIP treatment. The reasons for the reduction in the densification rate of nanocomposites, as already discussed in previous works [8,12,20], may be related to the presence of the nano-sized SiC particles on the grain boundaries of Al_2O_3 matrix.

Table 1. Relative density before and after HIP for Al_2O_3/SiC nanocomposites sintered at 1800 °C for 2h without and with sintering additives

SiC content (vol%)	Sintering additive (wt%)	Sintered density (%)	HIP density (%)
1	-	99.4	99.8
5	-	94.1	99.8
	0.1% MgO	98.4	99.9
	0.1% Y_2O_3	97.4	99.9
	0.1% MgO + 0.1% Y_2O_3	99.1	99.8
10	-	89.5	92.8
	0.1% MgO	93.5	99.8

It is clear from Table 1 that the MgO sintering additive enhances remarkably the densification of Al_2O_3/SiC nanocomposites. On the contrary, Y_2O_3 was less effective as a sintering additve than MgO. The nanocomposites codoped with MgO and Y_2O_3 showed more enhanced densification than the composite doped with only MgO. Some experimental studies clearly demonstrate that monolithic Al_2O_3 can be sintered to the full density[14,21]. For Al_2O_3/SiC nanocomposites, however, the full density could not be achieved by pressureless sintering, although sintering additives were added. Therefore, HIP was employed to achieve the full densification.

In sintered ceramic materials, open pores are generally eliminated when the bulk density has reached to 93 to 95% [22,23,24], which is essential to the post

HIP process. In Al_2O_3/SiC nanocomposite, however, the open pores disappeared at the bulk density of about 91%, regardless of the SiC content or addition of sintering additive. Therefore, the full densification was achieved by the HIP treatment for the Al_2O_3/SiC specimens sintered to the bulk density of above 91%[36].

3.2 Microstructural Observation

Figure 1 shows the thermally etched microstructures typical of Al_2O_3 monolith and Al_2O_3/SiC nanocomposites after HIP treatment. The microstructures exhibit the complete densification after HIP treatment, and show that SiC particles significantly inhibit grain growth of the Al_2O_3 matrix. This effect was more pronounced for the increased volume fraction of SiC particles, according to the pinning effect comprehensively described in the literature for composite systems containing second phases [25]. This pinning effect is also known from grain-boundary morphologies. Figure 1(a) shows the microstructure of Al_2O_3 with quite straight grain boundaries. On the other hand, the grain boundaries of the nonocomposites are irregular and restrained from migrating by SiC particles as shown in Figs. 1(c), (d) and (f).

Contrary to Al_2O_3/5 vol% SiC nanocomposites showing homogenous microstructure [Figs. 1(c), (d) and (f)], the microstructure of nanocomposites doped with 0.1% Y_2O_3 typify abnormal grain growth characteristics as shown in Fig. 1(e). Therefore, MgO as a sintering additive plays an important role in Al_2O_3/SiC nanocomposites for improving densification and controlling microstructure. When only Y_2O_3 was used as a dopant, on the contrary, negative impacts were observed for the microstructure.

TEM observation of Al_2O_3/5 vol% SiC nanocomposites after HIP shows a homogeneous distribution of the SiC particles present both within Al_2O_3 grains and at grain boundaries (Fig. 2), as was found for the hot-pressed Al_2O_3/SiC nanocomposites [7,8,26,27]. This figure also exhibits the distorted grain boundary due to the pinning effect of SiC second-phase particle.

3.3 Mechanical properties

3.3.1 Fracture strength

Figure 3 shows the fracture strength after HIP for Al_2O_3 monolith and Al_2O_3/SiC nanocomposites with different SiC contents. The fracture strength increased remarkably with increasing SiC content.

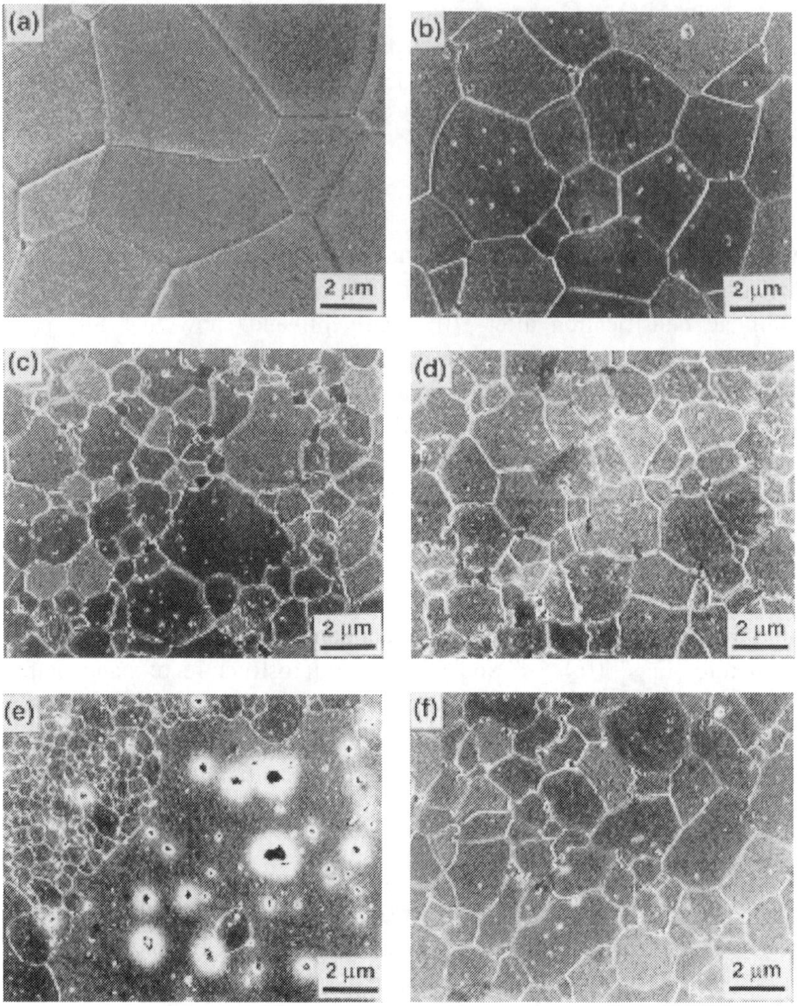

Fig. 1. SEM micrographs of thermally etched surfaces for Al_2O_3 monolith and Al_2O_3/SiC nanocomposites after HIP: (a) Al_2O_3, (b) 1 vol% SiC, (c) 5 vol% SiC, (d) 5 vol% SiC with 0.1 wt% MgO, (e) 5 vol% SiC with 0.1 wt% Y_2O_3 and (f) 5 vol% SiC with 0.1 wt% MgO + 0.1 wt% Y_2O_3 [(a) and (b): sintered at 1600 °C, (c) to (f): sintered at 1800 °C]

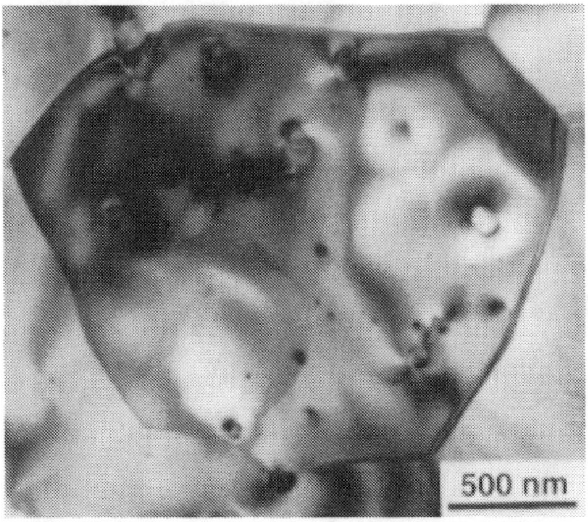

Fig. 2. TEM micrograph of Al_2O_3/5 vol% SiC with 0.1 wt% MgO nanocomposite sintered at 1800 °C and HIPed.

Figure 4 shows the fracture strength for Al_2O_3/5 vol% SiC nanocomposites without and with sintering additives after HIP. Al_2O_3/SiC nanocomposites without additive and with 0.1% MgO and with 0.1% MgO + 0.1% Y_2O_3 achieved the high fracture strength of 1 GPa, which is the same level with that obtained by the hot-pressing process [2,3,4]. Moreover, the fracture morphology was identical to that for the hot-pressed nanocomposites [7,9,10,28,29], in which the complete transgranular fracture mode was observed. The nanocomposites doped with 0.1% Y_2O_3, however, were not able to achieve high fracture strength because of the microstructural inhomogeneity such as abnormal grain growth during the pressureless sintering process, as shown in Fig. 1(e).

The notable strength improvement observed for the Al_2O_3/SiC nanocomposites can not be explained from only the moderate increase of fracture toughness. A variety of strengthening mechanisms have been proposed to explain the improved mechanical properties of nanocomposites [4,9,11,30,31,32], but none of those mechanisms has achieved general acceptance. However, it has been generally accepted that the fracture mode changes from intergranular in Al_2O_3 to transgranular in Al_2O_3/SiC nanocomposites with the presence of SiC particles, and this may be related with the strengthening of Al_2O_3/SiC nanocomposites.

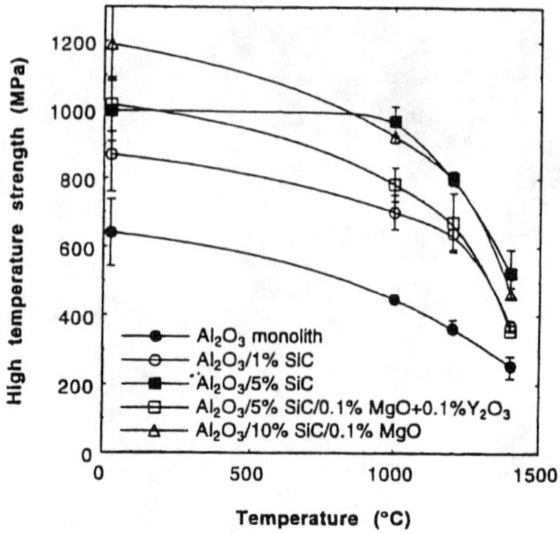

Fig. 3. Fracture strength as a function of SiC content for Al_2O_3/SiC nanocomposites after HIP.

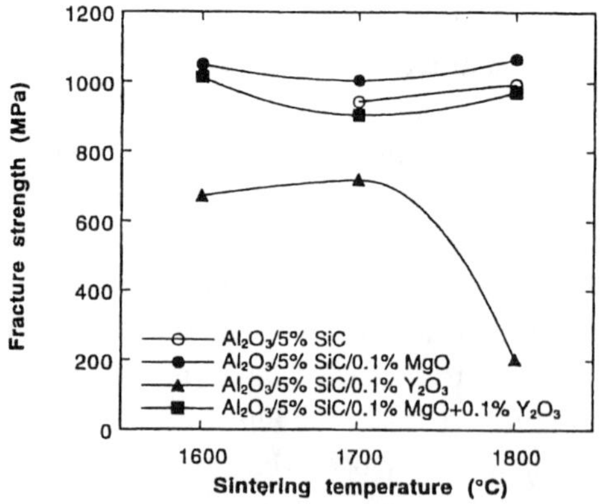

Fig. 4. Variations of the high-temperature strength for monolithic Al_2O_3 and Al_2O_3/SiC nanocomposites without and with sintering additives after HIP.

From the direct observation by HREM for Al_2O_3/5 vol% SiC nanocomposites doped with 0.1% Y_2O_3 + 0.1% MgO, it is revealed that there are no impurity phases, and the interfaces of Al_2O_3-Al_2O_3 and Al_2O_3-SiC are directly bonded. Consequently, the rigid bonding may play a role in suppressing grain boundary fractures and thus may increase the fracture strength of Al_2O_3/SiC nanocomposite.

Most elements present a low limit of solubility in Al_2O_3 and thus are segregated to grain boundaries. At high concentrations, some of them favor the formation of intergranular glassy phases that lead to striking modifications of Al_2O_3 properties [33]. For the nanocomposites doped with 0.1% Y_2O_3 + 0.1% MgO, however, we can not find out any intergranular phases. This may be attributed to the beneficial effect of MgO that is understood in terms of either a glass phase modification or a solid-liquid interface modification as suggested by Bae and Baik in Al_2O_3 monolith [34]. Clearly, further studies of grain boundary need to be done for determining a role of MgO additive.

3.3.2 High Temperature Strength

Figure 3 shows the fracture strength with temperature for Al_2O_3 monolith and Al_2O_3/SiC nanocomposites containing different SiC contents. Al_2O_3 monolith and Al_2O_3/1 vol% SiC nanocomposites without sintering additive exhibits the degradation in strength at high temperatures, whereas Al_2O_3/5 vol% SiC nanocomposites without sintering additive give the notable improvement in high temperature strength up to over 1000°C. On the other hand, Al_2O_3/10 vol% SiC nanocomposites doped with 0.1% MgO exhibit strength degradation at high temperatures, but high-temperature strength is same level as Al_2O_3/5 vol% SiC nanocomposites without sintering additive. Al_2O_3/5 vol% SiC nanocomposites doped with 0.1% MgO + 0.1% Y_2O_3 also exhibit the strength degradation at high temperatures and show lower high-temperature strength than that of nanocomposites without sintering additive.

It is well known that the grain boundary sliding and/or cavitation are responsible for the high-temperature strength degradation of the oxide ceramics [35]. For the Al_2O_3/SiC nanocomposites, SiC particles are dispersed at Al_2O_3-Al_2O_3 grain boundaries as well as within Al_2O_3 grains. Thus, it is believed that the observed increase in fracture strength of nanocomposites at high temperatures is mainly due to the prohibition of the grain boundary sliding or cavitation by SiC particles at grain boundaries. On the other hand, the decrease in high-temperature strength for the nanocomposites with sintering additives may be related to the change of grain boundary structure or the increase of grain boundary diffusion coefficient by addition of sintering additives.

4 Conclusions

Al_2O_3/SiC nanocomposites have been successfully fabricated by the pressureless sintering and subsequent HIP process. From the densification behaviors and microstructural observation, MgO sintering additive played an important role for improving densification and controlling microstructure. When only Y_2O_3 was used as a dopant, however, negative impacts were observed for the microstructure and the fracture strength.

Al_2O_3/5 vol% SiC nanocomposites without additive, with 0.1% MgO and with 0.1% MgO + 0.1% Y_2O_3 achieved the high fracture strength of 1 GPa. For nanocomposites doped with 0.1% Y_2O_3 + 0.1% MgO, the interfaces of Al_2O_3-Al_2O_3 and Al_2O_3-SiC were directly bonded. Consequently, the rigid bonding may play a role in suppressing grain boundary fractures and thus may increase the fracture strength.

The Al_2O_3/SiC nanocomposites without additive gave the notable improvement in high-temperature strength up to over 1000°C, whereas the Al_2O_3 monolith exhibited the degradation in strength at high temperature. The increase in strength at high temperature may be mainly due to the prohibition of the grain boundary sliding or cavitation by SiC particles at the grain boundaries.

References

1. Evans, A. G., Perspective on the development of high-toughness ceramics, *J. Am. Ceram. Soc.*, 73 (1990) 187-206.
2. Niihara, K., Nakahira, A., Sasaki, G. & Hirabayashi, M., Development of strong Al_2O_3/SiC composites. In *Proceedings of the International Meeting on Advanced Materials*, Vol. 4. Material Research Society, Tokyo, Japan, 1989, pp. 124-34.
3. Niihara, K. & Nakahira, A., Particle-strengthened oxide ceramic-nanocomposites. In *Advanced Structural Inorganic Composites*, Ed. P. Vincenzini. Advanced Structural Inorganic Composites, Elsevier Scientific Publishing Co., Trieste, 1990, pp. 637-64.
4. Niihara, K., New design concept of structural ceramics—ceramic nanocomposite. *J. Ceram. Soc. Jpn.*, **99** (1991) 974-82.
5. Niihara, K. & Nakahira, A., Stengthening and toughening mechanisms in nanocomposite ceramics. *Ann. Chim. (Paris)*, **16** (1991) 479-86.
6. Ohji, T., Nakahira, A., Hirano, T. & Niihara, K., Tensile creep behavior of alumina/silicon carbide nanocomposite. *J. Am. Ceram. Soc.*, **77** (1994) 3259-62.
7. Ohji, T., Hirano, T., Nakahira, A. & Niihara, K., Particle/matrix interface and its role in creep inhibition in alumina/silicon carbide nanocomposites. *J. Am. Ceram. Soc.*, **79** (1996) 33-45.

8. Borsa, C. E., Jiao, S., Todd, R. I. & Brook, R. J., Processing and properties of Al_2O_3/SiC nanocomposites. *J. Microsc.*, **177** (1995) 305-12.
9. Zhao, J., Stearns, L. C., Harmer, M. P., Chan, H. M., Miller, G. A. & Cook, R. F., Mechanical behavior of alumina-silicon carbide "Nanocomposite". *J. Am. Ceram. Soc.*, **76** (1993) 503-10.
10. Anya, C. C. & Roberts, S. G., Pressureless sintering and elastic constants of Al_2O_3-SiC nanocomposites. *J. Eur. Ceram. Soc.*, **17** (1997) 565-73.
11. Borsa, C. E., Jones, N. M. R., Brook, R. J. & Todd, R. I., Influence of processing on the microstructural development and flexure strength of Al_2O_3/SiC nanocomposites. *J. Eur. Ceram. Soc.*, **17** (1997) 865-72.
12. Walker, C. N., Borsa, C. E., T. I. Todd, Davidge, R. W. & Brook, R.J., Fabrication, characterisation and properties of alumina matrix nanocomposites. *Br. Ceram. Proc.*, **53** (1995) 249-64.
13. Coble, R. L., Sintering crystalline solids: II, Experimental test of diffusion models in porous compacts. *J. Appl. Phys.*, **32** (1961) 793-99.
14. Peelen, J. G. J., Influence of MgO on the evolution of the microstructure of Al_2O_3. *Mater. Sci. Res.*, **10** (1975) 443-53.
15. Harmer, M. P., Roberts, E. W. & Brook, R. J., Rapid sintering of pure and doped α-Al_2O_3. *Trans. Br. Ceram. Soc.*, **78** (1979) 22-25.
16. Delaunay, D., Huntz, A. M.& Lacombe, P., The Influence of yttrium on the sintering of Al_2O_3. *J. Less-Common Met.*, **70** (1980) 115-17.
17. Fang, J., Thompson, A. M., Harmer, M. P. & Chan, H. M., Effect of yttrium and lanthanum on the final-stage sintering behavior of ultrahigh-purity alumina. *J. Am. Ceram. Soc.*, **80** (1997) 2005-12.
18. Robertson, A. G., Wilkinson, D. S. & Cáceres, C. H. C., Creep and creep fracture in hot-pressed alumina. *J. Am. Ceram. Soc.*, **74** (1991) 915-21.
19. Lartigue, S. & Dupau, F., Grain boundary behavior in superplastic Mg-doped alumina with yttria codoping. *Acta Metall. Mater.*, **42** (1994) 293-302.
20. Stearns, L. C., Zhao, J. & Harmer, M. P., Processing and microstructure development in Al_2O_3-SiC 'Nanocomposites'. *J. Eur. Ceram. Soc.*, **10** (1992) 473-77.
21. Bae, S. I. & Baik, S., Critical concentration of MgO for the prevention of abnormal grain growth in alumina. *J. Am. Ceram. Soc.*, **77** (1994) 2499-504.
22. Härdtl, K. H., Gas isostatic hot pressing without molds. *Am. Ceram. Soc. Bull.*, **54** (1975) 201-207.

23. Tsukuma, K. & Shimada, M., Hot isostatic pressing of Y_2O_3-partially stabillized zirconia. *Am. Ceram. Soc. Bull.*, **64** (1985) 310-13.
24. Kwon, S.-T., Kim, D.-Y., Kang, T.-K. & Yoon, D. Y., Effect of sintering temperature on the densification of Al_2O_3. *J. Am. Ceram. Soc.*, **70** (1987) C-69-C-70.
25. Smith, C. S., Grains, phases and interphases: an interpretation of the microstructure, *Trans. Metal. Soc. AIME*, **175** (1948) 15-51.
26. Piciacchio, A., Lee, S.-H. & Messing, G. L., Processing and microstructure development in alumina-silicon carbide intragranular particulate composites. *J. Am. Ceram. Soc.*, **77** (1994) 2157-64.
27. Carroll, L., Sternitzke, M. & Derby, B., Silicon carbide particle size effects in alumina-based nanocomposites. *Acta. Mater.*, **44** (1996) 4543-53.
28. Nakahira, A. & Niihara, K., Sintering behaviors and consolidation process for Al_2O_3/SiC nanocomposites. *J. Ceram. Soc. Jpn.*, **100** (1992) 448-453.
29. Wang, J., Ponton, C. B. & Marquis, P. M., Thermal stability of Al_2O_3-5 vol% SiC nanocomposite. *J. Mater. Sci.*, **30** (1995) 321-33.
30. Thompson, A. M., Chan, H. M., Harmer, M. P. & Cook, R. F., Crack healing and stress relaxation in Al_2O_3-SiC 'nanocomposites'. *J. Am. Ceram. Soc.*, **78** (1995) 567-71.
31. Levin, I., Kaplan, W. D., Brandon, D. G.& Layyous, A. A., Effect of SiC submicrometer particle size and content on fracture toughness of alumina-SiC 'nanocomposites'. *J. Am. Ceram. Soc.*, **78** (1995) 254-56.
32. Ohji, T., Jeong, Y. K., Choa, Y. H. & Niihara, K., Strengthening and Toughening Mechanisms of Ceramic Nanocomposite. *J. Am. Ceram. Soc.* (in press)
33. Hansen, S. C.& Phillips, D. S., Grain boundary microstructures in a liquid-phase sintered alumina (α-Al_2O_3), *Philos. Mag. A*, **47** (1983) 209-34.
34. Bae, S. I. & Baik, S., Critical concentration of MgO for the prevention of abnormal grain growth in alumina, *J. Am. Ceram. Soc.*, **77** (1994) 2499-504.
35. Evans, A. G. & Blumenthal, W., High temperature failure in ceramics. In *Fracture Mechanics of Ceramics Vol. 6*, ed. R. C. Bradt, A. G. Evans, D. P. H. Hasselman & F. F. Lange. Plenum, New York, 1983, pp. 423-48.
36. Jeong, Y. K., Nakahira, A., and Niihara, K., Effects of Additives on Micro-structure and Properties of Alumina-Silicon Carbide Nanocomposites, *J. Mater. Sci. in print.*

MECHANICAL AND ELECTRICAL PROPERTIES OF 3Y-TZP/SiC AND 8YSZ/SiC NANOCOMPOSITES

Koichi Niihara, Noriko Bamba, Yong-Ho Choa, and Tohru Sekino
ISIR, Osaka University
8-1, Mihogaoka, Ibaraki, Osaka, 567-00, Japan

ABSTRACT

The nanocomposite techniques were applied in order to improve the ambient and elevated temperature mechanical properties of 3Y-TZP and 8YSZ. In 8YSZ system, the matrix grain growth was strongly inhibited by nano-sized SiC dispersoids, and then the fracture strength of the composites was remarkably improved. For instance, strength of 750 MPa was obtained by dispersing 20 vol% SiC particles. For 3Y-TZP/SiC nanocomposite system, high-temperature strength was also improved by incorporating SiC particles. The effect of SiC particles on ionic conductivity of 8YSZ was also investigated. The grain boundary conductivity decreased due to the decreasing grain size caused by the SiC dispersions, while an increase in lattice conductivity was observed by tensile residual stresses due to the thermal expansion mismatch between matrix and second phase for 8YSZ/20 vol% SiC nanocomposites.

INTRODUCTION

Pure zirconia has three polymorphisms, such as monoclinic (m-ZrO_2), tetragonal (t-ZrO_2) and cubic (c-ZrO_2)[1]. Pure ZrO_2 cannot be used for any industrial products due to the disruptive volume change accompanied to a transformation from t-ZrO_2 to m-ZrO_2 on cooling processes. However, the cubic or tetragonal structure can be stabilized by the addition of some oxides such as yttria (Y_2O_3), magnesia (MgO) or calsia (CaO). The high temperature cubic fluorite type phase can be maintained even at room temperature, dependent on the stabilizer content. The properties of each stabilized zirconia are controlled by the kind of stabilizer and its content.

In the Y_2O_3 stabilized zirconia (Y-ZrO_2) system which has been extensively studied, 2-3 mol% Y_2O_3 addition give only t-ZrO_2 and called as tetragonal zirconia polycrystal (2-3Y-TZP), while above 6 mol% Y-ZrO_2 completely stabilizes c-ZrO_2 (YSZ).

3Y-TZP is a representative structural material and has excellent mechanical properties due to the phase transformation toughening from t-ZrO_2 to m-ZrO_2[2]. However, the transformation toughening is only effective at ambient temperature. The rapid degradation of mechanical

To the extent authorized under the laws of the United States of America, all copyright interests in this publication are the property of The American Ceramic Society. Any duplication, reproduction, or republication of this publication or any part thereof, without the express written consent of The American Ceramic Society or fee paid to the Copyright Clearance Center, is prohibited.

properties occurs with increasing temperature, because the transformation toughening effect decreases at higher temperatures due to the increase in stability of t-ZrO_2[3,4]. Many attempts have been made to resolve this degradation of 3Y-TZP, for example, by dispersing second phase materials such as whiskers, platelets, fibers or particles with high Young's modulus[5-10]. In addition, composite techniques have been also applied to improve poor mechanical properties of 8 mol% Y_2O_3 fully stabilized ZrO_2 (8YSZ)[11-13], which is well-known as a solid electrolyte with high oxygen ionic conductivity over wide ranges of temperature and oxygen partial pressure[14-16]. Furthermore 8YSZ has not only poor mechanical properties, but also poor thermal conductivity. Considering some applications, not only the low mechanical properties, but also thermal imhomogeneity and thermal mismatch between electrolytes and other components of cell must be solved, because high mechanical stress is caused by those factors.

Recently, nanocomposite technique has been used to improve mechanical properties both at ambient and at elevated temperatures[17]. For example, the fracture strength[18] and creep resistance[19,20] of Al_2O_3 were improved by 3 to 5 times and by 3 to 4 orders by incorporating only 5 vol% nano-sized SiC particles, respectively. It is expected that the improvements in mechanical properties at both ambient and elevated temperatures expand the application of both 3Y-TZP and 8YSZ.

In this study, the improvements of high-temperature strength for 3Y-TZP and of mechanical/electric properties for 8YSZ were attempted by using the nanocomposite technique. Nano-sized SiC particulate was selected as a second phase, because SiC possesses high fracture strength even at elevated temperatures, high thermal shock resistance and a low coefficient of thermal expansion (CTE) which is an important factor in nanocomposites[17]. The purpose of the present work is to investigate the effects of nano-sized SiC dispersion on microstructure and mechanical properties for 3Y-TZP and 8YSZ. For 8YSZ, the effect of SiC addition on the ionic conductivity was also investigated.

EXPERIMENTAL PROCEDURE

3Y-TZP and 8YSZ with average crystalline sizes of 20 nm (Sumitomo Osaka Cement Co., Ltd., Osaka, Japan) and β-SiC with an average grain size of 150 nm (Mitsui Touatus Kagaku Ltd., Tokyo, Japan) were used as a matrix and a second phase, respectively. 3Y-TZP or 8YSZ and β-SiC (0, 5, 10 and 20 vol% with 3Y-TZP or 8YSZ content) were ball-milled in C_2H_5OH for 48 h. After drying, the mixed powders were dry ball-milled for 12 h and then were sieved under 150 μm to remove hard agglomerate particles. The powders were hot-pressed for 1 h under 30 MPa in Ar atmosphere. The hot-pressed bodies were cut and ground into rectangular bar specimens (4 x 3 x 36 mm) and polished using diamond paste with 0.5 μm to evaluate the microstructure and mechanical properties.

Density of the specimens was measured by the Archimedes method using toluene. Fracture strength was measured by the 3-point bending test (span: 30 mm, crosshead speed: 0.5 mm/min). Fracture toughness was evaluated by the indentation fracture (IF) method with 196 N load for 3Y-TZP and 49 N for 8YSZ. In the 3Y-TZP system, the samples were heated at 1300 °C for 10 min before IF measurements to remove m-ZrO_2 in polished surfaces which was introduced by machining. X-ray diffraction (XRD) analysis identified the phases in polished surface. For 3Y-TZP system, XRD analysis was also performed on fracture surfaces to assess the extent of m-ZrO_2 content. The volume fraction of m-ZrO_2, V_m, was calculated using following equation[21],

$$V_m = PX_m/(1+(P-1)X_m) \quad (1)$$

where X_m is the integrated intensity ratio and $P = 1.311$

$$X_m = (I_m(111) + I_m(\bar{1}11))/(I_m(111) + I_m(\bar{1}11) + I_t(111)) \quad (2)$$

I is the integrated intensity, and subscripts m and t indicate the m-ZrO_2 and the t-ZrO_2, respectively. The microstructure studies were performed by scanning electron microscopy (SEM) and transmission electron microscopy (TEM). The samples for SEM observation were thermally etched in Air for the monolith and in Ar atmosphere for the composites at 1300 to 1450 °C for 10 min. The grain size and its distribution were analyzed from SEM photographs by using an image analysis technique (NIH image written by Wayne Rasband at U.S. National Institutes of Health and available from the Internet by anonymous ftp from zippy.nimh.nih.gov or on floppy disk from NTIS, 5285 Port Royal Rd., Springfield, VA22161, part number PB93-504868).

Some 8YSZ and 8YSZ/SiC nanocomposite specimens were machined to platelets (5 x 5 x 0.5 mm) for measuring the impedance, Z, and the phase, θ. Impedance analysis was carried out in a frequency range of 20 Hz to 1 MHz with Ag electrodes in cooling from 802 °C to 200 °C in Ar. The grain boundary and lattice ionic conductivities were calculated by using a complex impedance plot.

RESULTS AND DISCUSSION

ZrO_2 (3Y-TZP and 8YSZ) monoliths and ZrO_2/5, 20 vol% SiC nanocomposites were fabricated by using hot-pressing technique. Hot-pressing temperatures for 3Y-TZP and 8YSZ systems were 1350 to 1850 °C and 1300 to 1900 °C, respectively. XRD analysis showed that all the composites were composed of only ZrO_2 (3Y-TZP: t-ZrO_2 and a little m-ZrO_2, 8YSZ: c-ZrO_2) and β-SiC, and free from reaction phases. Thus, the changes of microstructure and mechanical properties to be observed for ZrO_2/SiC nanocomposites should be mainly attributed to fine SiC particle dispersoids.

Mechanical Properties at Room Temperature

Generally, mechanical properties such as fracture strength and fracture toughness, greatly depend on the microstructure, for instance, the grain size. In addition, the transformation toughening also influences the properties in 3Y-TZP. At first, 8YSZ/SiC nanocomposites were focused on, in turn, 3Y-TZP/SiC nanocomposites were dealt with, because, in the case of 3Y-TZP system, an effect of transformation toughening must be taken into consideration in addition microstructure.

(a) 8YSZ/SiC Nanocomposites:

Fig. 1 shows the variation of fracture strength with hot-pressing temperature. The strength remarkably increased with SiC content and the hot-pressing temperature, and the hot-pressing temperature at which the maximum strength was observed became higher with increase in SiC content. The maximum strength of monolith was 300 MPa at 1400 °C. On the other hand, 8YSZ/20 vol% SiC nanocomposite hot-pressed at 1850 °C possessed the strength of 750 MPa which was about 2.5 times higher than that of the monolith. As mentioned before, the strength is influenced by the microstructure. The grain size distributions for composite materials that exhibited the maximum strength are demonstrated in Fig. 2, and the mean grain size is also indicated in this figure. The mean grain size of monolithic 8YSZ hot-pressed at 1400 °C was 7 μm, while those of nanocomposites dispersed 5 and 20 vol% SiC were 3 and 1 μm, respectively, even though they were hot-pressed at higher temperatures such as 1700 °C and 1850 °C. The width of grain size distribution was narrow by adding SiC particles. The grain size obviously became small with increasing SiC content. Furthermore, the homogeneity and distribution in the matrix grain size were also improved by the SiC dispersion. Those behaviors suggest strong inhibition of normal/abnormal grain growth by fine SiC particles, which should be attributed to inhibition of grain boundary diffusion and migration of matrix[22,23].

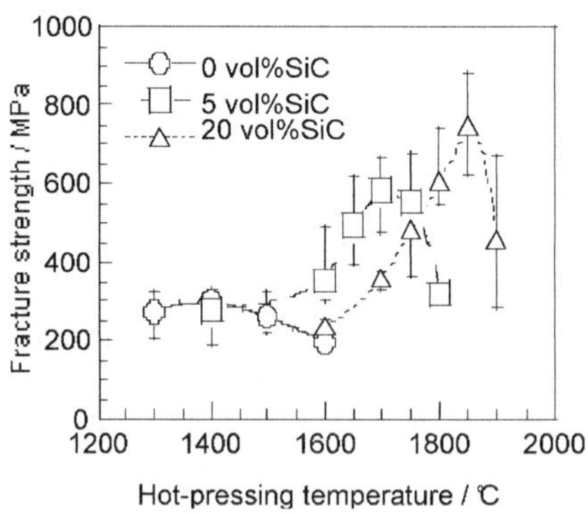

Fig. 1 The variation of fracture strength with hot-pressing temperature for 8YSZ/SiC nanocomposites.

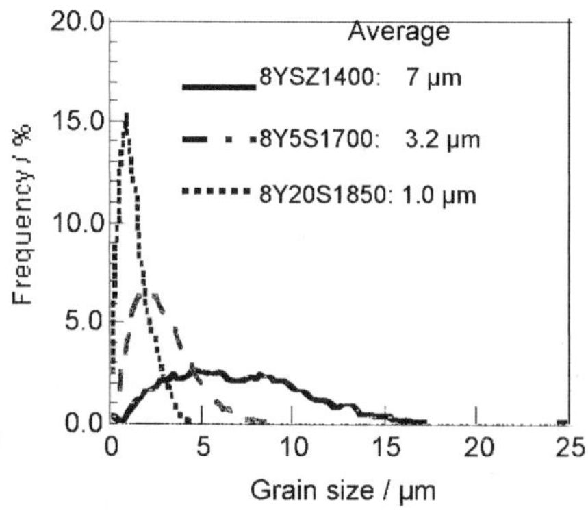

Fig. 2 Grain size distributions for 8YSZ monolith hot-pressed at 1400 °C, 8YSZ/5 vol% SiC nanocomposites at 1700 °C and 8YSZ/20 vol% SiC at 1850 °C.

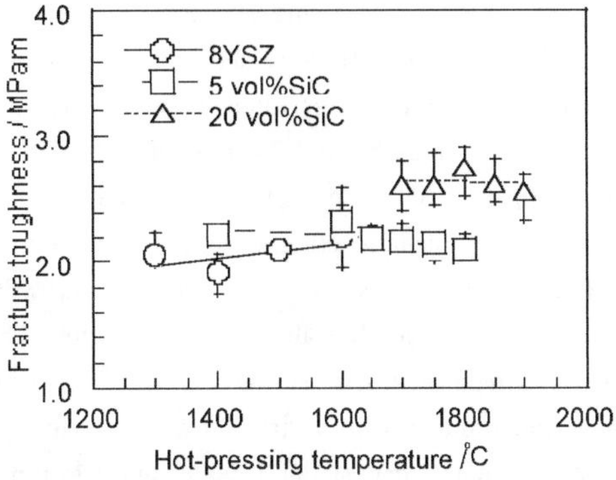

Fig. 3 The variation of fracture toughness as a function of hot-pressing temperature for 8YSZ/SiC nanocomposites.

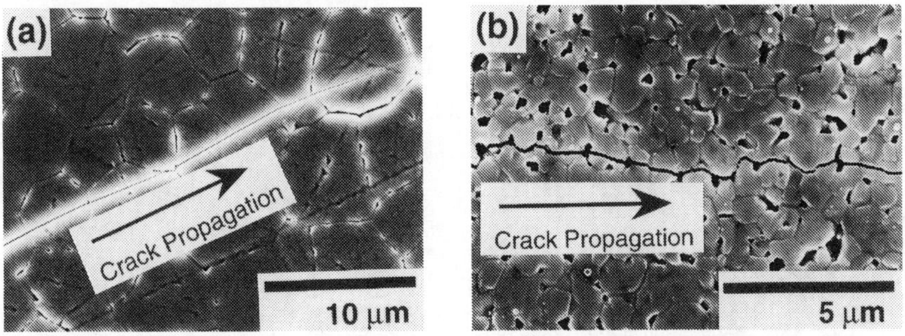

Fig. 4 SEM images of crack deflection for (a) 8YSZ monolith and (b) 8YSZ/20 vol% SiC nanocomposite.

Fine and homogeneous microstructure should make fracture origin small. Thus the strength of the composites could be expected to be remarkably improved. At low hot-pressing temperatures, the pores for the fracture origin should be remained, and then the strength was low.

Fig. 3 shows the variation of fracture toughness as a function of hot-pressing temperature. The toughness was also improved by dispersing SiC particles, which were 2.0, 2.2 and 2.6 $MPam^{1/2}$ for monolith and 5 vol% SiC and 20 vol% SiC nanocomposites, respectively. This should be because dispersed SiC particles deflected crack tip different from monolith as shown in Fig. 4.

(b) 3Y-TZP/SiC Nanocomposites:

Fig. 5 shows the fracture strength as a function of hot-pressing temperature. The maximum strength for monolith and nanocomposites was about 1.5 GPa, and little difference was obtained among them. Hot-pressing temperature for the maximum strength was 1500 °C, 1700 °C and 1750

°C for monolithic zirconia, 3Y-TZP/5 vol%SiC and 3Y-TZP/20 vol%SiC composites, respectively. 3Y-TZP monolith naturally had fine microstructure and its mean grain size was narrow, different from 8YSZ. Comparing monolith with the composites at the same hot-pressing temperature, such as 1700 °C, the mean grain size of monolith was 1.2 μm, while 5 and 20 vol% SiC nanocomposites gave 1.1 μm and 0.68 μm, respectively. SiC particles suppressed matrix grain growth at the same hot-pressing temperature. However, the large effect of inhibition of grain growth was not observed, probably because 3Y-TZP monolith does not have high grain growth rate. The SiC particles depressed not only the grain growth rate, but also the densification. Therefore, the hot-pressing temperature to have fully densified specimens became high with increasing SiC content. The grain size of specimen which exhibited the maximum strength were 0.5 μm for monolith hot-pressed at 1400 °C, 1.1 μm for 5 vol% SiC nanocomposite at 1700 °C and 0.75 μm for 20 vol% SiC at 1750 °C, respectively. In the 3Y-TZP system, thus the strength did not depend on only grain size, and then the transformation toughening effect should be also taken into consideration.

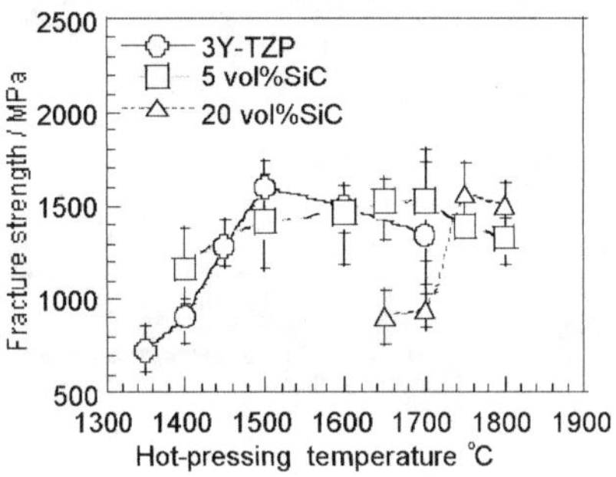

Fig. 5 The variation of fracture strength with hot-pressing temperature for 3Y-TZP /SiC composites.

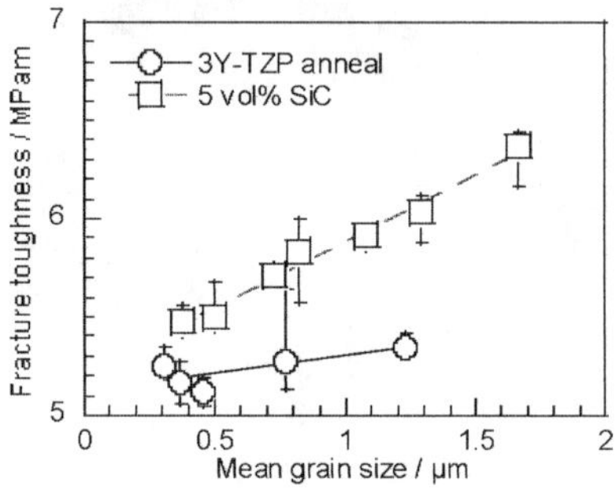

Fig. 6 The variation of fracture toughness as a function of mean grain size for 3Y-TZP and 3Y-TZP/5vol%SiC composite.

The variation of fracture toughness with the mean matrix grain size is shown in Fig. 6 for monolith and 5 vol% SiC nanocomposite. The toughness in both materials increased with increasing grain size, because transformation can occur easily when the grain size is large. The toughness increased also by dispersing SiC particles. Thus, m-ZrO_2 content on fracture surfaces was investigated by XRD method. As shown in Fig. 7, m-ZrO_2 content increased with increasing grain size and by dispersing SiC particles. Therefore, the improvement of toughness should be mainly attributed to enhanced transformation by dispersing SiC particles.

Recently, the residual stresses have been investigated for various kinds of nanocomposites[17,24,25]. In the ZrO_2/SiC composites, tensile and compressive residual stresses act for the ZrO_2 matrix and SiC particulate, respectively, as shown in Table 1.

Table I. Calculated residual stresses

	ZrO_2 matrix / MPa	SiC / MPa
Monolith	0	0
5 vol% SiC	63	1200
20 vol% SiC	250	1010

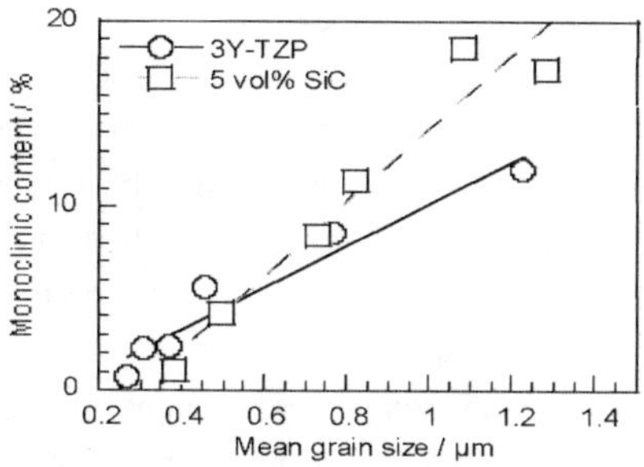

Fig. 7 m-ZrO_2 content at fracture surface as a function of mean grain size for3Y-TZP monolith and 3Y-TZP/5 vol% SiC composite.

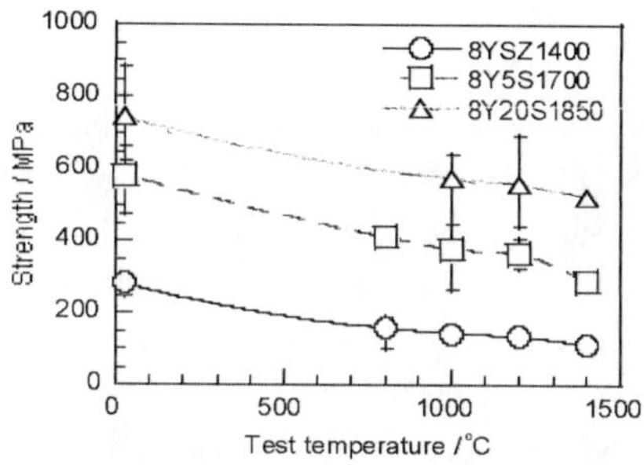

Fig. 8 The temperature dependence of fracture strength for 8YSZ monolith hot-pressed at 1400 °C, 8YSZ/5 vol% SiC composites at 1700 °C, and 8YSZ/20 vol% SiC at 1850 °C.

Those stresses were estimated by using the respective CTEs for ZrO_2 and SiC are 10×10^{-6} / K[26] and 4.45×10^{-6} / K[27], according to the model proposed by Taya *et al.*[28]. The tensile stress makes volume expansion accompanied transformation easy. In addition, defects, such as dislocation which is produced around SiC particle within matrix to reduce the high compressive stress, promotes the transformation[29]. In fact, dislocations were observed within matrix grains. Thus, the nanocomposites with high residual stresses by SiC particles exhibited the high toughness.

Mechanical Properties at High Temperatures

The samples employed for measurements of high-temperature fracture strength were as follows: 8YSZ monolith hot-pressed at 1400 °C (denoted by 8YSZ1400), 8YSZ/5 and 20 vol% SiC nanocomposites at 1700 °C and 1850 °C (8Y5S1700, 8Y20S1850), 3Y-TZP monolith at 1500 °C (3Y-TZP1500) and 3Y-TZP/5 and 20 vol% SiC nanocomposites at 1700 °C and 1750 °C (3Y5S1700 and 3Y20S1750), which exhibited the maximum strength as shown in Figs. 1 and 5

(a) 8YSZ/SiC Nanocomposites:

Fig. 8 shows the high-temperature strength for 8YSZ1400, 8Y5S1700 and 8Y20S1850. The strength for 8YSZ1400, 8Y5S1700 and 8Y20S1850 at 1000 °C in test temperature were 160 MPa, 380 MPa and 580 MPa, and thus, the strength for both monolith and composites decreased with increasing the temperature. However, the decreasing rate in strength with temperature from room temperature to 1000 °C decreased by dispersing SiC particles. And the high strength of above 500 MPa was kept at 1500 °C for the 8YSZ/20 vol%SiC nanocomposite system. These materials are expected to be used as a new structural material at high temperatures.

(b) 3Y-TZP/SiC Nanocomposites:

Fig. 9 shows the variation of fracture strength as a function of test temperature for 3Y-TZP1500, 3Y5S1700 and 3Y20S1750. These three kinds of specimens had similar strength at room temperature, about 1.5 Gpa, as shown in Fig. 5. However, the high-temperature strength for monolith and nanocomposites rapidly decreased with increasing test temperature. This decrease should be due to the decrease in transformation toughening effect. Although the monolith and composites exhibited similar strength at ambient temperature, the composites had higher strength than the monolith at elevated temperatures. On the other hand, plastic deformation behavior was observed for 3Y-TZP1500 at the temperature of 1200 °C, while the nanocomposites exhibited plastic deformation at 1400 °C. In general, plastic deformation behavior is related to the grain size, and the higher stress is necessary to be plastically deformed with increasing grain size. The grain size of the specimens was different from each other, and it became smaller in the order of 3Y5S1700, 3Y20S1750 and 3Y-TZP1500. From this result, the SiC particles are believed to suppress degradation in strength and plastic deformation.

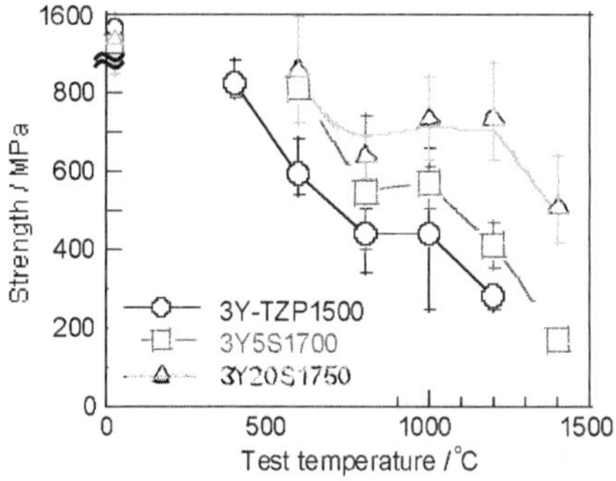

Fig. 9 The temperature dependence of fracture strength for 3Y-TZP monolith hot-pressed at 1500 °C, 3Y-TZP/5 vol% SiC nanocomposites at 1700 °C and 3Y-TZP/20 vol% SiC at 1750 °C.

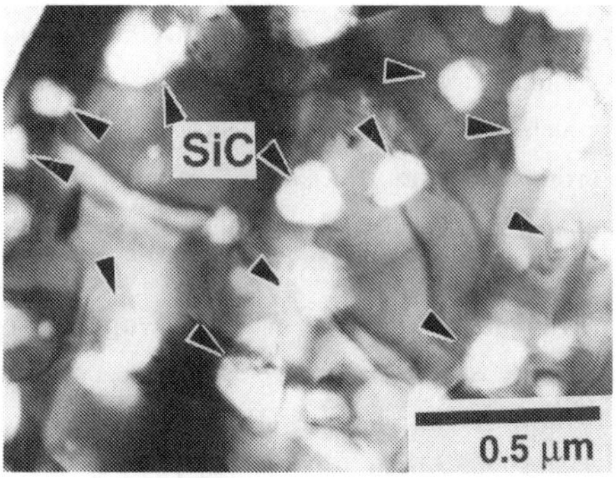

Fig. 10 TEM image of 8YSZ/20 vol% SiC nanocomposite hot-pressed at 1850 °C.

Ionic Conductivity

The ionic conductivity was measured for 8YSZ1400, 8Y5S1700 and 8Y20S1850. The measurement was also made for 8YSZ monolith hot-pressed at 1600 °C (8YSZ1600) to clarify the effects of the matrix grain size and the SiC dispersion on ionic conductivity. Electrical conductivity of SiC might be negligible in this study, because each SiC particle was isolated in matrix grains and/or at grain boundaries as shown in Fig. 10.

Firstly, the temperature dependence of grain boundary conductivity is shown in Fig. 11. The grain boundary conductivity obviously decreased in order of 8YSZ1600, 8YSZ1400, 8Y5S1700 and 8Y20S1850. The depression of the grain boundary conductivity increased with increasing SiC content. As shown in Fig. 11, some differences were observed for not only nanocomposites, but also monoliths. In general, the grain boundary resistance decreases with increasing matrix grain size[30], that is, the conductivity increases with the grain size. Therefore, the nanocomposites with fine microstructure had lower conductivity than monolith. In addition to the grain size, the effect of SiC particles located at grain boundary is considered to be another reason for the decrease in grain boundary conductivity. The effects of SiC particles may be explained as a following reason. SiC particles at grain boundary cause the ionic conduction process through the boundaries to be partially blocked. These effects are proportional to the volume of SiC particulates located at grain boundaries. In fact the content of SiC particles at grain boundaries essentially increased with the increase in the volume fraction of SiC.

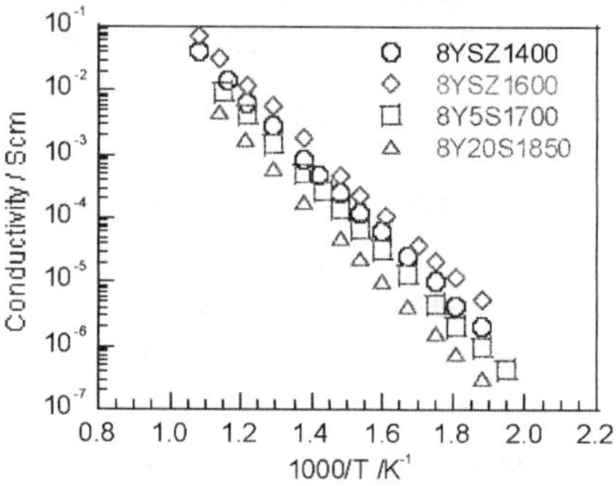

Fig. 11 The temperature dependence of grain boundary conductivity for 8YSZ monoliths hot-pressed at 1400 °C and 1600 °C, 8YSZ /5 vol%SiC nanocomposites at 1700 °C and 8YSZ/20 vol% SiC at 1850 °C.

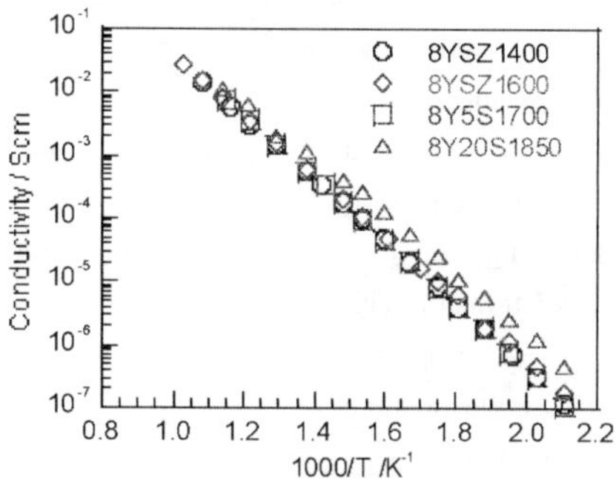

Fig. 12 The temperature dependence of lattice conductivity for 8YSZ monoliths hot-pressed at 1400 °C and 1600 °C, 8YSZ/5 vol% SiC nanocomposites at 1700 °C and 8YSZ/20 vol%SiC at 1850 °C.

The lattice ionic conductivity as a function of temperature is shown in Fig. 12. The temperature dependence of lattice conductivity for 8YSZ1400 and 8YSZ1600 seems to be almost same. 8Y5S1700 also had similar conductivity as both monoliths, in spite of incorporating SiC particulates. On the other hand, 8Y20S1850 had larger conductivity than monoliths, especially at low temperatures as shown in Fig. 12. According to Badwal et al. [31], the grain size made no effect on the lattice resistance, contrary to the grain boundary conductivity. Therefore, other factors should influence the conductivity. Generally, in oxides stabilized zirconia, the temperature dependence of conductivity depends on the diffusion coefficient of oxygen ion, which is varied by a temperature fixed solely by the activation energy required for oxygen ion mobility[31]. The increase in lattice conductivity at low temperatures for 8Y20S1850 might correspond to the increase in oxygen ion diffusion coefficient. This increase in oxygen diffusion coefficient will be explained by the residual stresses. The residual tensile stress shown in Table 1 expands the lattice spacing of 8YSZ matrix. The lattice expansion can make oxygen ion (O^{2-}) movement more easily within lattice. When the distance among zirconium ions (Zr^{4+}) or between Zr^{4+} and $Y^{3+}{}_{Zr}$ which substituted for Zr^{4+} and has a negative electric charge extended, the path for O^{2-} should become wide. The difference of lattice conductivity between 8Y5S1700 and 8Y20S1850 would be due to the amount of residual tensile stress. Compared with 8Y20S1850, the calculated residual tensile stress for 8Y5S1700 was much smaller.

CONCLUSIONS

Dispersing SiC particles gave better influences on mechanical properties of both 8YSZ and 3Y-TZP. The fracture strength at room temperature for 8YSZ system was significantly improved due to fine and homogeneous microstructure by dispersing SiC particles. 8YSZ/20 vol% SiC nanocomposite hot-pressed at 1850 °C exhibited the significantly improved strength, 750 MPa. The fracture toughness increased due to crack deflection by SiC particles. In addition, the SiC addition also greatly improved the high-temperature strength. For 3Y-TZP/SiC composites, the fracture toughness was increased without the decrease in strength, which should be attributed to enhanced transformability by the tensile residual stresses caused by the CTE mismatch between two phases. The high-temperature strength was improved, and the dispersed SiC particles suppressed plastic deformation.

Incorporating SiC particles caused the decrease in grain boundary conductivity. But the residual stress induced by the CTE mismatch between 8YSZ matrix and SiC dispersoid improved the lattice conductivity. Dispersing nano-sized second phase into the 8YSZ matrix grain gives a strong possibility to increase not only lattice conductivity but also total conductivity, when large amount of SiC particles can be dispersed within matrix grains.

ACKNOWLEGEMENTS

Appreciation is expressed to Dr. K. Ota and Dr. Y. Kobayashi for measurements of ionic conductivity and helpful discussions. Supported by the Japan Ministry of Education, Science,

Sports and Culture under Grant-in-Aid for Science Research Nos.08044150 and 09450262.

REFERENCES

[1] E.C. Subbarao, "Zirconia- an Overview"; pp. 1-24 in *Science and Technology of Zirconia Advance in Ceramics vol. 3*, Edited. by A.H. Heuer and L.W. Hobbs. Am. Ceram. Soc., Ohio, 1981.

[2] T. Masaki, "Mechanical Properties of Toughened ZrO_2-Y_2O_3 Ceramics," *J. Am. Ceram. Soc.*, **69** [8] 638-40 (1986).

[3] T. Masaki, K. Nakajima and K. Shinjo, "High Temperature Mechanical Properties of Y-PSZ HIPed Under an Oxygen-Containing Atmosphere," *J. Mater. Sci. Lett.*, **5** 1115-1118 (1986).

[4] K. Nakajima and T. Masaki, "High-Temperature Mechanical Properties of Y-PSZ Formed by Hot Isostatic Pressing in an Oxygen-Containing Atmosphere,"; pp. 625-633 in *Science and Technology of Zirconia Advance in Ceramics vol. 3*, Edited. by A. H. Heuer and L. W. Hobbs. Am. Ceram. Soc., Ohio, 1981.

[5] M. Hirano, H. Inada, "Fracture Toughness, Strength and Vickers Hardness of Yttria-Ceria-Doped Tetragonal Zirconia/Alumina Composites Fabricated by Hot Isostatic Pressing," *J. Mater. Sci.*, **27** 3511-3518 (1992).

[6] F. Meschke, G. D. Portu and N. Claussen, "Microstructure and Thermal Stability of Fine-Grained (Y, Mg)-PSZ Ceramics with Alumina Additions," *J. Eur. Ceram. Soc.*, **11** 481-486 (1993).

[7] N. Claussen, K. -L. Weisskopf and M RŸhle, "Tetragonal Zirconia Polycrystals Reinforced with SiC Whiskers," *J. Am. Ceram. Soc.*, **69** [3] 288-292 (1986).

[8] J. Hong, L. Gao, B.A. Show and D.P. Thompson, "SiC Platelet and SiC Platelet-Alumina Reinforced TZP Matrix Composites," *Br. Ceram. Transactions*, **94** [5] 201-204 (1995).

[9] S.E. Dougherty, T.G. Nieh, J. Wadsworth and Y. Akimune, "Mechanical Properties of a 20 vol% SiC Whisker-Reinforced, Yttria-Stabilized, Tetragonal Zirconia Composite at Elevated Temperature," *J. Mater. Res.*, **10** [1] 113-118 (1995).

[10] H.L. Lee and K.M. Lee, "Effect of SiC on the Mechanical Properties of 3Y-TZP/SiC Composites," *J. Mater. Sci. Lett.*, **13** 974-976 (1994).

[11] S. Terauchi, H. Takizawa, T. Endo and M. Shimada, "Ionic Conductivity and Mechanical Properties of Post-HIPed Cubic Zirconia/Alumina Composites," *Materials and Manufacturing Processes*, **9** [5] 965-974 (1994).

[12] L.M. Navarro, P. Recio, P. Duran, "Preparation and Properties Evaluation of Zirconia-Based Al_2O_3 Composites as Electrolytes for Solid Oxide Fuel Cell Systems," *J. Mater. Sci.*, **30** 1931-1960 (1995).

[13] K. Oe, K. Kikkawa, A. Kishimoto, Y. Nakamura and H. Yanagida, "Toughening of Ionic Conductive Zirconia Ceramics Utilizing a Nonlinear Effect," *Solid State Ionics*, **91** 131-136 (1996).

[14] Y. Saito, "Electrical Conduction of Stabilized Zirconia and Oxygen Sensor"; pp. 109-125 in *Zirconia Ceramics 1*, Edited by S. Somiya. Uchida Rokakuho, Tokyo, 1983.

[15] J.F. Baumard, B. Cales and A.M. Anthony, "Stabilized Zirconia : Basic Science and Applications"; pp. 127-144 in *Zirconia Ceramics 1*, Edited by S. Somiya. Uchida Rokakuho, Tokyo, 1983.

[16] M. Takeuchi, "O_2 Sensor" *Bull. Ceram. Soc. Jpn.*, **17** [6] 433-438 (1982).

[17] K. Niihara, "New Design Concept of Structural Ceramics-Ceramic Nanocomposites," *J. Ceram. Soc. Jpn.*, **99** [10] 974-982 (1991).

[18] K. Niihara, A. Nakahira, T. Uchiyama and T. Hirano, "High-Temperature Mechanical Properties of Al_2O_3-SiC Composites"; pp. 103-116 in *Fracture Mechanics of Ceramics*, vol. 7 Edited by R.C. Bradt, A.G. Evans, D.P.H. Hasselman and F.F. Lange. Plenum Press, New York, 1986.

[19] A. Nakahira and K. Niihara, "Microstructure and Fracture Bbehaviors at High Temperatures for Al_2O_3-SiC Nanocomposites"; pp. 165-178 in *Fracture Mechanics of Ceramics*, vol. 9 Edited by M. Sasaki, R.C. Bradt, D.P.H. Hasselman and D. Munz. Plenum Press, New York, 1992.

[20] T. Ohji, A. Nakahira, T. Hirano and K. Niihara, "Tensile Creep Behavior of Alumina/Silicon Carbide Nanocomposite," *J. Am. Ceram. Soc.*, **77** [12] 3259-3262 (1994).

[21] H. Toraya, M. Yoshimura and S. Somiya, "Calibration Curve for Quantitative Analysis of the Monoclinic-Tetragonal ZrO_2 System by X-Ray Diffraction," *J. Am. Ceram. Soc.*, **67** [6] C119-C121 (1984).

[22] N. Bamba, Y. H. Choa, T. Sekino, and K. Niihara, "Microstructure and Mechnical Properties of Yttria Stabilized Zirconia/Silicon Carbide Nanocomposites", *J. Euro. Ceram. Soc,* in print

[23] S. Hori, R. Kurita, M. Yoshimura and S. Somiya, "Suppressed Grain Growth in Final Stage Sintering of Al_2O_3 with Dispersed ZrO_2 Particles," *J. Mater. Sci. Let.*, **4** 1067-1070 (1985).

[24] M. Sternitzke, "Structural Ceramic Nanocomposites," *J. Eur. Ceram. Soc.*, **17** 1061-1082 (1997).

[25] Y. H. Choa, A. Nakahira, T. Ohji and K. Niihara, *Acta Materialia*, in contribution.

[26] *Cyclopedia of Fine Ceramics*, p. 595, Edited by H. Yanagida *et al.*, Gihodo Press, Tokyo, 1987.

[27] *Cyclopedia of Fine Ceramics*, p. 228, Edited by H. Yanagida *et al.*, Gihodo Press, Tokyo, 1987.

[28] M. Taya, S.Hayashi, A.S. Kobayashi and H.S. Yoon, "Toughening of a Particulate-Reinforced Ceramic-Matrix Composites by Thermal Residual Stress," *J. Am. Ceram. Soc.*, 73 [5] 1382-1391 (1990).

[29] I.W. Chen, Y.H. Chiao and K. Tsuzaki, "Statistics of Martensitic Nucleation," *Acta Metall.* **33** [10] 1847-1859 (1985).

[30] S. P. S. Badwal and J. Drennan, "Yttria-Zirconia: Effect of Microstructure on Conductivity," *J. Mater. Sci.*, **22** 3231-3239 (1987).

[31] W. D. Kingery, H. K. Bowen and D. R. Uhlmann, p. 239 in *Introduction to Ceramics*, 2nd ed. Edited by, John Wiley and Sons, NY, 1975.

MECHANICAL AND ELECTRICAL PROPERTIES OF ZnO/Ag NANOCOMPOSITES

Y. Hayashi, Y. H. Choa, J. P. Singh* and K. Niihara
The Institute of Scientific and Industrial Research , Osaka University
Ibaraki, Osaka 567-0047, Japan
*Argonne National Laboratory
Argonne, Illinois 60439, U.S.A.

ABSTRACT

Effects of Ag particle dispersions on microstructural development and some properties were investigated for ZnO/Ag nanocomposites. They were fabricated by Pulse Electric Current Sintering (PECS) Process to achieve finer and densified microstructure. ZnO/Ag nanocomposites with novel microstructure which were prepared by a reduction process using Ag_2O fine powders were compared with microcomposites prepared by mixing of Ag and ZnO powders. SEM observation indicated that fine Ag particles were homogeneously dispersed within the ZnO matrix grains and at the grain boundaries for ZnO/Ag nanocomposites prepared by the reduction process using Ag_2O fine powder. Hardness and fracture toughness increased with increasing the Ag volume fraction. Linear resistivity was decreased with increasing Ag volume fraction. However, the mechanical and electrical properties appeared to the significantly different for composites prepared by two different powder processes.

INTRODUCTION

Zinc oxide (ZnO) having wurtzite structure is a variable resistor and is used widely as a gas sensor, absorber of abnormal voltage, piezo-electric material, etc.. In these applications, ZnO is required to give better performances not only in electrical properties, but also in mechanical, chemical and electrical stability in special environments. However, ZnO displays poor mechanical properties. There are some material designs to strengthen and toughen ceramics by using composite techniques to incorporate particulate, whisker or platelet reinforcement.[1,2] Recent investigations have shown that ceramic composites having nano-sized metal particulate dispersions show excellent mechanical properties like hardness, Young's

To the extent authorized under the laws of the United States of America, all copyright interests in this publication are the property of The American Ceramic Society. Any duplication, reproduction, or republication of this publication or any part thereof, without the express written consent of The American Ceramic Society or fee paid to the Copyright Clearance Center, is prohibited.

modulus, fracture strength and toughness even at high temperatures.[3-7] In this study, therefore, the nanocomposite techniques were applied to improve the mechanical properties and electrical properties of ZnO. Ag particle was selected as a dispersion phase in this work, because Ag gives good sinterability[8] and high electrical conductivity which gives to wide applications in high voltage regions to ZnO materials.

Recently, pulse electrical current process (PECS) has received much interest as a new process for sintering various kinds of powder materials.[9,10] The process is a type of solid compression sintering process that is similar to hot-pressing (HP) because some pressure is applied during heating. In a special case, a process commercially called Spark Plasma Sintering (SPS) which enables us to sinter at considerably rapid heating rates.

The main effects for sintering enhancement in hot-pressing process are thermal diffusion and plastic flow due to high pressure. It is indicated in the PECS process that self-exothermic reactions due to electric discharge between particles at an early stage of ON-OFF d.c. pulse application enhances sintering processes, as well as effects of high pressure and rapid sintering. Recent reports support these advantages of the PECS process for obtaining highly densified and strengthened ceramic composites.[11,12]

In this study, ZnO/Ag nanocomposites were fabricated by PECS Process to achieve finer microstructure with the high density. Effects of Ag particle dispersions on microstructural development and some properties were investigated. ZnO/Ag nanocomposites with a novel microstructure which were prepared by a reduction process using Ag_2O fine powder are compared with microcomposites prepared by mixing the Ag and ZnO powders. Furthermore, we attempted to understand the role of finer Ag dispersions on microstructure evolution, and electrical and mechanical property improvements for the ZnO/Ag nanocomposites from PECS method.

EXPERIMENTAL PROCEDURE

ZnO (Sakai Chemical Industry Co. Ltd., Osaka, Japan) was used as a matrix and Ag and Ag_2O (Kojundo Chemical Lab. Co. Ltd., Tokyo, Japan) as dispersion to fabricate the composite materials. Average particle size of ZnO is 0.28μm and Ag and Ag_2O are 3μm. the powder mixture was made by ball-milling ZnO and Ag powders in ethanol. Another powder mixture was prepared from ZnO and Ag_2O powders, and this composite powder was calcined in air atmosphere to get fine Ag after ball-milled in ethanol. ZnO/Ag_2O mixture powder were evaluated by thermogravimetry (TG) / differential thermal analysis (DTA) (DTG-50, Shimadzu Co., Kyoto, Japan) and high temperature X-ray diffraction (X1, Sintag Inc., USA) in order to select the calcination temperature required to decompose Ag_2O into Ag and O_2. The mixture was packed into carbon dies, and sintered at 700 to 900 °C under an applied pressure of 40 MPa for 1 to 20 minutes in vacuum by PECS (SPS-2080, Sumitomo Coal Mining Co., Tokyo, Japan). The heating rate is 200°C / min below 600°C and is 100°C / min at 600°C to sintering temperature in PECS method. Under these sintering conditions, the maximum current value was about 2500 A and the maximum voltage value was about 3.5V. The sintered disks, 30 mm in diameter and 4.0 mm in thickness, were cut ground and polished rectangular bars 2×3×25 mm in size. The phase composition of the powders and the sintered composites was examined by X-ray diffraction (RU-200B, Rigaku Co., Tokyo, Japan). The bulk density was measured by the Archimedes method in water. Fracture toughness was estimated by the indentation fracture (IF) method using Vickers diamond indenter. To evaluate linear resistivity, In-Ga electrode was printed on both sides of specimen. The linear resistivity was measured with a digital multi-meter (Model 7522, Yokogawa Electric Co., Tokyo, Japan). The microstructure was observed with field emission scanning electron microscopy (FE-SEM) (Model S-5000, Hitachi, Japan) and transmission electron microscopy (TEM) (Model H-8100, Hitachi, Japan).

RESULTS AND DISCUSSION

Reduction of Ag_2O

Ag_2O powder was subjected to thermogravimetry (TG) / differential thermal analysis (DTA) in order to know the reduction temperature for decomposing Ag_2O into Ag and O_2 in air. In TG/DTA, weight change (at 160°C to 240°C) represented an exothermic reaction with a corresponding decrease in weight. This weight change is due to Ag_2O decomposition forming Ag and O_2. There is no reaction taking place at temperatures higher than 240 °C. Fig. 1 shows high temperature X-ray diffraction (HT-XRD) patterns for the Ag_2O powder. Pt peaks come from a Pt HT-XRD stage. For powders reduced below 150°C, Ag peaks are not observed, and only Ag_2O peaks are clearly seen. However, broader and weaker Ag peaks for powders calcined at 200°C are observed in addition to Ag_2O peaks. For powders reduced over 250 ° C, only Ag peaks are observed clearly. This observation suggests that decomposition of Ag_2O is completed at 300°C. Therefore, 300°C was selected as the reduction temperature to obtain the ZnO/Ag composite powder. Fig. 2 shows TEM images of reduction powder. After reduction treatment at 300 °C, Ag_2O particles were reduced to fine spherical Ag particles. The diameter of Ag particles was 30-70 nm.

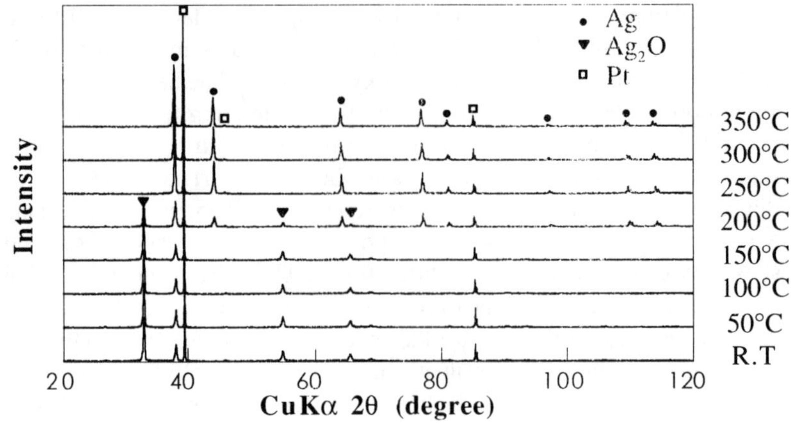

Fig.1 High temperature X-ray analysis of Ag_2O in air.

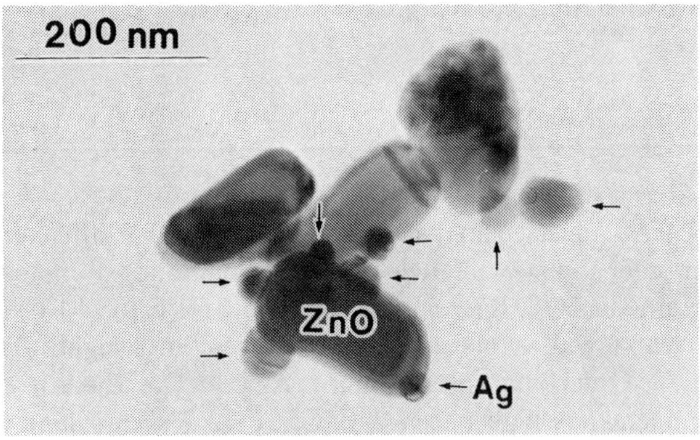

Fig.2 TEM image of ZnO/5vol% Ag reduction powder.

Microstructure of ZnO/Ag composites

Fig. 3 shows SEM micrographs of ZnO/Ag composites with varying Ag contents. The Ag particles are spherical in shape, and are mainly dispersed homogeneously at grain boundaries and triple grain junctions. It suggests that Ag particles migrated during sintering and grain growth with moving grain boundaries. It can also be seen that the Ag particle size increased with increasing Ag contents. Similar observations were also made for PZT/Ag composites.[8] It is believed that the Ag particles become soft and then agglomerates to form large particles at higher

Ag content. The matrix grain size decreases with increasing Ag content as observed in Fig. 3. Micrographs in Fig.4 show the effect of the source of Ag (Ag powder or Ag_2O powder) addition on microstructure development. It was observed that composites from the Ag_2O powder and its reduction had Ag particles primarily less than 0.3 μm, dispersed in the ZnO matrix, while the Ag particles were larger than 30 μm in composites fabricated from Ag powders. These large Ag dispersions are attributed to agglomeration during ball milling due to high ductility of Ag and to poor dispersibility due to the density differences between ZnO and Ag.

Average grain size of the ZnO matrix in ZnO/5 vol% Ag composites prepared using Ag_2O and reduction process is smaller than that of ZnO/5 vol% Ag composites prepared by conventional mixing. This is attributed to the stronger pinning effect of Ag particles on the grain boundary migration of ZnO by dispersing larger number of Ag Particles with the smaller particle size, Ag particle. Also, as expected, both families of composites showed increase in grain size with increasing sintering temperature. It can be also seen that the Ag particle size increased with increasing sintering temperature. It is believed that the Ag particles become softer at higher sintering temperatures and form large particles.

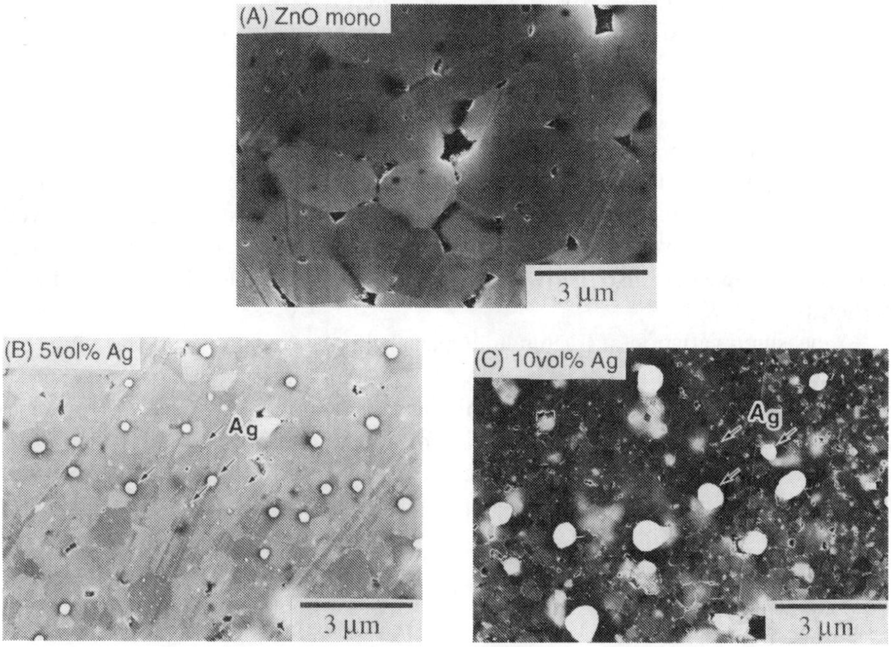

Fig.3 SEM micrographs of ZnO/ Ag composites with varying Ag contents. These composites were sintered at 800°C for 5 min in vacuum.

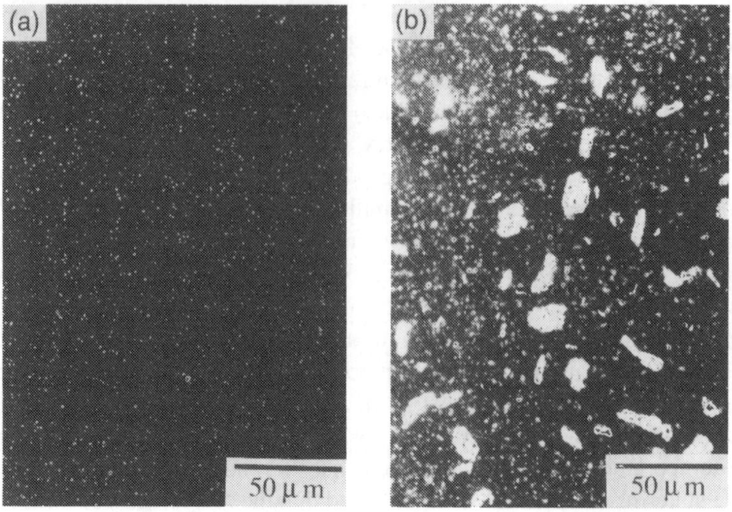

Fig.4 SEM images for ZnO/ 5vol%Ag composites prepared from Ag_2O (a) and Ag (b) starting powders. These composites were sintered at 800°C for 5 min in vacuum.

Mechanical Properties

In general, mechanical properties of ceramics and composites are greatly controlled by their microstructure. Ceramic/metal composites have been observed to display big improvements in mechanical properties.[3-6,8,13] The metal particulate dispersed composites also derive their toughness from the plasticity of metal particles.

A measured variation of fracture toughness with Ag content is shown in Fig. 5. The fracture toughness of ZnO with 10 vol% Ag additions increased to a value of ~2.77MPa$\sqrt{m}$ from a value of ~0.95 MPa$\sqrt{m}$ for monolithic ZnO. This represents an approximately three-folds increase in fracture toughness due to 10 vol% Ag additions. This increase in fracture toughness is believed to be results from crack deflection and bridging provided by Ag particle dispersions.

Fig. 6 shows the dependence of hardness on Ag content. As shown in this figure, the hardness increased with increasing Ag content. In general, the hardness of ceramics and composites is affected by density, grain size, grain boundary and second phase additions. In the present ZnO/Ag composites, the second phase Ag has high ductility and high malleability. In ZnO/Ag composites, the density increased with increasing Ag content and grain size decreased with increasing Ag content. Therefore, it is believed that the increase in hardness of ZnO/Ag composites is a result of the increasing density and the decreasing matrix grain size.

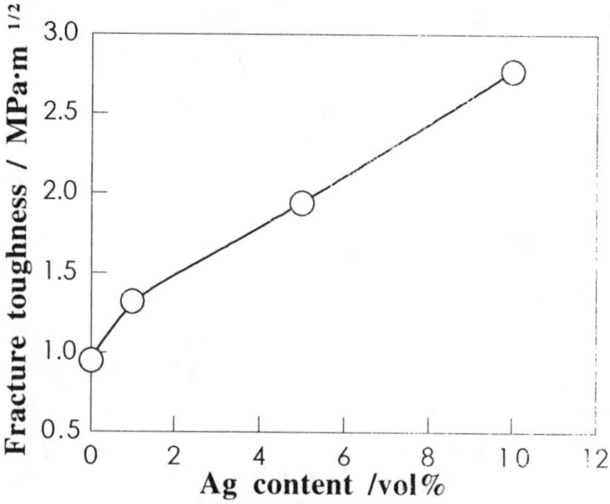

Fig.5 Variation of fracture toughness as a function of Ag content for ZnO and ZnO/Ag composites.

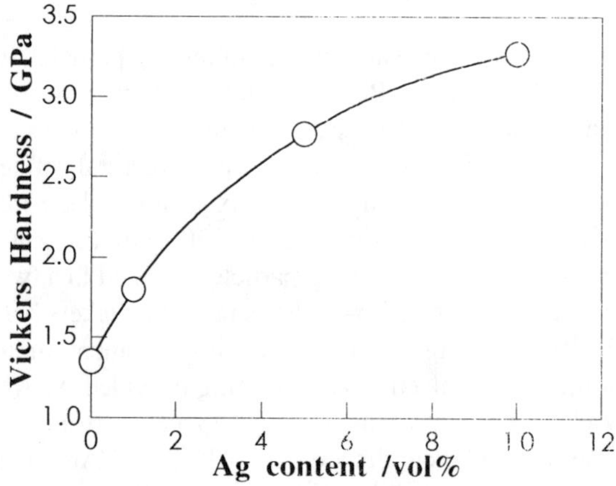

Fig.6 Variation of Vickers hardness as a function of Ag content for ZnO and ZnO/Ag composites.

Linear resistivity

Fig. 7 shows the dependence of linear resistivity on Ag content for ZnO/Ag composites. As shown in this figure, the linear resistivity decreased with increasing Ag content. The linear resistivity of the ZnO/Ag composites prepared using Ag_2O and reduction process is smaller than that of conventionally prepared. This change of linear resistivity can be explained by the effects of size and volume of dispersed Ag particles as well by an effective medium theory and percolation theory.[14,15]

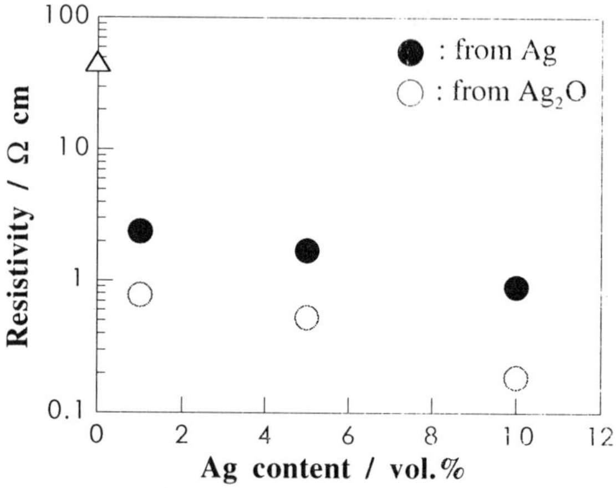

Fig.7 Variation of resistivity as a function of Ag content for ZnO and ZnO/Ag composites. These composites were sintered at 800°C for 5 min in vacuum.

SUMMARY

Two types of ZnO-based composites incorporating Ag particles were fabricated by Pulse Electric Current Sintering (PECS) method. Homogeneous dispersion of fine Ag particles was achieved using Ag_2O and reduction process. The uniform dispersion of Ag in ZnO matrix resulted in the microstructural refinement. The Ag particles were uniformly dispersed along the matrix grains and the matrix grain size decreased from ~3 μ m for monolithic ZnO to ~300 nm for ZnO/ 10vol% Ag composites. The uniform distribution of Ag particles resulted in a two-fold increase in fracture toughness. In addition, a 2.5-folds increase in Vickers hardness for ZnO was observed with 10 vol% Ag addition. The improvement in toughness was mainly caused from the crack deflection and bridging provided by Ag particles, and the increment in Vickers hardness was believed to cause from the matrix grain refinement by the fine metal particle dispersion. Also, ZnO/Ag composite resulted in substantial increase in linear resistivity with increasing Ag content.

REFERENCES

[1]C. Fan and M. N. Rahaman,"Factors Controlling the Sintering of Ceramic Particulate Composites: I , Conventional Processing," Journal of the American Ceramic Society, **75** [8] 2056-65 (1992)

[2]Y. Nakada and T. Kimura,"Effects of Shape and Size of Inclusions on the Sintering of ZnO-ZrO_2 Composites," Journal of the American Ceramic Society, **80** [2] 401-06 (1997)

[3]T. Sekino, T. Nakajima, S. Ueda, and K. Niihara,"Reduction and Sintering of a Nickel-Dispersed-Alumina Composite and Its Properties," Journal of the American Ceramic Society, **80** [5] 1139-48 (1997)

[4]T. Sekino and K. Niihara, "Microstructural Characteristics and Mechanical Properties for Al_2O_3/Metal Nanocomposites," Nanostructured Materials, **6** [5-8] 663-66 (1995)

[5]T. Sekino, T. Nakajima and K. Niihara, "Mechanical and Magnetic Properties of Nickel Dispersed Alumina-Based Nanocomposite," Material Letters, **29**, Nos.1-3, 165-69 (1996)

[6]M. Nawa, K. Yamazaki, T. Sekino, K.Niihara, "Microstructure and Mechanical Behavior of 3Y-TZP/Mo Nanocomposites Possessing a Novel Interpenetrated Intragranular Nicrostructure," Journal of Material Science, **31** [11] 2849-858 (1996)

[7]K. Niihara, "New design Concept of Structural Ceramics -Ceramic Nanocomposites-," Journal of the Ceramic Society of Japan, **99** [10] 974-82, (1991)

[8]H. J. Hwang, K. Watari, M. Sando, M. Toriyama and K. Niihara, "Low-temparature Sintering and High-Strength Pb(Zr,Ti)O_3-Matrix Composites Incorporating Silver Particles," Journal of the American Ceramic Society, **80** [3] 791-93 (1997)

[9]M. Tomota,"Mechanism of Spark Plasma Sintering"; pp. 69-76 in Microwave, Plasma and Thermochemical Processing of Advanced Materials, Edited by S. Miyake and M. Samandi. JWIR Osaka University, Osaka, 1997.

[10]M. Omori and T. Hirai,"Materials and Composite Formation by Spark Plasma System" ; pp. 50-55 in Microwave, Plasma and Thermochemical Processing of Advanced Materials, Edited by S. Miyake and M. Samandi. JWIR Osaka University, Osaka, 1997.

[11]N. Tamari, T. Tanaka, K.Tanaka, M. Kawahara and M. Tokita,"Effect of Spark Plasma Sintering on densification and Mechanical Properties of Silicon Carbide," Journal of the Ceramic Society of Japan, **103** [7] 740-42 (1995)

[12]T. Nishimura, M. Mitoki, H. Hirotsuru, M. Kawahara,"Fabrication of Silicon Nitride Nano-ceramics by Spark Plasma Sintering," Journal of Materials Science Letters, **14** [15] 1046-047 (1995)

[13]H. Hyuga, Y. Hayashi, T. Sekino and K. Niihara,"Fabrication Process and Electrical Properties of $BaTiO_3$/Ni Nanocomposites." Journal of Nanostructured Materials, **9** [1-8] 547-550 (1997)

[14]S. Kirkpatrick, "Percolation and Conduction," Reviews of Modern Physics, **45** [4] 574-88 (1973)

[15]R. Landauer,"The Electrical Resistance of Binary Metallic Mixtures," Jounal of Applied Physics, **23** [7] 779-84 (1952)

MECHANICAL PROPERTIES AND THERMAL SHOCK RESISTANCE OF SiO_2-Si_3N_4 COMPOSITE MATERIALS

Li Baoshun, Yao Junjie and Zhang Yufeng
Shanghai Institute of Ceramics, Chinese Academy of Sciences
1295 Dingxi Road
Shanghai 200050, P. R. China
Huang Xueren
Chinese Academy Launching Vehicles Technology
P. O. Box 9200-76
Beijing 100076, P. R. China

ABSTRACT

SiO_2-Si_3N_4 composites were obtained by hot pressing at 1350℃ for 0.5h. Flexural strength and fracture toughness of the composites were 143MPa and 1.7MPa·m$^{1/2}$ respectively. The composites were heated at 1000℃ for 0.5h and rapidly quenched in water, then residual flexural strength of the composites measured. The composites gave excellent thermal shock resistance. Toughening mechanism of the composites is discussed.

INTRODUCTION

SiO_2 glass is one of candidates for antenna window materials because of its excellent dielectric properties, outstanding thermal shock resistance and low density. Since 1956 researchers at Georgia Institute of Technology have thoroughly investigated fused slip-casting SiO_2 and successfully applied SiO_2 as antenna window material[1]. However, the lower mechanical strength and poor rain erosion resistance make it difficult for SiO_2 material to withstand the severe conditions experienced by high-speed vehicles. Therefore, it is intended to toughen the fused SiO_2 by introducing SiO_2 and Al_2O_3 fibers[2]. However, because of the poor combination between the fibers and the matrix, the mechanical properties of SiO_2 matrix material can not be improved by introducing fibers.

According to multi-phase ceramic theory[3,4], flexural strength and fracture toughness of the matrix material can be enhanced by introducing second-phase particles. The main toughening mechanism is the difference in thermal expansion

To the extent authorized under the laws of the United States of America, all copyright interests in this publication are the property of The American Ceramic Society. Any duplication, reproduction, or republication of this publication or any part thereof, without the express written consent of The American Ceramic Society or fee paid to the Copyright Clearance Center, is prohibited.

coefficient and elastic modulus between the particles and the matrix giving rise to residual stress fields around the particles, resulting in microcrack deflection[5]. In this paper Si_3N_4 particles with high elastic modulus, high thermal expansion coefficient and good dielectric properties were introduced into the SiO_2 matrix material as the second phase. It was expected that flexural strength and fracture toughness of the SiO_2 matrix material would be significantly improved when simultaneously retaining excellent dielectric properties and thermal shock resistance of the SiO_2 matrix material[6].

EXPERIMENTAL

The SiO_2-Si_3N_4 composites, with various concentrations of Si_3N_4 of 0, 5, 10, 20, 30, 40, and 50vol% respectively, were prepared from the fused SiO_2 and Si_3N_4 particles as starting materials. The chemical compositions of these starting materials are presented in Table Ⅰ and Table Ⅱ. XRD analysis showed that the SiO_2 was amorphous with an average diameter of 3.4μm. Si_3N_4 was mainly composed of α-Si_3N_4, accoMPanying with minor amount of β-Si_3N_4, and the average diameter 1.7μm. The composites were prepared according to the following procedure. First, the batch with various contents of Si_3N_4 were ball-mixed for 24h using distilled water as ball media. The resultant mixtures were then dried. Finally the mixtures were hot-pressed in an N_2 atmosphere at 1250, 1300 and 1350℃ with 20MPa hot pressure for o.5hr. after sifting.

Table Ⅰ. Chemical analysis of SiO_2

Composition	SiO_2	Fe_2O_3	K_2O	CaO	MgO	Na_2O	Al_2O_3
Content (wt%)	99.49	0.014	0.035	0.004	0.007	0.017	0.30

Table Ⅱ. Chemical analysis of Si_3N_4

Composition	Si_3N_4	Si_{free}	C_{free}	SiO_2	Fe_2O_3	MgO
Content (wt%)	94.35	0.07	0.19	3.15	0.049	0.011

CHARACTERIZATION

Mechanical properties of the composites were determined by three-point bend test using an Instron 1195 universal test machine. The sample sizes were 2.5×5×30mm for both flexural strength and fracture toughness. The support span was 20mm and the loading speed was $0.25mm.min^{-1}$ for flexural strength testing and $0.05mm.min^{-1}$ for fracture toughness testing. The composites, after being heat-treated at 610, 810, 1010 and 1210℃ for 0.5hr., were quenched by immersion into cool water. Residual flexural strength of the composites after drying were measured by three-point bend tests. The crystalline phase of the composites was examined by X-ray diffractometry (Model RAX-10). The bulk density of the composites was determined by Archimede method. SEM (Model EPMA-8705QH II) and TEM (Model JEM-200CXJ) analysis were employed to observe microstructure and fractograph of the composites. The Vickers indentation crack propagation morphology was observed from SEM analysis.

RESULT AND DISCUSSION

Sintering and phase compositions

Fig.1 presents XRD patterns of some composites. It can be seen that a minor amount of β-cristobalite formed in SiO_2 matrix material hot-pressed at 1250℃ (Fig.1a), while a large amount appeared when hot-pressed at 1350℃ (Fig.1b). When adding 10vol% Si_3N_4 into SiO_2 matrix material, the composite contained β-cristobalite when hot-pressed at 1350℃ (Fig.1c). Further additions of Si_3N_4 particles, up to 30vol%, showed that a trace amount of β-cristobalite was detected (Fig.1d). Therefore, it was concluded that the precipitation of cristobalite can be significantly reduced by adding Si_3N_4 particles. The formation of cristobalite depends on many factors such as the presence of other forms of quartz, impurities, atmosphere, holding time and temperature. The desired concentration of cristobalite precipitated was 12vol%[2].

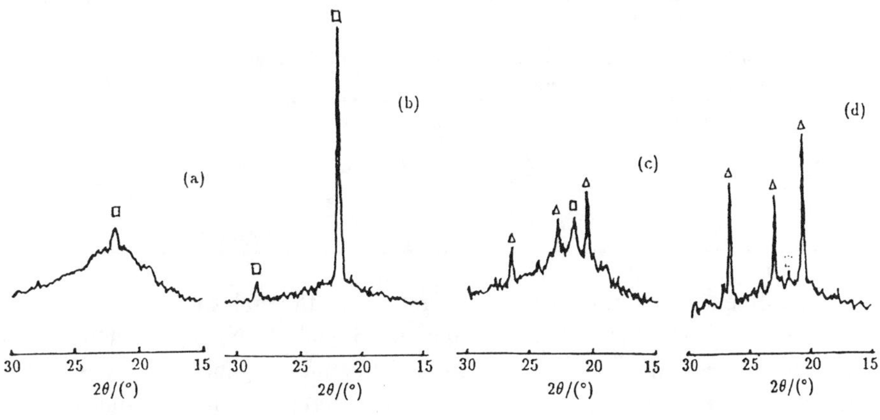

Fig. 1. XRD patterns of some composites
(a) SiO_2 hot-pressed at 1250℃; (b) SiO_2 hot-pressed at 1350℃; (c) SiO_2-10vol%Si_3N_4 hot-pressed at 1350℃; (d) SiO_2-30vol%Si_3N_4 hot-pressed at 1350℃
Legend: □: β-cristobalite; Δ: α-Si_3N_4

In addition, it can also be found from XRD patterns that the peak corresponding to Si_2N_2O did not appear, indicating that the composites were composed of only SiO_2 and Si_3N_4 without formation of Si_2N_2O. The absence of Si_2N_2O showed that SiO_2 and Si_3N_4 could retain their distinctive chemical properties with good chemical compatibility[7].

The dependence of relative density of the composite hot pressed at 1350℃ on the concentration of Si_3N_4 is illustrated in Fig.2, indicating that the relative density of the composite decreased with increasing Si_3N_4 concentration. Reason for this might be that the sintering activation energy increased with a second phase, impeding the densification process of the composites.

Mechanical properties and microstructure

The influence of hot pressing temperature on flexural strength and fracture toughness of the SiO_2 matrix material is showed in Table III. The data shows that flexural strength decreases with increasing hot-pressing temperature, which may be due to the appearance of microcracks caused by cristobalite precipitation. The appearance of microcracks could reduce the flexural strength, but improve fracture toughness because of its toughening mechanism. Thus, fracture toughness reaches a maximum value when hot-pressed at 1300℃. However at 1350℃ the excessive microcracks resulting from large amounts of cristobalite precipitated (perhaps exceeding the desired value 12vol%) may give rise to cleavage mechanisms inside the composite, causing both flexural strength and fracture toughness to decrease.

Table III. Flexural strength and fracture toughness of SiO_2 matrix material vs. hot-pressing temperature

Hot-pressing temperature (℃)	σ (MPa)	K_{IC} (MPa·m$^{1/2}$)
1250	69±8	1±0.1
1300	43±3	1.2±0
1350	34±3	0.8±0

The influence of hot-pressing temperature on flexural strength and fracture toughness of SiO_2-Si_3N_4 composites containing 30vol% Si_3N_4 is presented in Table Ⅳ. It can be seen that flexural strength and fracture toughness are considerably enhanced with increasing hot-pressing temperature. The matrix material was reinforced and strengthened by adding the Si_3N_4 particles, and the precipitation of β-cristobalite suppressed and reduced below the desired value. However, at higher hot-pressing temperatures, much more cristobalite would probably precipitate; in addition, SiO_2 could react with Si_3N_4 to produce Si_2N_2O, which would lead to the incoMPatibility between SiO_2 and Si_3N_4. Further studies regarding Si_2N_2O on mechanical properties and dielectric properties are in process.

Table IV. Flexural strength and fracture toughness of SiO_2-30vol%Si_3N_4 composite vs. hot-pressing temperature

Hot-pressing temperature (℃)	σ (MPa)	K_{IC} ($MPa{\cdot}m^{1/2}$)
1250	77±6	0.7±0
1300	117±12	1.4±0
1350	143±9	1.7±0

The effect of concentration of Si_3N_4 particles on flexural strength and fracture toughness of the composite hot-pressed at 1350℃ is illustrated in Fig.3. Flexural

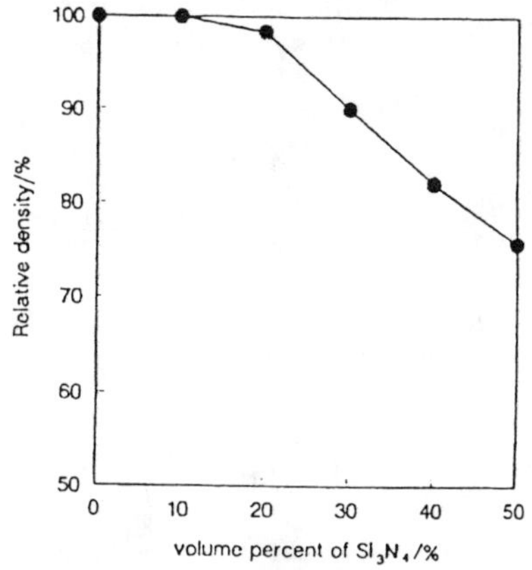

Fig. 2. Relative density of composites hot pressed at 1350℃ vs. vol% Si_3N_4

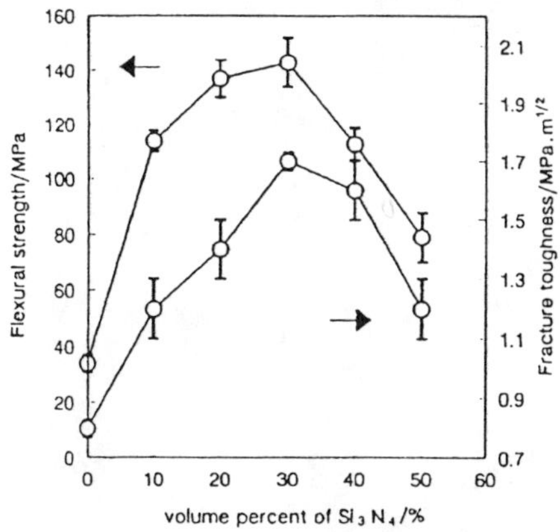

Fig. 3. Flexural strength and fracture toughness of SiO_2-Si_3N_4 composites hot-pressed at 1350℃ vs. vol%Si_3N_4

strength and fracture toughness increases with increasing amount of Si_3N_4. When adding 30vol% Si_3N_4, flexural strength and fracture toughness of the composite reaches the maximum value of 143MPa and $1.7MPa{\cdot}m^{1/2}$ respectively. These values are equal to 4.2 and 2.5 times greater than those of SiO_2 matrix material. However further additions of Si_3N_4 into the SiO_2 matrix material show the flexural strength and fracture toughness to decrease. This might be due to the high concentration of Si_3N_4 makes it impractical for the composites to completely sinter.

Fractographs of some composites are showed in Fig.4. It can be observed that the SiO_2 matrix material hot pressed at 1250℃ appears smooth (Fig.4.a) and the densification process completes. However, when introducing Si_3N_4, the cross-section of the composites became more rough with increasing Si_3N_4 concentration (Fig.4b, 4c, and 4d). A reason for this might be that microcracks would deflect along Si_3N_4 particles when fracturing, and the more the addition of Si_3N_4 particles, the more the microcracks would deflect allowing the cross-section to take on the gradient forms. Therefore, during fracture, the length of microcracks increases and more energy is consumed resulting in fracture toughness enhancement.

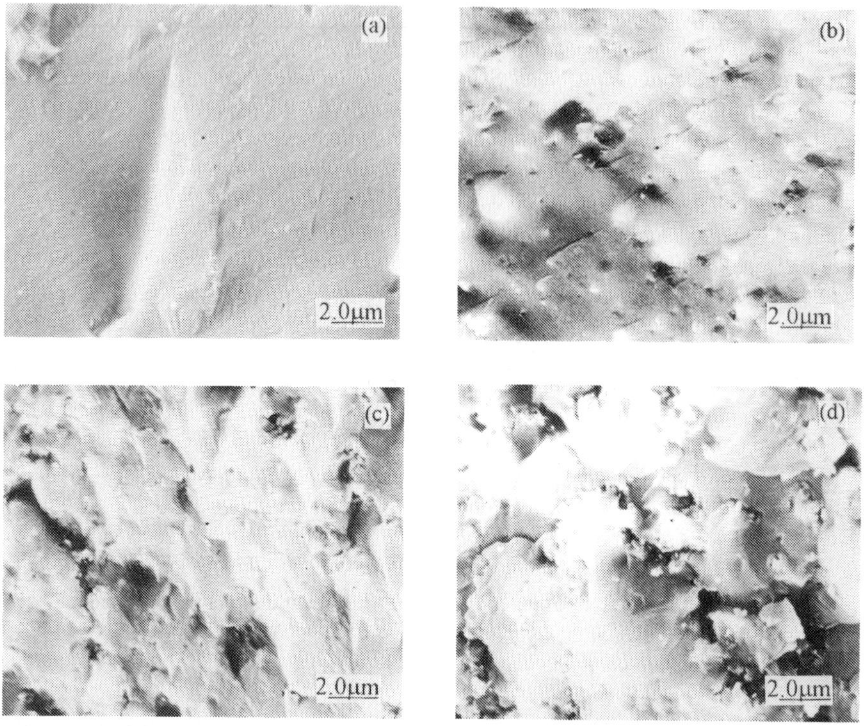

Fig. 4. SEM fractographs of composites
(a)SiO_2 sintered at 1250℃; (b)SiO_2-10vol%Si_3N_4 sintered at 1350℃;
(c)SiO_2-20vol%Si_3N_4 sintered at 1350℃;(d)SiO_2-30vol%Si_3N_4 sintered at 1350℃

Fig.5 shows TEM micrographs of the composites. For the composite with 30vol% Si_3N_4 hot-pressed at 1250℃, as illustrated in Fig.5a, SiO_2 region exists alone and are separated from each other by Si_3N_4 particles. The diameter of SiO_2 is about 6μm and larger than the starting size. Si_3N_4 particles are dispersed around the SiO_2, which make it difficult for SiO_2 to form the continuous phase because of the impractical agglomeration process. Thus, there exists a large amount of porosity in the composite resulting in lower the degree of densification along with poor flexural strength and fracture toughness. When hot-pressed at 1350℃(Fig.5b) the amorphous SiO_2 forms a continuous phase resulting in the enhancement of density. In this case, although Si_3N_4 particles are not homogeneously dispersed in the SiO_2 matrix and there is still some porosity, flexural strength and fracture toughness are enhanced.

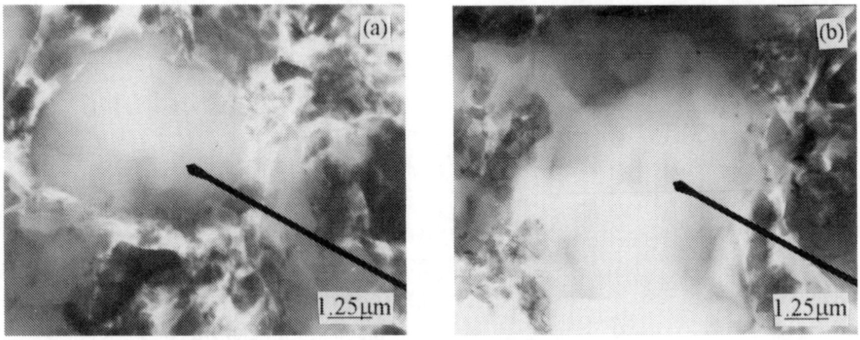

Fig. 5. TEM micrographs of composites (indication is SiO_2)
(a) SiO_2-30vol%Si_3N_4 sintered at 1250℃;(b)SiO_2-30vol%Si_3N_4 sintered at 1350℃

Thermal shock resistance

The change in flexural strength of the SiO_2 matrix material and SiO_2-Si_3N_4 composites after thermal shock are presented in Fig.6. Flexural strength of composites with 5 and 10vol% Si_3N_4 are higher than those of the SiO_2 matrix material. When quenching in water, residual flexural strengths were still high. The change in flexural strength for all composites vs. quenching temperature was analogous. When the temperature difference ΔT was below 1000℃, flexural strength decreases slightly; however, when the ΔT is above 1000℃, flexural strength reduces abruptly. This suggests that the critical quenching temperature of the composites is 1000℃, which means that the 5 and 10vol% Si_3N_4 composites have the same excellent thermal shock resistance as the pure SiO_2 matrix material.

Thermal shock resistance of the composites depends on two main factors: thermal expansion coefficient and thermal conductivity coefficient. For the composite with 5vol% Si_3N_4, thermal expansion and conductivity coefficients were nearly consistent with those of the SiO_2 matrix material; while thermal expansion and conductivity coefficients of the composite doped with 10vol% Si_3N_4 increased the same way compared with those of the SiO_2 matrix material. As a result thermal shock resistance has not perceptibly changed for the composites with various amounts of Si_3N_4. In addition, thermal shock resistance damage parameter R^{IV} was directly proportional to the square of fracture toughness and inversely proportional to the square of the flexural strength. In this case, because flexural strength and fracture toughness improved to the same level due to the addition of the second phase Si_3N_4, the SiO_2-Si_3N_4 composites could retain the same thermal shock resistance with the SiO_2 matrix material.

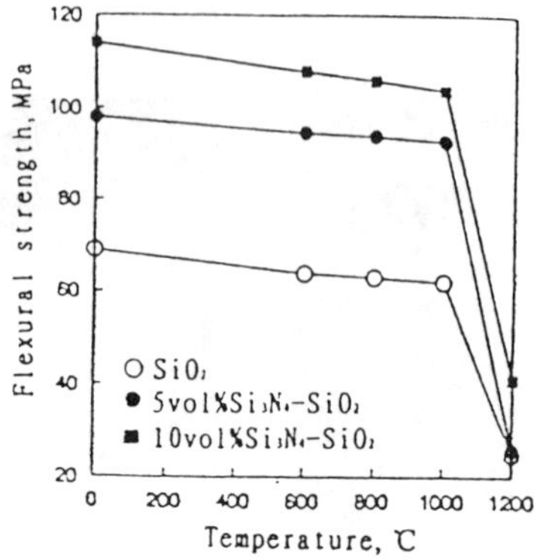

Fig. 6. Flexural strength after quenching of the pure SiO_2 matrix and the SiO_2-Si_3N_4 composites vs. quenching temperature

Fig.7 shows SEM fractographs of the composites after thermal shock. For the case of the SiO_2 matrix material without thermal shock (Fig.7a) the cross-section appeared smooth with no microcracks. When ΔT=1000℃(Fig.7b) some smaller microcracks were formed, which did not cast much effect on flexural strength of the matrix material. However, at ΔT=1200℃(Fig.7c), the microstructure has changed and small microcracks caused by thermal stress had coalesced into large microcracks which ultimately give rise to the decrease in flexural strength. For the case of the composite with 10vol% Si_3N_4(Fig.7d), the cross-section of the composite seems rougher than that of the SiO_2 matrix material because of the Si_3N_4 particles. There was not much difference of the cross-section when ΔT=0℃ and ΔT=1000 ℃ (Fig.7e). However, at ΔT=1200 ℃ (Fig.7f), the microstructure has changed i.e., some microcracks formed resulting in the decrease of flexural strength.

Toughening mechanism analysis

When the composites were cooled from hot-pressing temperature to room temperature, a draw stress is produced inside the Si_3N_4 particles because the thermal expansion coefficient, α_P, of the Si_3N_4 particles is larger than that of the SiO_2 matrix[8],

$$P = \Delta\alpha\Delta T/[(1+\nu_m)/2E_m+(1-2\nu_p)/E_p] \quad (1)$$

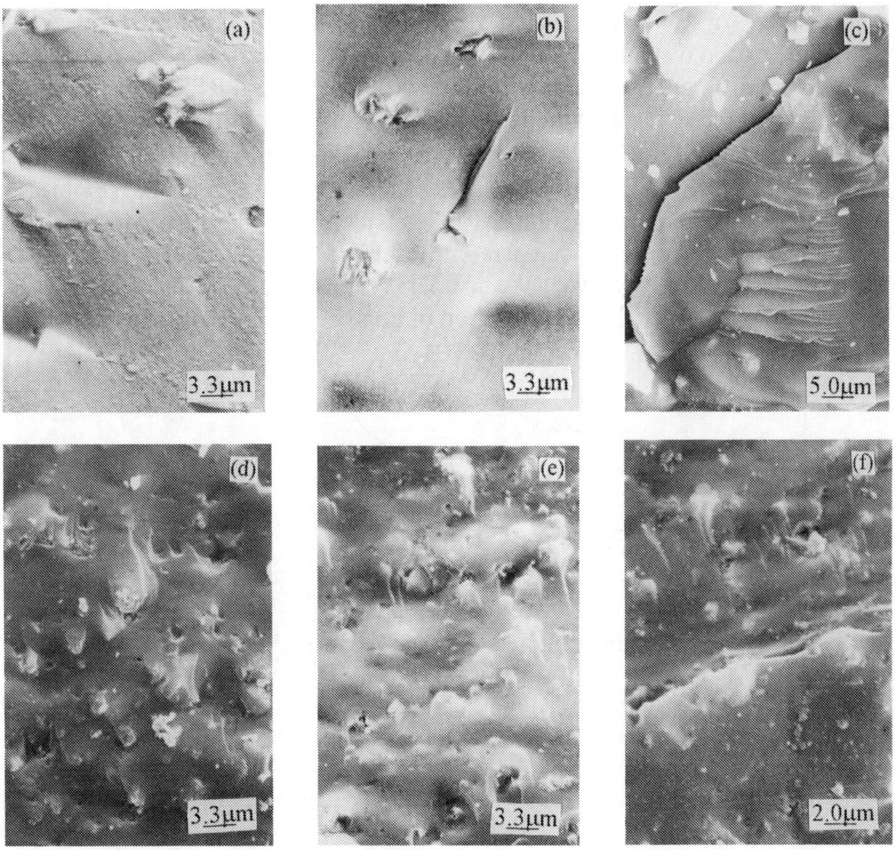

Fig. 7. SEM fractographs of composites after quenching
(a) Unshocked SiO_2 matrix material; (b) SiO_2 matrix material after 1000℃ thermal shock; (c) SiO_2 matrix material after 1200 ℃ thermal shock; (d) Unshocked SiO_2-10vol%Si_3N_4 composites; (e) SiO_2-10vol%Si_3N_4 composites after 1000 ℃ thermal shock; (f) SiO_2-10vol%Si_3N_4 composites after 1200℃ thermal shock

Meanwhile, a radius draw stress, σ_r, and a tangential compressed stress, σ_f, are also be produced in the SiO_2 matrix,

$$\sigma_r = P(r/R)^3 \tag{2}$$

$$\sigma_f = -1/2\ P(r/R)^3 \tag{3}$$

where:$\Delta\alpha=\alpha_p-\alpha_m$. v, E are Poisson ratio and elastic modulus respectively. Subscripts m and p stand for SiO_2 matrix and Si_3N_4 particles respectively; ΔT was temperature difference; r is the radius of Si_3N_4 particles; and R is the distance from some point of the stress field to the center of Si_3N_4 particles. According to equation (2) and the data in Table Ⅳ, P was 276MPa on the assumption that ΔT was equal to 1200℃, indicating that a radius draw stress of 276MPa and a tangential compressed stress of 138MPa would be produced in one side of the interface. Moreover, the as-produced stresses would dampen with triple of the ratio of r to R. Thus, the microcracks would deflect when the microcracks propagated around the stress field as illustrated in Fig.8[9]. The deflection of microcracks increases the expanding

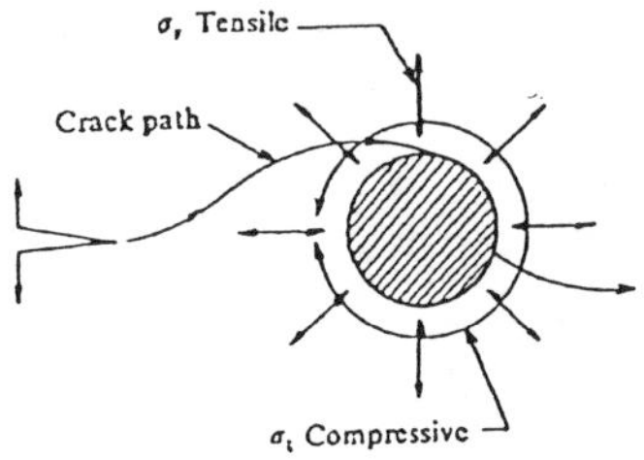

Fig. 8. Crack deflection by residual stress in particulate composites (α_p-α_m)

path which improves the expanding hindrance resulting in fracture toughness enhancement. Fig.9 shows SEM micrographs of Vickers indentation crack propagation of composites. It can be seen that, for the 30vol% Si_3N_4 composite(Fig.9b), the microcracks deflect when encountering with Si_3N_4 particles;

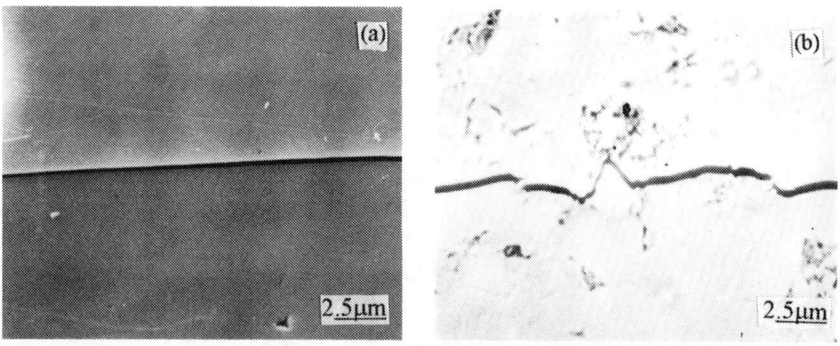

Fig. 9. SEM micrograph of Vickers indentation crack propagation of composites
(a) SiO_2 hot-pressed at 1250℃; (b) SiO_2-30vol% Si_3N_4 hot-pressed at 1350℃

while deflection of microcracks has not taken place in the SiO_2 matrix material(Fig.9a).

Microcrack deflection resulting from the interaction between microcracks and the surrounding stress field is different from that directly caused by the interaction between the microcracks and particles, in that the toughness values of the latter increased by 12-15% [5]. When the SiO_2 matrix material is heat-treated at 1250℃, small amounts of cristobalite precipitate and fracture toughness was measured as 1MPa·m$^{1/2}$. After adding Si_3N_4 particles, the maximum fracture toughness of the composites was 1.7MPa·m$^{1/2}$, thereby increasing 70%, indicating that microcrack deflection caused by the residual stress field is the main toughening mechanism.

Another reason for improvement of mechanical properties is the difference in physical properties such as elastic modulus and thermal expansion coefficient between the adding agent and the matrix. Since the elastic modulus E_p of the adding agent, Si_3N_4 particles, is several times greater than that of the SiO_2 matrix E_m, the composites could withstand more load causing the flexural strength to increase.

In addition, a compressive stress is formed in the SiO_2 matrix during cooling due to the difference in thermal expansion coefficient. When the composites are loaded the external tensile force, the compressive stress will firstly be relieved so that the presence of the compressive stress raises the flexural strength of the composites.

CONCLUSIONS

The mechanical properties of SiO_2-Si_3N_4 composites were strongly affected by Si_3N_4 particles and hot-pressing temperature. Flexural strength and fracture toughness of the composite with 30vol% Si_3N_4 hot-pressed at 1350℃ were 143MPa and 1.7 MPa·m$^{1/2}$. The precipitation of cristobalite in the composites could be reduced by introducing Si_3N_4. SiO_2-Si_3N_4 composites retained the excellent thermal shock resistance of the SiO_2 matrix material. The experimental results showed that microcrack deflection induced by the residual stress field created from the thermal expansion mismatch of Si_3N_4 particles and the SiO_2 matrix was the main toughening mechanism. Introduction of Si_3N_4 particles with high elastic modulus into the SiO_2 matrix and the resultant compressive stresses induced in the SiO_2 matrix can significantly enhance flexural strength of the composites.

REFERENCES

[1]J. D. Walton, Jr., "Reaction Sintered Silicon Nitride for High Temperature Radome Applications," *American Ceramic Society Bulletin*, 53[3] 255-8 (1974).

[2]F. P. Meyer, *"Proceedings of the 15th Symposium on Electromagnetic Windows*," pp.170, Georgia Institute of Technology, Atlanta, Georgia, 1980.

[3]D. L. Jiang and J. K. Guo, "Multiphase Ceramics," *Journal of the Chinese Ceramics Society*, 19[3] 258-68 (1991).

[4]G. Pezzotti and T. J. Nishida, "Effect of Size and Morphology of Particulate SiC Dispersions on Fracture Behaviour of Si_3N_4 without Sintering Aids," *Journal of Material Science*, 29[7] 1765-72 (1994).

[5]M. Taya, S. Hayachi, A. S. Kobayashi and H. S. Yoon, "Toughening of a Particulate-Reinforced Ceramic-Matrix Composite by Thermal Residual Stress," *Journal of American Ceramic Society*, 73[5] 1382-91 (1990).

[6]J. J. Yao, B. S. Li, X. X. Huang and J. K. Guo, "Research on Dielectric Properties of SiO_2-Si_3N_4 Antenna Window Material," *Journal of Functional Materials and Devices*, 2[2] 65-70 (1996).

[7]W. Y. Shun, W. L. Li and D. S Yang, "Control of Phase Compositions in o-β Multiphase Sialon Ceramic," *Scientia Sinica (series A)*, [10] 1093-1100 (1987).

[8]R. W. Davidge, *"Mechanical Behaviour of Ceramics*," pp.86, Cambridge University Press, Cambridge, 1979.

[9]G. C. Wei and P. F. Becher, "Improvements in Mechanical Properties in SiC by the Addition of TiC Particles," *Journal of American Ceramic Society*, 67[8] 571-74 (1984).

REACTION-BONDED MULLITE-ZrO_2 COMPOSITES WITH OXIDE ADDITIVES

E.Rocha-Rangel[a], H.Balmori-Ramírez[a] and S.D.De la Torre[b]
[a] Department of Metallurgical Engineering, ESIQIE-IPN, A.P. 75-872, 07300 Mexico City, México
[b] Centro de Investigación en Materiales Avanzados, CIMAV, CP 31109, Chihuahua, México

ABSTRACT

Highly densified mullite-ZrO_2, mullite-ZrO_2-Al_2MgO_4 and mullite-ZrO_2-Al_2TiO_5 were prepared by reaction-bonding a mixture of $ZrSiO_4$, Al_2O_3, and Al with MgO or TiO_2. In contrast to the conventional mullite, these materials display high density, substantial strengthening and toughening, particularly in the composites with TiO_2 additions. The mechanical properties and the microstructure evolution of these composites are discussed.

INTRODUCTION

Mullite has attracted the scientists attention due to its good physical properties. However, it is difficult to obtain dense mullite bodies without a long treatment at high sintering temperatures (>1700^{0}C) [1, 2]. Except for the above disadvantage, mullite also shows low fracture toughness. Dispersing ZrO_2 particles has been used to toughen mullite [2-4]. Dense mullite-ZrO_2 has been tried by electro-fusion [5], sintering [2], reaction sintering [3] and reaction bonding [6] methods. In this work, the reaction-bonding method is modified by introducing the sintering additives such as MgO and TiO_2 [8], to produce mullite-ZrO_2 (MZ), mullite-ZrO_2-Al_2MgO_4 (MZM) and mullite-ZrO_2-Al_2TiO_5 (MZT). The related reactions are listed as follows:

a) Oxidation of aluminum [6]:

$$4Al + 3O_2 \rightarrow 2Al_2O_3 \quad (1)$$

b) Decomposition of zircon [7]:

$$ZrSiO_4 \rightarrow ZrO_2 + SiO_2 \quad (2)$$

c) Formation of mullite-zirconia composites:

$$3Al_2O_3 + ZrO_2 + 2SiO_2 \rightarrow \{3Al_2O_3 \cdot 2SiO_2\} + ZrO_2 \quad (3)$$

d) Reactions from the sintering additives MgO and TiO_2 [8]:

$$2ZrSiO_4 + 3Al_2O_3 + x(Al_2O_3 + MgO) \rightarrow \{3Al_2O_3 \cdot 2SiO_2\} + 2ZrO_2 + xMgOAl_2O_4 \quad (4)$$

$$2ZrSiO_4 + 3Al_2O_3 + x(Al_2O_3 + TiO_2) \rightarrow \{3Al_2O_3 \cdot 2SiO_2\} + 2ZrO_2 + xAl_2TiO_5 \quad (5)$$

To the extent authorized under the laws of the United States of America, all copyright interests in this publication are the property of The American Ceramic Society. Any duplication, reproduction, or republication of this publication or any part thereof, without the express written consent of The American Ceramic Society or fee paid to the Copyright Clearance Center, is prohibited.

EXPERIMENTAL PROCEDURE

Powder mixtures of $ZrSiO_4$ (64%), Al (18.5%) and Al_2O_3 (17.5%) with or without the sintering additives (1.8% MgO or 7% TiO_2) were attrition-milled at 400 rpm for 0.5, 6 and 12 h in an stainless steel container. The container was filled up with 125 ml of isopropyl alcohol and 3 kg of 3 mm diameter zirconia balls. The ball powder ratio was fixed at 36:1. After milling and drying in air at 70^{0}C, the powder was configured into cylindrical specimens (20 mm diameter x 3 mm thickness) by uniaxially pressing at 300 MPa followed by cold isostatically pressing at 400 MPa. The green samples were heated up in air by according to the heating cycles depicted in Fig. 1. Specific surface area of the milled powder was measured. Phase evolution and microstructure were characterized by X-ray diffraction (XRD) and by scanning electron microscopy (SEM), respectively. Density was measured by the Archimedes method. Indentation-fracture toughness was measured at 500 N [9]. 3-point bend strength was determined in 18x4x3 mm^3 specimens (14 mm load-point span). Samples for mechanical testing were polished down to 0.3 μm with alumina and then were annealed at 1000°C for 1h.

RESULTS AND DISCUSSION

Sintering and Phase Identification

The densities of sintered samples are shown in Fig. 2. Sintered density increases with the milling time, although milling for more than 6 h only produces small differences. Milling is believed to be beneficial for densification because it increases the powder reactivity by reducing the particle size, breaking the agglomerates, and increasing the specific surface area, to produce diffusion for better consolidation of the bodies.

The XRD patterns of samples sintered at 1550°C for 2 h without sintering additives are shown in Fig. 3. The XRD patterns of powders milled for 6 and 12 h show diffraction peaks of mullite and ZrO_2 (monoclinic and tetragonal), as well as zircon and traces of α-Al_2O_3. It indicates that reactions (2) and (3) are not terminated even at these high temperatures. However, the formation of mullite-ZrO_2 is nearly completed for the powders milled for 12 h because the mullite and the

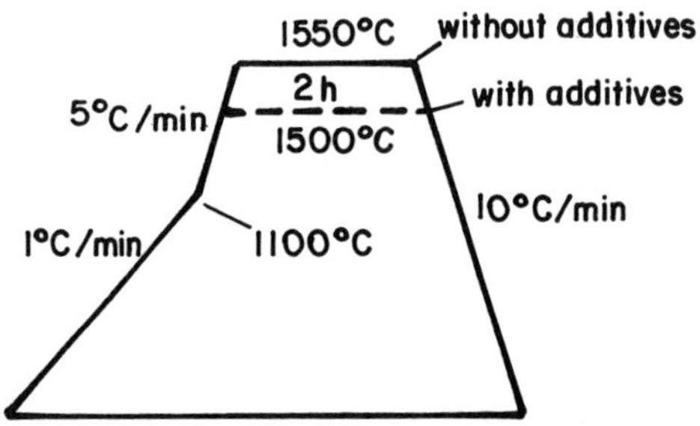

Figure 1. Heating cycle of samples.

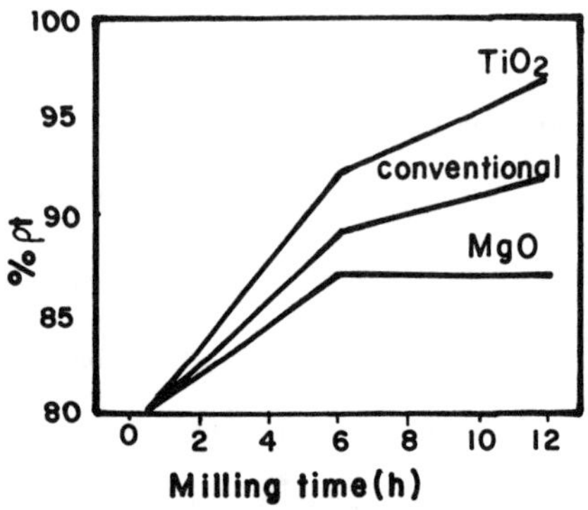

Figure 2. Relative density of sintered samples as a function of milling time

ZrO_2 intensity peaks are higher, whereas the zircon intensity peaks are smaller with the traces of α-Al_2O_3. The XRD pattern for the powder milled for 0.5 h only shows traces of mullite and zirconia. This pattern does not show any Al signals, but from SEM micrographs, it shows surface of the sintered sample containing some Al gray spots.

In Fig. 4 the XRD patterns of samples with 1.8 % MgO are shown. In the powders milled for 6 and 12 h, zircon is decomposed but mullite is formed because practically the presence of zircon is not observed. A higher amount of tetragonal-ZrO_2 (45.6 %) is retained in these samples than in the samples without additives (7.01 %). Diffusion of MgO into the ZrO_2 precipitates is responsible for this difference. The presence of two diffraction peaks of spinel Al_2MgO_4 at $2\theta = 37^0$ and 45.2° indicate that the reaction (4) has also occurred.

In Fig. 5 the XRD patterns of samples prepared with 7 % of TiO_2 are shown. In the powders milled for 6 and 12 h, reactions (2) to (4) were completed as it is evident from the presence of mullite and zirconia, while the alumina and zircon peaks are not present. Here zirconia is mainly in the monoclinic form. The formation of Al_2TiO_5 is referred to the diffraction peak at 2θ=34°, which implicates that reaction (5) has been developed. On the other hand, in the powder milled for 0.5 h without additives and with MgO, reactions (1) to (3) are only in the incipient stage.

From figures 3-5, it indicates reactions (1) to (5) are completed as the milling time increases. The powder homogeneity is improved at longer milling times. As particle size reduces, the diffusion mechanism enhances. In samples with MgO additions the content of t-ZrO_2 is higher than the other two cases. It is probably because MgO acts as the stabilizer of the t-ZrO_2 phase[11].

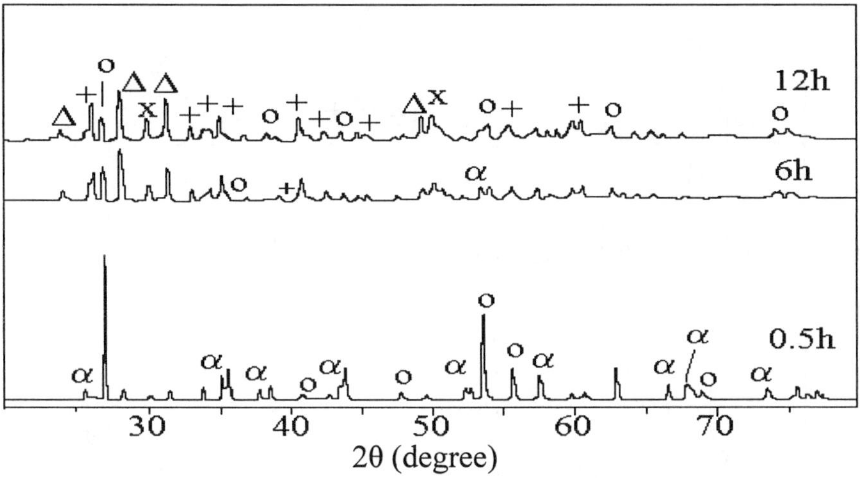

Figure 3. XRD patterns of sample without sintering additives, sintered at 1550 ^{0}C for 2h. o: zircon, +: mullite, Δ: monoclinic ZrO_2, x: tetragonal ZrO_2, α: α-Al_2O_3.

Microstructure

The microstructure of three different samples without additives, with MgO, and with TiO_2, are presented in figure 6. These powders have been milled for 12 h. Zirconia grains (white) are homogeneously dispersed in the mullite matrix (dark). Samples with Mg0 (Fig. 6b) disclose

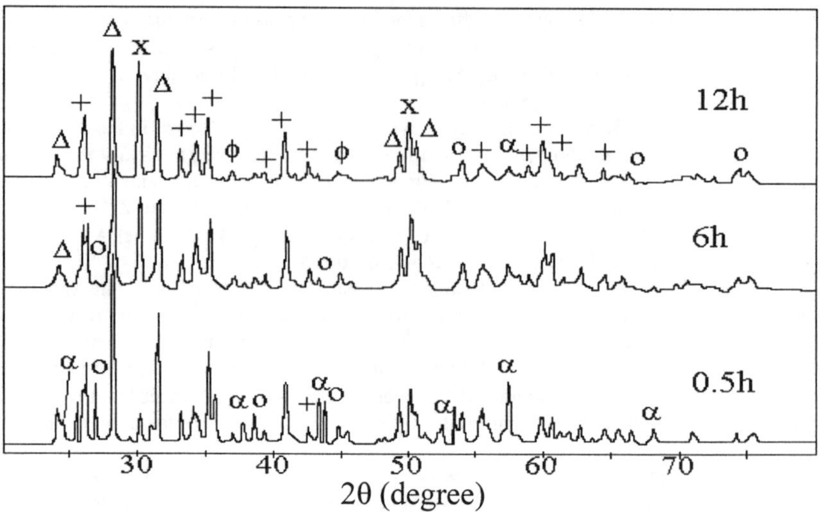

Figure 4. XRD patterns of samples with addition of 1.8% of MgO sintered at 1500^{0}C, 2h. ϕ: Al_2MgO_4, 0: $ZrSiO_4$, +: mullite, Δ: monoclinic ZrO_2, x:: tetragonal ZrO_2, α: α-Al_2O_3.

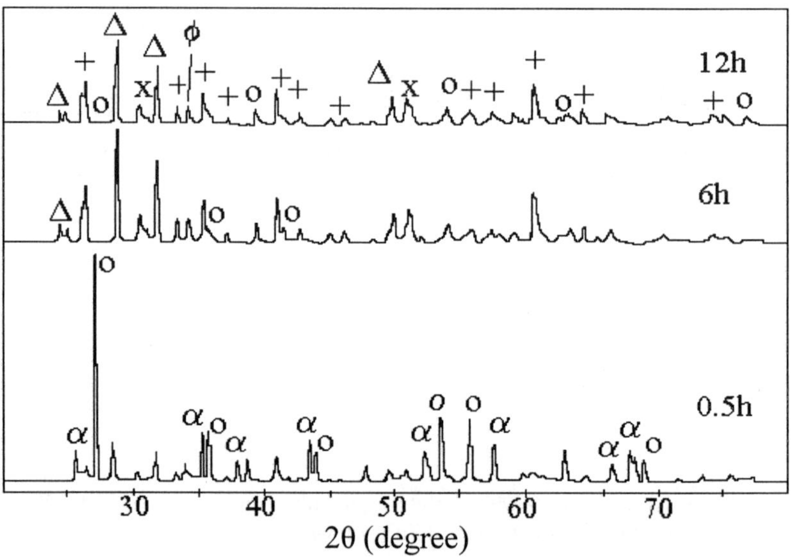

Figure 5. XRD patterns of samples with addition of 7% TiO_2 sintered at 1500^{0}C, 2h. ϕ: Al_2TiO_5, 0: $ZrSiO_4$, +: mullite, Δ: monoclinic ZrO_2, x:: tetragonal ZrO_2, α: α-Al_2O_3.

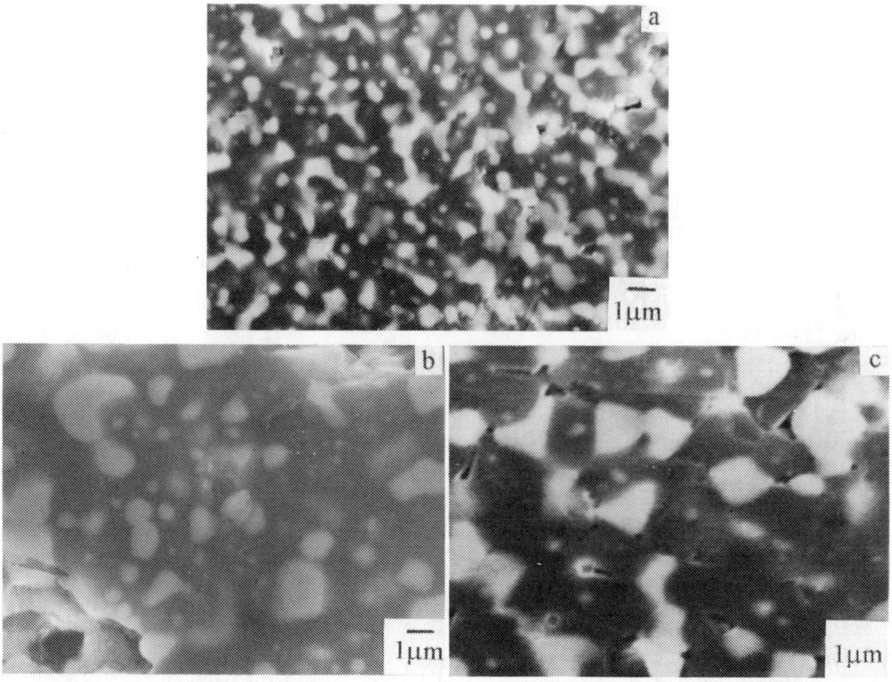

Figure 6. Microstructure of reaction-bonded samples a) Mullite-ZrO_2, b) Mullite-ZrO_2-MgO and c) Mullite-ZrO_2-TiO_2

larger porosity. The ZrO_2 precipitates in samples with MgO (Fig. 6b) and TiO_2 (Fig. 6c) look coarser than the sample without additives (Fig. 6a). Materials with TiO_2 are also characterized by the presence of few ZrO_2 agglomerates consisting of joined particles. In figure 6a it is possible to see one phase more (light grey) that corresponds to zircon. Some porosity is also present in that sample. The ZrO_2 particle size is > 1 μm in all the materials. It is larger than the critical particle size for t → m ZrO_2. This is the reason why ZrO_2 is mainly in the monoclinic phase. As mentioned above, in the materials containing MgO, the tetragonal phase has been partly stabilized because the ZrO_2 particles contain MgO.

Mechanical Properties

Mechanical properties of the reaction-bonded materials are listed in Table 1. From Table 1, it is evident that these mechanical properties increase with milling time. As the milling time increases the finer and denser sintered microstructures are obtained. On the other hand, the use of sintering additives leads to better densification of the composites via formation of a transitory liquid phase[8] which contributes to the increment of mechanical properties. Although MgO- ceramics have higher t-ZrO_2 content than TiO_2 ceramics, the properties in the latter are better. Therefore, the toughening mechanism in the mullite-ZrO_2-TiO_2 materials is needed to further study. The higher density obtained in these samples can be one reason for the higher toughness values measured. Moya [12] also suggested that microcracking rather than the stress induced martensitic transformation of t-ZrO_2 might help to explain the increases in K_{IC} of mullite-ZrO_2-Al_2TiO_5 composites.

Table 1. Mechanical properties of reaction-bonded materials.

Sample	HV (GPa)	E (GPa)	K_{IC} ($MPa \bullet m^{1/2}$)	MOR (MPa)
MZ (6h)	518	177	2.29	276
MZ (12h)	1047	276	3.29	305
MZM (6h)	440	101	4.03	163
MM (12h)	574	137	4.40	209
MZT (6h)	1143	212	4.14	363
MZT (12h)	1194	241	4.54	372

MZ: Sample without additives; MZM: Sample with MgO; MZT: Sample with TiO_2
Milling time in parenthesis.

CONCLUSIONS

1. Longer milling times lead to faster reaction-bonding and finer mullite-ZrO_2 microstructures.
2. Additions of TiO_2 and MgO enhance the reaction decomposition rate of zircon.
3. The addition of titania enhances the densification and improves the mechanical properties of reaction-bonded mullite-ZrO_2 composites.

ACKNOWLDEGEMENTS

E. R.R. thanks UAM and IPN for financial support. Work supported by IPN (DEPI-962503).

REFERENCES

1. M.D.Sacks and J.A.Pask, "Sintering of Mullite-Containing Materials: Effect of Composition", J. Am. Ceram. Soc., 65 (1982)65-70.
2. J.S. Moya and M.I. Osendi, "Effect of ZrO_2 (ss) in Mullite on the Sintering and Mechanical Properties of Mullite/ZrO_2 Composites", J. Mater. Sci. Lett., 2(1983)599-601.
3. T. Koyama, S.Hayashi, A. Yasumori and K.Okada, "Microstructure Properties of Mullite/Zirconia Composites Prepared From Alumina and Zircon Under Various Firing Conditions", J. Eur. Ceram. Soc., 16(1996)231-237.
4. A. Leriche, "Mechanical Properties and Microstructures of Mullite-Zirconia Composites", pp. 541-552, Ceram. Trans., Vol. 6, Mullite and Mullite Matrix Composites, 1991.
5. S. Prochazka, J.S. Wallace and N. Claussen, "Microstructure of Sintered/Mullite Zirconia Composites", J. Am. Ceram. Soc., 66(1983)C-125.
6. N. Claussen, S. Wu and D. Holz, "Reaction Bonding of Aluminum Oxide (RBAO) Composites: Processing, Reaction Mechanisms and Properties", J. Eur. Ceram. Soc., 14(1994)97-109.
7. E. Di Rupo and M.R. Anseau, "Solid State Reaction in The ZrO_2-SiO_2-α-Al_2O_3 System", J. Mater. Sci., 15(1980)114-118.
8. K.Srikrisna, G.Thomas and J.S.Moya, "Sintering Additives for Mullite-Zirconia Composites", Adv. Ceramics, Vol. 24, Science and Technology of Zirconia III, (1988), 276-286.
9. K. Niihara, R. Morena and D.P.H. Hasselman, "Evaluation of K_{IC} of Brittle Solids by the Indentation Method with Low Crack-to-Indent Ratios", J. Mater. Sci. Lett.., 1(1982)13-16.
10. P. Peña, J.S. Moya, S. De Aza, E. Cardinal, F. Cambier, C. Leblud, M.R. Anseau and A. Leriche, "Effect of Magnesia Additions on the Reaction Sintering of Zirconia-Alumina to Produce Zirconia Toughened Mullite", J. Mater. Sci., 2(1983)772-774.
11. D.L. Porter and A.H. Heuer, "Microstructural Development in MgO Partially Stabilized Zirconia (Mg-PSZ) ", J. Am. Ceram. Soc., 62(1979)298-305.
12. J.S. Moya, "Mechanical Properties of Sintered, In-Situ reacted Mullite-Zirconia Composites", J. Am. Ceram. Soc., (1981)C-78.

PREPARATION, MICROSTRUCTURAL AND MECHANICAL PROPERTIES OF Ti(C,N) WHISKER / Al_2O_3 COMPOSITES

P. Pettersson, M. Johnsson and M. Nygren
Department of Inorganic Chemistry, Stockholm University,
S-106 91 Stockholm, Sweden

ABSTRACT

A study of Ti(C,N) whisker / alumina composites was performed with focus on different whisker content. The whiskers were produced *via* carbothermal reduction of TiO_2 involving a vapour-liquid-solid growth mechanism. The whiskers have a diameter of 1-3 μm and a length of 30-40 μm. Composites with 10, 15, 20, 25 and 30 vol.% Ti(C,N) whiskers were prepared. The Al_2O_3 was dispersed in water with Dispex A40 added as a dispersing agent and MgO as an alumina grain growth inhibitor. All composites were hot-pressed (HP) at 1700 °C, 28 MPa for 1½ h in argon atmosphere. The samples were found to be fully dense by use of Archimedes method. The microstructure of both whiskers and composites were investigated in a scanning electron microscope (SEM). The whiskers were found to survive sintering. The Vickers hardness (Hv_{49}) and the fracture toughness (K_{1c}) values were measured for all compositions. The Vickers hardness increase from approximately 16.4 GPa for the composite with 10 vol.% until it reaches the highest value (18 GPa) at a whisker amount of 25 vol.%. The fracture toughness increase with increasing amount of whiskers, from approximately 3 $MPa{\cdot}m^{1/2}$ for pure Al_2O_3 to almost 5.5 $MPa{\cdot}m^{1/2}$ for the composite with 30 vol.% whiskers.

To the extent authorized under the laws of the United States of America, all copyright interests in this publication are the property of The American Ceramic Society. Any duplication, reproduction, or republication of this publication or any part thereof, without the express written consent of The American Ceramic Society or fee paid to the Copyright Clearance Center, is prohibited.

INTRODUCTION

A common way to reinforce alumina is to disperse SiC whiskers into the matrix material. SiC_w has many benefits such as high toughness, strength and resistance to oxidative environments. SiC is one of few commercially available whisker materials and is consequently the most studied.

During the last few years a number of new carbide and carbonitride whiskers has been developed such as TiC and Ti(C,N) [1-2] that may be interesting alternatives to SiC.

The aim of this work was to test if Ti(C,N) whiskers can be used as a reinforcing material for Al_2O_3. It was thus necessary to ensure that the whiskers survive the sintering temperature and to obtain a good dispersion of whiskers into the alumina matrix. Effort were also made to study how the microstructure and the mechanical properties vary with the volume percentage of whiskers in the matrix.

EXPERIMENTAL

Whisker synthesis

The Ti(C,N) whiskers were synthesised carbothermally *via* a VLS (vapour-liquid-solid) growth mechanism where the reacting elements are present in all three states of aggregation during the synthesis [1-2]. The starting material (a mixture of TiO_2, C, NaCl and Ni, all commercially available) was heated in a graphite furnace (Thermal Technology) in nitrogen atmosphere at 1250 $^{\circ}$C for 6 h. In those studies Ni was mainly used as catalyst, however, Fe has proved to give coarser Ti(C,N) whisker [3] and was used in this study.

The whisker product contains carbon residues that must be removed. The whisker-carbon mixture was shaken with water and an organic solvent, the carbon residue accumulates in the organic solvent while the whiskers remain in the water phase. Residual Fe was washed away in 1 vol.% HCl. The refined whiskers were analysed chemically for Ti, N, C and O from this data the bulk composition was estimated to be $TiC_{0.21}N_{0.63}O_{0.16}$. The whisker morphology was investigated both in an optical microscope and in a scanning electron microscope (SEM, JEOL 820). The whisker yield is estimated to be 80 vol.%, the whiskers have a length of 30 - 40 μm and a diameter in the range 1 - 3 μm (see Figure 1).

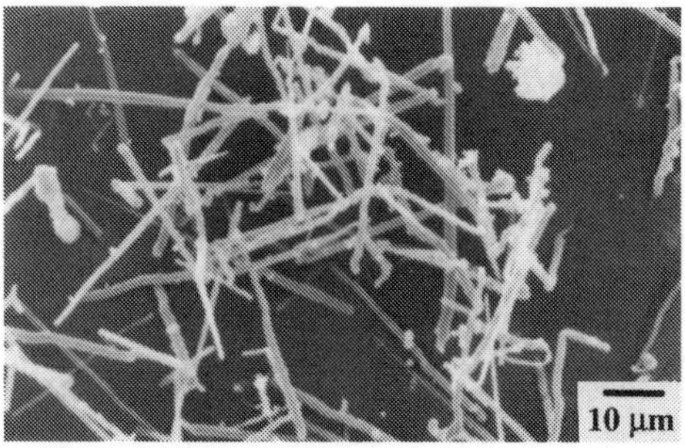

Figure 1. Ti(C,N) whiskers having a length of 30 - 40 μm and a diameter in the range 1 - 3 μm.

Composite production

A series with composites containing 10, 15, 20, 25 and 30 vol.% Ti(C,N) whiskers and alumina were produced. The composite matrix material was Al_2O_3 (AKP-30, Sumitomo Chemical, Inc.).

Some MgO was added to the composite mixture (0.25 wt% by Al_2O_3 dry weight) to prevent alumina grain growth [4]. The dispersion is made in water and to improve the dispersion of MgO it was added as the easily soluble salt $Mg(NO_3)_2 \cdot 6H_2O$ which decompose to MgO at elevated temperatures.

A mixture of Al_2O_3, $Mg(NO_3)_2 \cdot 6H_2O$, Dispex A40 (a dispersing agent) were first ball milled with Al_2O_3 cylpebs as milling media in water for approximately 18 h, then Ti(C,N) whiskers were added to the slurry and the milling proceeded for another 5 h. The slurry was instantly frozen in liquid nitrogen and subsequently freeze dried.

Thermogravimetric measurements (TG, Perkin Elmer TGA7) were performed on $Mg(NO_3)_2 \cdot 6H_2O$, Dispex A40 and a composite mixture in Ar and it was found that oxygen from dispex oxidise the Ti(C,N) whiskers and consequently the procedure was repeated in Ar-5%H_2 (see Figure 2). From these studies it was decided to heat treat the composite mixtures at 600 °C (1 h) in Ar-5% H_2 to remove residues from Dispex A40 and crystal water from $Mg(NO_3)_2 \cdot 6H_2O$ before hot-pressing.

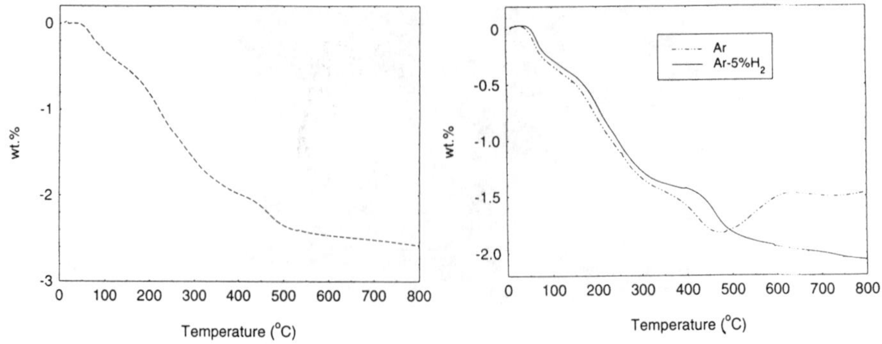

(a) **(b)**

Figure 2. Weight loss versus temperature for starting mixtures containing
(a) Al_2O_3 - $Mg(NO_3)_2 \cdot 6H_2O$ - Dispex A40
(b) Al_2O_3 - 20 vol.% $Ti(C,N)_w$ - $Mg(NO_3)_2 \cdot 6H_2O$ - Dispex A40.

Green bodies were made with a diameter of 12 mm. The green bodies were sintered in a hot-press furnace (Thermal Technology) at 1700 °C, 28 MPa for 90 min. The density was measured on all sintered samples by use of Archimedes principle.

The microstructure for sections both parallel to and perpendicular to the pressure directions were investigated with a scanning electron microscope SEM (JEOL 820).

Mechanical properties

The hardness (Hv_{49}) and fracture toughness (K_{1c}) were evaluated by the Vickers indentation technique at a load of 49 N for all compositions. This technique is simple and fast, however, it is rather sensitive to geometry and surface properties such as residual stress. Five indents were made on a row in the middle (to minimise near-surface effects) on each sample piece. Mean values for both orientations are used. The hardness was calculated from the following relation:

$$H = P/\alpha_0 a^2 \qquad (1)$$

Where **P** is the load in N; for Vickers indenters $\alpha_0 = 2$; ***a*** is the mean value of the indentation diagonals. The error in the measured values is estimated to be ±0.5 GPa.

The fracture toughness was calculated according to Anstis *et al.* [5]:

$$K_{Ic} = A \cdot \left(E/H\right)^{1/2} \cdot \left(P/c\right)^{3/2} \tag{2}$$

Where **A** is a constant (0.017); **E** is Youngs modulus; **H** is the hardness and **c** the crack length. Youngs modulus was estimated for the different composites by assuming a linear relation between the values for Al_2O_3 (380 GPa) and Ti(C,N). For Ti(C,N) the Youngs modulus was approximated to the value for TiC (448 MPa). The error in the measured K_{Ic} values is estimated to be ±0.4 $MPa \cdot m^{1/2}$.

RESULTS

From the results of the TG experiments it was decided that the residual products after dispersion of the composite materials could be burned off at 600 °C in Ar-5%H_2 (see Figure 2(b)). After hot-pressing all samples were fully dense (see Table I) which reveal that the hot-press furnace program was well suited for this kind of material.

Table I. Density of the composites after hot-pressing

Amount of whiskers (vol.%)	Density (% of theoretical density)
10	98.7
15	99.7
20	99.0
25	99.2
30	98.5

The microstructure of the five different whisker composites oriented perpendicular to the pressure direction are shown in Figures 3(a-e). The white needle shaped areas are whiskers and the dark grey areas are alumina. From these pictures it is evident that the whiskers survived the sintering process and kept their shape and size, the whiskers appear to be homogeneously dispersed over the composite volume.

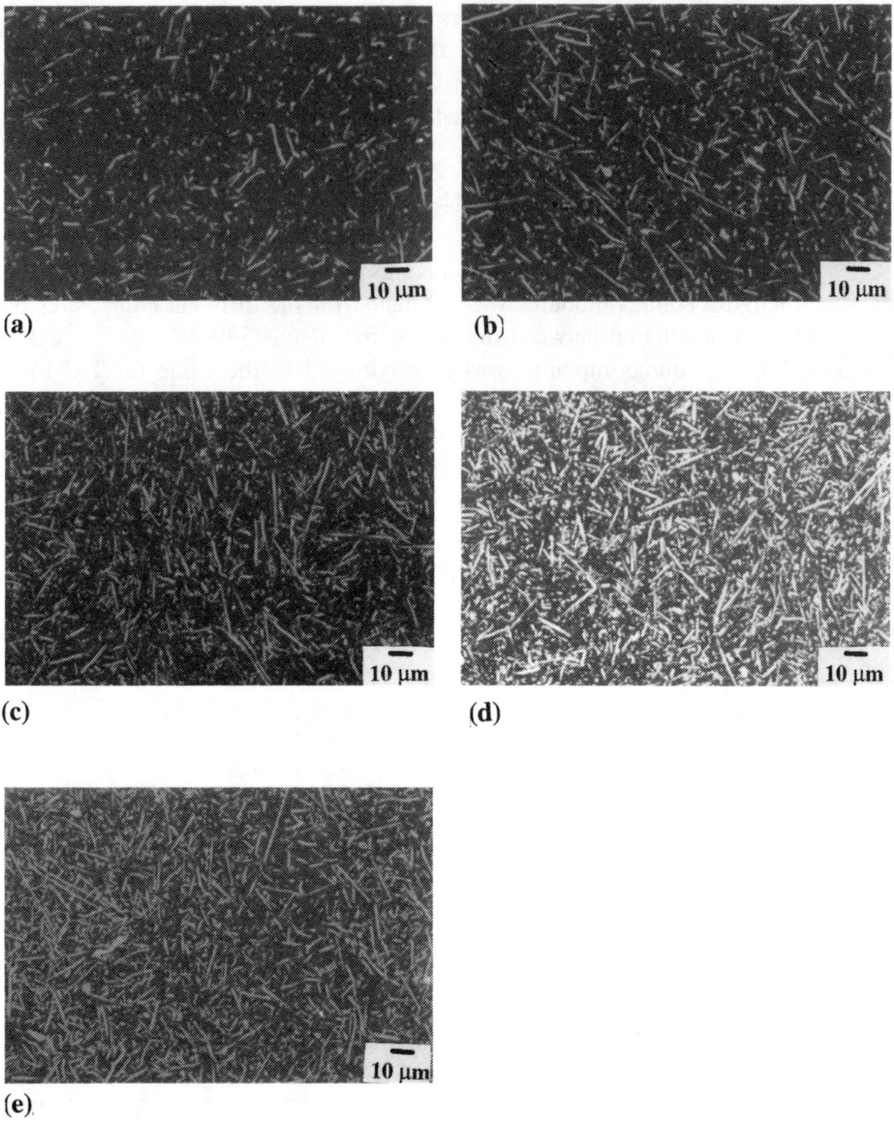

Figure 3. SEM micrographs of the different composites. All micrographs are from sections oriented perpendicular to the pressure direction **(a)** 10 vol.% whiskers **(b)** 15 vol.% whiskers **(c)** 20 vol.% whiskers **(d)** 25 vol.% whiskers **(e)** 30 vol.% whiskers.

The microstructure for the orientation parallel to the pressure direction for the composite sample with 15 vol.% whiskers is shown in Figure 4. It is obvious from figures 3 and 4 that the whiskers arrange themselves with the long side perpendicular to the pressure direction.

Figure 4. SEM micrograph of the section parallel to the pressure direction for the composite containing 15 vol.% Ti(C,N).

There is a tendency for the hardness and fracture toughness to increase with increasing amount of whiskers for the studied composites. The Vickers hardness increase with increasing amount of whiskers to a value of about 18 GPa for 25 vol.% whiskers. The fracture toughness increase from about 3 MPa·m$^{1/2}$ for pure Al_2O_3 to about 5.5 MPa·m$^{1/2}$ for the composite sample with 30 vol. % whiskers.

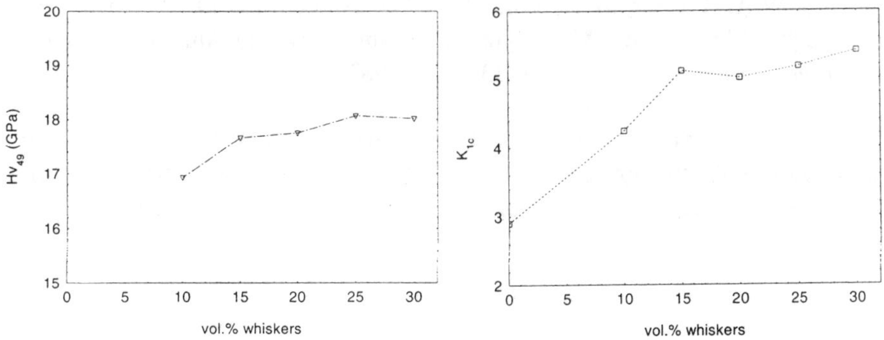

(a) **(b)**

Figure 5. **(a)** Hv_{49} for the different composites. **(b)** K_{1c} for the different composites and a reference sample without any whiskers.

CONCLUSIONS

Composites with different volume percentages of Ti(C,N) whiskers dispersed in Al_2O_3 matrix was investigated. It was found that it was possible to disperse the whiskers homogeneously in the matrix and that the whiskers survived the sintering procedure. The fracture toughness increase with increasing amount of whiskers while the Vickers hardness increase until it reaches the highest value at a whisker amount of 25 vol.%.

ACKNOWLEDGEMENT

The Brinell Competence Centre and the Swedish Foundation for Strategic Research are acknowledged for financial support.

REFERENCES

[1] N. Ahlén, M. Johnsson and M. Nygren, "Carbothermal Synthesis of TiC Whiskers via a Vapour-Liquid-Solid Growth Mechanism," *Journal of American Ceramic Society*, **79** [11] 2803-2808 (1996).

[2] N. Ahlén, M. Johnsson and M. Nygren, "Synthesis and characterization of $TiN_{1-X}C_X$ whiskers", *Processing and Handling of Powders and Dusts,* pp. 57-67 Eds. T.P. Battle and H. Henein, The Minerals, Metals & Materials Society, Orlando, Florida, 1997.

[3] N. Ahlén, M. Johnsson, and M. Nygren: Unpublished results.

[4] W.A. Kaysser, M. Sprissler, C.A. Handwerker and J.E. Blendell: "Effect of a Liquid Phase on the Morphology of Grain Growth in Alumina", *Journal of American Ceramic Society*, **70** [5] 339-343 (1987).

[5] G. R. Anstis, P. Chantikul, B. R. Lawn and D. B. Marshall, "A Critical Evaluation of Indentation Techniques for Measuring Fracture Toughness: I, Direct Crack Measurements," *Journal of American Ceramic Society*, **64** [9] 533-538 (1981).

MICROSTRUCTURAL MODIFICATION OF 70% Si_3N_4-30% BAS CERAMIC MATRIX COMPOSITE BY VARYING THE PARTICLE SIZE DISTRIBUTION OF Si_3N_4 STARTING POWDERS

Feng Yu and Kenneth W. White
Department of Mechanical Engineering,
University of Houston, Houston, TX 77204-4792

ABSTRACT

The development of bimodal microstructures was observed in the 70% Si_3N_4-30% BAS self-reinforced composite pressureless sintered at 1930°C by using Si_3N_4 starting powders with different particle size distribution. It was found that the addition of coarse silicon nitride powder encourages the abnormal growth of β grains. Subsequently, the influence of the sintering time on grain growth was considered for the various Si_3N_4 starting powder size distributions.

I. INTRODUCTION

Silicon nitride-based ceramics have attracted interest for use in engineering applications because of their excellent performance at both room temperature and elevated temperatures. However, despite their excellent strength, significant improvements in fracture resistance are required to satisfy reliability and design criteria for practical applications.

Beyond the classical issues of density and phase composition, the fracture toughness, R-curve behavior and other mechanical properties of silicon nitride ceramics also sensitively depend upon the size distribution and aspect ratio of β-Si_3N_4 grains. A number of research activities [1-8] have been conducted to develop microstructures with better fracture resistance. Reported fracture toughness values between 4 and 6 MPa√m most frequently corresponded to relatively uniform microstructures consisting of β whiskers of several microns in length [4,9-14]. Although this fine microstructure usually shows a characteristically high flexural strength, normally ranging from 800 to 1100 MPa, little or no R-curve behavior is observed. Fracture toughness improvements to between 7 and 10 MPa√m are observed through increases in the average whisker diameter [3,4,9-14]. Rising R-curves were often detected from the silicon nitride ceramics with coarse microstructures. However, because large β grains may act as crack initiation sites, coarse microstructures generally result in a relatively low flexural strength, typically ranging from 500 to 800 MPa. It has been found that coarse β whiskers (usually larger than 15 μm in length) surrounded by a fine matrix of short whiskers provide an ideal compromise between toughness and high strength [3-14]. Increased fracture toughness is attributed mainly to crack deflection and crack bridging by

To the extent authorized under the laws of the United States of America, all copyright interests in this publication are the property of The American Ceramic Society. Any duplication, reproduction, or republication of this publication or any part thereof, without the express written consent of The American Ceramic Society or fee paid to the Copyright Clearance Center, is prohibited.

abnormally grown β grains [3,4]. High α content Si_3N_4 powders are usually used as starting powders when developing bimodal microstructures.

Sintering of these materials generally follows three stages: nucleation of the β phase in the α rich matrix, growth of β grains during the α to β phase transformation, and subsequently grain growth of β-Si_3N_4. According to Emoto *et al.* [15], the driving force of β grain growth, following completion of the α to β phase transformation, depends on the grain size difference. For the case of abnormal grain growth, a critical minimum grain size difference is required. This minimum requirement may be suppressed by high-pressure operations, such as hot pressing, and hot-isostatic pressing. The onset temperature of abnormal grain growth during hot pressing was reported as about 1850°C by several authors [16,17]. The relatively small driving force for grain growth in pressureless sintered microstructures typically suppresses a bimodal distribution. Seeding with large β particles often accommodates this shortcoming through enhanced abnormal grain growth [18-20].

Recent work [21-26] on barium aluminum silicate (BAS), of the composition $BaO•Al_2O_3•2SiO_2$, shows promise as a refractory matrix material. Our previous study [26] demonstrated that silicon nitride with 30 vol.% BAS matrix can be processed to full density without high pressure. The as-sintered material showed an average flexural strength of about 963 MPa and an average fracture toughness of 5.36 MPa√m at room temperature. BAS almost completely crystallizes into hexacelsian during cooling, contributing to good high temperature strength. However, to fully realize the engineering benefits of this high strength, appropriate improvements in the fracture toughness and damage tolerance of this material must be obtained.

It has been shown [1,2,4-9] that the microstructural control of silicon nitride ceramics depends upon: the properties of the Si_3N_4 starting powder, the composition of sintering additives and the sintering conditions. In this study, the size distribution of Si_3N_4 starting powders was modified by blending with coarse Si_3N_4 powders, with goal of obtaining bimodal sintered microstructures. The microstructural development with sintering conditions was also explored through quantitative methods.

II. EXPERIMENTAL PROCEDURE

Characterization of Raw Si_3N_4 Powders

Because of the difficulties in obtaining high β content powders [19,27], two α-rich silicon nitride powders, UBE-E10 and UBE-ESP, were used in this study. Phase characterization by X-ray diffraction was performed on a Siemens Diffrac-5000 X-ray diffractometer. A peak breadth (B(2θ)) was determined by measuring the half maximum intensity breadth of the (201) peak (2θ=30.992°) of α phase from the XRD spectrum. The average particle size could be estimated according to Scherrer equation [28],

$$L = \frac{0.94\lambda}{B(2\theta)\cos\theta}, \tag{1}$$

where λ is the wavelength of the X-ray radiation (λ=0.154nm). The evaluation of the average β particle size was performed by using the (101) reflection of β-Si_3N_4. The observations of particle morphology were carried out in a JEM-2000FX transmission electron microscope fitted with an energy dispersive X-ray spectrometer (EDS). The average particle size was also estimated from measurements made on the TEM micrographs. The particle size distribution was determined by sedimentation techniques.

Processing

Table I. The calculated β-Si_3N_4 nucleus density in the powder compact and the measured β-Si_3N_4 grain density in the as-sintered samples

Sample code	E10:ESP (by volume) (with 30vol.%BAS)	β nucleus density (N/μm^2)	β grain density (after 30 minutes at 1930°C) (N/μm^2)	β grain density (after 60 minutes at 1930°C) (N/μm^2)
100E10	100:0	1.24	1.23	1.25
85E10	85:15	1.09	1.14	1.09
70E10	70:30	0.94	0.82	0.89
50E10	50:50	0.74	0.79	0.78
100ESP	0:100	0.24	0.22	0.22

30 vol.% BAS constituent powders were mixed with 70 vol.% Si_3N_4 powders, which were prepared by mixing E10 and ESP in ratios of 100:0, 85:15, 70:30, 50:50 and 0:100 (Table 1), in iso-propyl alcohol for 48 hours in a ball mill using the silicon nitride grinding media. The 30 vol.%BAS-70 vol.%Si_3N_4 composition, selected for the present study, corresponds to that developed previously **[23,26]** for ease of sinterability and acceptable mechanical properties and electromagnetic performance. Samples were sintered at 1930°C for 5, 30, 60, 120, 210, and 270 minutes respectively in nitrogen.

Microstructural Characterization

Sintered materials were sectioned, ground and etched by immersion in molten NaOH for 2 minutes at 350°C, to characterize the whisker morphology. Quantitative microstructural evaluation was performed on SEM micrographs, where the diameter, length, and area of each grain were determined through commercially available image analysis software (NIH Image, National Institute of Health, USA). At least 700 whiskers were characterized for each processing condition. The randomly oriented grains prompted us to use the method proposed by Braue *et al.* **[16,27,29]** to calculate the real average aspect ratio as the mean of the largest 10% observed aspect ratios. This aspect ratio, along with the grain diameter, which could be revealed completely by overetching, served as our two key microstructural indicators.

III. RESULTS AND DISCUSSION

Characterization of Raw Si_3N_4 Powders

The content of the β phase in each raw silicon nitride powder was calculated from the XRD spectrum in a manner described by Gazzara **[30]**, as presented in Table II. The E10 powder in this study has more β phase than the ESP powder. Also from this table, the average α and β crystallite sizes estimated from the XRD spectrum are very close in each product, and agree quite well with the specific surface area data provided by the manufacturer.

Table II. Characterization of Si_3N_4 starting powders

Powder	Average particle size (From TEM) (μm)	Average α particle size (From XRD) (μm)	Average particle size (Converted from SSA*) (μm)	β-Si_3N_4 crystallite diameter (From XRD) (μm)	β content (vol.%)
E10	0.17	0.29	0.22	0.31	5.76
ESP	0.35	0.50	0.43	0.53	3.65

*provided by the manufacturer

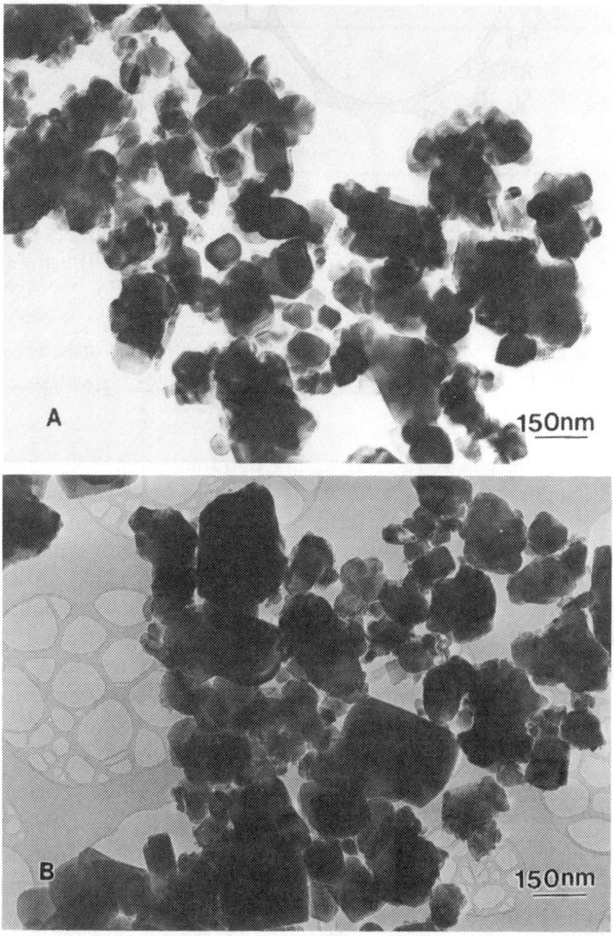

Figure 1. TEM micrographs showing the morphology of (a) E10 and (b) ESP powders

Figure 1(a) and (b) are TEM micrographs showing the morphology of E10 powder and ESP powder, respectively. Both powders are in almost spherical forms. The particle-size distribution and morphology of α and β, therefore, are assumed to be the same in each powder sample. The average particle size calculated from TEM micrograph measure-ments is about 0.19 μm for E10 and about 0.35 μm for ESP, slightly larger than that of XRD analysis. The appearance of either E10 or ESP did not change much after milling for 48 hours on the ball mill.

The particle size distribution curves for both powders (Figure 2), obtained from sedimentation techniques, appears larger than the TEM and the XRD results. The average particle size, which corresponds to 50% cumulative weight, was 0.8 mm for E10 and 1.05 mm for ESP. The aggregation of powders during sedimentation test is attributed to this difference. Figure 3 also shows that the size distribution of E10 is narrower than that of ESP.

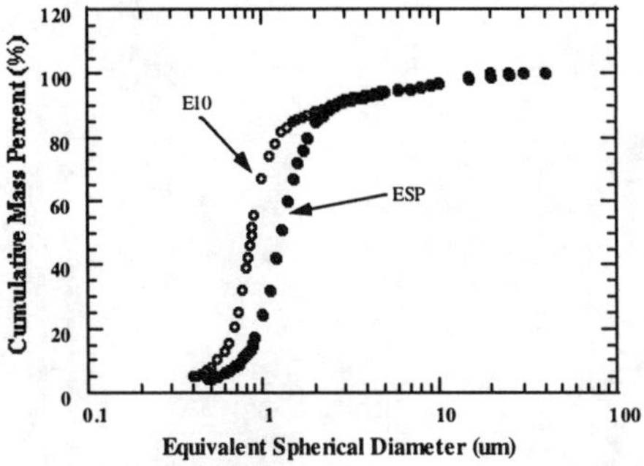

Figure 2. The particle size distribution of E10 and ESP powders

Characterization of Microstructure

Figures 3(a) and (b) show the microstructure of 100E10 and 50ESP samples, respectively, which were sintered at 1930°C for 210 minutes. They represent two typical microstructures, with and without abnormal grain growth. Figures 4(a) and (b) show the average whisker width and the average aspect ratio of samples sintered from different powders as functions of sintering time at 1930°C. There is not much difference in the average whisker width and average aspect ratio between samples sintered for 30 minutes. Based on the solution-reprecipitation mechanism, the nucleation and growth of β depends upon the mass transport of SiN_4 tetrahedra in the liquid, but is driven by the supersaturation [2]. Therefore, increase in the local supersaturation resulting from dissolution of the thermodynamically unstable α phase at the onset of α to β transformation can accelerate the development of elongated β grain [16]. The rapid increase of both width and aspect ratio between 30 and 60 minutes most likely results from the corresponding β phase development. After completion of the phase transforma-tion, grain coarsening in each sample is evident with sintering time, but the increased rate of grain diameter is small for 100E10, implying that variations of the sintering parameters cannot effectively modify the whisker morphology, due, apparently, to the insufficient driving force for grain growth during pressureless sintering. The higher fraction of coarse particles (ESP) promoted doubling of the whisker width in the 50E10 sample when the sintering time was extended from 30 to 270 minutes. We, therefore, cite the starting β size and morphology as one of the critical features for microstructural control.

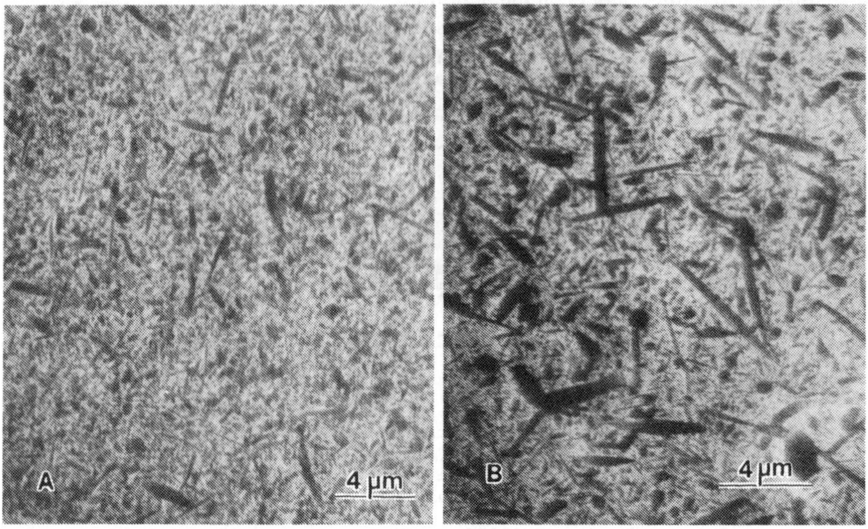

Figure 3. Micorstructure of (a) 100E10 sample and (b) 50ESP sample (1930°C-210min)

Characterization of Abnormal Grain Growth

As suggested by Hirosaki, *et al.* **[18]**, the diameter/cross-sectional area, which could better describe the importance of coarse grains, was used to statistically characterize the grain morphology instead of diameter/population parameters. Figures 5 (a) to (c) clearly illustrate the development of bimodal microstructures with sintering time at a temperature of 1930°C.

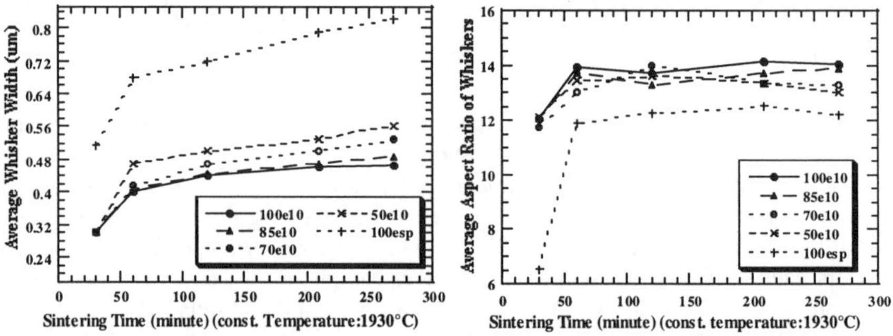

(a) (b)

Figure 4. Average whisker width (a) and average whisker diameter (b) vs. sintering time

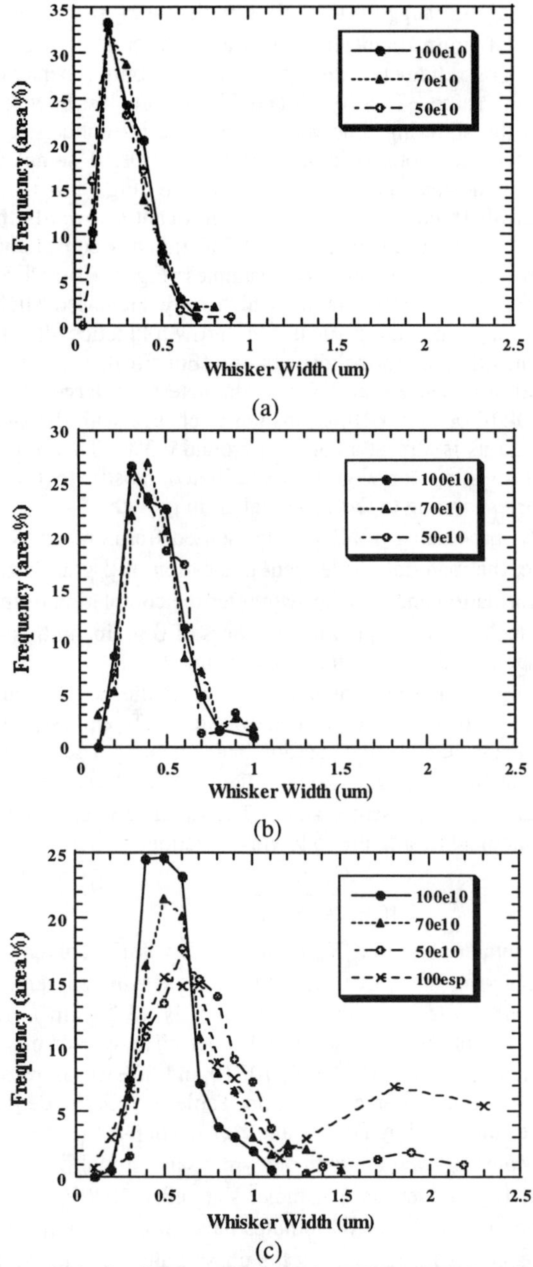

Figure 5. Whisker width distribution of 100E10, 70E10, 50E10 and 100ESP samples after sintering at 1930°C for (a) 30min, (b) 60min, and (c) 210min

Statistical analyses suggest that all samples start from microstructures with normal distributions of grain diameter (Fig. 5(a)). After sintering at 1930°C for 30 minutes, the average grain diameters and grain-size distribution behaviors of these three samples are almost the same. The 70E10 and 50E10 samples, however, do have some large grains. When the sintering time was extended to 60 minutes, clear indication of abnormal grain growth was observed in 70E10 and 50E10 samples, while 100E10 sample still retained normal grain-size distribution (Fig. 5(b)). In this period, distribution curve of all three samples shifted significantly towards the large size end with sintering time. A longer sintering time of 210 minutes considerably promoted the abnormal grain growth in 70E10 and 50E10 samples (Figs. 2(b) and 5(c)). In contrast, the distribution curve of 100E10 sample, which only increased slightly with a little broader distribution range indicating small grain growth kinetics after the completion of phase transformation, still kept normal distribution (Fig. 5(c)).

The population of large grains whose diameters are larger than three times the average, either in 70E10 or in 50E10, seems not to change with the sintering time. The frequency of those grains is almost a constant around 0.3%. Therefore, the accelerating growth of previous existing large grains contributed mostly to the evolution of the distribution peak corresponding to the abnormal grain growth.

In this study, processing of all samples started from one or two kinds of high α powders. Therefore, the materials underwent nucleation of β grains, growth of β grains during α to β transformation and grain growth after the completion of the transformation. Understanding the nucleation and growth behaviors of β grains in this material becomes critical for development of the bimodal microstructure.

By comparing the initial β nucleus density in the powder compact and the β grain density in the as-sintered material, Petzow and Hoffmann [31] concluded that E10 and ESP powders resulted in the predominant heterogeneous nucleation on the preexisting β-Si_3N_4 in the oxynitride liquid. Same method was applied to identify the nucleation mechanism in the present system. The initial β nucleus densities of 100E10 and 100ESP were calculated using the following equation,

$$D_p = \frac{3}{2}\frac{V_\beta\, \rho_s}{100\pi\, (R_N)^2\, \rho_{SN}}. \qquad (2)$$

D_p is the number of β nuclei per μm^2. V_β is the initial volume content of β in the Si_3N_4 starting powder. ρ_s is the theoretical density of the as-sintered sample, which is 3.25 g/cm^3 for this material, while ρ_{SN} is the density of Si_3N_4 (3.2 g/cm^3). R_N corresponds to the mean particle radius of the Si_3N_4 starting powder. The rule-of-mixtures calculations of the initial β nucleus densities of 85E10, 70E10 and 50E10 samples as well as the β grain density of as-sintered sample appear in Table 1. Grain density determination followed the analyses proposed by Petzow and Hoffmann [31].

All five samples showed approximately the same initial β nucleus density in the powder compact and β grain density in samples sintered at 1930°C for 30 minutes (Table I). Samples sintered at 1930°C for 60 minutes have similar β grain density as well. It seems that no new β grains formed during the phase transformation stage suggesting the predominant heterogeneous nucleation on the pre-existing β in the BAS liquid. TEM observation also revealed that β grains commonly had dark cores in the center of the grains, as shown in Figure 6. Although the contrast of this dark core was weak in most situations, it could be identified in most β grains by adjusting the orientation. This core structure, also reported by Dressler *et al.* [17,32], is generally believed to be pre-existing β nuclei.

During transformation, α particles, which are more thermodynamically unstable, have higher solubility than β nuclei with similar size and shape. They tend to dissolve into the liquid phase to set up supersaturation that encourages the growth of pre-existing β nuclei. High concentration of α in the liquid promotes the growth of all β grains but benefits more to the growth of large grains [16]. Certainly, some small β grains may dissolve back into the liquid during this stage.

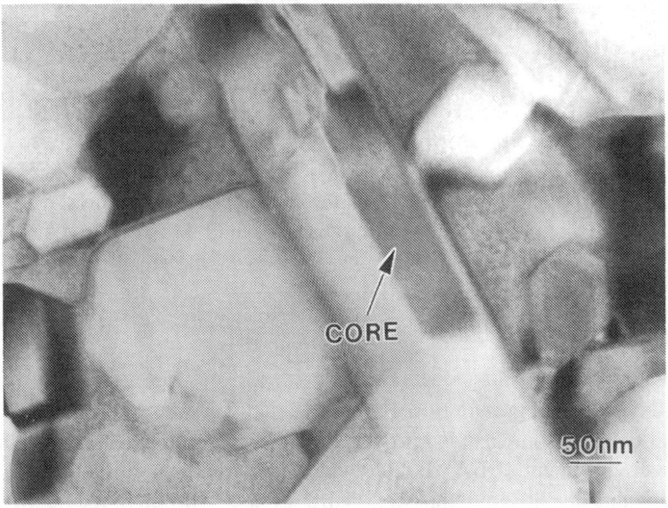

Figure 6. TEM micrograph showing the core structure within a β-Si_3N_4 grain

After completion of the phase transformation, the driving force for β grain growth depends solely on the difference in grain size. Emoto *et al.* [15] suggested that abnormal grain growth occurs only when the size difference exceeds a characteristic critical value. In the mean time, those grains that are smaller than a critical size will redissolve into the liquid, contributing to the growth of other grains [5,6]. Therefore, the grain size distribution at that time, which depends basically upon the β nucleus size distribution and expands during phase transformation, decides the development of bimodal microstructure. A narrow grain size distribution results in slow grain growth kinetics, most possibly without forming the bimodal microstructure, because the number of grains which will redissolve into the liquid is small while imposing an excessive demand upon the supply of parent material. A broad grain size distribution, on the contrary, favors the abnormal grain growth that results in the increasing average grain size with sintering time.

We conclude, therefore, that the microstructure of this material depends sensitively upon the density of initial β nuclei besides their size difference. The microstructural development during the α to β phase transformation becomes critical in deciding the final microstructure. A lower number of nuclei result in a larger distance between β particles in the powder compact, permitting more grain growth during phase transformation without significant impingement by other grains. For a fixed Si_3N_4 content, we could also assume that a smaller number of β grains in the liquid means more sufficient materials, which is contributed from the dissolution of α phase, for the growth of each grain if we exclude the consideration of kinetic issues like diffusion rate and interface reaction etc.

Based on the above ideas, it becomes relatively easy to explain the different microstructures obtained in this study. All samples contain 30 vol.% BAS and were sintered under the same conditions. Only the density and size distribution of initial β nuclei vary between these samples (Table 1). As expected, the material with low β content and large nucleus size like in 100ESP and 50E10 samples reveals the coarse microstructure. On the other hand, the material with relatively high nucleus density and narrow nuclei size distribution, like in 100E10 and 85E10 samples, shows the fine grained and homogenous microstructure. 100E10 sample, with a narrow grain size distribution and large β grain density, shows a slow grain growth rate without abnormal grain. With the introduction of certain volume fraction of ESP to the starting powder, as 70E10 and 50E10 samples, the resulting broader and larger β nucleus distribution encourages the abnormal grain growth. The growth of the smaller β nuclei retains the importance in developing a fine-grained matrix.

IV. CONCLUSIONS

The addition of some coarse silicon nitride powder to the starting powders encourages the abnormal growth of β grains in the 70 vol.% Si_3N_4-30 vol.% BAS system. Heterogeneous nucleation was identified as the predominant nucleation mechanism for β grains in the environment of BAS liquid. The microstructure of this material is strongly influenced by the density and the size distribution of initial β nuclei in the powder compact. Low density of β nuclei with the broader size distribution tends to promote coarse microstructure with a bimodal grain distribution.

REFERENCES

1. F. F. Lange, "Fracture Toughness of Si_3N_4 as a Function of the Initial α-Phase Content", J. Am. Ceram. Soc. **62[7-8]** 428(1979)
2. G. Wotting, B. Kanka and G. Ziegler, "Microstructural Development, Microstructural Characterization and Relation to Mechanical Properties of Dense Silicon Nitride", *Non-oxide Technical and Engineering Ceramics* Elsevier Applied Science, London, England, 83(1986)
3. P. F. Becher,, S. L. Hwang, H. T. Lin and T. N. Tieg, "Microstructural contributions to the fracture resistance of silicon nitride ceramics", *Tailoring of Mechanical Properties of Si_3N_4 Ceramics* M. J. Hoffmann and G. Petzow (eds) 87(1994)

4. E. Sun, P. F. Becher, C. H. Hsueh, K. P. Plucknett and S. B. Waters, "Microstructural Design of Silicon Nitrides with Improved Fracture Toughness" (SXVI-011-97) 99th Annual Meeting of American Ceraimc Society, May 4-7, 1997, Cincinnati, OH, USA
5. M. J. Hoffmann and G. Petzow, "Microstructural Design of Si_3N_4 Based Ceramics", *Silicon Nitride Ceramics-Scientific and Technological Advances* Mater. Res. Soc. Symp. Proc. **287** 3(1993)
6. M. J. Hoffmann and G. Petzow, "Tailored Microstructures of Silicon Nitride Ceramics", Pure & Appl. Chem. **66[9]** 1807(1994)
7. M. J. Hoffmann, "Analysis of Microstructural Development and Mechanical Properties of Si_3N_4 ceramics", *Tailoring of Mechanical Properties of Si_3N_4 Ceramics,* M. J. Hoffmann and G. Petzow (eds), Kluwer Academic Publishers, 59(1994)
8. M. J. Hoffmann, "Relationship Between Microstructure and Mechanical Properties of Silicon Nitride Ceramics", Pure & Appl. Chem. **67[6]** 939(1995)
9. A. J. Pyzik and D. R. Beaman, "Microstructure and Properties of Self-Reinforced Silicon Nitride", J. Am. Ceram. Soc. **76[11]** 2737(1993)
10. P. Sajgalik, J. Dusza and M. J. Hoffmann, "Relationship Between Microstructure, Toughening Mechanisms, and Fracture Toughness of Reinforced Silicon Nitride Ceramics", J. Am. Ceram. Soc. **78[10]** 2619(1995)
11. K. Matsuhiro and T. Takahashi, "The Effect of Grain Size on the Toughness of Sintered Si_3N_4", Ceram. Eng. Sci. Proc. **10[7-8]** 807(1989)
12. A. Okada and N. Hirosaki, "Effects of Microstructure on the R-curve Behavior of Sintered Silicon Nitride", J. Mater. Sci. **25** 1656-1661(1990)
13. T. Kawashima,H. Okamoto,H. Yamamoto and A. Kitamura, "Grain Size Dependence of the Fracture Toughness of Silicon Nitride Ceramics", J. Ceram. Soc. Jpn. **99[4]** 310(1991)
14. N. Hirosaki and Y. Akimune, "Effect of Grain Growth of β-Silicon Nitride on Strength, Weibull Modulus, and Fracture Toughness", J. Am. Ceram. Soc. **76[7]** 1892(1993)
15. H. Emoto, and M. Mitomo, "Control and Characterization of Abnormally Grown Grains in Silicon Nitride Ceramics", J. Eur. Ceram. Soc. **17** 797(1997)
16. M. Mitomo and S. Uenosono, "Microstructural Development During Gas-Pressure Sintering of α-Silicon Nitride", J. Am. Ceram. Soc. **75[1]** 103-8(1992)
17. W. Dressler, H. Kleebe, M. J. Hoffmann, M. Ruhle and G. Petzow, "Model Experiements Concerning Abnormal Grain Growth in Silicon Nitride", J. Eur. Ceram. Soc. **16** 3(1996)
18. N. Hirosaki, Y. Akimune and M. Mitomo, "Quantitative Analysis of Microstructure of Self-Reinforced Silicon Nitride Ceramics", J. Ceram. Soc. Jpn., Int. Edition **101** 1209(1993)
19. K. Hirao, T. Nagaoka, M. E. Brito, and S. Kanzaki, "Microstructure Control of Silicon Nitride by Seeding with Rodlike β-silicon Nitride Particles", J. Am. Ceram. Soc. **77[7]** 1857(1994)

20. N. Hirosaki and Y. Akimune, "Microstructure Characterization of Gas-pressure-sintered β-silicon nitride Containing Large β-silicon nitride Seeds", J. Am. Ceram. Soc. **77[4]** 1093(1994)
21. H. Pickup and R. J. Brook, "Barium Oxide as a Sintering Aid for Silicon Nitride", Br. Ceram. Soc. Proc. **39** 69(1987)
22. C. J. Hwang, and R. A. Newman, "Silicon Nitride Ceramics with Celsian as an Additive", J. Mater. Sci. **31** 150(1996)
23. K. K. Richardson, D. W. Freitag and D. Hunn, "Barium Aluminosilicate Reinforced In Situ with Silicon Nitride", J. Am. Ceram. Soc. **78[10]** 2662(1995)
24. S. W. Quander, A. Bandyopadhyay and P. B. Aswarth, "Synthesis and Properties of in situ Si_3N_4-reinforce $BaO.Al_2O_3.2SiO_2$ Ceramic Matrix Composites", J. Mater. Sci. **32** 2021(1997)
25. F. Yu, C. R. Ortiz-Longo, K. W. White and D. Hunn, "The Microstructural Characterization of In Situ Grown Si_3N_4 Whisker-Reinforced Barium Aluminum Silicate Ceramic Matrix Composite", Accepted for publication by J. Mater. Sci.
26. F. Yu, N. Nagarajan, Y. Fang and K. W. White, "Microstructural Control of a 70% Si_3N_4-30% Barium Aluminum Silicate (BAS) Self-Reinforce Composite", submitted to J. Am. Ceram. Soc.
27. M. Mitomo, M. Tsutsumi, H. Tanaka, S. Uenosono and F. Saito, "Grain Growth During Gas-Pressure Sintering of β-Silicon Nitride", J. Am. Ceram. Soc. **73[8]** 2441(1990)
28. B. E. Warren, X-ray Diffraction Dover Publications, New York, 1990, P.253
29. W. Braue, G. Wotting, and G. Ziegler, *Ceramic Microstructure 86*, Plenum Press, New York, (1987)
30. C. P. Gazzara and D. R. Messier, Am. Ceram. Soc. Bull. **56[9]** 777(1977)
31. G. Petzow and M. J. Hoffmann, "Grain Growth Study in Si_3N_4-Ceramics", Mater. Sci. Forum **113-115** 91(1993)
32. M. Kramer, M. J. Hoffmann and G. Petzow, "Grain Growth Studies of Silicon Nitride Dispersed in an Oxynitride Glass", J. Am. Ceram. Soc. **76[11]** 2778(1993)

MECHANICAL BEHAVIOR OF THE SILICON NITRIDE-BARIUM ALUMINOSILICATE CERAMIC MATRIX COMPOSITE WITH REDUCED LIQUID PHASE

N.Nagarajan, F.Yu, Y.Fang and K.W.White
Dept. of Mechanical Engineering, University of Houston, Houston, TX.

ABSTRACT

The silicon nitride-barium aluminosilicate ceramic composite offers great advantages in structural applications, where a combination of high strength and fracture resistance are of primary importance. In this work, pressure-less sintering of SN-BAS composite is carried out with low volume fractions of the liquid phase. The effect of reducing the BAS phase on the densification, and the inter-relationship between densification and phase transformation is examined. These studies are then applied to processing of low liquid volume fraction containing SN-BAS composite, by using a two-step sintering technique. The flexure strength and fracture toughness measured are then correlated with the microstructural changes.

INTRODUCTION

The Silicon nitride-barium aluminosilicate (SN-BAS) Ceramic Matrix Composite processed by the pressure-less sintering technique, exhibits good mechanical properties, comparable to those of the hot-pressed and hot-isostatic pressed silicon nitride based ceramic components, with the added advantage of low manufacturing costs. Previous work on the pressure-less sintering of the SN-BAS system [(1)] has shown that beta silicon nitride whiskers can be grown in situ, as a result of the alpha-beta phase transformation occurring in the presence of liquid BAS at temperatures above 1760°C. Further work, by way of microstructural modifications on this composite[(2)], with a composition of 70% silicon nitride-30% BAS, by volume, improved the flexural strength of this material, up to 960MPa, with a fracture toughness of 5.9MPa.m$^{1/2}$. The composite also exhibits fairly good R-Curve behavior[(3)] with whisker bridging, pullout, and crack deflection being the main toughening mechanisms. Improved toughness values were also obtained, by the development of the bimodal microstructures in this material[(4)]

To the extent authorized under the laws of the United States of America, all copyright interests in this publication are the property of The American Ceramic Society. Any duplication, reproduction, or republication of this publication or any part thereof, without the express written consent of The American Ceramic Society or fee paid to the Copyright Clearance Center, is prohibited.

The high temperature properties of silicon nitride ceramics are generally dictated by the secondary phases, formed by the sintering additives. In the SN-BAS system, the complete crystallization of the BAS liquid phase on cooling, offers an added advantage in the use of this material. The high temperature properties of the composite are also improved because of the crystalline BAS refractory phase, in comparison to the amorphous or partially crystallized secondary phases in other Si_3N_4 systems. There is however, a very thin layer of the amorphous glassy phase in between beta silicon nitride grains, which is thought of as a thermodynamic constraint(5). It has been observed that the high temperature strength of SN-BAS composite, is excellent up to 1200°C, after which it suffers a drop in strength. The softening of the BAS phase is believed to be the reason for the drop in flexural strengths, in addition to the residual glassy phase. In order to improve the high temperature properties of this composite, it thus becomes necessary to reduce the amount of the liquid BAS phase while sintering.

The role of the liquid volume fraction assumes greater importance in the case of pressure-less sintering, as the driving forces for sintering are reduced in this case, in comparison to hot pressing(HP) or hot isostatic pressing(HIP)(6). The volume of the liquid phase plays a key role in the densification of the material. The three basic stages of liquid phase sintering are re-arrangement, solution-precipitation, and grain coalescence, and the volume of the liquid phase governs the extent of the re-arrangement process, and the subsequent pore filling process(7). The effect of reduced liquid phase contents is to decrease the extent of re-arrangement, thereby necessitating other mechanisms, like solution – precipitation to be operable for complete densification.

In this work, the inter-relationship between the densification and the transformation in Si_3N_4, and the effect of a reduced liquid phase, on the densification of the composite, are studied, and these results are applied in developing sintering techniques for the pressure-less sintering of SN-BAS systems with reduced liquid phase.

EXPERIMENTAL PROCEDURE

Processing

Batches of four different compositions of the SN-BAS composite were prepared by varying the amount of the liquid phase constituents, so as to obtain 60%SN-40%BAS, 70%SN-30%BAS, 80%SN-20%BAS, and 90%SN-10%BAS stoichiometry. Each batch was prepared by mixing the constituent powders with 2-propanol for 72 hours, on a ball mill using silicon nitride grinding media. The powders were dried, and compacted under a uni-axial pressure of 50MPa. The green disks were then baked at 450°C in order to evaporate the organic lubricants.

Samples were packed in a powder bed, and sintered at temperatures from 1740°C through 1950°C, with sintering times varying from 30 minutes to 4 hours. Nitrogen was passed continuously throughout the sintering process, so as to minimize the surface decomposition of the samples. Samples of the type 80SN-20BAS and 90SN-10BAS were also sintered using a different sequence of sintering, the two-step sintering scheme, besides the one step sintering procedure used for samples of type 60SN-40BAS and 70SN-30BAS In the two step sintering scheme, samples were first sintered at 1800°C for a dwell time of 4 hours, and then sintered at 1950°C for varying dwell times. The heating and cooling rates were the same as that of the one-step sintering scheme.

Microstructural and Mechanical Characterisation

Microstructures were observed on a scanning electron microscope. Each sample was etched in molten NaOH at 350°C for 2-3 minutes, in order to observe whisker morphology. Fracture surfaces were observed for closed porosity, and possible toughening mechanisms. Bulk densities were measured by water immersion method. Phase characterization by x-ray diffraction was done on a D-5000 Siemens Diffractometer using Ni filtered CuKa radiation at a scan rate of 0.02°/sec. The percentages of α and β silicon nitride, were calculated using the method described by Gazzara and Messier.[8]:

The flexural strengths of the samples were measured by using the three-point bend tests. Indentation method was used to measure the fracture toughness values for each material. The K_{IC} measured was averaged for different contact loads applied.

RESULTS AND DISCUSSION

The liquid phase sintering process is visualized by the classical Kingery model[9] , to consist of three distinct stages, particle re-arrangement, solution-precipitation, and coalescence. The particle re-arrangement process itself consists of primary and secondary re-arrangement, and both stages are requisite for full densification. The solution-precipitation process includes Ostwald ripening of the grains, at the later stages of the process, and has no effect on the densification, unless, grain coarsening induces changes in the curvature of the liquid meniscus, which facilitates pore filling [10]. Grain coalescence, leading to the formation of a solid skeleton, inhibits densification. However, when certain morphological and thermodynamic criteria are fulfilled, mechanisms like pore migration, can cause densification even in this stage [11].

In the case of the SN-BAS system with low volume fractions of BAS, mechanisms of densification like pore filling, and pore migration are inoperable, and hence, the primary mechanisms of densification are only re-arrangement and solution-precipitation. This places further constraint on the amount of the liquid phase that is essential for complete densification of the composite. Fig 1 shows the densification curves at 1800°C at different times of sintering, for the 60SN-40BAS, 70SN-30BAS, 80SN-20BAS and the 90SN-10BAS samples. While the 60SN-40BAS samples attain complete densification even before the start of the dwell time, the 80SN-20BAS, and the 90SN-10BAS samples, do not attain full densification even after prolonged sintering at 1800°C. Since, transformation, that accompanies the solution-precipitation process in this composite, is minimum, this shows that particle re-arrangement, which is rapid, is sufficient to densify the composite that has more than 30% of the liquid phase. However, particle re-arrangement alone, is not sufficient to densify the 80SN-20BAS and 90SN-10BAS samples. The lower amount of liquid phase in both these cases results in an increased resistance to particle migration, and hence the re-arrangement stage is not rapid, but slow. Hence, solution-precipitation, which in effect, is the alpha-beta transformation, and grain growth seem to influence the achievement of higher densities in this case.

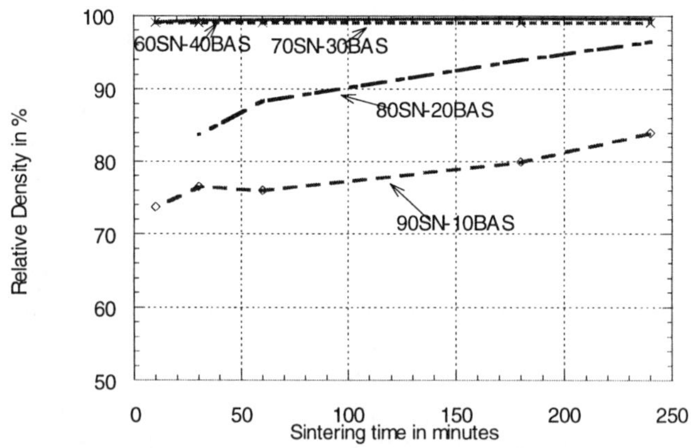

Fig 1 Densification curves for varying liquid volume fractions of the composite at 1800°C

The role of transformation in the densification of silicon nitride systems is a subject of much controversy. Two different opinions have been proposed to explain the influence of transformation. Based on the classical Kingery model of liquid phase sintering, one opinion is that the transformation is beneficial to densification, because of its grain refinement, and the other is that transformation does not lead to grain refinement, and hence is unrelated to densification[12] . It becomes critical to understand the effect of transformation on the densification in the SN-BAS system, if improvement in the densities of the lower liquid phase containing materials is to be made.

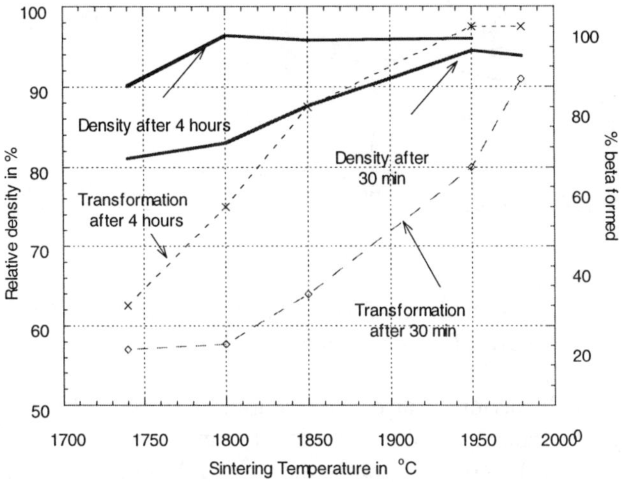

Fig 2. Densification and transformation of the 80SN-20BAS composite with varying temperatures of sintering, at 30 minutes, and 4hours of sintering.

Fig 2 shows the densification and transformation of the 80SN-20BAS composite, at two different sintering times for different sintering temperatures. It is very clear that the densification at higher temperatures, is faster initially, and does not increase much with time, whereas, at lower temperatures, for e.g., 1800°C, the density improves a great deal, and exceeds values of density at 1950°C, at longer times of sintering.

In Fig 3(a) and 3(b), the densification and transformation of the 80SN-20BAS samples at different sintering times, is shown at two temperatures 1800°C, and 1950°C, respectively. It is evident that, at 1800°C, densification increases with the transformation, whereas, at 1950°C, the transformation increases with time, while the densification remains a constant. This implies that the transformation after a certain stage prevents any further densification. This stage corresponds to the formation of the rigid solid skeleton, which suppresses further re-arrangement, and shrinkage.

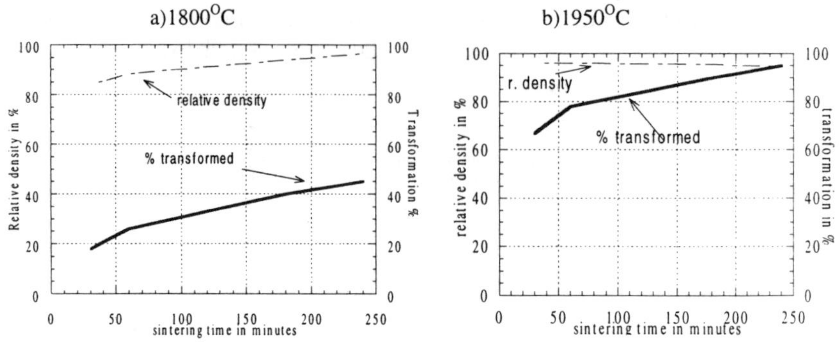

Fig 3. The densification and transformation of the 80SN-20BAS composite, at two different temperatures, for varying sintering times.

The same behavior is also observed in the 90SN-10BAS samples, as shown in Fig 4(a) and 4(b). The amount of transformed beta can be directly related to the size of the growing grains, because, it has been observed(13) that the transformation occurs via heterogeneous nucleation, and the beta transformed precipitates on the preexisting beta nuclei(14). Hence, impingement and interlocking of the growing beta grains, would be a feature in the later stages of transformation. It is this impingement and interlocking, that causes the formation of a rigid skeleton, and hinders any further densification. Therefore, we can conclude that grain growth leading to grain impingement is the factor that needs to be related to densification.

It is clear that when continued transformation leads to grain impingement it inhibits densification in low liquid phase volume materials. It would thus be beneficial to densify the material first, by controlling the transformation, and then completing the transformation at higher temperatures. This idea led us to applying the two step sintering technique on the 80SN-20BAS samples. From Fig 2, the ideal conditions for densification was found to be a sintering time of 4hrs at 1800°C, and the ideal temperature for transformation, 1950°C

A comparison between the densities of the one step and two step sintered samples, at different sintering dwell times at 1950°C, is shown in Fig 5. The two step sintered samples have higher densities, than the one step sintered samples, which is predicted by our analysis of the inter-relationship between densification and transformation.

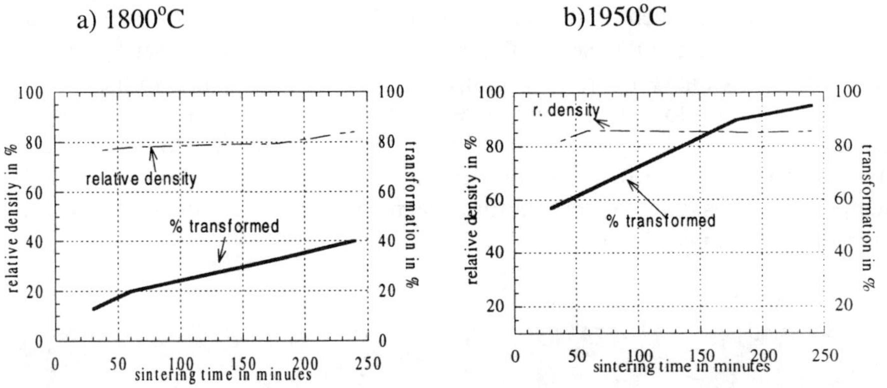

Fig 4. The densification and transformation of the 90SN-10BAS composite at two different temperatures, for varying sintering times.

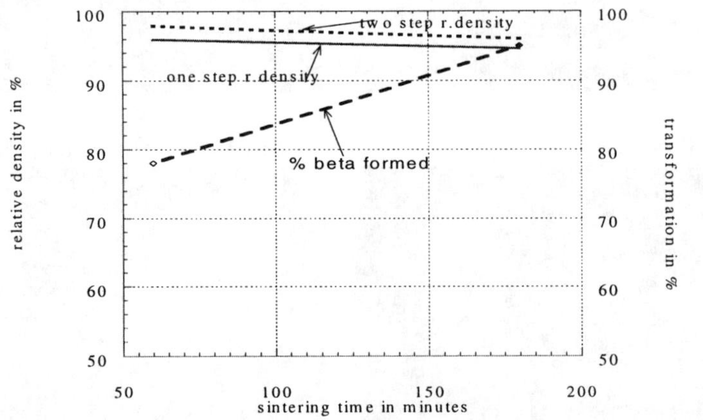

Fig. 5. A comparison of the densities of the one-step and two-step sintered 80SN-20BAS samples.

The microstructures developed out of the one step and two step sintered samples are shown in Figs. 6(a), (b) (c) and (d). The two step sintered samples seem to have a uniform microstructure, while the one step samples seem to have a broader distribution of grain sizes. This difference could only be caused during the first step of the sintering. In the first step, at 1800°C, the viscosity is higher than at 1950°C. A strong relationship has been found between the viscosity of the liquid and the aspect ratio of the grains[15]. Generally, grains of higher aspect ratios are obtained, with higher viscosity liquid phases. In the case of the two step sintered samples, much of the nucleation probably occurs in the high viscosity melt, while this is not the case in the one step sintered samples. Hence, two step sintered samples have finer, more uniform, grain sizes compared to the one step samples. Also, for longer sintering times at lower temperatures, the microstructure could reach a thermodynamic equilibrium. As a result, the driving force for accelerated grain growth is decreased, and further sintering at higher temperatures, results in a more uniform microstructure, in comparison to the one step sintered samples.

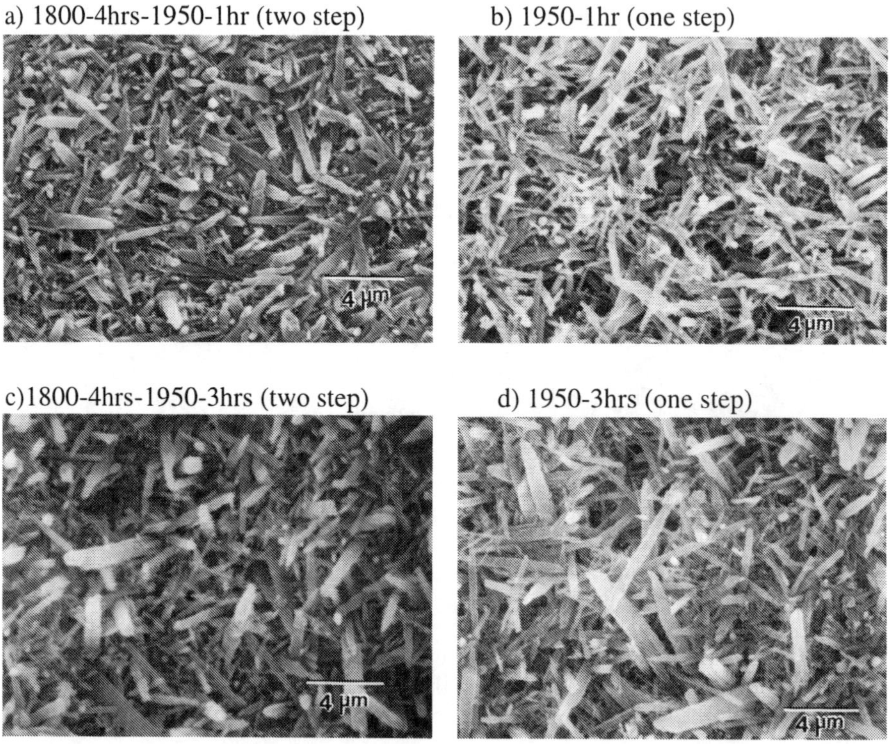

Fig 6. Morphology of the one step and two step sintered 80SN-20BAS sample

The flexural strengths of the one step and two step sintered 80SN-20BAS samples, are plotted in Fig 7, along with strengths of the 70SN-30BAS samples, for comparison. The strengths of the two step sintered samples are higher, when compared to the one step sintered samples for the same amount of transformation. This can be attributed to the development of finer, and more uniform microstructures, in the case of the two step sintered samples. The maximum strength reached by the two step sintered 80SN-20BAS sample, which was sintered at 1950°C for 3hours, was about 960MPa,.

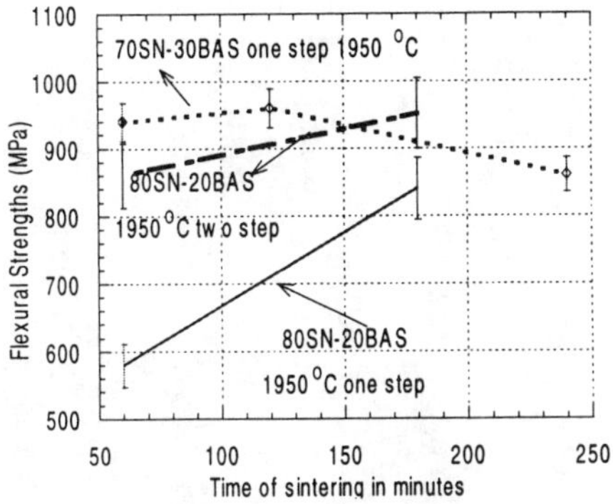

Fig 7 Flexural Strengths of the 80SN-20BAS one step and two step samples, in comparison with 70SN-30BAS

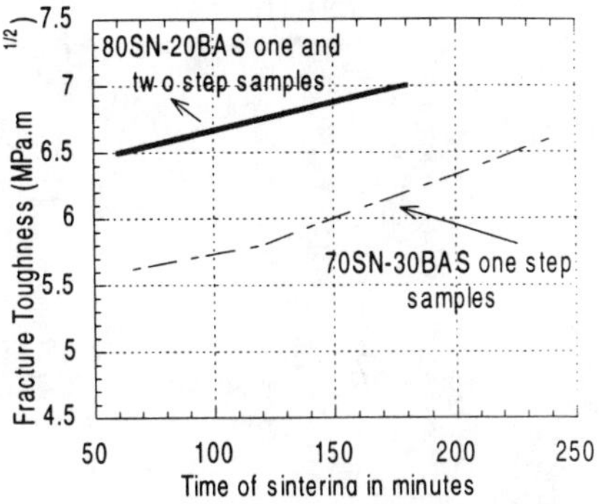

Fig 8.Fracture toughness of 80SN-20BAS one step and two step samp--les, compared to 70SN-30BAS

The critical fracture toughness from indentation measurements for the one step and two step sintered 80SN-20BAS samples are shown in Fig 8, in comparison with that of the 70SN-30BAS samples. The 80-20 samples have higher K_{IC} than the 70SN-30BAS samples. This is expected, because of the greater volume of beta whiskers that is present in the 80SN-20BAS, than in that of the 70SN-30BAS samples. However, the K_{IC} numbers for 80SN-20BAS one step and two step samples are almost the same. The volume of whiskers are the same in both the cases,. and we do not observe any abnormal growth of whiskers, in the one step sintered samples, even though they have a broader distribution of grain size Hence, it is possible that the critical fracture toughness does not change much with the type of processing. Electron micrographs of the fracture surfaces of the one step and two step samples are shown in Fig 9(a), (b), (c) and (d).

Crack deflection seems to be a prevalent mechanism of toughnening, as seen from the micrographs. The crack deflection sites per unit area increase with the sintering times for both types of samples, which is reflected by the increasing K_{IC} values with sintering time. The fracture surfaces of the one step and two step samples do not show much of a variation, which agrees with the K_{IC} values obtained. Whisker pullout, which seemed to be the dominant mechanism of toughening in the 70SN-30BAS samples [(3)], is not prevalent in the 80SN-20BAS samples. This may be due to the fact that the whiskers are more tangled, and interlocked in the 80SN-20BAS, than in the case of 70-30 material.

a) 1800-4hr-1950-1hr (two step)

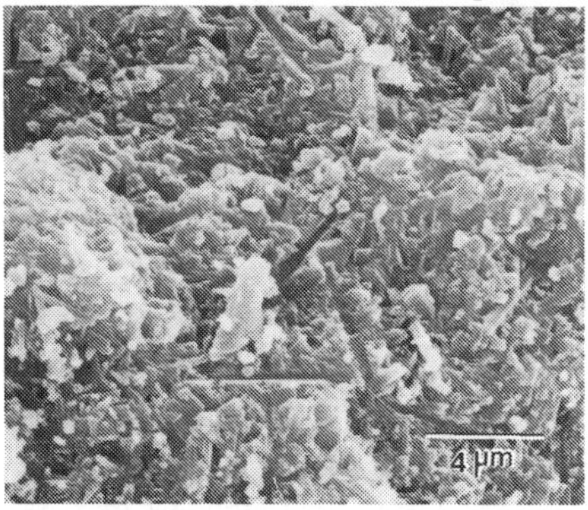

b)1950-1hr (one step)

c) 1800-4hr-1950-3hr (two step)

d) 1950-3hr (one step)

Fig 9. Fracture surfaces of the 80SN-20BAS one step and two step sintered samples.

CONCLUSIONS

A clearer perception of the inter-relationship between the densification and the alpha-beta transformation, was achieved, by studying the densification and transformation of SN-BAS composite with varying volume fractions of the liquid phase. The transformation is not essential for densification, if there is sufficient liquid in the composite, to attain full densification by particle re-arrangement. However, when low liquid volume fraction does not permit densification, then solution-precipitation is needed for densification, as a result of which, transformation, which occurs by a solution-precipitation, aides densification. Transformation inhibits densification in the solid skeleton formation stage, but only when mechanisms like pore migration are not active.

Based on our understanding of the densification-transformation relationship, the densification of the 80SN-20BAS material was achieved up to 97% TD, by the two step sintering technique, which helped achieve high densities by controlled transformation. Complete densification was not achieved, however, in the 90-10 samples, under these processing conditions.

High strengths, up to 950MPa and high fracture toughness values, up to 7MPa.$m^{1/2}$ were obtained by the two step sintering of 80SN-20BAS composite.The flexural strength does not show a great increase with the increase in silicon nitride whisker content, but there was a rise in the K_{IC} values

REFERENCES:

1. Kerry K.Richardson et. al., in "Barium Aluminosilicate Reinforced In Situ with Silicon Nitride". J Am. Cer. Soc. 78 [10] 2662-2668 1995
2. F. Yu Ph.D. Thesis, University of Houston, Houston, TX., May 1998.
3. Y. Fang et. al., in "R-Curve Behavior in Silicon Nitride-BAS Ceramic Matrix Composite" Presented at the 99th Annual Meeting of the Am. Cer. Soc, May 1997.
4. F. Yu et.al., work in progress.
5. D. R. Clarke in "On the Equilibrium Thickness of Intergranular Phases in Ceramic Materials" J. Am. Cer. Soc. 70 [1] 15-22, 1987

6. Thomas M. Shaw in "Model for the Effect of Powder Packing on the Driving Force for Liquid Phase Sintering" J. Am. Cer. Soc. 76 [3] 664-70 1993.
7. Randall M. German in "Liquid Phase Sintering" Plenum Press, NY, 1985
8. C. P. Gazzara et. al., in "Quantitative Determination of Phase Content of Silicon Nitride by X-Ray Diffraction Analysis". Cer. Bull. [50] 777-780 1977.
9. W. D. Kingery, in " Densification during Sintering in the Presence of a Liquid Phase. I. " J. App. Physics. [30] No.3. 301-306, 1959.
10. Suk-Joong L. Kang et.al., in "Densification and Shrinkage During Liquid Phase Sintering" J. Am. Cer. Soc., 74 [2] 425-427, 1993
11. T. H. Courtney, in "Densification and Structural Development in Liquid Phase Sintering" Metall. Trans. [15A] 1065-1074, 1985.
12. Lee, Dong-Duk et.al., in "Effect of α to β(β') Phase Transition on the Sintering of Silicon Nitride Ceramics". J. Am. Cer. Soc. 73 [3] 767-769, 1990.
13. F. Yu, unpublished work, University of Houston, TX
14. M. Kramer et. al., in "Grain Growth Kinetics of Silicon Nitride during alpha-beta Transformation" Acta Metall. Vol. 41 No. 10, pp 2939-2947 1993.
15. Ziegler G., et.al., in "Relationships between Processing, microstructure and properties of dense and reaction bonded silicon nitride" J. Mater. Sci. 22[1987] 3041-3046.

Phase Relationships and Microstructures in Composite Systems

PHASE RELATIONSHIPS AND MICROSTRUCTURAL CONTROL OF ISOTHERMALLY SOLIDIFIED YAG-SPINEL-ALUMINA COMPOSITES

Shuqiang Wang
New Energy and Industrial Technology Development Organization(NEDO), National Industrial Research Institute of Nagoya, Nagoya 462-8510, Japan
Takashi Akatsu, Yasuhiro Tanabe and Eiichi Yasuda
Research Center for Materials Design, Materials and Structures Laboratory, Tokyo Institute of Technology, 4259 Nagatsuta, Midori, Yokohama 226, Japan

ABSTRACT

Phase relationships and microstructures of YAG-spinel-alumina system were studied through isothermal solidification. It was revealed that the ternary system consists of an Al_2O_3-rich spinel solid solution-YAG binary eutectic system and an Al_2O_3-rich spinel solid solution-YAG-alumina ternary eutectic system. For the former, it was verified by the solidified sample that $YAlO_3$ metastable phase could be precipitated during rapid cooling process. When reheating the solidified sample containing the $YAlO_3$ precipitates, however, YAG phase was reprecipitated in the form of nano-sized grains. For the solidified sample with alumina primary phase, the metastable precipitation of $YAlO_3$ phase did not occur at the same cooling rate, instead, layered YAG precipitates were formed between the alumina primary phase and the characteristic eutectics consisting of alumina-rich spinel solid solution and YAG phases. Similar microstructure was observed in the ternary-phase composite with YAG-rich composition. The formation of the solidified microstructure was explained based on the divorced eutectic precipitation of ternary eutectic system with off-eutectic composition.

INTRODUCTION

Oxide/oxide composites have been recognized to be promising candidate for high-temperature structural applications above 1500°C considering the coherent chemical stability of the oxide materials in oxidizing atmosphere[1,2]. The current studies are focused on developing novel methods to prepare the oxide/oxide composites with improved high temperature mechanical properties such as creep resistance and fracture toughness. It is known that unidirectionally solidified

To the extent authorized under the laws of the United States of America, all copyright interests in this publication are the property of The American Ceramic Society. Any duplication, reproduction, or republication of this publication or any part thereof, without the express written consent of The American Ceramic Society or fee paid to the Copyright Clearance Center, is prohibited.

$YAG-Al_2O_3$ eutectic composites can exhibit excellent creep resistance[3,4], moreover, it was revealed that coherent interfaces were present between the eutectic phases[5]. The presence of the low-energy interfaces was proved to be also effective to the enhancement of high-temperature fracture toughness in Al_2O_3/c-ZrO_2 (Y) duplex ceramics[6] and of creep resistance in Al_2O_3/SiC nanocomposites[7]. In addition, it is also noted that spinel ceramics can exhibit more excellent mechanical performances such as creep resistance[8,9] and high-temperature hardness[1,2], in comparison with alumina ceramics. Moreover, both spinel and YAG crystals exhibit increased fracture toughness with increase in temperature[10,6]. Therefore, the authors suggest that YAG-spinel cubic system composites may have better high-temperature mechanical behaviours, in which the interface stresses should be thermally uniform and isotropic in comparison with YAG-alumina system.

Recently, the authors examined the phase relation of YAG-spinel system through isothermal solidification, revealing that this system exhibits a binary eutectic phase relation[11,12], which could be maintained with formation of Al_2O_3-rich spinel solid solution (MA_{ss}) [13]. Consequently, it was suggested that a binary eutectic phase region YAG-MA_{ss} and a ternary eutectic phase region YAG-MA_{ss}–alumina may exist in the YAG-spinel-alumina ternary system[13]. In this work, the composites with compositions situated in the YAG-MA_{ss} phase region and YAG-MA_{ss}–alumina phase region were prepared to verify the phase relations and especially to investigate the changes in phase compositions and microstructures in the isothermally solidified composites with emphasis on divorced eutectic precipitation.

EXPERIMENTAL PROCEDURE

High purity Al_2O_3 (99.99%,TM-100, Taimei Chem. Co., Ltd.), MgO (99.98%, Ube Ind. Ltd.) and Y_2O_3 (99.9%, Shin-Etsu Chemical Co., Ltd.) powders were used as starting materials to prepare the composites with compositions designated by C_M, C_A and C_Y, as shown in Fig. 1. The powder mixtures were prepared by ball-milling with plastic-coated steel balls in ethanol for 24h. The slurries were dried be means of evaporator and oven, then the dried mixtures were calcined at 700°C for 1h in air. Solid state reaction and densification of the samples were conducted simultaneously by hot pressing with graphite dies at 1600°C in vacuum for 1h, respectively. The hot pressed samples with diameter 16mm and thickness 2-3mm were cut into several smaller pieces to prepare for subsequent annealing treatments. The annealing processes proceeded in argon atmosphere through heating the specimens at 10°C/min up to the temperature at which the specimen was melted substantially. At each annealing temperature, the specimen was held for 1h and then cooled down at 100°C/min or 0.5°C/min. The temperature measurement was carried out by optical pyrometer (IR-Q2C, Chino Works, Ltd.) and calibrated by Pt-30Ph/Pt-6Rh thermocouple. The mean error of the calibrated temperatures was about ±5°C in the whole temperature range. In addition, part of the solidified specimens were reheated at 1500°C for 2h to examine the change in microstructure and phase composition.

The microstructures and phase compositions of the specimens before and after the annealing treatments were characterized by SEM, EPMA and XRD.

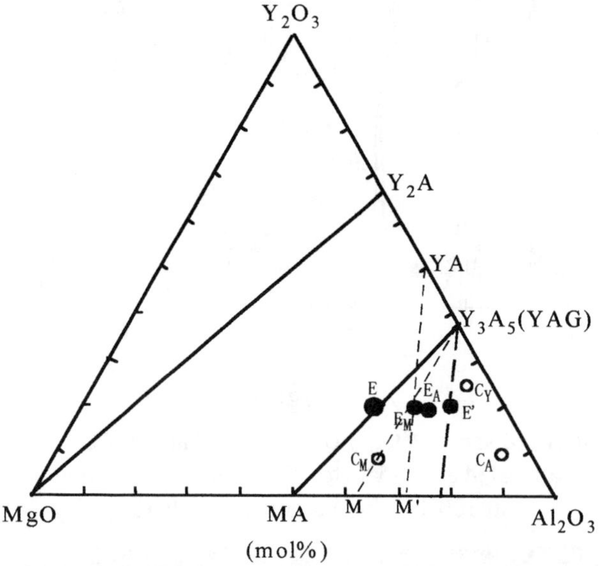

Fig. 1 Diagram of partial phase relations of system MgO-Al_2O_3-Y_2O_3.

RESULTS AND DISCUSSION

Metastable Precipitation in MA_{ss}-YAG Composite

Figure 2(a) shows the XRD pattern of the sample, C_M, hot pressed at 1600°C, indicating that only two phases, YAG and Al_2O_3-rich spinel solid solution were present in the composite. Subsequently, it was identified that the hot pressed sample was melted significantly up to 1795°C. However, it was found from XRD analysis (Fig. 2b) that the phase composition of sample annealed at 1795°C for 1h and cooled down at 100°C/min was changed, in which the initial YAG phase disappeared, instead $YAlO_3$ phase precipitated. Fig. 3 shows a typical microstructure of this solidified specimen. It is observed from Fig. 3 that the solidified microstructure consists of large grain-sized and faceted spinel solid solution (dark contrast) and intergranular precipitates in which $YAlO_3$ precipitates are involved, the average composition of the intergranular precipitates, determined by EPMA quantitative analysis, being E_M (Al_2O_3: 64±1mol%, MgO: 17±3mol% and Y_2O_3: 19±3mol%), as shown in Fig. 1.

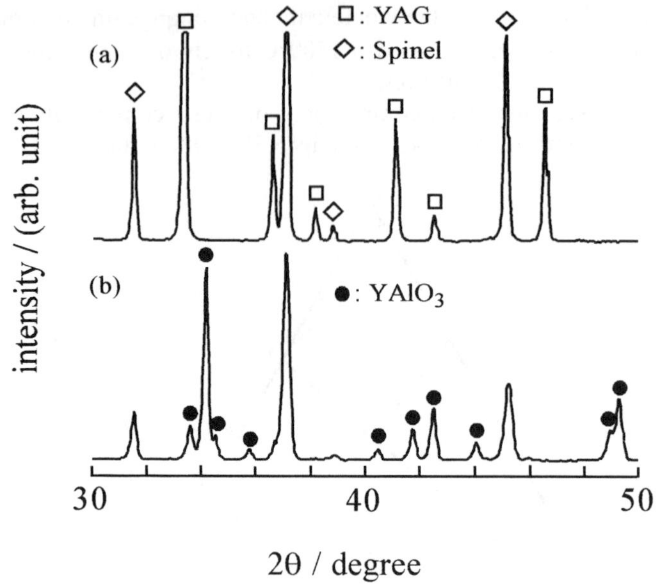

Fig. 2 XRD patterns of samples C_M, (a) hot pressed at 1600°C for 1h in vacuum, and (b) annealed at 1795°C for 1h followed by cooling down at 100°C/min in argon atmosphere.

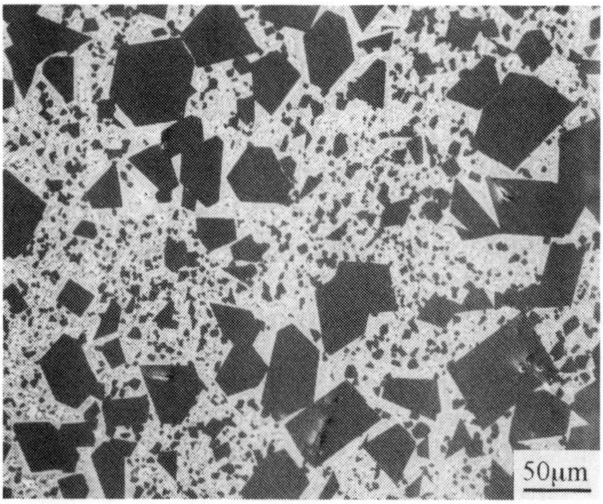

Fig. 3 SEM micrograph of samples C_M, annealed at 1795°C for 1h followed by cooling down at 100°C/min in argon atmosphere.

The decreased melting temperature of the YAG-MA_{ss} system in comparison with YAG-MA system (about 1830°C[11]) indicates that the eutectic temperature of YAG-MA_{ss} system decreases with increase in Al_2O_3/MgO molar ratios in the spinel solid solution phase, in which the dependence of solubility of Al_2O_3 in spinel on temperature may be determined by the phase diagram of MgO-Al_2O_3 binary system. In addition, it is known from comparing the compositions between C_M and E_M (Fig. 1) that the spinel solid solution should behave as primary phase. In other words, the spinel solid solution phase may coexist with the eutectic melt at the annealing temperature of 1795°C. Thus, the formation of the faceted and large grains of the spinel solid solution phase in the solidified sample(Fig. 3) may be attributable to the primary growth of the MA_{ss} phase during the annealing and solidification processes. On the other hand, the formation of the $YAlO_3$ precipitates may imply that the following reaction,

$$xY_3Al_5O_{12} + MA_{ss} \rightarrow 3xYAlO_3 + xAl_2O_3 \cdot MA_{ss} \qquad (1)$$

took place during the solidification process, indicating that a spinel solid solution (MA_{ss}') phase with higher content of Al_2O_3 (or $xAl_2O_3 \cdot MA_{ss}$) should be coexist with the $YAlO_3$ phase in the intergranular precipitates (Fig. 3). It could be thought from Fig. 1 that the initial composition of the spinel solid solution phase in this composite should be situated at M based on the sample composition C_M, while the composition of MA_{ss}' in the intergranular precipitate with composition E_M should be at M'. However, XRD analyses (Fig. 2) showed that the positions of the intensities corresponding to spinel solid solution phases in as hot pressed and solidified specimens are not changed evidently. A possible reason may be due to the MA_{ss}' precipitate in the intergranular phases(Fig. 3), which was present in the form of an amorphous phase. As suggested in the work on rapidly solidified YAG-spinel and YAG-alumina composites with off-eutectic composition[14,15], the formation of the amorphous spinel solid solution, if any, could be attributed to the lower melting temperature of the metastable system $YAlO_3$-MA_{ss} (or MA_{ss}' in this work) than that of the equilibrium system YAG-spinel solid solution, below which nucleation of spinel phase was depressed.

Further experiments indicated that the phase compositions of the solidified composites could be changed when solidification and annealing processes were changed. For example, when the sample solidified from 1795°C at initially 0.5°C/min to about 1770°C and then continuously at 100°C/min, $YAlO_3$ phase would not precipitate, while the equilibrium YAG phase was maintained, as verified by the XRD patterns shown in Fig. 4(a). On the other hand, when the solidified sample containing $YAlO_3$ precipitates (Fig. 3) was reheated at 1500°C for 2h in air , the $YAlO_3$ phase was transformed into YAG , which were proved by

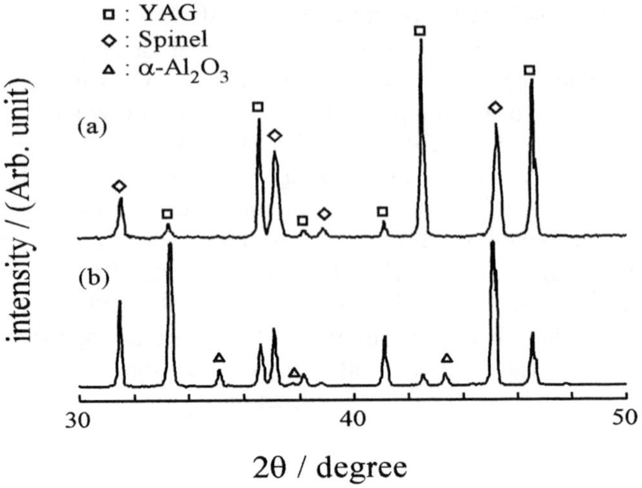

Fig. 4 XRD patterns of sample C_M, (a) annealed at 1795°C for 1h and cooled down slowly at initially 0.5°C/min to 1770°C and then quickly at 100°C/min in Ar, (b) annealed at 1795°C for 1h and cooled down at 100°C/min in Ar and then reheated from room temperature to 1500°C, keeping for 2h in air.

XRD analysis, as shown in Fig. 4(b).

Figures 5(a) and 5(b) show the microstructures of the specimens corresponding to Figs. 4(a) and 4(b), respectively. It was observed from Fig. 5(a) that the YAG phase (bright contrast) was present as a continuous phase between spinel solid solution primary phase which exhibited dendritic morphologies. In contrast, it was clearly seen from Fig. 5(b) that YAG precipitates existed as fine-grained structures, and the faceted interface characteristic was maintained between the YAG precipitates and the large sized primary phase of spinel solid solution. In addition, it was also observed that some dendritic precipitates of spinel solid solution, demonstrated by EPMA, were formed with random distributions in YAG precipitates (Fig . 5b) . Furthermore, it may be noted that the interface characteristics between YAG and spinel solid solution phases in the two specimens shown in the SEM micrographs (Fig. 5) seem to be different. The interfaces shown in Fig. 5(a) exhibited brighter contrast than the surrounding phases, which was not observed in Fig. 5(b).

The precipitation of $YAlO_3$ and spinel solid solution in rapid solidification process indicated that short-range order structure of $YAlO_3$ and MA_{ss} were present in the eutectic melt at the beginning of the solidification. However, due to the decreased solubility of Al_2O_3 in spinel with decrease in temperature , Al_2O_3 would

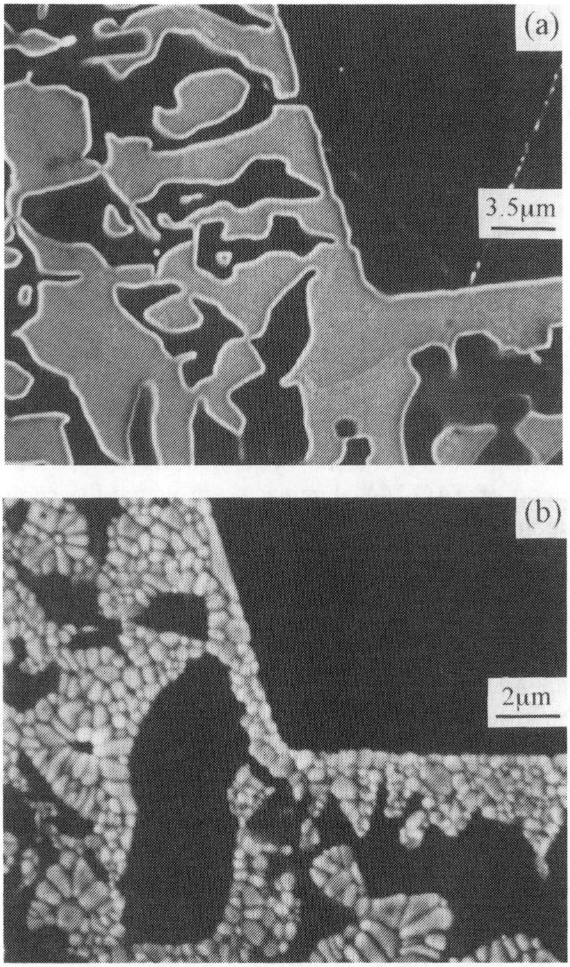

Fig. 5 SEM micrographs of the sample C_M, corresponding to Figs. 4(a) and (b), respectively.

precipitate gradually from the spinel solid solution under certain degree of supercooling. Therefore, it may be suggested that during slow cooling process, the basic structural units of Al_2O_3 exist throughout the supercooled melt, and at the same time, nuclei of YAG may be generated through the adverse reaction of Eq. (1) at relatively higher temperature. The interface characteristics of the specimen shown in Fig. 5(a) may imply that the eutectic nucleation of YAG and spinel solid solution occurred at the solid/liquid interfaces during the slow solidification process. Similarly, for the sample reheated at 1500°C (Fig. 5b), large amount of Al_2O_3 should be precipitated from the initial spinel solid solution, which, then reacted with $YAlO_3$ by solid state reaction to produce YAG. Evidently, nucleation process of the YAG phase and formation of the YAG/spinel interfaces by solid state reaction in the reheating process at relatively low temperature (1500°C) were different from the precipitation process during solidification, in which YAG phase should precipitate by heterogeneous nucleation at the interfaces between primary phase spinel solid solution and the supercooled eutectic melt. It can be expected that the interfaces formed in solidification should possess lower interface energies than those formed by solid state reaction.

Microstructural Characteristics of YAG-MA_{ss}-A eutectic Composites with off-Eutectic Compositions

It has proved that binary eutectic system, YAG-MA_{ss}, is involved in the ternary phase region of YAG-MA-alumina(A), in which the composition of the spinel solid solution (MA_{ss}) should be dependent on the phase relation of binary MgO- Al_2O_3 system. Subsequently, it can reasonably be expected that a ternary eutectic system, YAG-MA_{ss}-A, is present. The presence of the ternary system was verified by the composites with compositions C_A and C_Y, XRD patterns of the hot-pressed specimens being shown in Fig. 6. It was further identified through annealing experiments that the melting temperature of sample C_A is about 1775°C, indicating that the lowest melting temperature or eutectic point in the entire phase region of YAG-MA-A is situated in the sub-ternary phase region of YAG-MA_{ss}-A. Furthermore, it was indicated that equilibrium ternary phase relations in samples

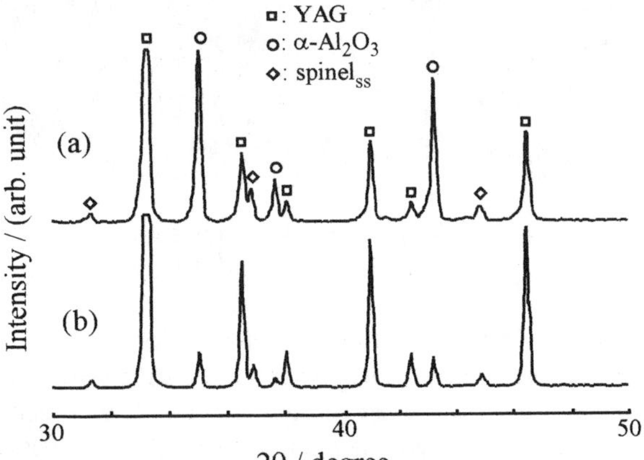

Fig. 6 XRD patterns of the composites with compositions C_A and C_Y, hot pressed at 1600°C for 1h in vacuum.

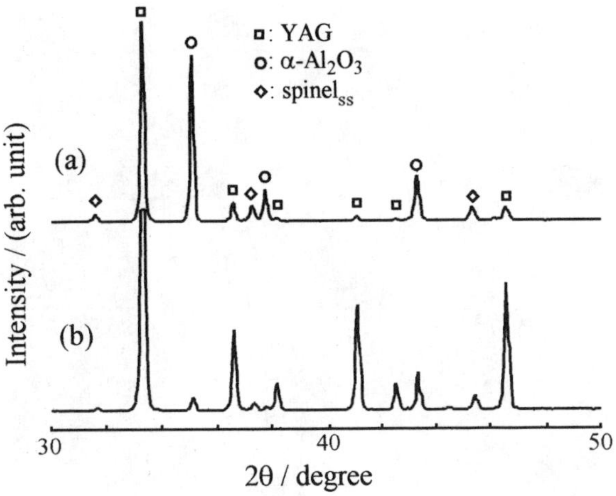

Fig. 7 XRD patterns of samples C_A (a) and C_Y (b), annealed at 1795°C for 1h and cooled down at 100°C/min in Ar.

C_A and C_Y were maintained during the annealing process at the temperatures higher than melting temperature, as proved by XRD patterns shown in Fig. 7. It may be noted that under the identical annealing process, metastable precipitation of $YAlO_3$ occurred only in sample C_M (Fig. 2b).

Figure 8 shows the typical microstructures of the specimens C_A and C_Y annealed at 1795°C for 1h and cooled down at 100°C/min in argon atmosphere, corresponding to Fig. 7. For specimen C_A (Fig. 8a), characteristic eutectic precipitates were observed, whose composition was detected by EPMA to be Al_2O_3: 66.5mol%, MgO: 15.5mol%, Y_2O_3: 18mol%, as expressed by E_A in Fig. 1. Moreover, it is noted that a layered precipitate exhibiting bright contrast, which was determined by EPMA to be YAG precipitate, was formed between the primary phase alumina and the characteristic eutectic precipitates. In contrast, for specimen C_Y (Fig. 8b), intergranular Al_2O_3 precipitates which exhibited a continuous phase were formed surrounding the primary phase YAG, while the spinel composition appeared only locally in the Al_2O_3 precipitates, which was identified by EPMA mapping analysis.

Formation of the microstructural characteristics shown in Fig. 8 may be attributable to the divorced eutectic precipitation in ternary eutectic system with off-eutectic composition P, as illustrated schematically in Fig. 9. According to the formation principle of the divorced eutectic precipitates (DEPs) in binary eutectic system[16], the composition of the ternary eutectic melt would change from E to E' due to metastable primary growth of phase A. The subsequent precipitation

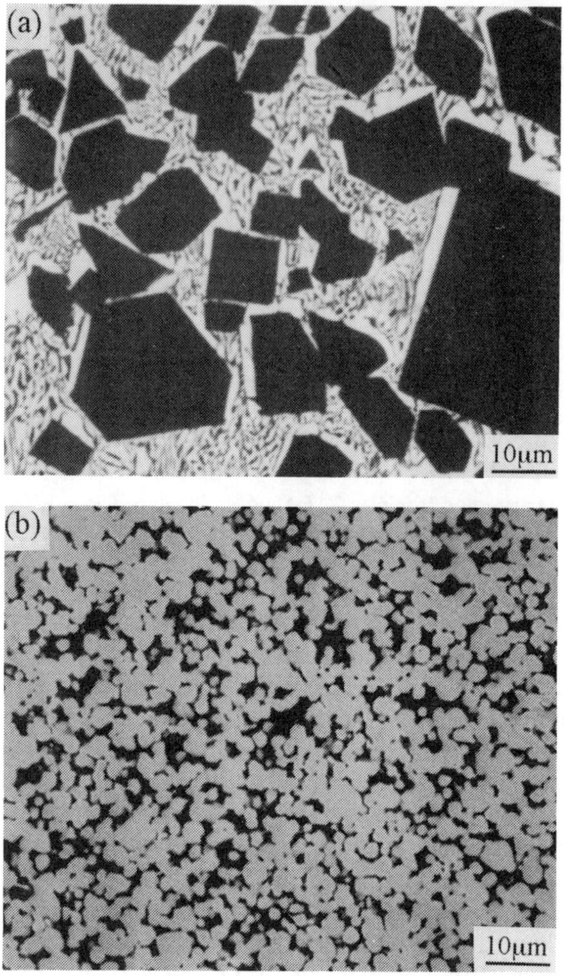

Fig. 8 SEM micrographs of the samples C_A (a) and C_Y (b), annealed at 1795°C for 1h and cooled down at 100°C/min in Ar.

of the liquid phase of composition E' would depend on the nucleation barriers between homogeneous nucleation within the liquid phase and the heterogeneous nucleation on the primary phase A. Certainly, the heterogeneous nucleation would more possibly occur before the homogeneous nucleation, in which two kinds of nuclei with compositions B and C would be involved in the heterogeneous nucleation processes on the primary phase A. In this case, if phase B nucleated prior to phase C on the primary phase A, the liquid phase composition would tend to change metastably from E' to E" with the decrease in temperature. At this moment, if phase C nucleated within the supercooled melt, the metastable eutectic precipitates with composition eutectic E" would be obtained which might correspond to the characteristic eutectics with composition E_A in the solidified specimen C_A containing primary phase of alumina (Fig. 8a). If it is true, it will imply that in the ternary eutectic melt of system YAG-MA_{ss}-A with Al_2O_3 as primary phase, YAG (regarding as phase B in Fig. 9) can precipitate secondarily on the primary phase Al_2O_3, indicating that the nucleation free energy of YAG is lower than that of MA_{ss} (regarding as phase C in Fig. 9) on the identical Al_2O_3 primary phase. Similarly, the nucleation free energy of Al_2O_3 should be lower than

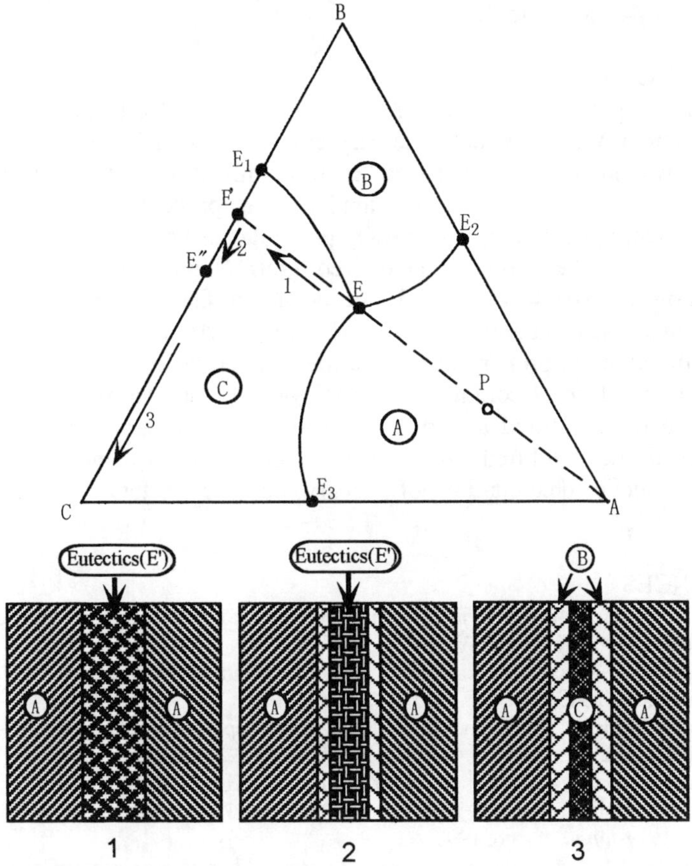

Fig. 9 Schematic illustration of the formation process of the divorced eutectic precipitates (DEPs) in a ternary eutectic system with off-eutectic composition P.

that of MA_{ss} on the identical YAG primary phase, as identified by the solidified specimen C_Y with YAG as primary phase (Fig. 8b).

Furthermore, it can certainly be thought from Fig. 9 that if the secondary precipitation of phase B could be controlled to precipitate continuously, finally only component C would be remained in the supercooled melt or precipitate, corresponding to the microstructure as illustrated in Fig. 9. It is seen from Fig. 9 that there are three layers of interphases could be formed in the solidified ternary eutectic composites with off-eutectic composition. Because the formation processes of the intergranular precipitates are assumed to be similar to the formation of the DEPs in binary eutectic system, the intergranular precipitates could form as nano-size crystallites and the interfaces or sub-interfaces could behave as low-energy interfaces.

CONCLUSIONS

The YAG-spinel(MA)-alumina(A) system were revealed to consist of a binary eutectic system YAG-MA_{ss} and a ternary eutectic system YAG-MA_{ss}-A. For the former, it was identified by the isothermal solidification that $YAlO_3$ metastable phase could be precipitated during rapid cooling process. When reheating the solidified sample containing the $YAlO_3$ precipitates, however, YAG phase was reprecipitated in the form of nano-sized grains. In contrast, the metastable precipitation of $YAlO_3$ phase did not occur in the alumina-rich composite solidified at the same cooling rate, instead, layered YAG precipitates were formed between the alumina primary phase and the characteristic eutectics consisting of Al_2O_3-rich spinel solid solution and YAG phases. Similar microstructure was observed in the solidified ternary composite with YAG as primary phase. The formation of the solidified microstructures could be explained based on the divorced eutectic precipitation of ternary eutectic system with off-eutectic composition.

REFERENCES

[1]R. Raj, "Fundamental research in Structural ceramics for Service Near 2000°C", *J. Am. Ceram. Soc.*, **76** [9] 2147-74 (1993).

[2]W. B. Hillig, "A Methodology for Estimating the Mechanical Properties of Oxides at High Temperatures", *J. Am. Ceram. Soc.*, **76** [1] 129-38 (1993).

[3]T. A. Parthasarathy, T. Mah, and L. E. Matson, "Creep Behavior of an Al_2O_3-$Y_3Al_5O_{12}$ Eutectic Composite," *Ceram. Eng. Sci. Proc.*, **11** [9-10] 1628-38 (1990).

[4]Y. Waku, N. Nakagawa, T. Wakamoto et al., "High Temperature Properties of Unidirectionally Solidified Al_2O_3/YAG Composites"; pp.170 in *Proceedings of Fall Meeting of Ceram. Soc. of Japan*, Kyoto, 1995.

[5]R. S. Hay and L. E. Matson, "Alumina/Yttrium-Aluminum Garnet Crystallographic Orientation Relationships and Interphase Boundaries: Observations and Interpretation by Geometric Criteria", *Acta Mettal. Mater.*, **39** [8] 1981-94 (1991).

[6]J. D. French, H. M. Chan, M. P. Harmer, and G. A. Miller, "High-Temperature Fracture Toughness of Duplex Microstructures", *J. Am. Ceram. Soc.*, **79** [1] 58-64 (1996).
[7]T. Ohji, T. Hirano, A. Nakahira, and K. Niihara, "Particle/Matrix interface and Its Role in Creep Inhibition in Alumina/Silicon Carbide Nanocomposites", *J. Am. Ceram. Soc.*, **79** [1] 33-45 (1996).
[8]R. W. Davidge, *Mechanical Behaviours of Ceramics*, Eds. R. W. Cahn, M. W. Thompson and I. M. Ward, Cambridge Univ. Press, London, 1979.
[9]T. Shiono, H. Ishitomi, Y. Okamoto and T. Nishida, "Deformation Behavior of Polycrystalline $MgAl_2O_4$ Spinel at High Temperature", p. 78, in *Proceedings of Fall Meeting of Ceram. Soc. of Japan*, Kyoto, 1995.
[10]R. L. Stewart and R. C. Bradt, "The Brittle to Ductile transition in $MgAl_2O_4$ Spinel", pp. 21-29, Plastic Deformation of ceramics, Eds. R. C. Cradt, C. A. Brookes and J. L. Routbort, plenum Press, New York, 1995.
[11]S. Wang, T. Akatsu, Y. Tanabe, Z. Nakagawa and E. Yasuda, "Identification of Eutectic Phase Relation and Microstructures of Isothermally Solidified YAG-Spinel Composite", to be published.
[12]S. Wang, T. Akatsu, Y. Tanabe and E. Yasuda, "Microstructural Characteristics of $Al_2O_3/Y_3Al_5O_{12}/MgAl_2O_4$ Composites with Intergranular Eutectics", *Proceedings of The 2nd International Meeting of Pacific Rim Ceramic Societies*, Cairns, Australia, 15-17 July 1996, in press.
[13]S. Wang, T. Akatsu, Y. Tanabe, Z. Nakagawa and E. Yasuda, " Eutectic Precipitation of Spinel Solid Solution-YAG System", *J. Am. Ceram. Soc.*, **81** [1] (1996).
[14]S. Wang, T. Akatsu, Y. Tanabe and E. Yasuda, "Metastable Precipitation of $YAlO_3$ in Isothermally Solidified YAG-Spinel Composite with YAG-Rich Composition", unpublished work.
[15]S. Wang, F. Yamamoto, T. Akatsu, Y. Tanabe and E. Yasuda, "Metastable Precipitation in Isothermally Solidified YAG-Alumina Composites with off-Eutectic Composition", unpublished work.
[16]S. Wang, T. Akatsu, Y. Tanabe and E. Yasuda, "Divorced Eutectic Precipitation in Spinel-YAG Composite with Spinel-Rich Composition", in this proceedings.

PRECIPITATION OF $YAlO_3$ IN YAG-SPINEL EUTECTIC COMPOSITE WITH YAG-RICH COMPOSITION

Shuqiang Wang
New Energy and Industrial Technology Development Organization(NEDO), National Industrial Research Institute of Nagoya, Nagoya 462-8510, Japan
Takashi Akatsu, Yasuhiro Tanabe and Eiichi Yasuda
Research Center for Materials Design, Materials and Structures Laboratory, Tokyo Institute of Technology, 4259 Nagatsuta, Midori, Yokohama 226, Japan

ABSTRACT

Isothermally solidified microstructure of YAG-spinel composite with YAG-rich composition was investigated. It was found from XRD analysis that spinel phase disappeared and $YAlO_3$ phase was formed significantly in the solidified composite. Microstructural observations showed that the $YAlO_3$ phase with relatively higher hardness was present as needle-like precipitates grown in a phase with spinel rich composition. Formation of the $YAlO_3$ precipitates was attributed to the metastable precipitation of melting YAG component during the cooling process, which resulted in the formation of a metastable system of $YAlO_3$ and Al_2O_3-rich spinel solid solution with lower melting temperature. It was suggested that the Al_2O_3-rich spinel solid solution was present as a metastable amorphous phase due to the lower nucleation temperature.

INTRODUCTION

Divorced eutectic precipitation has been investigated in isothermally solidified $Y_3Al_5O_{12}$(YAG)-$MgAl_2O_4$(spinel) composite with spinel-rich composition[1,2]. It was indicated that the divorced eutectic precipitates (DEPs) of YAG composition could be present as nano-sized intergranular precipitates in the solidified composite. Moreover, it was revealed that specific crystallographic orientation relationships were present at the interfaces between the YAG precipitates and the primary phase spinel, indicating that strong or low-energy interfaces were formed to some extent in the solidified composite. As a result, it was expected that high temperature mechanical properties such as creep resistance of oxide-oxide composites could be enhanced through formation of the divorced eutectic

To the extent authorized under the laws of the United States of America, all copyright interests in this publication are the property of The American Ceramic Society. Any duplication, reproduction, or republication of this publication or any part thereof, without the express written consent of The American Ceramic Society or fee paid to the Copyright Clearance Center, is prohibited.

precipitates. However, it was known that the formation of the DEPs depended on the metastable primary growth below eutectic temperature[1]. For the isothermally solidified YAG-spinel composite with spinel composition, the metastable primary growth was verified by the acicular spinel precipitates present in the solidified composite[2]. Obviously, growth rate of the primary phase would be an important factor to determine the metastable change in composition and melting temperature of supercooled melt during the cooling process, which should be related not only to the cooling rate but also to the crystal structure of the primary phase. It may be imagined that under the same solidification conditions, the composition of the divorced eutectic precipitates and even the solidified microstructure would be changed with the change in the primary phases with different crystal structures.

Therefore, in this work, the isothermally solidified YAG-spinel composite with YAG-rich composition was investigated in comparison with the composite with spinel-rich composition. It was found that needle-like precipitates of $YAlO_3$ were formed significantly in the solidified microstructure of the YAG-rich composite. The formation process and the growth morphology of the metastable $YAlO_3$ precipitates in the YAG-spinel system are discussed in this paper.

EXPERIMENTAL PROCEDURE

High purity Al_2O_3 (99.99%,TM-100, Taimei Chem. Co., Ltd.), MgO (99.98%, Ube Ind. Ltd.) and Y_2O_3 (99.9%, Shin-Etsu Chemical Co., Ltd.) powders were used as starting materials to prepare the sample with composition C_Y (YAG : spinel = 11 : 4 molar ratio), as shown in Fig. 1, in which C_S expresses the

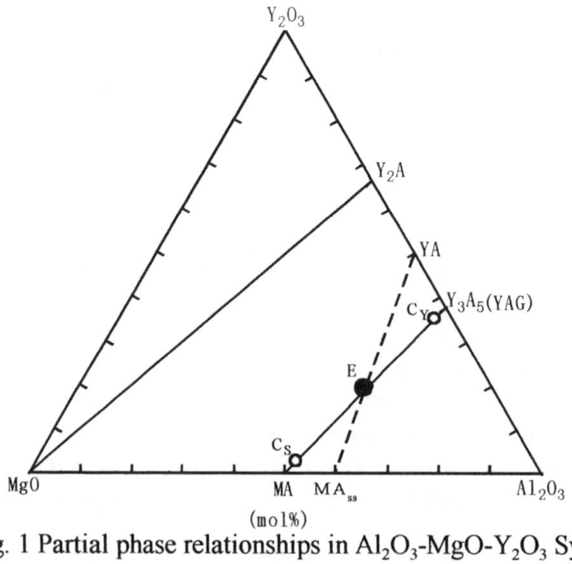

Fig. 1 Partial phase relationships in Al_2O_3-MgO-Y_2O_3 System.

composition of the spinel-rich (YAG : spinel = 1 : 30 molar ratio) sample, as reported previously[1]. Solid state reaction and densification of the powder mixture was carried out by hot pressing at 1700°C for 1h in argon atmosphere. The hot pressed sample with diameter 16mm and thickness 2mm was then cut into smaller pieces for subsequent annealing treatment which was conducted at higher temperature up to the eutectic point about 1830°C of this system[3] for 1h in Ar and cooled down quickly at about 100°C/min.

Phase compositions and microstructures of polished surfaces of the hot pressed and annealed specimens were analyzed by means of x-ray diffraction (XRD), electron probe micro analyzer (EPMA, Shimadzu Mfg. Co., Ltd.) equipped with SEM and back scattering electron(BSE) detectors, and transmission electron microscopy (TEM, JEM-2000EX). The specimen for the TEM analysis was prepared by Ar ion-thinning. In addition, polished surfaces of the annealed specimen were analyzed by atomic force microscope (AFM) to examine the growth morphology of the precipitates.

RESULTS AND DISCUSSION

Figure 2 shows the XRD patterns of the specimen hot pressed at 1700 °C for 1h and the specimen annealed at 1830°C for 1h followed by cooling down at 100°C/min. It was found that the spinel phase present initially in the hot pressed sample disappeared, while $YAlO_3$ phase was formed in the annealed specimen. A typical SEM micrograph of the annealed specimen are shown in Fig. 3. It can be

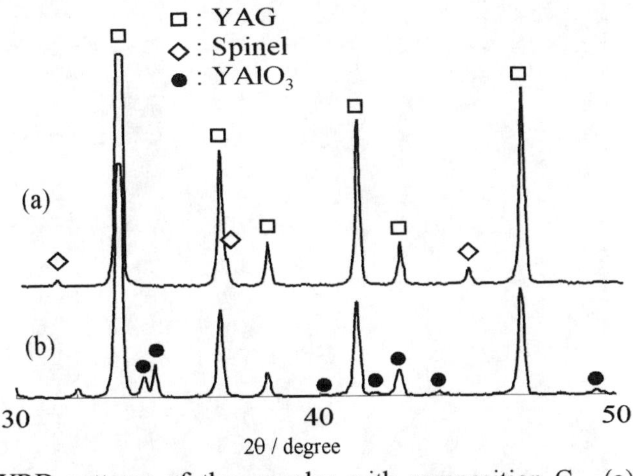

Fig. 2 XRD patterns of the samples with composition C_Y, (a) hot pressed at 1700°C for 1h, (b) annealed at 1830°C for 1h and cooled down at 100°C/min in argon atmosphere.

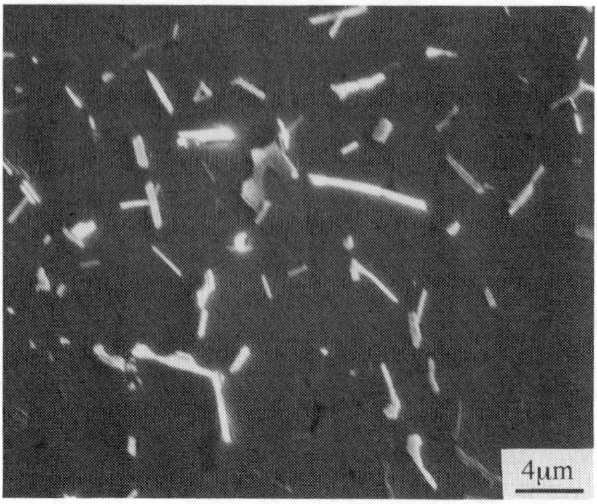

Fig. 3 A typical SEM micrograph of the polished surface of the solidified specimen containing $YAlO_3$ precipitate.

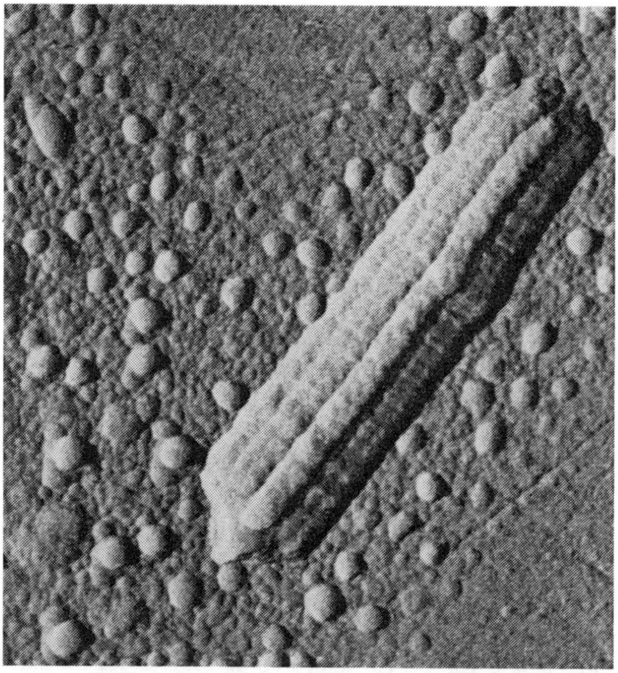

Fig. 4 AFM deflection image of a $YAlO_3$ precipitate shown in Fig. 3.

seen from the SEM picture that large amount of acicular precipitates are produced significantly in the solidified specimen, which were proved to possess $YAlO_3$ composition. The growth morphologies of the $YAlO_3$ precipitates observed on the polished surface were verified by AFM analysis. Fig. 4 shows an AFM deflection image of a $YAlO_3$ precipitate on the polished surface (Fig. 3). It was clearly indicated that the needle-like precipitate exhibited the height raised from the surrounding surface, which may imply that the $YAlO_3$ precipitates possess higher hardness than the surrounding phases. In addition, it was noted in Fig. 4 that many small particles with similarly higher hardness were present around the acicular precipitate. Furthermore, the acicular precipitates in the annealed specimen were observed by TEM, as shown Fig. 5, the single crystal feature of the precipitate

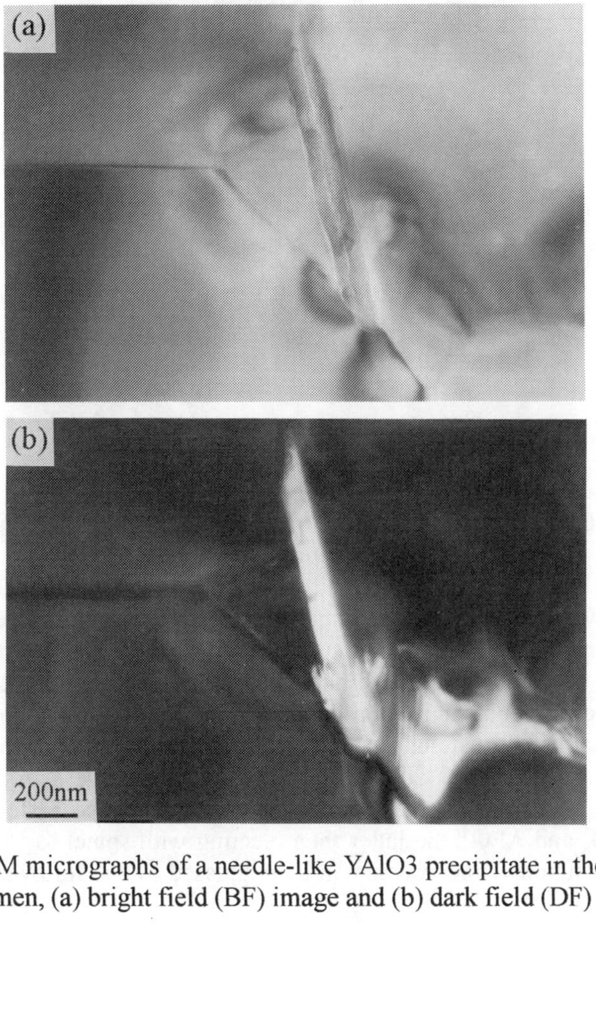

Fig. 5 TEM micrographs of a needle-like YAlO3 precipitate in the solidified specimen, (a) bright field (BF) image and (b) dark field (DF) image.

being identified by the dark-field image and selected area diffraction (SAD) analysis of the $YAlO_3$ precipitates. In addition, the TEM picture of Fig. 6 shows that the needle-like precipitate grows intra-granularly. The phase surrounding the precipitate may correspond to the phase in which the nano-sized particles or nuclei were present (Fig. 4). It was further identified by XRD analysis that the metastable $YAlO_3$ precipitates was still present after the reheating treatment at 1400°C for 1h in air.

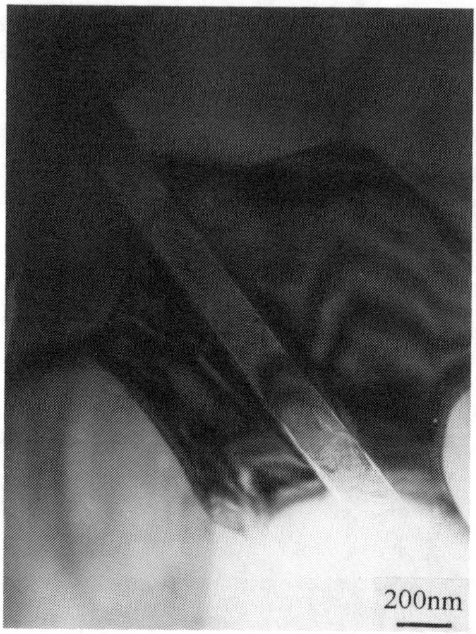

Fig. 6 TEM bright field image of a needle-like $YAlO_3$ precipitate grown intragranularly.

It is known that YAG is the primary phase which should co-exist with the eutectic melt at the eutectic temperature. Moreover, primary growth of the YAG phase should occur before eutectic nucleation during cooling process because of the lower heterogeneous nucleation barrier. However, due to the complex crystal structure containing 160 atoms per unit cell, the equilibrium precipitation of YAG phase would be depressed during the rapid cooling process. As a result, the melting YAG component in the supercooled melt could precipitate as metastable phases, $YAlO_3$ and Al_2O_3, the latter then reacting with spinel to form Al_2O_3-rich spinel solid solution. This process can be expressed by the following continuous reaction equations.

$$Y_3Al_5O_{12} = 3YAlO_3 + Al_2O_3 \quad (1)$$
$$mAl_2O_3 + nMgAl_2O_4 = nMgO{\cdot}(n{+}m)\ Al_2O_3 \quad (2)$$

Consequently, a metastable system, $YAlO_3$-spinel solid solution or YA-MA_{ss}, would result from the eutectic melt during the rapid solidification process. Assuming that the metastable YA-MA_{ss} system could still maintain a eutectic phase relationship with eutectic composition E', its eutectic temperature ($T_{E'}$) would be lower than T_E, as illustrated schematically in Fig. 7. If it was true, $YAlO_3$ phase would precipitate preferentially as a primary phase from the melt with composition E in the metastable YA-MA_{ss} system, resulting in the change in composition of the supercooled melt from E towards E' with decrease in temperature until the temperature $T_{E'}$ was achieved, below which YA-MA_{ss} eutectic might precipitate . Otherwise , the $YAlO_3$ phase would continue to

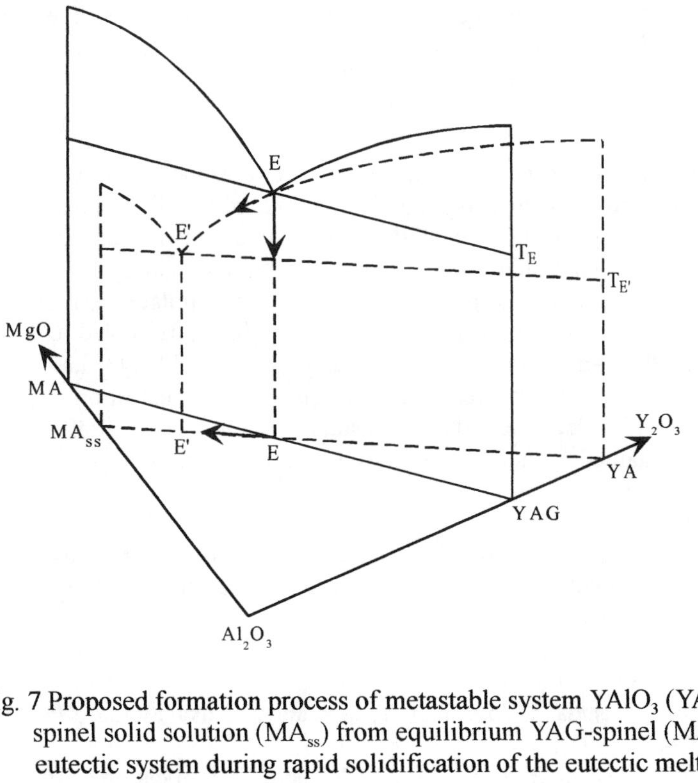

Fig. 7 Proposed formation process of metastable system $YAlO_3$ (YA)-spinel solid solution (MA_{ss}) from equilibrium YAG-spinel (MA) eutectic system during rapid solidification of the eutectic melt.

precipitate with further decrease in temperature, and finally, only spinel solid solution (MA_{ss}) would be present in the supercooled melt.

Based on the above analysis, the acicular precipitates formed in the specimen annealed at 1830°C for 1h and cooled down at 100°C/min could be attributed to the metastable precipitation of $YAlO_3$ phase due to metastable decomposition of YAG component in the supercooled melt during rapid cooling process. According to the illustration in Fig. 7, the $YAlO_3$ phase precipitates at the temperature above $T_{E'}$, behaving as primary phase growth in a supercooled melt with changed composition, and therefore exhibiting acicular growth morphology. Below $T_{E'}$, however, not only the primary growth but also nucleation of $YAlO_3$ phase would occur in the metastable melt with composition E'. The nano-sized particles, as observed in Fig. 4, might be the $YAlO_3$ nuclei formed in the metastable melt below $T_{E'}$. Evidently, it may be thought that spinel solid solution would also nucleate or precipitate from the metastable melt below $T_{E'}$. However, it was known from the XRD analysis of the solidified specimen (Fig. 2b) that spinel (or solid solution) phase was not precipitated, indicating that an amorphous phase with MA_{SS} composition in the solidified specimen might be formed due to the significantly decreased nucleation temperature below $T_{E'}$, at which the nucleation of MA_{ss} phase was depressed.

CONCLUSIONS

Isothermally solidified microstructure of YAG-spinel composite with YAG-rich composition was investigated. It was identified that metastable $YAlO_3$ precipitates were present as needle-like precipitates surrounded by a phase with composition of spinel solid solution which exhibit an amorphous phase in the solidified microstructure. Formation of the $YAlO_3$ precipitates could be attributed to the metastable decomposition of YAG in the supercooled eutectic melt, resulting in the formation of the metastable system of $YAlO_3$ and spinel solid solution with lower melting temperature, below which nucleation of spinel solid solution might be depressed. The metastable phases in the solidified composite could be maintained to some extent during reheating treatment.

ACKNOWLEDGEMENT

The authors are very grateful to Mr. F. Yamamoto for his contribution to this work in sample preparation. We also thank Dr. W. Suchanek and Dr. J. –F. Ma for their helps in AFM analysis.

REFERENCES

1. S. Wang, E. Yasuda, T. Akatsu, Y. Tanabe and Z. Nakagawa, *J. Mater. Sci. Lett.*, **16**, 1580-83 (1997).
2. S. Wang, T. Akatsu, Y. Tanabe and E. Yasuda, "Identification of Eutectic phase

relation and Microstructures of Isothermally solidified YAG-Spinel Composite", to be published work.
3. S. Wang, T. Akatsu, Y. Tanabe and E. Yasuda, "Divorced Eutectic Precipitation in Isothermally Solidified YAG-Spinel Eutectic Composite with Spinel-Rich Composition", to be published work.
4. S. Wang, T. Akatsu, Y. Tanabe and E. Yasuda, "Acicular Primary Growth in Isothermally Solidified YAG-Spinel Composite with Spinel-Rich Composition", to be published work.

MICROSTRUCTURAL CHARACTERISTICS OF YAG-SPINEL EUTECTIC COMPOSITES WITH CHANGED COMPOSITIONS

Shuqiang Wang
New Energy and Industrial Technology Development Organization(NEDO), National Industrial Research Institute of Nagoya, Nagoya 462-8510, Japan
Takashi Akatsu, Yasuhiro Tanabe and Eiichi Yasuda
Research Center for Materials Design, Materials and Structures Laboratory, Tokyo Institute of Technology, 4259 Nagatsuta, Midori, Yokohama 226, Japan

ABSTRACT

Eutectic microstructures of YAG-spinel composites prepared by isothermal solidification were examined. It was identified that the composition of the eutectic precipitates in the YAG-spinel system could be changed with formation of Al_2O_3-rich spinel solid solution. The change in composition resulted in the precipitation of YAG dendrites in the solidified composite. It was observed that the divorced eutectic precipitates (DEPs) with spinel-rich composition were formed significantly between the YAG dendrites and the characteristic eutectic precipitates. Furthermore, it was revealed that the microstructure of the characteristic eutectics with Al_2O_3-rich spinel solid solution and YAG composition was composed of dendritic YAG grains and fine-grained matrix of spinel solid solution.

INTRODUCTION

Divorced eutectic precipitates (DEPs) of $Y_3Al_5O_{12}$ garnet (YAG)-$MgAl_2O_4$ (spinel) system were investigated by isothermal solidification[1]. Subsequently, the binary eutectic phase relation of this system was verified by the isothermally solidified microstructure of the YAG-spinel composite[2], the eutectic temperature of the characteristic eutectics with composition E (Figure 1) being about 1830°C. Furthermore, it was found that Al_2O_3-rich spinel solid solution (MA_{ss}) would be formed due to the vaporization of MgO component in the spinel phase during high temperature annealing process in low partial pressure of oxygen, e.g., in argon atmosphere. It is known from the phase diagram of Figure 1 that the formation of MA_{ss} implies that the eutectic composition of the YAG-spinel system

To the extent authorized under the laws of the United States of America, all copyright interests in this publication are the property of The American Ceramic Society. Any duplication, reproduction, or republication of this publication or any part thereof, without the express written consent of The American Ceramic Society or fee paid to the Copyright Clearance Center, is prohibited.

would change from E into the ternary system YAG-spinel-alumina. It has been identified that the ternary system is composed of YAG-MA_{ss} binary eutectic system and YAG-MA_{ss}-alumina ternary system[3]. In this study, the microstructural change of the characteristic eutectics in YAG-spinel system, accompanied with composition change, was investigated through employing the high-temperature vaporization of MgO component in spinel phase. It was indicated that the change in composition resulted in the precipitation of YAG dendrites in the solidified composite. It was observed that the DEPs with composition rich in MA_{ss} were formed significantly between the YAG dendrites and the characteristic eutectic precipitates of YAG-MA_{ss} composition. Furthermore, it was revealed by thermally etched specimen that the characteristic eutectics of YAG-MA_{ss} composition was composed of dendritic YAG grains and fine-grained matrix of spinel solid solution.

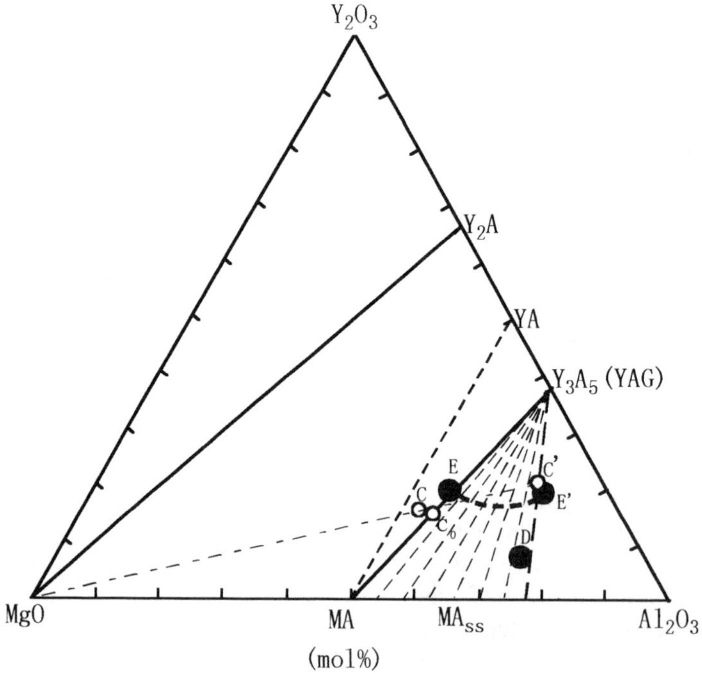

Fig.1 Diagram of partial phase relations in MgO-Al_2O_3-Y_2O_3 ternary system.

EXPERIMENTAL PROCEDURE

The samples with composition about YAG : spinel = 1: 3 (molar ratio) were prepared by ball-milling of the high-purity powder mixture of Al_2O_3, MgO and Y_2O_3 in ethanol for 24h, as described in the paper[2]. Solid state reactions and densification of the sample were carried out simultaneously during hot pressing with graphite dies at 1700°C for 1h under 33MPa in Ar atmosphere. Subsequently, the as hot pressed specimens were annealed at 1830°C for 1h, then two kinds of solidification processes were conducted using the identical graphite dies, one is that the annealed specimen was cooled down quickly at 100°C/min in Ar atmosphere, and the other is that the annealed specimen was cooled down initially 0.5°C/min to 1770°C and then continuously at 100°C/min to lower temperatures in vacuum. The microstructures and the phase compositions of the solidified specimens were analyzed by electron probe micro analyzer (EPMA-1400, Shimadzu Manufacturing. Co. Ltd., Kyoto, Japan) and XRD. The standard samples of Al_2O_3, MgO and Y_2O_3 used in the EPMA quantitative analysis were prepared by hot pressing of the identical high-purity powders, respectively.

RESULTS AND DISCUSSION

Microstructure of Rapidly Solidified Specimen

Figure 2 shows a SEM picture of the as hot pressed sample. It was clearly indicated that the composite was composed of two phases, YAG (bright contrast) and spinel (dark), which was verified by XRD analysis, the chemical composition being checked by EPMA , as expressed by C (52 mol% Al_2O_3 +

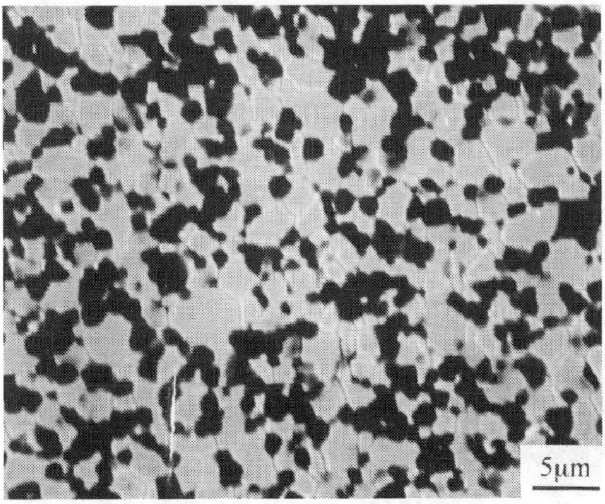

Fig. 2 SEM micrograph of the as hot pressed specimen with composition C.

32 mol%MgO + 16 mol%Y_2O_3) in Fig. 1, in which C_0 corresponds to the composition of YAG : spinel = 1 : 3 molar ratio. Subsequently, the solidified microstructures of the composite were examined. It was found that all of the solidified specimens exhibited spherical shapes, indicating that a sufficient liquid phase was formed during the high-temperature annealing processes. Figure 3 shows the typical microstructures of the specimen annealed at 1830°C for 1h and cooled down at 100°C/min in Ar atmosphere. It was observed that the characteristic eutectics and the primary phases contained microstructures were

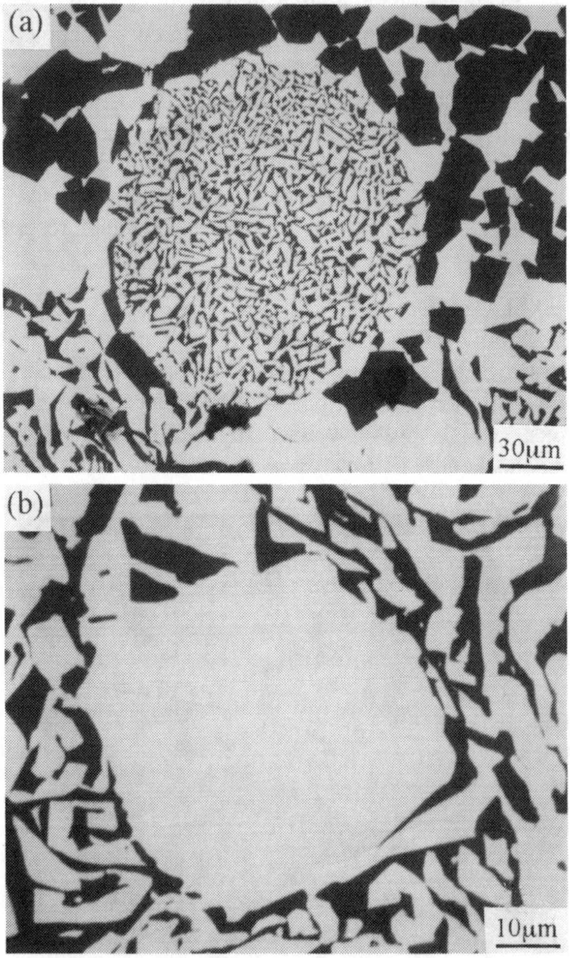

Fig. 3 Typical microstructures in the composite annealed at 1830°C for 1h and cooled down at 100°C/min in Ar, (a) a middle region, (b) a bottom region.

involved in the solidified microstructure, in which the presence of YAG and spinel phases in the solidified specimen was confirmed by XRD analysis, as described together in next section (Figure 6a). The chemical composition of the characteristic eutectic precipitates shown in Figure 3(a) was determined by EPMA quantitative analysis, as expressed by point E in Figure 1. The presence of the faceted spinel grains (dark contrast) shown in Figure 3(a) was consistent with the composition of the composite containing spinel primary phase, comparing compositions C with E (Figure 1). Whereas the formation of the primary phase-like YAG precipitate surrounded directly by the characteristic eutectics (Figure 3b) may be related to the presence of residual unmelted YAG solid phase or the composition deviation caused by the difference in densities of the components in the eutectic melt during the annealing process.

Furthermore, it was identified through thermal etching treatment that the YAG precipitates surrounding the spinel primary phase grains were formed mostly like a continuous phase exhibiting little grain boundaries (Figure 4). In addition, it may be noted from the SEM picture shown in Figure 4 that specific interfaces exhibiting raised morphology (bright contrast) were formed between primary phase spinel and precipitates YAG which may possess low interface energy. The formation of the microstructure shown in Figure 4 may be attributed to the divorced eutectic precipitation caused by the metastable primary growth of spinel below eutectic temperature during cooling process[4].

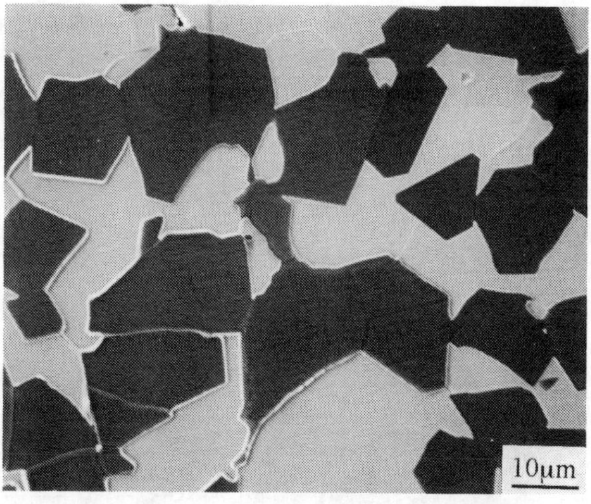

Fig. 4 SEM micrograph of the solidified composite containing primary phase spinel (dark phase) and YAG precipitates (bright phase), thermally etched at 1500°C for 1h in air.

Eutectic Precipitation and Microstructural Change of Slowly Cooled Specimen

In order to understand the changes in composition and microstructure of the YAG-spinel eutectic precipitates due to the vaporization of MgO component in spinel phase, the specimen annealed at 1830°C for 1h and cooled down initially 0.5°C/min to 1770°C and then continuously at 100°C/min to lower temperatures in vacuum was examined. It was observed that most of the solidified microstructures exhibited the characteristic eutectics and locally the dendritic precipitates (Figure 5). It was identified by EPMA analysis that the dendritic precipitates possess the composition of YAG and the composition of the characteristic eutectic precipitates is Al_2O_3: 71 mol%, MgO: 11 mol% and Y_2O_3: 19 mol%, as expressed in Figure 1 by the point E'. The XRD analysis verified that the phase compositions of this solidified specimen were composed of YAG and Al_2O_3-rich spinel solid solution (Figure 6b). The presence of the spinel solid solution was proved by the peak positions corresponding to spinel, shifting significantly toward high angle direction, comparing with the specimen solidified quickly in Ar (Figure 6a). Thus, it was concluded that the eutectic precipitates with composition E' were composed of YAG and Al_2O_3-rich solid solution of spinel (MA_{ss}). This result indicates that the binary eutectic phase relation of the YAG-spinel system can be maintained with the formation of Al_2O_3-rich spinel solid solution, as expressed by tie lines in Figure 1. It was known that the eutectic temperature of YAG-spinel system tended to decrease with the formation of Al_2O_3-rich spinel solid solution[1,5].

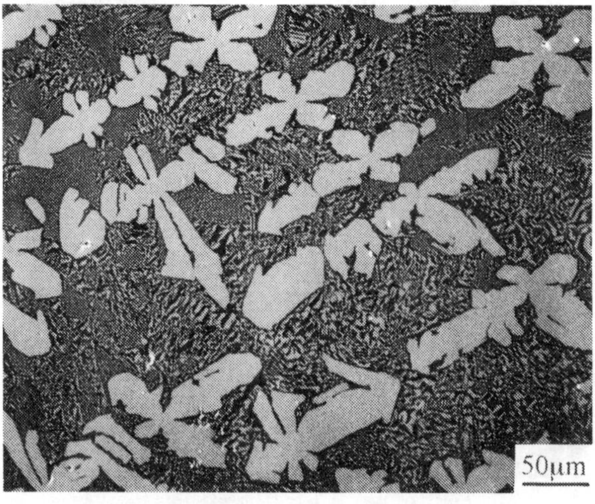

Fig. 5 Typical microstructure of the specimen annealed at 1830°C for 1h followed by cooling down slowly at 0.5°C/min to 1770°C and then quickly at 100C/min in vacuum.

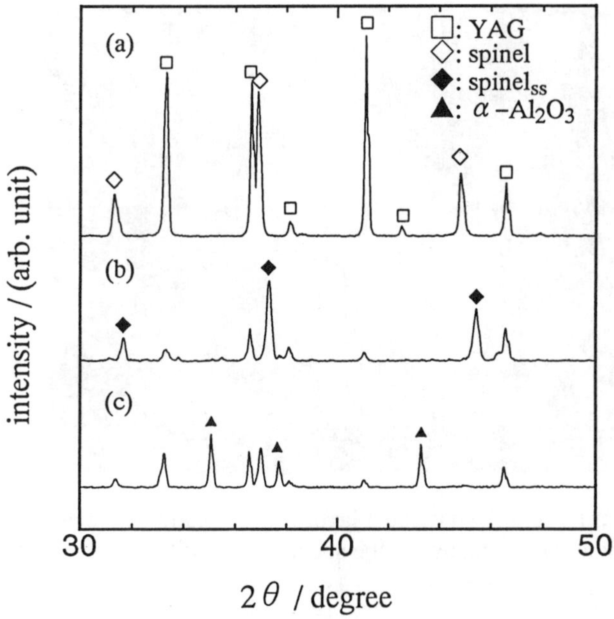

Fig. 6 XRD patterns of the YAG-spinel composite, (a) annealed at 1830°C for 1h followed by cooling at 100°C/min in Ar, (b) annealed at 1830°C for 1h followed by cooling down slowly at 0.5°C/min to 1770°C and then quickly at 100°C/min in vacuum, and (c) reheated at 1500°C for 2h in air for the specimen solidified in vacuum (b).

It was estimated from the above analyses that the composition of the sample shifted from C to C', accompanying the changes in eutectic composition from E to E' (Figure 1), due to the vaporization of MgO component in spinel during the annealing process. If it is the case, it can be expected that the primary phase was changed from spinel to YAG, comparing the compositions C and E with C' and E'. The presence of the YAG primary phase in this solidified specimen was verified by the formation of YAG dendrites (Figure 5).

In addition, it was noted from Figure 5 that there was a layer of precipitate (black phase) between the YAG dendrites and the characteristic eutectics, as seen more clearly in Figure 7. Similar phenomenon can be observed at the interdendritic regions where the characteristic eutectics were not exhibited. Figure 8 shows a high- magnification SEM micrograph of the interdendritic precipitate, whose composition determined by EPMA is situated at point D (Al_2O_3: 73 mol%, MgO: 20 mol% and Y_2O_3: 8 mol%) (Figure 1). The formation of the layered or the interdendrtic precipitates surrounding the YAG dendrites was probably

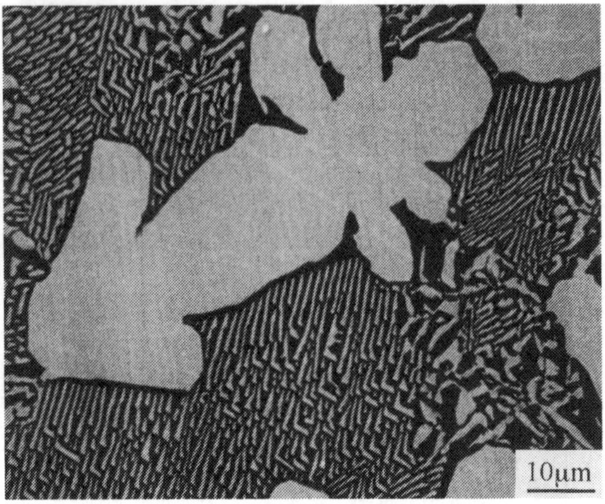

Fig. 7 SEM micrograph showing a layered precipitate between YAG dendrites and characteristic eutectics in the specimen solidified in vacuum (Fig. 5).

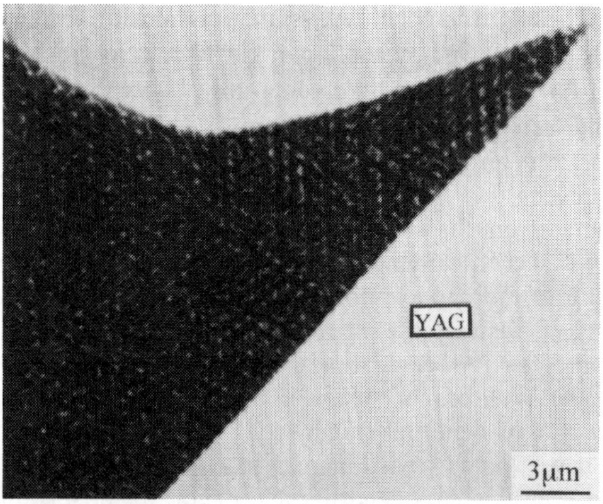

Fig. 8 SEM micrograph showing the growth morphology of the interdendritic precipitate with composition D.

associated with divorced eutectic precipitation originated from the metastable primary growth of YAG in the supercooled eutectic melt[4]. With the metastable precipitation of YAG which acts as primary phase, the spinel component in the supercooled melt was rejected at the liquid/solid interface, resulting in the spinel-rich composition in the melt near the interfaces. The presence of the metastable melt with spinel-rich composition was evidenced by the composition of the interdendritic precipitate and the morphology of the layered precipitate surrounding the YAG dendrites. The crystallography of the divorced eutectic precipitates (DEPs) and the interface character between the DEPs and the primary phase are to be clarified.

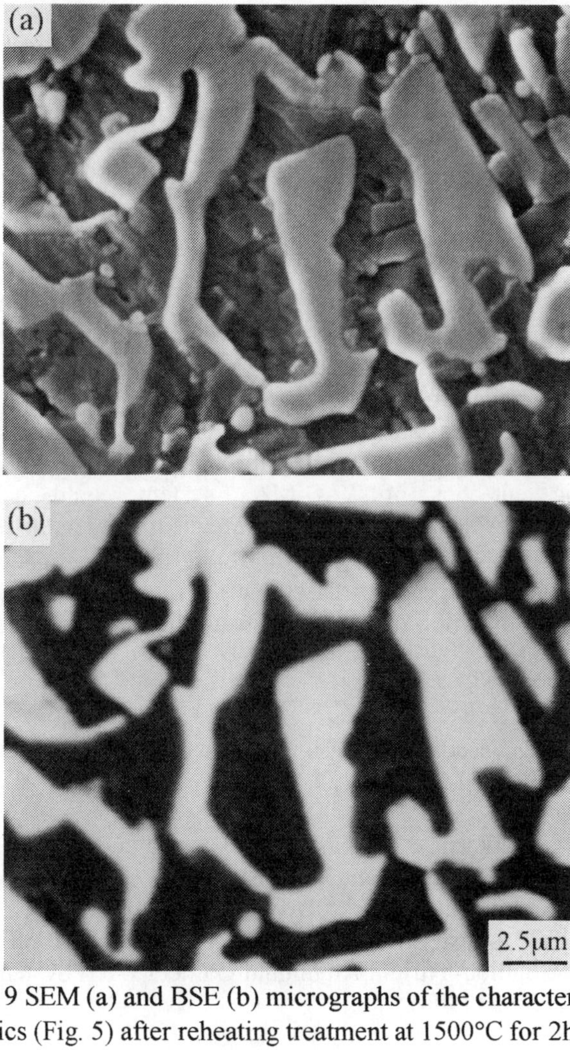

Fig. 9 SEM (a) and BSE (b) micrographs of the characteristic eutectics (Fig. 5) after reheating treatment at 1500°C for 2h in air.

On the other hand, it was thought that α-Al_2O_3 phase would precipitate from the Al_2O_3-rich spinel solid solution (MA_{ss}) formed at high temperatures during annealing at lower temperatures according to MgO-Al_2O_3 binary phase diagram, which was verified by reheating the slowly solidified specimen at 1500°C for 2h in air. It was evidenced by XRD analysis (Figure 6c) of the specimen reheated in air that α-Al_2O_3 phase appeared, moreover, the peak positions corresponding to MA_{ss} shifted left, comparing with Figure 6(b). Microstructural observation revealed that the characteristic eutectics were still maintained in the reheated specimen, however, high-magnification observation (Figure 9) showed that the phase regions of MA_{ss} (dark) exhibited a polycrystalline morphology, while the YAG phase (bright) exhibited single crystal-like dendritic growth morphologies. The reason for the formation of the fine-grained microstructure of the MA_{ss} phase in the characteristic eutectics (Figure 9) may be related to the precipitation of α-Al_2O_3 from the matrix of MA_{ss} by solid state reaction.

CONCLUSIONS

The eutectic phase relation of YAG-spinel system can be maintained with the formation of Al_2O_3-rich spinel solid solution (MA_{ss}) caused by vaporization of MgO component in spinel phase. Moreover, the divorced eutectic precipitates (DEPs) with YAG or spinel-rich composition were formed surrounding the primary phase spinel or YAG dendrites, respectively, whose formation could be attributable to the metastable primary growth in the supercooled melt below the eutectic temperature. It was revealed through thermal etching process that the characteristic eutectics of YAG-MA_{ss} system could exhibit a specific microstructure in which YAG dendrites were distributed in the fine-grained matrix composed of spinel and α-Al_2O_3 phases. The further investigation on the interface character between the primary phases and the intergranular DEPs is necessary to understand the formation of interfaces in the liquid-phase sintering ceramics or composites.

REFERENCES:

[1]S. Wang, E. Yasuda, T. Akatsu, Y. Tanabe and Z. Nakagawa, “Intergranular Precipitation in $Y_3Al_5O_{12}/MgAl_2O_4$ System”, *J. Mater. Sci. Lett.*, **16**, 1580-83 (1997).

[2]S. Wang, T. Akatsu, Y. Tanabe, Z. Nakagawa and E. Yasuda, "Identification of Eutectic Phase Relation and Microstructural Characteristic of Isothermally Solidified YAG-Spinel Composite", to be published work.

[3]S. Wang, T. Akatsu, Y. Tanabe and E. Yasuda, "Eutectic Precipitation of the Spinel Solid Solution-Yttrium Aluminum Garnet (YAG) System", *J. Am. Ceram. Soc.*, **81** [1] 263-65 (1998).

[4]S. Wang, T. Akatsu, Y. Tanabe and E. Yasuda, "Divorced Eutectic Precipitation in Spinel-YAG Composite with Spinel-rich Composition", presented at *Symposium III- Ceramic Matrix Composites, SIII-059-98, The 100th Annual Meeting & Exposition of the American Ceramic Society.*

[5]S. Wang, T. Akatsu, Y. Tanabe and E. Yasuda, "Phase Relationships and Microstructural Control of Isothermally Solidified YAG-Spinel-Alumina Composites", presented at *Symposium III- Ceramic Matrix Composites, SIIIP-016-98, The 100th Annual Meeting & Exposition of the American Ceramic Society.*

GRAIN BOUNDARY PHASE IN THE α/β-SIALON COMPOSITE

Tzer-Shin Sheu
Department of Materials Science and Engineering
I-Sou University
Kaohsiung, Taiwan

ABSTRACT

Sintering behaviors of (Ca, Li, Y)-SiAlON composites had been studied in our previous study. The results showed that the sintering additives CaO and Li_2O had excellent sinterability for this α/β-SiAlON composite. In this study, cations (Ca^{+2}) and anions (N^{-3} or O^{-2}) were found segregated into the grain boundary. From TEM microstuctural observations, a minor glassy phase which contained Ca^{+2} and Y^{+3} was present at some grain boundary triple junctions. This segregation phenomenon of large cations in the grain boundary phase was used to explain why the α/β-SiAlON composite had a higher content of β-SiAlON in the mixed oxide system. Furthermore, the oxidation behaviors of these α/β-SiAlON composites at 1400-1500°C were observed.

INTRODUCTION

The α/β-SiAlON composite has been found to have excellent mechanical properties at high temperatures.[1-2] With the sintering additive Y_2O_3, the α/β-SiAlON composite is fully densified at 1780°C by hot pressing in our previous study. For this Y-SiAlON composite, the optimum flexural strength and fracture toughness are located in the phase content with 30-40% α-SiAlON.[1] From crystal chemistry,[3-4] like Y^{+3}, cations Li^{+1} and Ca^{+2} are in the interstitial site of the Si-N lattice; therefore Li_2O and CaO are introduced to substitute Y_2O_3 as the sintering additives to fabricate α/β-SiAlON composite in this study.

The systems Ca-Si-Al-O-N and Li-Si-Al-O-N have much lower eutectic temperatures than the system Y-Si-Al-O-N.[5-7] α-SiAlON is expected to form at lower temperature in the former two systems. In our recent studies,[8] the α/β-SiAlON composite has been observed to have excellent sintering behaviors in the mixed oxide system such as Y_2O_3+Li_2O in Fig. 1. Alternatively CaO+Li_2O is also used to replace Y_2O_3 as the sintering additives to obtain the fully densified

To the extent authorized under the laws of the United States of America, all copyright interests in this publication are the property of The American Ceramic Society. Any duplication, reproduction, or republication of this publication or any part thereof, without the express written consent of The American Ceramic Society or fee paid to the Copyright Clearance Center, is prohibited.

α/β-SiAlON composite as shown in Fig. 2.[8] The phase contents of these hot-pressed samples are listed in Table I.[8] It indicates that the mixed oxide system has a higher β-SiAlON content than the single oxide system; however, this phenomenon has not been fully explained. In this study, through the observation of grain boundary phase, it is hoped that the above controversies are well understood. Furthermore, the oxidation behaviors of these α/β-SiAlON composites at high temperatures are also investigated.

Table 1. Phase content and composition of hot-pressed samples[8]

Sample	Composition	β-SiAlON (wt%)
C1	$Ca_{0.15}Si_{11.55}Al_{0.45}N_{15.85}O_{0.15}$	26
L1	$Li_{0.20}Si_{11.70}Al_{0.30}N_{15.90}O_{0.10}$	31
L2	$Li_{0.30}Si_{11.55}Al_{0.45}N_{15.85}O_{0.15}$	12
L3	$Li_{0.40}Si_{11.40}Al_{0.60}N_{15.80}O_{0.20}$	0
Y1	$Y_{0.10}Si_{11.55}Al_{0.45}N_{15.85}O_{0.15}$	67
CL1	$Ca_{0.05}Li_{0.05}Si_{11.775}Al_{0.225}N_{15.925}O_{0.075}$	100
CL2	$Ca_{0.075}Li_{0.15}Si_{11.55}Al_{0.45}N_{15.85}O_{0.15}$	94
CL3	$Ca_{0.10}Li_{0.20}Si_{11.25}Al_{0.75}N_{15.75}O_{0.25}$	75
CL4	$Ca_{0.15}Li_{0.15}Si_{11.325}Al_{0.675}N_{15.775}O_{0.225}$	50
CL5	$Ca_{0.20}Li_{0.10}Si_{11.25}Al_{0.75}N_{15.75}O_{0.25}$	20
CL6	$Ca_{0.20}Li_{0.10}Si_{11.05}Al_{0.95}N_{15.68}O_{0.15}$	26
CY1	$Ca_{0.05}Y_{0.05}Si_{11.625}Al_{0.375}N_{15.875}O_{0.125}$	100
CY2	$Ca_{0.075}Y_{0.025}Si_{11.6625}Al_{0.3375}N_{15.8875}O_{0.125}$	97
CY3	$Ca_{0.075}Y_{0.05}Si_{11.55}Al_{0.45}N_{15.85}O_{0.15}$	84
LY1	$Li_{0.025}Y_{0.075}Si_{11.625}Al_{0.375}N_{15.875}O_{0.125}$	95
LY2	$Li_{0.05}Y_{0.05}Si_{11.70}Al_{0.30}N_{15.90}O_{0.10}$	88
LY3	$Li_{0.10}Y_{0.025}Si_{11.775}Al_{0.225}N_{15.925}O_{0.075}$	88
LY4	$Li_{0.15}Y_{0.05}Si_{11.55}Al_{0.45}N_{15.85}O_{0.15}$	--
CLY1	$Ca_{0.05}Li_{0.10}Y_{0.033}Si_{11.55}Al_{0.45}N_{15.85}O_{0.15}$	--

EXPERIMENTAL PROCEDURES

Starting powders Si_3N_4 (UBE, SN-E10, 95%β), AlN(Tokuyama), Y_2O_3 (99.99% pure, Cerac Inc.), LiO_2 (97% pure, Aldrich), and $CaCO_3$ (99% pure, Ruedel-de Haen) were used to prepare the different compositions of α/β-SiAlON composite. According to the composition $Y_{k/3}Ca_{m/2}Li_nSi_{12-3(k+m+n)/2}Al_{3(k+m+n)/2}N_{16(k+m+n)/2}O_{(k+m+n)/2}$, these starting powders were mixed with the isopropyl alcohol by using a homogenizer or an attritor. After removing the isopropyl alcohol, the dried mixing powders were hot-pressed.

A batch of 15 g mixed powder was put inside the graphite die, of which its inside cross section was 5.2 x 2.6 cm^2. During hot pressing, the specimen was pressed uniaxially under 18 MPa, with 1.5 atm of N_2 flowing gas inside the

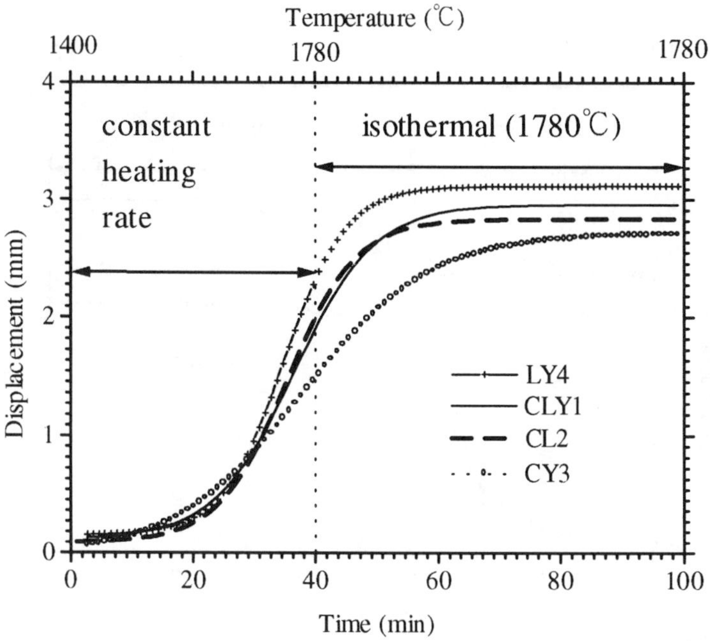

Fig.1. Sintering behavior of α/β-SiAlON composite.[8] The indices for the specimens are referred to Table I.

heating chamber. Temperature was increased from 25°C to 1780°C with a heating rate 10°C/min, and then was isothermally kept at 1780°C for 1 h. The sintering curve was observed from the displacement of the pushing ram of the hot-pressing unit.

Phase contents of the α- and β-SiAlONs were determined by an X-ray diffraction method.[9] An analytical electron microscopy (Jeol AEM 3100) was used to observe the microstructure of the sintered specimen and also was used to determine the composition of grain boundary phase. ESCA was used to analyze the composition of the fracture surface. Oxidation resistance of the sintered α/β-SiAlON composite was conducted at 1400-1500°C in air.

RESULTS AND DISCUSSION

A TEM micrograph for sample CY3 is shown in Fig. 3. The sintering additives for sample CY3 are CaO and Y_2O_3. The glassy phase pocket as marked by the symbol "A" in Fig. 3 is occasionally observed in some grain boundary triple junctions. From the energy dispersive X-ray spectrum (EDS), this minor glassy phase is identified to contain Ca^{+2} & Y^{+3}. The presence of large cations

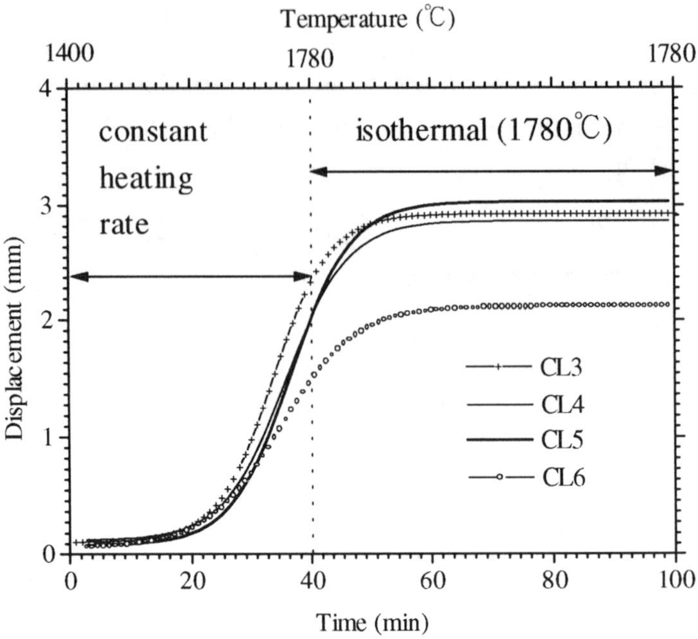

Fig. 2. Sintering behavior of hot-pressed samples.[8] Sample indices are referred to Table I.

in the grain boundary phase causes the phase content of α-SiAlON much lower than the original design. This is one of resasons for the α/β-SiAlON composite having a higher β-SiAlON content in the mixed oxide sintering additive system. A crystalline phase marked by the symbol "B" in Fig. 3 is probably α-SiAlON but not β-SiAlON, because Ca^{+2} was found in its energy dispersive X-ray spectrum.

Fig. 4 shows that several chemical species Si, N, O, and Ca exist on the fracture surface for sample CL3. The presence of these chemical species is determined by ESCA. Chemical element C is believed from the contamination during the sample preparation. The sintering additives for sample CL3 are CaO and Li_2O. The content of Ca on the fracture surface is much higher than its bulk composition. It suggests that Ca be segregated into the grain boundary because the fracture surface of α/β-SiAlON is intergranular. However, Li and Al are not detectable on the fracture surface. This no significant segregation of Al to the grain boundary is related to the cooperative formation of α- and β-SiAlONs during sintering.[1] As more β-SiAlON is formed, more Al is in the β-SiAlON crystal structure. Consequently, without enough Al, large cations such as Ca^{+2} are unable to locate themselves inside the Si-N interstitial site to form α-SiAlON.

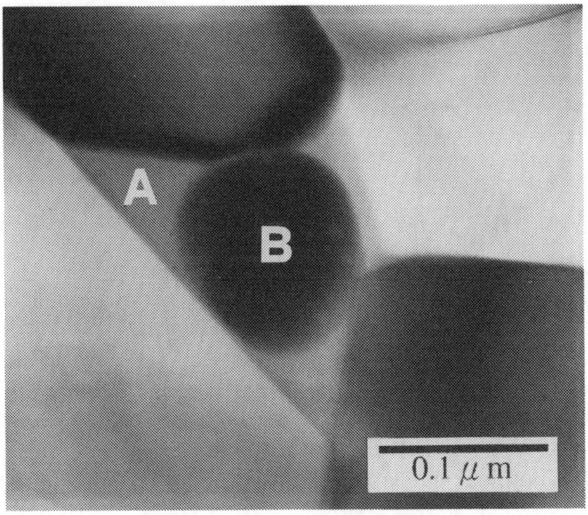

Fig. 3. TEM micrograph for sample CY3. Symbol "A" is for a grain boundary phase. Symbol "B" is for a crystalline phase.

Therefore large cations are segregated into the grain boundary but Al is not.

Fig. 5 shows the fractograph of the oxidized α/β-SiAlON. This specimen is oxidized at 1500°C in air. Several pores are found located at the interface between the oxidation layer and the α/β-SiAlON. These porosities are probably from the interactions between the evolved gases and the locally melted oxide layer during oxidation. Moreover, the oxidation layer also contains several microcracks as shown in Fig. 5. These phenomena indicate that α/β-SiAlON composite does not have excellent oxidation resistance at high temperatures. Further studies have been undertaken to tailor grain boundary phase for improving high-temperature oxidation resistance and mechanical properties.

CONCLUSIONS

1. Different combinations of oxides $CaO+Li_2O$, $CaO+Y_2O_3$, $Li_2O+Y_2O_3$, and $CaO+Li_2O+Y_2O_3$ have been used to fabricate the fully densified α/β-SiAlON composite. Li_2O and CaO are suggested to partially or completely substitute Y_2O_3 as the sintering additives.
2. Some large cations such as Ca^{+2} & Y^{+3} are segregated to the grain boundary phase. This phenomenon causes α/β-SiAlON composite in the mixed oxide system to have a higher β-SiAlON content than that in the single oxide system.
3. The (Ca, Li,Y)-SiAlON composite does not show excellent oxidation resistance at 1500°C. A further study to tailor the grain boundary phase is

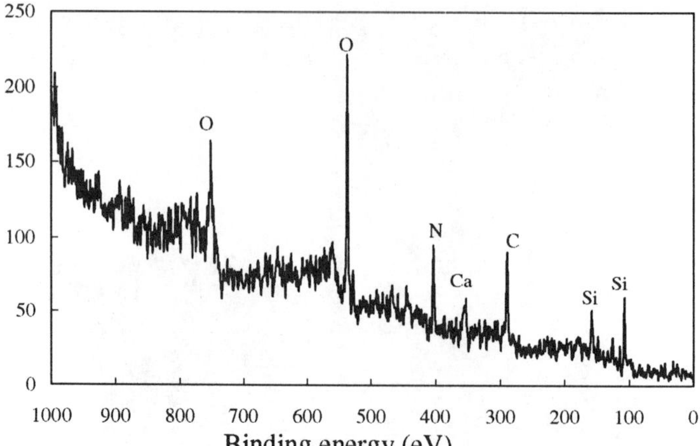

Fig. 4. Chemical species on the fracture surface of sample CL3.

undertaken to further improve its high-temperature mechanical properties under the oxidizing conditions.

ACKNOWLEDGEMENT

The author would like to thank Taiwan National Science Foundation for financial support for this research and oral presentation, and Mr. C. S. Tsai, Mr. C. C. Tsai, and Mr. D. Y. Chen at I-Sou University for their research assistance.

REFERNCES

[1]T. S. Sheu, "Microstructure and Mechanical Properties of the In-Situ β-Si_3N_4/α'-SiAlON Composite," J. Am. Ceram. Soc., 77 [9] 2345-53 (1994).

[2]Y. Ukyo, and S. Wada, "High Strength Si_3N_4 Ceramics," Nippon Seramikkusu Kyikai Gakujutsu, 97 [8] 872-74 (1989).

[3]Z. K. Huang, T. Y. Tien, and T. S. Yen, "Subsolidus Phase Relationships in Si_3N_4-AlN-Rare-Earth Oxide System," J. Am. Ceram. Soc., 69 [10] C-241-C-242 (1986).

[4]K. H. Jack, "The Characterization of α'-SiAlONs and α–β Relationships in SiAlONs and Silicon Nitride," pp.45-60 in NATO ASI series E65, Progress in Nitrogen Ceramics. Edited by F. L. Riley. Martinus Nijhoff, The Hague, Netherlands, 1983.

[5]I. A. Bondar, and F. Ya. Galakhov, Izv. Akad. Nauk SSSR, Ser. Khim., 7, 1325 (1963); or Fig. 2586 in Phase Diagram for Ceramists, Vol. II. Edited by E. M. Levin, C. R. Robbins, and H. F. McMurdie. The American Ceramic Society, Inc. Ohio, U.S.A. 1969.

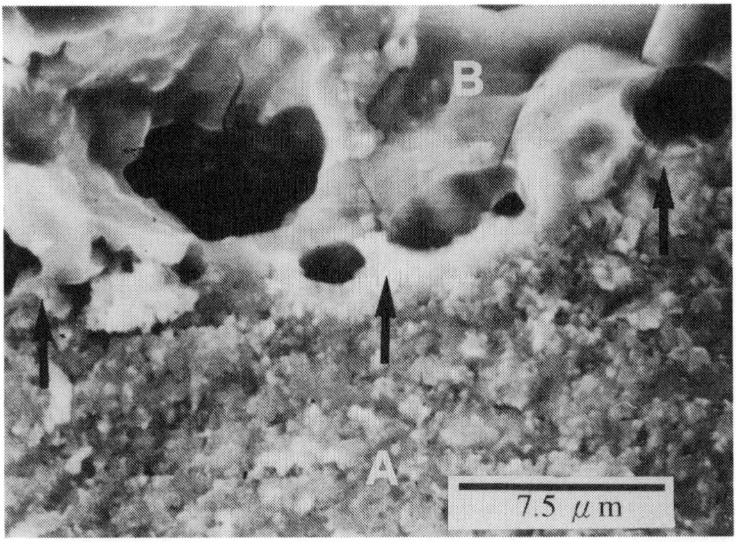

Fig. 5 Fractograph of the oxidized α/β-SiAlON sample. Arrows indicate the interface between the oxidation layer and the α/β-SiAlON composite. Symbol "A" is for the oxidation layer; symbol "B" is for the α/β-SiAlON composite.

[6]E. F. Osborn, and Arnulf, revised and redrawn "Phase Diagrams of Oxide Systems," Plate, published by the American Ceramic Society and Edward Orton, Jr., Ceramic Foundation, 1960; or Fig. 630 in Phase Diagram for Ceramists, Vol. I. Edited by E. M. Levin, C. R. Robbins, and H. F. McMurdie. The American Ceramic Society, Inc. Ohio, U.S.A. 1964.

[7]R. Roy, and E. F. Osborn, J. Am. Chem. Soc., 71 [6] 2092 (1949); or Fig. 450 in Phase Diagram for Ceramists, Vol. I. Edited by E. M. Levin, C. R. Robbins, and H. F. McMurdie. The American Ceramic Society, Inc. Ohio, U.S.A. 1964

[8]T. S. Sheu, "Phase Transformation and Microstructure in the Mixed Cations of α'/β'-SiAlON composite," pp. 449-456 in Ceramic Engineering & Science Proceedings, Vol. 18, Issue 4, 1997. 21th Annual Conferece on Composite, Advanced Ceramics, Materials, and Structures—B. The American Ceramic Society.

[9]C. P. Gazzara and D. R. Messier, "Determination of Phase Content of Si3N4 by X-ray Diffraction Analysis," Am. Ceram. Soc. Bull., 56 [9] 777-80 (1977).

EUTECTIC PRECIPITATION IN SPINEL-YAG COMPOSITE WITH OFF-EUTECTIC COMPOSITION

Shuqiang Wang
New Energy and Industrial Technology Development Organization(NEDO), National Industrial Research Institute of Nagoya, Nagoya 462-8510, Japan
Takashi Akatsu, Yasuhiro Tanabe and Eiichi Yasuda
Research Center for Materials Design, Materials and Structures Laboratory, Tokyo Institute of Technology, 4259 Nagatsuta, Midori, Yokohama 226, Japan

ABSTRACT

Formation mechanism and possible microstructural characteristics of divorced eutectic precipitates in oxide eutectic composite with off-eutectic composition were investigated, based on which, a spinel-YAG composite with spinel-rich composition was prepared through isothermal solidification. It was verified that the eutectic melt formed during high temperature annealing process could precipitate as fine-grained intergranular precipitates with YAG composition, accompanied with specific acicular primary growth of spinel phase during cooling process. Furthermore, a set of crystallographic orientation relationships between YAG and spinel phases in the solidified composite were confirmed by TEM and structural analyses, indicating that low-energy interfaces were formed to some extent in the solidified composite containing the intergranular eutectic precipitates.

INTRODUCTION

Oxide/oxide composites have increasingly been recognized to be promising candidate for long-term services at high temperature above 1500°C and in oxidizing atmosphere, based on the inherent properties of oxides such as oxidation resistance, chemical stability and low volatility[1]. However, it has to be admitted that the performances of structural oxide ceramics and their composites are much far from the requirements for high-temperature applications, such as fracture stress and toughness, resistance to creep deformation and cavitation at grain boundaries or interfaces, thermal shock resistance. For the long-term load bearing applications at high temperatures, fractures of the structural ceramic materials are

To the extent authorized under the laws of the United States of America, all copyright interests in this publication are the property of The American Ceramic Society. Any duplication, reproduction, or republication of this publication or any part thereof, without the express written consent of The American Ceramic Society or fee paid to the Copyright Clearance Center, is prohibited.

often caused by creep fracture due to crack extension accompanied with creep deformation which usually results from grain-boundary sliding. Consequently, grain boundary strengthening should have become one of basic and necessary considerations in microstructural design of high-temperature engineering ceramics. However, it is known that a variety of approaches including particulate, whisker, fiber and platelet reinforced composites are being utilized to improve the mechanical properties of the brittle oxide ceramics, in which most studies are focused on the morphology (equiaxed, platelet and elongated), size (micro and nano scaled) and distribution position (intragranular and intergranular) of the second-phase reinforcements or the matrix. Few studies are associated with the dependence of grain boundary or interface structural characteristics on the high-temperature performances. Moreover, it is indicated that weakly bonded interfaces are preferred for whisker or fiber reinforced composites in order to obtained high toughness or fracture energy[2-4]. It was believed that the weak interfaces directly or indirectly conflict with the creep and cavitation resistance needed for long-term service at high temperatures[1].

Recently, it has been reported that nanocomposites and micro-grained duplex composites can improve not only room-temperature but also high-temperature mechanical properties such as strength, toughness and creep fracture resistance[5-7]. Particularly, it is found that unidirectionally solidified eutectic composites can exhibit much more excellent high-temperature strength and creep resistance than the solid phase sintered materials with identical composition[8]. It has been identified that specific crystallographic orientation relationships are usually present between the two phases of the eutectic composites, indicating that the eutectic interfaces possess to some extent low-energy features. The enhanced creep resistance for the nanocomposites and the eutectic composites may be related to the depression of grain-boundary sliding due to the presence of the strong interfaces in the composites.

Generally, the oxide eutectic composites are prepared through unidirectional (UD) solidification of eutectic melts to control the growth direction of the eutectic precipitates so as to obtain the anisotropic microstructure[9-12]. Due to the particular processing and the limitation in composition (eutectic point), the UD method has not been applied to prepare the general ceramic composites with various compositions. In contrast, the sintering methods widely used for ceramic materials are generally carried out isothermally, while little was known about the eutectic composites prepared by isothermal solidification. On the other hand, liquid-phase sintering methods have been employed extensively in preparation of ceramic materials for improving sintering densification, in which a limited amount of liquid phase is usually situated at the grain boundaries of main crystalline phase during the sintering process. However, in most cases, compositions of the sintered materials (oxides and/or nonoxides) possess eutectic or peritectic phase relationships in which the main crystalline phases behave frequently as the primary phases. For a eutectic system with off-eutectic composition, the interfaces in the solidified composites involve not only the eutectic interfaces formed within the intergranular eutectic melt but also the interfaces between the intergranular precipitates and the primary phase. Crystallization of the intergranualr precipitates and the growth morphology of the main crystalline phase have been investigated extensively in, e.g., the studies of high-performance silicon nitride ceramics with oxide additives[6,13]. However, the interface characteristics between the precipitates and the primary phase and the effect of primary growth on the crystallization and morphology of the precipitates are little reported.

In this study, the solidified microstructure of the eutectic composite with off-eutectic composition is investigated in principle based on the formation process of the so-called divorced eutectic precipitates[14]. Then, a practical high temperature eutectic system, $MgAl_2O_4$(spinel) and $Y_3Al_5O_{12}$(YAG)[15], was selected. A spinel-YAG composite with spinel-rich composition was prepared by isothermal solidification and the solidified microstructures were examined to verify the designed microstructure.

MICROSTRUCTURAL DESIGN OF OXIDE/OXIDE COMPOSITES BASED ON DIVORCED EUTECTIC PRECIPITATION

Liquid-Solid Phase Transformation in Primary Phase Contained Eutectic Melt

It is known that eutectic precipitation will result in not only the reduction in chemical free energy but also the increase in free energies due to the generation of new interfaces. Therefore, the driving force for an A-B eutectic phase transformation to proceed spontaneously is usually proportional to the reduction in total Gibbs free energy ΔG given by

$$\Delta G = \Delta G_{chem} + \Delta G_{inter} + \Delta G_{st} \quad (1)$$

where ΔG_{chem} is the reduction in total chemical free energy; ΔG_{inter} is the increase in total free energy due to the formation of new interfaces, e.g., the interfaces between solid phase A (or B) and liquid phase and the interfaces between solid phases A and B; ΔG_{st} is the strain energy term due to the formation of the A/B interfaces, which can be neglected if the interfaces are formed from heterogeneous nucleation at the liquid-solid interfaces. For the eutectic melt containing primary phase A, there are two kinds of possible nucleation ways, i.e., homogeneous or heterogeneous nucleation. The nucleation barriers for the homogeneous and heterogeneous nucleation of a spherical nucleus of phase B, for example, can be given respectively as follows,

$$\Delta G^*_{ho} = \frac{16\pi(\gamma_B)^3}{3(\Delta G_B)^2} \quad (2)$$

$$\Delta G_{he}^{*} = \Delta G_{ho}^{*} f\ (\theta_{B/A}) = \frac{16\pi(\gamma_B f^{\frac{1}{3}}(\theta_{B/A}))^3}{3(\Delta G_B)^2} \qquad (3)$$

where γ_B and ΔG_B are the liquid/solid interface energy and the chemical free-energy change per unit volume of phase B nucleated from the eutectic melt; $f(\theta_{B/A})$ is the contact coefficient corresponding to contact angle $\theta_{B/A}$ of B nucleus with the primary phase A, which can be derived by[16]

$$f\ (\theta_{B/A}) = \frac{(2+\cos\theta_{B/A})(1-\cos\theta_{B/A})^2}{4} \qquad (4)$$

$$\cos\theta_{B/A} = \frac{\gamma_A - \gamma_{A/B}}{\gamma_B} \qquad (5)$$

where γ_A and $\gamma_{A/B}$ express the interface energies at the interfaces of primary phase A with eutectic melt and nuclei B, respectively. It is noted from Eqs. (2) and (3) that the difference in nucleation barriers corresponding to homogeneous and heterogeneous nucleation of phase B lies only in the decrease in interface energies from γ_B to $\gamma_B f^{1/3}(\theta_{B/A})$. Consequently, the lowest supercooling degrees required for the heterogeneous nucleation of both phase B and phase A in the eutectic melt on primary phase A could be derived respectively by

$$\Delta T_B = -T_E(S_B\gamma_B f^{1/3}(\theta_{B/A}) + S_{B/A}\gamma_{B/A})/V_B\Delta H_E \qquad (6)$$

and

$$\Delta T_A = -T_E(S_A\gamma_A f^{1/3}(\theta_{A/A}) + S_{A/A}\gamma_{A/A})/V_A\Delta H_E \qquad (7)$$

where S_B, S_A, $S_{B/A}$ and $S_{A/A}$ express the areas of the interfaces of nuclei B and A with eutectic liquid phase and with primary phase A, respectively; γ_A and $\gamma_{A/A}$ the interface energy between phase A and eutectic melt and grain-boundary energy between nuclei A and primary phase A; V_B and V_A the volumes of nuclei B and A respectively; ΔH_B and ΔH_A the transformation heat per unit volume of nuclei B and A at eutectic temperature T_E, respectively. The contact coefficient $f(\theta_{A/A})$ is similar to Eq. (4), in which $\cos(\theta_{A/A})$, however, is given by

$$\cos\theta_{A/A} = \frac{\gamma_A - \gamma_{A/A}}{\gamma_A} \qquad (8)$$

Because the grain-boundary energy $\gamma_{A/A}$ could be zero, corresponding to the case of primary growth of phase A in eutectic melt, there would be $f(\theta_{A/A})\to 0$, and consequently, $\Delta T_A \to 0$. In other words, due to the pre-existence of primary phase A, the primary growth of phase A would occur prior to heterogeneous nucleation of phase on phase A at much smaller supercooling degree, especially for larger interface energy $\gamma_{B/A}$. In this case, it can be thought that if the supercooling lies between ΔT_A and ΔT_B, i.e., $\Delta T_A < \Delta T < \Delta T_B$, only primary growth of phase A would appear in the solidification process and finally result in the formation of the divorced eutectic precipitates.

Formation and Microstructure of Divorced Eutectic Precipitates

As analyzed above, for an A-B eutectic system with A-rich composition (x_0), component A in the eutectic melt could be controlled to precipitate heterogeneously through primary growth on the preexisting primary phase A prior to the nucleation of the second phase B. According to Gibbs's phase rule, if it happened, the composition of the liquid phase would change metastably from initial eutectic composition x_E toward B-rich composition with decreasing temperature until some supercooling required for the precipitation of solid phase B which may be metastable phase with smaller liquid/solid interface energy was achieved, as illustrated by the Gibbs energy-composition curves (Fig. 1), the correspondent phase diagram being illustrated in Fig. 2 .   It can be thought that if

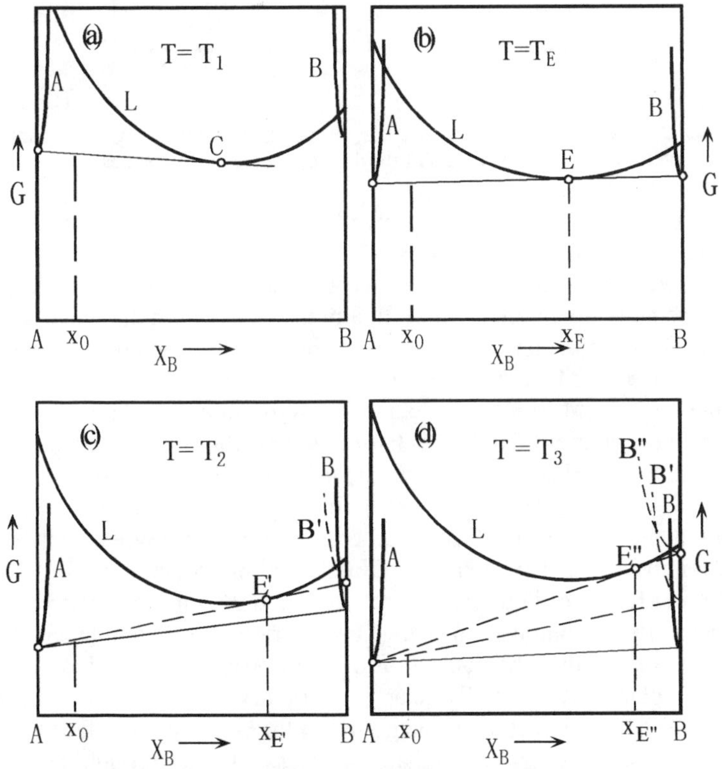

Fig. 1 Schematic G-x_B curves of A-B eutectic system with composition x_0 at different temperatures, $T_1 > T_E > T_2 > T_3$.

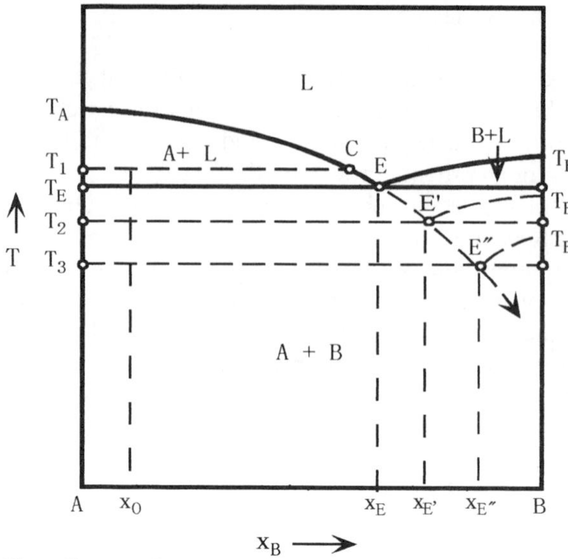

Fig. 2 Phase diagram of A-B eutectic system corresponding to Fig. 1, showing the formation process of the divorced eutectic precipitates.

phase B even in the form of metastable phase could not be precipitated from the supercooled melt, the composition of the supercooled melt would change continuously from x_E up to pure B composition along the metastable liquidus line with decreasing temperature (Fig. 2). In a broad sense, both the metastable phases with pure B composition and the metastable eutectic precipitates with B-rich composition are called the divorced eutectic precipitates (DEPs).

On the other hand, similar to the redistribution of solute during solidification, a concentration gradient exhibiting B-rich composition should be present in the supercooled liquid phase near the liquid/solid interface as the primary growth proceeds, as shown in Fig. 3, depending on the diffusion kinetics in the supercooled melt[17]. The nucleation process in the supercooled melt exhibiting changed compositions can be predicted by the thermodynamic driving forces, that is, the driving forces of both the entire system and the nucleation of phase B tend to increase with the composition change toward richer in B component. Therefore, it can be suggested that the nucleation of solid phase B in the supercooled melt would take place preferentially near the primary phase A, and consequently resulting in the microstructure as suggested in Fig. 4. If the interface between the primary phase A and the precipitates B was formed through heterogeneous nucleation of phase B on phase A due to the lower nucleation barrier, it might exhibit low-energy feature according to the relations of interface energy with supercooling and nucleation barrier, as described above.

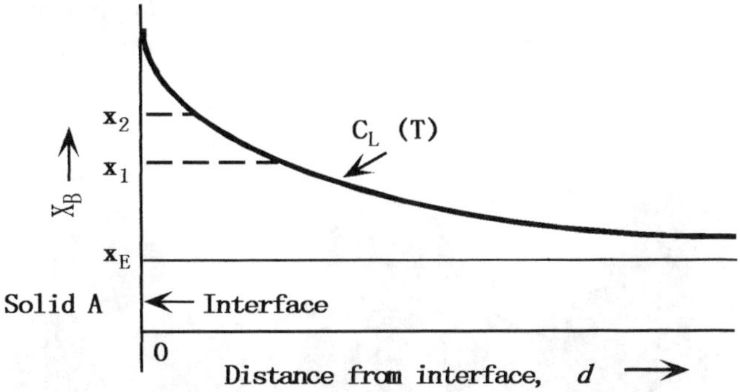

Fig. 3 Proposed concentration profile present near the interface between the primary phase and the supercooled melt caused by metastable primary growth during solidification.

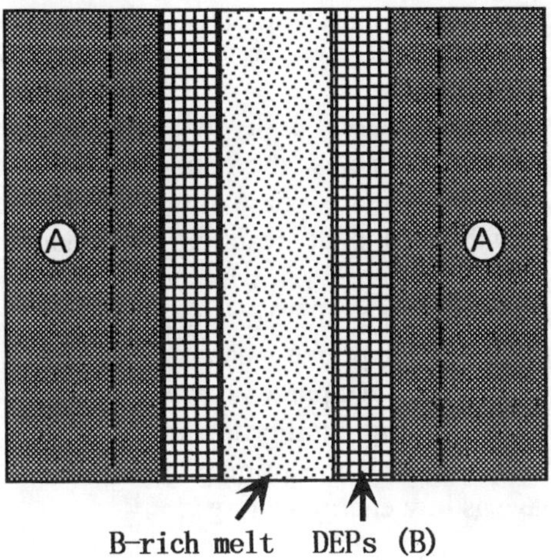

Fig. 4 Illustration of the proposed interface structure containing divorced eutectic precipitates (B) in A-B eutectic composites with A-rich composition.

Microstructural Design Based on the Divorced Eutectic Precipitation

In general, it is believed that the solidified microstructure of a eutectic system with off-eutectic composition is like the one shown in Fig. 5(a). Considering the formation of the divorced eutectic precipitates, as described above, the solidified microstructure may be formed like the one shown in Fig. 5(b), in which a layered precipitates of composition B could be present between the primary phase and the characteristic eutectics. Furthermore, it could be thought that with decrease in content of the liquid phase, the precipitation of characteristic eutectics would not appear, and the resulting microstructure would only exhibiting primary phase and the intergranular divorced eutectic precipitates (B), as shown in Fig. 5(c).

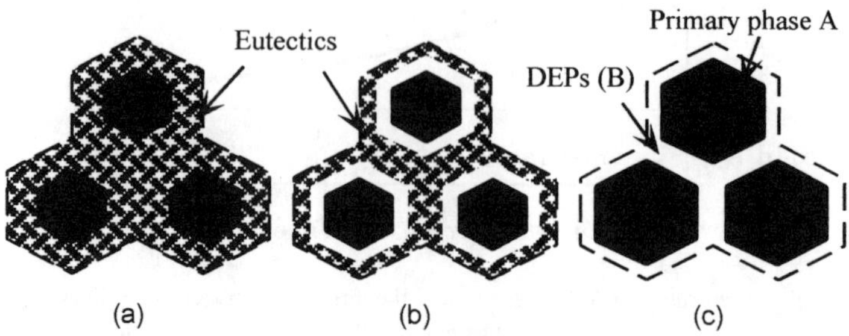

Fig. 5 Three kinds of intergranular precipitates proposed for solidified A-B eutectic composites with A-rich composition, (a) characteristic eutectics, (b) DEPs (B) and characteristic eutectics, and (c) only DEPs (B) coexisting with primary phase A.

In addition, because the formation of divorced eutectic precipitates depends on the metastable primary growth below the equilibrium eutectic temperature, the growth morphology of the primary phase might be controlled by the kinetic process of crystal growth from melt, resulting in dendritic or cellular growth morphologies during solidification[18,19], in which cooling rate may be an important parameter controlling the resulting microstructure containing the divorced eutectic precipitates. It is predicted that primary phase A grows like dendrites and the intergranular divorced eutectic precipitates of component B may be present with fine-grained structures exhibiting equilibrium or metastable phase composition, moreover, the interfaces between the primary phase and the divorced eutectic precipitates may behave as low-energy features.

EXPERIMENTAL PROCEDURE

High purity Al_2O_3 (99.99%,TM-100, Taimei Chem. Co., Ltd.), MgO (99.98%, Ube Ind. Ltd.) and Y_2O_3 (99.9%, Shin-Etsu Chemical Co., Ltd.) powders were used as starting materials to prepare the spinel/YAG diffusion couple[20] and the composite sample (1Y30S) with composition of YAG : spinel = 1 : 30 molar ratio. Solid state reaction and densification of the sample was carried out by hot pressing at 1700°C for 1h in argon atmosphere. Subsequently, the couple sepcimens were annealed at the temperatures up to melt appearing; the hot pressed composite specimen (1Y30S) was annealed at the eutectic temperature about 1830°C [14] for 1h in Ar followed by cooling down quickly at 100°C/min.

Microstructures of the hot pressed and solidified specimens were characterized by means of x-ray diffraction (XRD), micro Raman spectroscopy (MRS, T-64000 micro-Raman system), electron probe micro analyzer (EPMA, Shimadzu Mfg. Co., Ltd.), scanning electron microscope (SEM), and transmission electron microscopy (TEM, JEM-2000EX). The specimens used in the TEM study were prepared through crushing and grinding the annealed bulk sample into micro-sized particles in an agate mortar. After ultrasonic dispersion in ethanol for a few minutes, the sufficiently dispersed particles were taken by a meshed holder as the specimen for TEM observation. Selected area diffraction (SAD) was used in the TEM analysis to determine the crystal structures and the orientation relationships (ORs) between the two phases at their interfaces, in which the chemical composition of the observed phases were detected qualitatively by the energy disperse X-ray spectroscopy (EDX, TN-2000) equipped in the TEM.

RESULTS AND DISCUSSION

Figure 6 shows a typical solidified microstructure in spinel side of the spinel/YAG couple annealed at 1815°C for 1h in argon atmosphere (cooling rate: 10°C/min), in which the dark phase is spinel primary phase and the intergranular bright phase is the precipitates with YAG composition[20], whose crystallinity was examined together with YAG polycrystal by MRS, as shown in Fig. 7. It was indicated that the YAG precipitates exhibited lower crystallinity or much finer grain structures, whose formation may be attributable to the divorced eutectic precipitation, as described above.

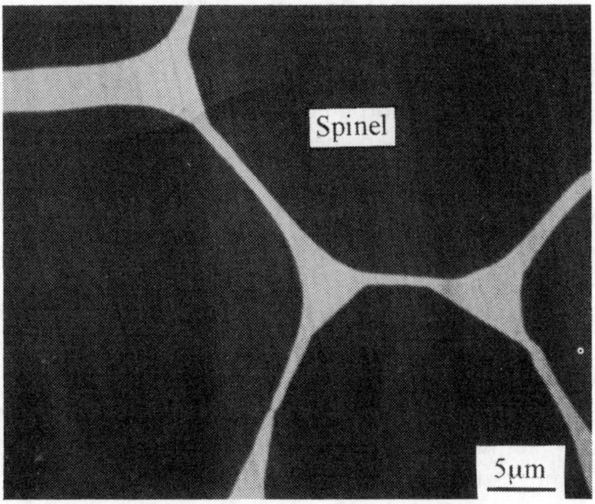

Fig. 6 Typical microstructure of solidified spinel/YAG diffusion couple, in spinel-rich side.

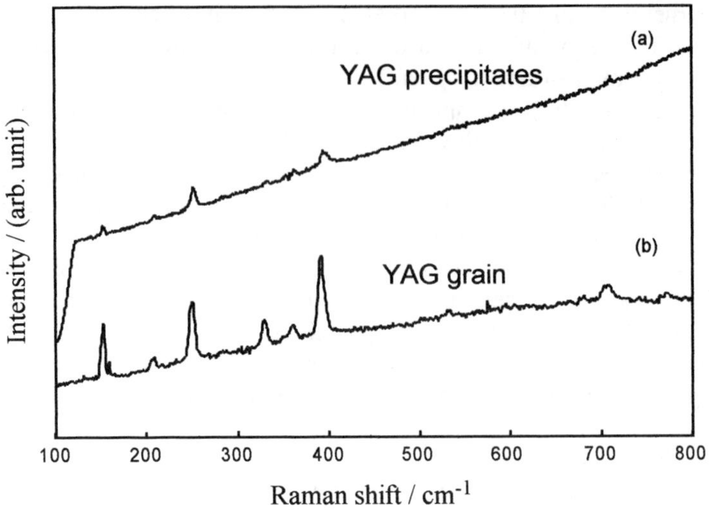

Fig. 7 Micro Raman spectra of the intergranular precipitates with YAG composition (a) and correspondent YAG polycrystals in the solidified spinel/YAG diffusion couple.

The presence of the divorced eutectic precipitates in the spinel-YAG system was verified using the spinel-rich composite (1Y30S). Fig. 8 shows the SEM microstructures of the specimen hot pressed at 1700°C for 1h (a) and annealed at 1828°C for 1h followed by cooling down at 100°C/min (b). It is seen that the phase distribution in the two-phase composite (dark phase is spinel and bright phase is YAG) was changed significantly after the annealing process although the phase compositions were maintained, as identified by XRD analysis[21]. Furthermore, the crystallization morphology of the intergranular YAG precipitates was observed using thermally etched specimen, as shown in Fig. 9, indicating that the divorced eutectic precipitates of YAG composition possess fine-grained structure. Subsequently, Fig. 10 shows a typical SEM micrograph of a fracture surface of the annealed specimen. It may be noted that the fracture surface exhibit transgranular fracture feature, moreover, it was found that some needle-like precipitates appeared to form in the annealed specimen. It was identified by TEM analysis that the needle-like precipitates were spinel crystal, as shown in Fig. 11, whose formation process was attributed to the metastable primary growth[22].

Furthermore, the interfaces between primary phase spinel and the intergranular YAG precipitates in the annealed specimen were examined by TEM-SAD technique, the detailed experimental results being described in the paper[21] . It was identified that specific crystallographic orientation relationships

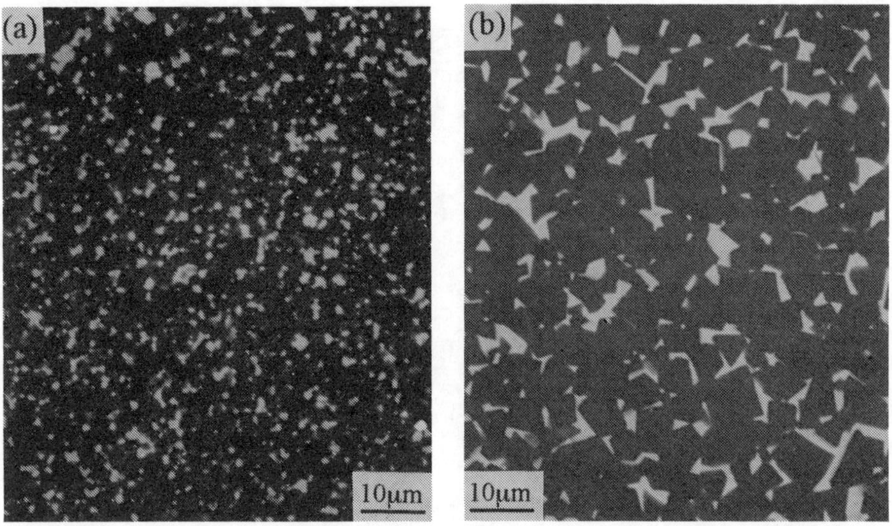

Fig. 8 SEM microstructures of the specimen (1Y30S) hot pressed at 1700°C for 1h (a) and annealed at 1828°C for 1h followed by cooling down at 100°C/min (b).

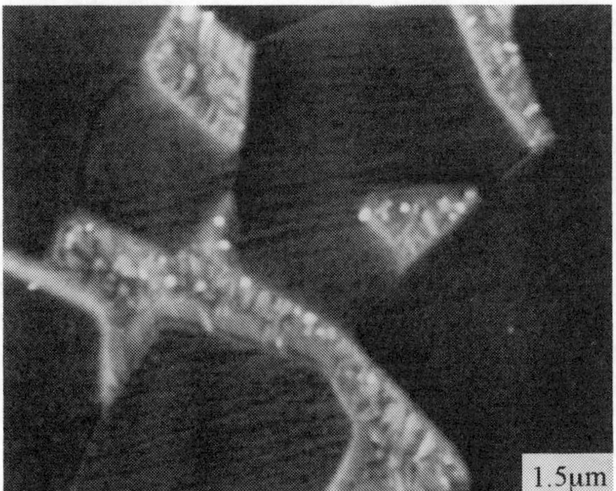

Fig. 9 SEM micrograph of the specimen (1Y30S) annealed at 1828°C for 1h followed by cooling down at 100°C/min, thermally etched at 1400°C for 30min in air.

Fig. 10 A typical room-temperature fracture surface of the specimen (1Y30S) annealed at 1828°C for 1h followed by cooling down at 100°C/min in Ar.

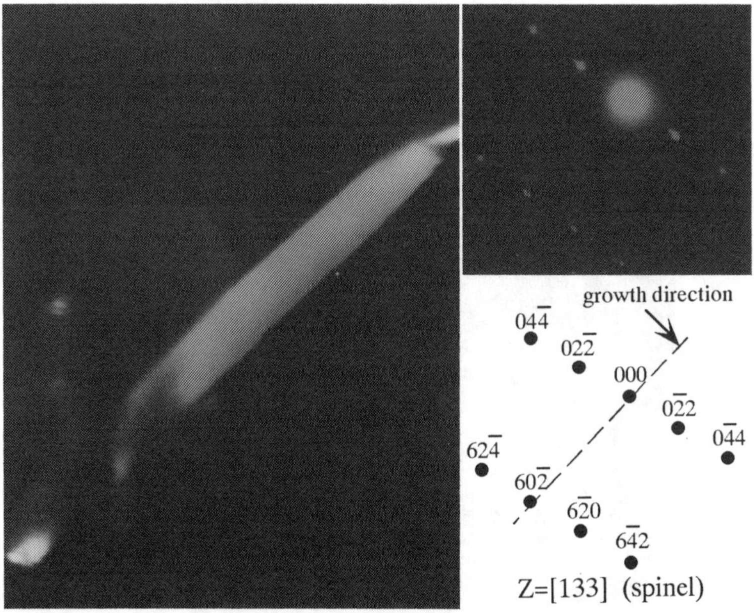

Fig. 11 TEM analysis (dark-field image and diffraction pattern) of a needle-like precipitate in the specimen (1Y30S) annealed at 1828°C for 1h followed by cooling down at 100°C/min in Ar.

(ORs) are present at the YAG/spinel interfaces, the ORs being determined to be

$$(640)[001][\bar{2}32][\bar{2}36]_{YAG} // (111)[2\bar{1}\bar{1}][11\bar{2}][10\bar{1}]_{spinel}$$

, indicating that low-energy interfaces similar to eutectic interfaces may be formed between the spinel and YAG phases in the spinel-rich composite.

CONCLUSIONS

The formation process of the divorced eutectic precipitates was investigated by means of thermodynamic analysis of eutectic precipitation of a binary system with off-eutectic composition, based on which, a microstructural design of high-temperature structural oxide/oxide composites was proposed. It was verified by spinel-YAG system with spinel-rich composition that the divorced eutectic precipitates with YAG composition could be formed as fine-grained intergranular precipitates, accompanying with acicular primary growth of spinel phase during annealing process followed by isothermal solidification. Furthermore, a set of crystallographic orientation relationships between YAG and spinel phases in the solidified composite were identified, indicating that low-energy interfaces may exist to some extent in the solidified composite containing the divorced eutectic precipitates.

ACKNOWLEDGEMENT

The authors are very grateful to Mr. F. Yamamoto for his contribution to this work in sample preparation and thank Prof. M. Yoshimura for his helpful discussion on metastable precipitation.

REFERENCES

[1]R. Raj, "Fundamental Research in Structural Ceramics for Service Near 2000°C", *J. Am. Ceram. Soc.*, **76** [9] 2147-74 (1993).

[2]P. E. D. Morgon and D. B. Marshall, "Ceramic Composites of Monazite and Alumina", *J. Am. Ceram. Soc.*, **78** [6] 1553-63 (1995).

[3]P. F. Becher, "Microstructural Design of Toughened Ceramics", *J. Am. Ceram. Soc.*,**74** [2] 255-69 (1991).

[4]T. Akatsu, M. Ohshima, Y. Tanabe, E. Yasuda, and T. Yano, "The Effect of Whisker-coating on the Work-of-fracture of SiC-whisker/Al_2O_3 Composites at High Temperature", p. 231 in *Proceedings of Annual Meeting of the Ceramic Society of Japan*, Yokohama, April 2-4, 1996.

[5]T. Ohji, T. Hirano, A. Nakahira, and K. Niihara, "Particle/Matrix interface and Its Role in Creep Inhibition in Alumina/Silicon Carbide Nanocomposites", *J. Am. Ceram. Soc.*, **79** [1] 33-45 (1996).

[6]S. Inone, T. Ono, A. Nakahira and K. Niihara, "Impromement of High temperature Strength by Grain Boundary Crystallization in Silicon Nitride", *Proc. Conf. '1st Intern. Symposium on the Science of Engineering Ceramics'*, Koda, October 1991, S. Kimura and K. Niihara, p. 129-134.

[7]J. D. French, H. M. Chan, M. P. Harmer, and G. A. Miller, "High-Temperature Fracture

Toughness of Duplex Microstructures", *J. Am. Ceram. Soc.*, **79** [1] 58-64 (1996).
[8]Y. Waku, N. Nakagawa, T. Wakamoto et al., "High Temperature Properties of Unidirectionally Solidified Al_2O_3/YAG Composites"; pp.170 in *Proceedings of Annual Meeting of Ceram. Soc. of Japan*, Kyoto, 1995.
[9]V. S. Stubican and R. C. Bradt, "Eutectic Solidification in Ceramic Systems," *Ann. Rev. Mater. Sci.*, **11**, 267-97 (1981).
[10]V. S. Stubican, R. C. Bradt, F. L. Kennard, et al., "Ceramic Eutectic Composites," *Mater. Sci. Res.*, **20**, 103-114 (1986).
[11]R. S. Hay and L. E. Matson, "Alumina/Yttrium-Aluminum Garnet Crystallographic Orientation Relationships and Interphase Boundaries: Observations and Interpretation by Geometric Criteria", *Acta Mettal. Mater.*, **39** [8] 1981-94 (1991).
[12]T. A. Parthasarathy, T. Mah, and L. E. Matson, "Creep Behavior of an Al_2O_3-$Y_3Al_5O_{12}$ Eutectic Composite," *Ceram. Eng. Sci. Proc.*, **11** [9-10] 1628-38 (1990).
[13]W. E. Lee, C. H. Drummond, G. E. Hilmas and S. Kumar, "Microstructural Evolution in Near-eutectic Yttrium Silicate Compositions Fabricated from a Bulk Melt and as an Intergranular Phase in Silicon Nitride", *J. Am. Ceram. Soc.*, **73** [12] 3575-79 (1990).
[14]B. Chalmers, *Principles of Solidification*; p. 218. John Wiley & Sons, Inc., New York, 1964.
[15]S. Wang, T. Akatsu, Y. Tanabe and E. Yasuda, "Identification of Eutectic phase relation and Microstructures of Isothermally solidified YAG-Spinel Composite", unpublished work.
[16]W. D. Kingery, H. K. Bowen and D. R. Uhlmann, pp.334 in *Introduction to Ceramics*, second edition, John Wiley & Sons, New York, 1976.
[17]B. Chalmers, *Principles of Solidification*; p. 132. John Wiley & Sons, Inc., New York, 1964.
[18]Ibid, p. 91-185.
[19]R. Doherty, Dendritic Growth, Chapter 15 in *Crystal Growth*, ed, B. R. Pamplin, Pergamon Press, Oxford, 1975.
[20]S. Wang, E. Yasuda, T. Akatsu, Y. Tanabe and Z. Nakagawa, "Intergranular Precipitation in $Y_3Al_5O_{12}/MgAl_2O_4$ System", *J. Mater. Sci. Lett.*, **16**, 1580-83 (1997).
[21]S. Wang, T. Akatsu, Y. Tanabe and E. Yasuda, "Divorced Eutectic Precipitation in Isothermally Solidified YAG-Spinel Composite with Spinel-Rich Composition", unpublished work.
[22]S. Wang, T. Akatsu, Y. Tanabe and E. Yasuda, "Acicular Primary Growth in Isothermally Solidified YAG-Spinel Composite with Spinel-Rich Composition ", unpublished work.

THERMODYNAMICS OF THE Mg-C-N-O SYSTEM FOR EVALUATION OF MgO-C BRICKS

Siang-Chun Liu
Department of Materials and Mineral Resources Engineering
National Taipei University of Technology
Taipei, Taiwan 10643

ABSTRACT

Based on the thermodynamic properties, MgO melts and decomposes at 3105 and 3697K, respectively. Through thermodynamic calculations, MgO is reduced and vaporized into Mg and CO gases by carbon at 2036 K in MgO-C compacted bricks. If the oxygen pressure could possibly be reduced to 3.425×10^{-35} atm, MgO can be decomposed and vaporized at 1366K. Under one atm pressure, magnesium carbides can not be formed at any temperature, unless the atmospheric pressure is increased. Mg_3N_2 will be stable under a nitrogen atmosphere below 1703K. It is impossible for MgO to be carbothermic and nitrided into Mg_3N_2. Finally, a phase stability diagram for Mg-N system and a space diagram for the Mg-C-N-O was constructed.

INTRODUCTION

The knowledge of the vaporization chemistry of MgO and MgO-C is very important for understanding the practice of MgO-C bricks as a furnace lining in the metallurgieal industry. The study of complex vapor species need to be done. There are eight condensed phases and twenty-eight gaseous species in the Mg-C-N-O system, as shown in Table I. The relationships between these phases will always depend on the equilibrium of vapor pressure. Kinetic and mass-transfer factors will also influence practical behavior, but these will not be considered here. The phase stability for the Mg-C-N-O system has been calculated and analyzed using JANAF[1] thermodynamic properties.

PHASE EQUILIBRIA AND DISCUSSION

Mg-C-O System

Four condensed phases and ten gaseous species can be formed in the Mg-C-O

To the extent authorized under the laws of the United States of America, all copyright interests in this publication are the property of The American Ceramic Society. Any duplication, reproduction, or republication of this publication or any part thereof, without the express written consent of The American Ceramic Society or fee paid to the Copyright Clearance Center, is prohibited.

Table I. Condensed phases and gas species in the Mg-C-N-O system

Condensed Phases	Gas Species			
Mg	Mg_1	C_4	CNO	N_2O
C	Mg_2	C_5	N	N_2O_3
MgC	MgN	CO	N_2	N_2O_4
Mg_2C_3	MgO	CO_2	N_3	N_2O_5
MgC_2	C_1	C_3O_2	NO	O
$MgCO_3$	C_2	C_2O	NO_2	O_2
Mg_3N_2	C_3	CN_2	NO_3	O_3
MgO				

system using JANAF[1] data, as shown in Table II. The MgC, Mg_2C_3 and MgC_2 crystals, shown in Fig. 1, could barely be presented under standard conditions, since their Gibbs free energy of formation, ΔG_f^o, are indicated as "positive" in the JANAF table and they can only be formed under pressure as high as 7.7 Gpa, as shown in Fig. 8940 of the Phase Equilibria Diagram.[2]

Table II. Condensed phases and gas species in the Mg-C-O system

Condensed Phases	Gas Species		
C	Mg_1	CO_2	C_4
Mg	Mg_2	C_1	C_5
MgO	MgO	C_2	
$MgCO_3$	CO	C_3	

When MgO is mixed with carbon and compacted into bricks or used for the linning of furnace in the iron and steel industry, the following reaction will occur in the MgO-C bricks at higher temperatures:

$$MgO_{(s)} + C_{(s)} \Leftrightarrow Mg_{(g)} + CO_{(g)} \quad (1)$$

Based on the JANAF data, the Gibbs free energy change for reaction (1) will be zero ($\Delta G_r^o = 0$) at 2119K, where MgO will decompose and vaporize when Mg and

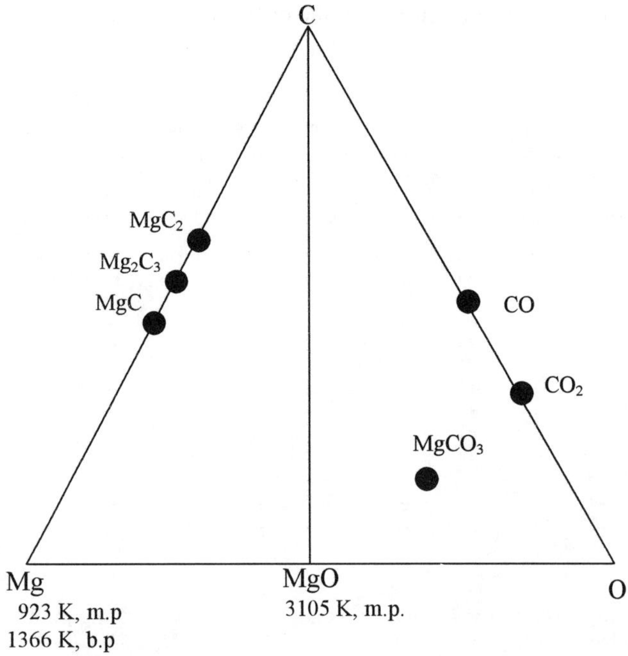

Fig. 1. Ternary diagram for Mg-C-O system. *$MgCO_3$ decomposed into MgO and CO_2 at 680 K.

CO gases each reach 1 atm separately. Other gaseous species can exist and are calculated as shown in Table III. In reality, the dissociation and vaporization pressure for the MgO-C brick should reach 1 atm at 2036K (1236℃), also shown in Table III. In comparison with Al_2O_3-C, Al_2O_3-SiC-C and MgO-Cr_2O_3 as composite refractories, the MgO-C brick has been verified by Asano's laboratory studies[3] to have the least corrosion resistance to the slag in the iron-making process. The total pressure of dissociation and vaporization for MgO-C brick is established as a function of temperature in the following equation:

$$\log P_t = -15824/T + 7.769 \quad \text{..........(2)}$$

Mg-O System

According to the Gibbs free energy of formation for MgO, ΔG_f^o, in the JANAF table, MgO crystals will melt and decompose into Mg vapor and O_2 gas at

Table III. Equilibrium vapor pressure with the coexistance of MgO-C phases. $P_T = 1 atm$

T,K	1600	1700	1800	1900	2000	2100	2119	2200	2300	2400	2500
log Po_2	-21.360	-19.732	-18.284	-16.990	-15.828	-14.776	-14.588	-13.882	-12.952	-12.154	-11.422
log PMg	-2.445	-1.853	-1.330	-0.863	-0.444	-0.067	0	0.275	0.586	0.869	1.129
PMg	0.0036	0.0140	0.0468	0.137	0.360	0.858	1	1.884	3.855	7.396	3.459
log Pco	-2.445	-1.853	-1.330	-0.863	-0.444	-0.067	0	0.275	0.586	0.869	1.129
Pco	0.0036	0.0140	0.0468	0.137	0.360	0.858	1	1.884	3.855	7.396	13.459
log Pco_2	-8.423	--	--	--	--	--	--	--	--	--	-3.417
Pco_2	3.776×10^{-9}	--	--	--	--	--	--	--	--	--	7.128×10^{-4}
log PMgO	-11.964	--	--	--	--	--	--	--	--	--	-4.227
PMgO	1.086×10^{-12}	--	--	--	--	--	--	--	--	--	5.280×10^{-5}
PMg_2	2.891×10^{-9}	--	--	--	--	--	--	--	--	--	0.014
PC_1	6.280×10^{-16}	--	--	--	--	--	--	--	--	--	1.0×10^{-7}
PC_2	5.930×10^{-18}	--	--	--	--	--	--	--	--	--	3.390×10^{-8}
PC_3	2.642×10^{-16}	--	--	--	--	--	--	--	--	--	5.0×10^{-7}
PC_4	1.420×10^{-21}	--	--	--	--	--	--	--	--	--	2.570×10^{-10}
PC_5	2.244×10^{-21}	--	--	--	--	--	--	--	--	--	5.0×10^{-7}
P_t	0.0072	0.028	0.0936	0.274	0.720	1.716	2	3.786	7.710	17.792	26.918
log P_t	-2.143	-1.553	-1.029	-0.562	-0.143	0.235	0.301	0.576	0.887	1.170	1.430

3105 and 3696.671 K, respectively. If the oxygen potential can be lowered to a condition of $pO_2 = 3.425 \times 10^{-35}$ atm, MgO can be reduced into Mg vapor at 1366 K. The relationship of the equilibrium oxygen pressure with respect to temperature between $Mg_{(s/l)}$ and $MgO_{(s)}$ are established as:

$$\log pO_2,\ Mg_{(l)} / MgO_{(s)} = -63570/T + 12.072 \ (923\text{-}1366 \text{ K}) \quad (3)$$

$$\log pO_2,\ Mg_{(s)} / MgO_{(s)} = -62764/T + 11.202 \ (298\text{-}923 \text{ K}) \quad (4)$$

where 923 and 1366 K are the melting and boiling points of magnesium, respectively. The phase stability diagram for Mg-O system was constructed as shown in Fig. 2.

The vaporization characteristic of the Mg-O system is calculated using the following equations:

$$Mg_{(s/l)} \Leftrightarrow Mg_{(g)} \quad (5)$$

$$2Mg_{(s/l)} \Leftrightarrow Mg_{2(g)} \quad (6)$$

$$Mg_{(s/l)} + 1/2O_{2(g)} \Leftrightarrow MgO_{(g)} \quad (7)$$

Table IV. Vapor pressures for different species under varies pO_2 at 1300 K, P_T = 1 atm.

		$Mg_{(l)} \leftrightarrow$	$MgO_{(s)}$				
log Po_2	-50	-40	-36.828	-30	-20	-10	0
log P_{Mg}	-0.249	-0.249	0.249	-3.663	-8.663	-13.663	-18.663
P_{Mg}	0.564	0.564	0.564	0.0002	2.173×10^{-9}	2.173×10^{-14}	2.173×10^{-19}
log P_{Mg2}	-3.925	-3.925	-3.925	-10.753	-20.753	30.753	-40.753
P_{Mg2}	0.00012	0.00012	0.00012	1.766×10^{-11}	1.766×10^{-21}	1.766×10^{-31}	1.766×10^{-41}
log P_{MgO}	-23.507	-18.507	-16.921	-16.921	-16.921	-16.921	-16.921
P_{Mgo}	3.111×10^{-24}	3.111×10^{-19}	1.199×10^{-17}	1.199×10^{-17}	1.199×10^{-17}	1.199×10^{-17}	1.199×10^{-17}
P_t	0.56412	0.56412	0.56412	0.0002	2.173×10^{-9}	2.174×10^{-14}	1.221×10^{-17}
log P_t	-0.249	-0.249	-0.249	-3.663	- 8.663	-13.663	-16.913

$$MgO_{(s/l)} \Leftrightarrow Mg_{(g)} + 1/2O_{2(g)} \quad \text{.......(8)}$$
$$2MgO_{(s/l)} \Leftrightarrow Mg_{2(g)} + O_{2(g)} \quad \text{.......(9)}$$
$$MgO_{(s/l)} \Leftrightarrow MgO_{(g)} \quad \text{.......(10)}$$

The results of the calculations at 1300 K are listed in Table IV and plotted in Fig. 3. Table IV and Fig.3 indicates that (1) the vapor pressure of pure magnesium will be 0.564 atm at 1300 K and reaches 1 atm at 1366 K, the boiling point of Mg, (2) the total pressure of the Mg-bearing gas reaches the maximum with the coexistence of $Mg_{(s/l)}$ and $MgO_{(S)}$ at any given temperature, and (3) the small amount of vapor consisting of $Mg_{2(g)}$ and $MgO_{(g)}$ is insignificant and could not influence the total pressure.

However, the following differences are found when comparing Fig 6258 from the published Phase Diagrams for Ceramists [4], Hultgren [5] and JANAF data : (1) log pMg ≅ -1 (0.1 atm) reaches equilibrium with pure $Mg_{(l)}$ at 1327℃(1600 K) in Fig 6258(B). But according to Hultgren, beyond the boiling point of Mg(1366 K), the vapor pressure of magnesium reaches 5.380 atm at 1600 K. (2) Base on ΔG_f^o, MgO = -402,026 j at 1600K as indicated in JANAF, thermodynamic calculation shows log pO_2 = -26,250 when $MgO_{(s)}$ becomes equilibrium with pMg = 1atm. But Fig. 6258(B) illustrates that MgO crystal becomes equilibrium with liquid magnesium at 1327℃ (1600K) when log pO_2 ≅ -24, which is incorrect. Since 1600 K is beyond the boiling point of Mg, Mg should not be in a liquid phase. (3) Beyond 1366 K, b.p. of Mg there are three variables, temperature, pMg and pO_2, should be taken into consideration as shown in the following equation:

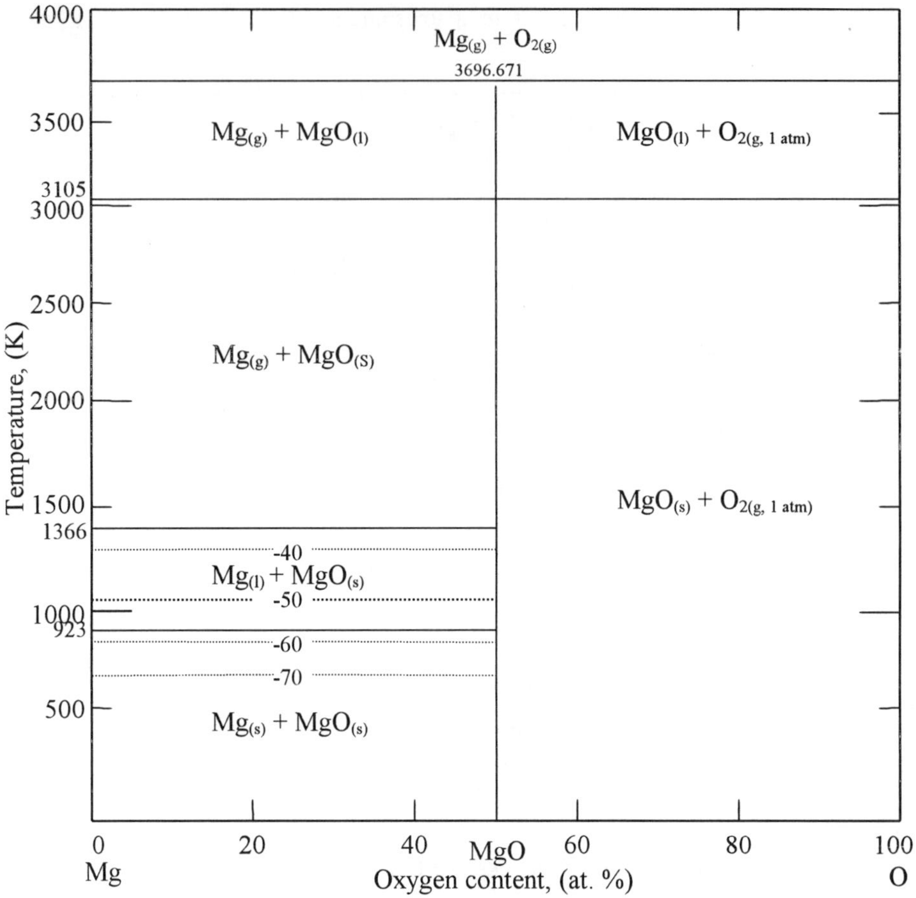

Fig. 2. Phase stability diagram for the Mg-O system. Dashed lines are oxygen isobars. $P_T = 1$ atm.

$$Mg_{(g)} + 1/2O_{2(g)} \Leftrightarrow MgO_{(s/l)} \qquad (11)$$

But these three variables were not being considered in Fig.6258(A).

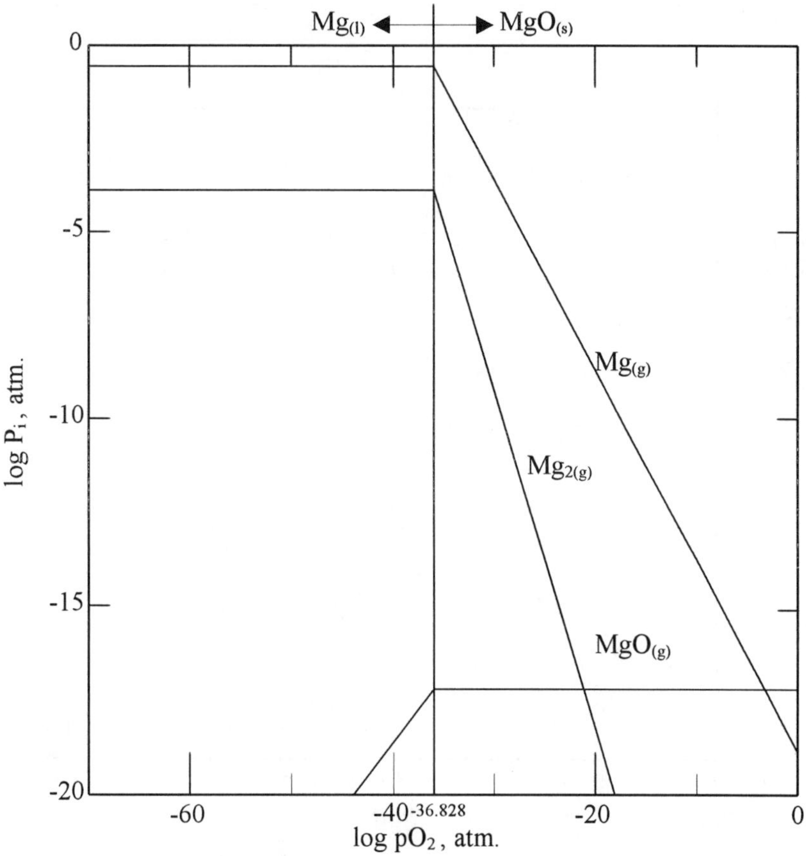

Fig. 3. Vapor pressures for different species in Mg-O system at 1300 K.

Mg-N System

Based on the Gibbs free energy formation for Mg_3N_2 in JANAF table, Mg_3N_2 crystals will decompose and vaporize into Mg and N_2 gases at 1703.75 K. The relationship for the nitrogen equilibrium pressure with respect to temperature between $Mg_{(s/l/g)}$ and $Mg_3N_{2(S)}$ was established with the following equations and plotted as Fig. 4.

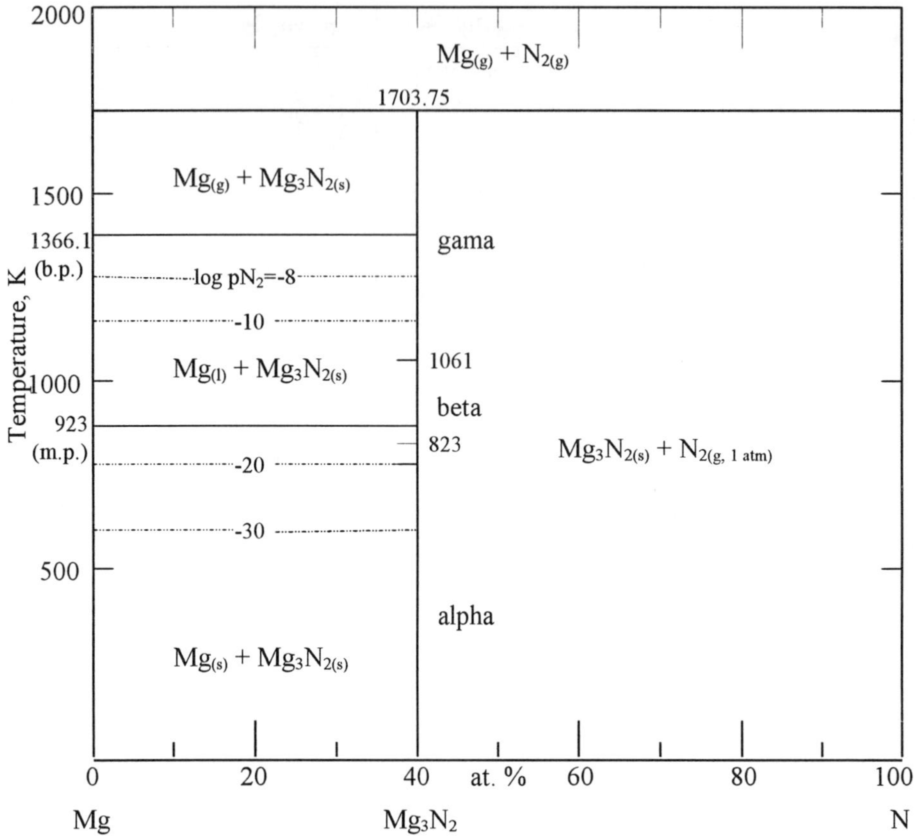

Fig. 4. Phase relations in the Mg-N system. Dashed lines are nitrogen isobars, P_T = 1atm.

$$\log pN_2\ (Mg_{(s)} / \alpha - Mg_3N_2) = -24085/T + 10.543 \qquad (12)$$

$$\log pN_2\ (Mg_{(l)} / \gamma - Mg_3N_2) = -25522/T + 12.085 \qquad (13)$$

$$\log pN_2\ (Mg_{(g)} / \gamma - Mg_3N_2) = -45434/T + 26.667 - 3 \log pMg \qquad (14)$$

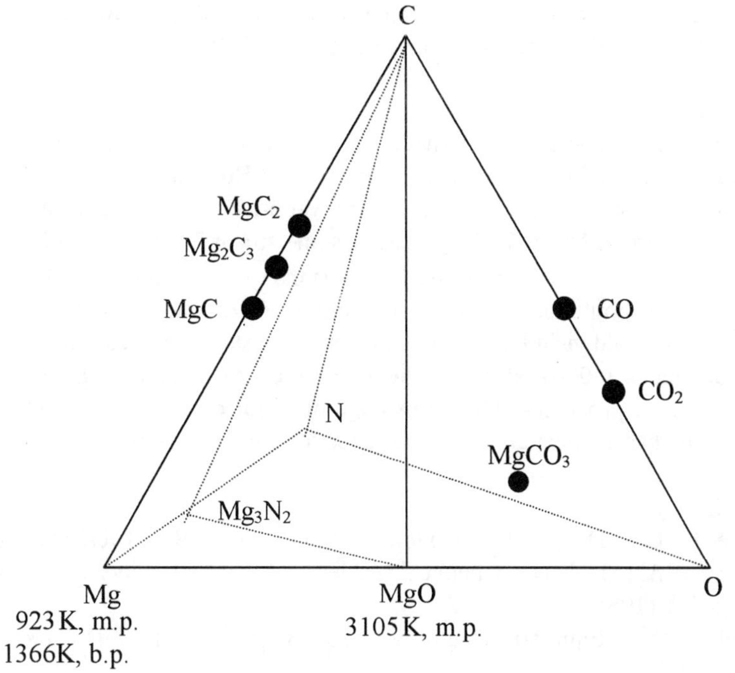

Fig. 5. Space diagram for the system of Mg-C-N-O.

Mg-C-N-O system

Magnesium carbides could not be formed as described in the Mg-C-O system. Mg_3N_2 is very stable under N_2 atmosphere and it will decompose and vaporize into Mg and N_2 gases at 1703.75 K as stated in Mg-N system. MgO either with or without the presence of carbon, could not form Mg_3N_2 when reacting with pure nitrogen gas, since the change of Gibbs free energy for the following reactions are "positive", $\Delta G_{\gamma}^{o} = +$, from room temperature to the decomposition temperature of Mg_3N_2, 1703.75 K.

$$3MgO_{(s)} + N_{2(g)} \Leftrightarrow Mg_3N_{2\,(s)} + 3/2O_{2(g)} \quad (15)$$

$$3MgO_{(s)} + 3C_{(s)} + N_{2(g)} \Leftrightarrow Mg_3N_{2(s)} + 3CO_{(g)} \quad (16)$$

The space diagram of the Mg-C-N-O system is constructed as shown in Fig. 5, which shows all phase relationships we have discussed this far.

CONCLUSIONS

When we use pure MgO as a furnace lining, it is very stable, neither carbide nor nitride can be formed under normal air operations. But when it is mixed with carbon as MgO-C bricks, it will decompose and vaporize into Mg vapor and CO gas, the total pressure of Mg and CO gas reaches one atm at 2036 K (1763℃).

In most calculations, 1 atm is used as a standard of evaluation. A vapor pressure of 0.1 atm should also result in the rapid vaporization of gas stream. But 0.01 or 0.001 atm would yield kinetically slow or a very slow vaporization process. Any condition under 0.0001 atm may be considered to yield negligible vapor transport in practical processes. The knowledge of vapor chemistry is worthwhile to understand and has an even greater importance for commercial exploitation.

REFERENCES

[1] M.W Chase, Jr., C.A. Davies, T.R. Downey, JR., D.J. Fruip, R.A. McDonald, and A.N. Syverud, "JANAF Thermochemical Tables," 3rd., ed., J. Phys.Chem. Ref. Data, 14, Suppl. 1 (1985).

[2] A.E.McHale, "Phase Equilibria Diagrams," Vol. X, pp.254, Fig. 8940, *the Am. Ceram. Soc.* (1994).

[3] K. Asano,A. Ishii, and J.Nakao, "Laboratory Study on Refractories for Iron-Bath Smelting Reduction Furnance," *Ceramics*, 43[2], pp. 74-79 (1991).

[4] R.S. Roth, J.R. Dennis, and H.F. McMurdie, "Phase Diagrams for Ceramists," Vol. VI, pp. 3-4. Fig,6258, *the Am. Ceram. Soc.* (1987).

[5] R.Hultgren, P.D. Desai, D.T. Hawkins, M. Gleiser, and K.K. Kelley, "Selected Values of the Thermodynamic Properities of the Elements," *Am. Soc. for Metals*, pp. 294-300 (1973).

Environmental Effects and Creep Behavior

STABILITY OF BN INTERPHASES AND THEIR EFFECT ON THE DEFORMATION OF FIBER REINFORCED CERAMIC MATRIX COMPOSITE MATERIALS

Margaret M. Stackpoole[1], Rajendra K. Bordia[1],
Charles H. Henager, Jr.[2], Russell H. Jones[2],

[1]Department of Material Science and Eng,, University of Washington, Seattle, WA,
[2]Pacific Northwest National Laboratories, Richland, WA

ABSTRACT

Time-dependent deformation behavior, in bending, of notched SiC/SiC ceramic matrix composite materials was investigated at elevated temperatures under varying pO_2. Materials investigated consisted of a chemical vapor infiltrated (CVI) β-SiC matrix reinforced with Nicalon® SiC fibers with a 0.4μm BN interphase. From these results, activation energies for composite deformation were obtained as a function of oxygen partial pressure. Finally Thermo-Gravimetric experiments were conducted to obtain the oxidation kinetics of the BN interphase. The activation energy for oxidation from the TGA data was correlated to the activation energy for damage evolution and also compared to activation energies for fiber creep taken from the literature. From these comparisons it is shown that in the inert environment, deformation kinetics is controlled be fiber creep while in oxidizing environments, it is controlled by the interface oxidization kinetics.

INTRODUCTION

Many of the most promising ceramic matrix composites have been optimized for high toughness at relatively low temperatures. However, the most promising applications of continuous fiber reinforced composites are at high temperatures, e.g. boilers, turbine combustors and heat exchangers. It is realized that superior composite properties are dominated by the interphase and for these composites high ambient temperature toughness is obtained by coating the fibers with a material that allows weak bonding between the fiber and the matrix. This leads to crack deflection, debonding at the fiber matrix interface, crack wake fiber bridging and fiber pull out which contribute to high toughness and damage tolerance.[(1)] Since these materials have been optimized for high toughness at relatively low temperatures the long term, high temperature performance of these composites needs further study. It is expected that time dependent crack growth and deformation will control the lifetime of these materials at elevated temperatures.[(2)] The interphase of these composite materials has been specifically tailored so that crack wake processes lead to debonding and fiber pull-out giving a damage tolerant material, therefore, crack wake bridging by the fibers dictates their desirable mechanical behavior. If the temperature is high enough, then the bridging fibers could creep and relax the bridging forces. Thus, at elevated temperatures the time

To the extent authorized under the laws of the United States of America, all copyright interests in this publication are the property of The American Ceramic Society. Any duplication, reproduction, or republication of this publication or any part thereof, without the express written consent of The American Ceramic Society or fee paid to the Copyright Clearance Center, is prohibited.

dependent response of the fibers in the bridging zone greatly influences the deformation of these composites. The crack velocity-stress intensity factor relationship of these composites has been previously investigated and higher crack velocities were observed in Ar with up to 20% oxygen than in pure Ar. Properties of the C interphase deteriorated when oxygen was present in the environment and this leaded to a further decrease in the effectiveness of bridging fibers.[3]

Fiber creep reasonably predicts the correct crack velocity and time dependence in pure Ar, however, other factors need to be considered when oxygen is introduced into the environment. In these environments interphase removal or reaction must also be considered. Recent investigations[4] have shown that oxygen is detrimental to the tensile stress rupture properties of a BN interphase composite due to glass formation in the 950°C - 1200°C region. At intermediate temperatures the BN reacted with water vapor from the atmosphere and vaporized. Other investigators[5,6] concluded that at intermediate temperatures (~600°C) a reduction in flexural strength was observed. The strength reduction was due to the oxidation product not forming fast enough to seal the stress induced cracks and this lead to internal oxidation of the sample. At this temperature the BN oxidized via volatilization and did not result in borosilicate formation. At higher temperatures (950°C) the reaction product formation was faster thus sealing the interior of the sample from further oxidation. From this brief review of the literature, it is apparent that the effect of pO_2 on the oxidation of the interphase and the mechanical properties of composites is rather complex and depends on many factors. These include sample geometry, interphase layer thickness, pO_2 and temperature. The present work will investigate the effect of pO_2 and temperature on the time dependent deformation of the composite and compare these results with data taken from an inert environment. The effect of pO_2 and temperature on interface reaction is also investigated on the same materials and an attempt made to correlate interface reaction with composite deformation.

MATERIALS USED AND EXPERIMENTAL PROCEDURE

The composite materials in this investigation consisted of a woven Nicalon fiber cloth (0°/90°) which was 8 plies in thickness. The matrix material, β-SiC, was applied by chemical vapor infiltration. The composite had a residual porosity of between 10% and 15%. The interface present was a 0.4μm Boron Nitride (BN) layer which was deposited on the fibers before the CVI step. The as received composites were 4mm x 13mm x 50mm in dimensions and were fabricated by RCI of Whittier, CA. Single-edge-notched bend bars (SENB) with dimensions 3.5 mm x 5 mm x 50 mm were prepared from the as received bars. The bars had a blunt U shaped notch and an initial a/w ratio of ~ 0.25. These specimens were loaded in four-point bending using a fully articulated SiC bend fixture. All tests were performed inside a sealed mullite tube and a split clamshell furnace. Further details of the experiment set-up are provided in Ref. 7.

All the experiments were conducted under static loading for times up to 1×10^5 seconds giving long term displacement/time data. The test temperatures varied between 800°C to 1100°C. Samples were tested in pure Ar and in Ar with varying oxygen partial pressures. The specimens were loaded such that the outer fiber stress was slightly higher than the matrix cracking stress. Testing continued until sample failure. Finally, samples that were tested in an oxygen environment were brought up to temperature in pure Ar and the gas mixture was switched over just prior to testing. The gas flow rate through the mullite tube holding the specimen was set to ~ 1.6×10^{-6} m^3/s. Time, load, displacement, oxygen partial pressure and temperature were recorded for each experiment. All the load/displacement data were corrected for the compliance of the SiC four point bend fixture used in the experiments. At least two tests were conducted under each set of conditions to check for reproducibility.

Thermo-Gravimetric Analysis (TGA) was used to monitor the oxidation behavior of the composite at temperatures between 800°C and 1100°C in varying oxygen environments for periods up to 12 hours (Netzsch Instrument STA409). Gas flow was maintained at 1 SCFH throughout the experiment. Before each test the sample was placed in an Al_2O_3 pan in the furnace. The furnace was then sealed and the gas flow was allowed to equilibrate before heat up. The furnace was then ramped up to the required temperature and held for the specified time. Mass change was recorded as a function of time. A sample was also run in pure Ar and this showed little mass change indicating that all mass changes are attributed to reactions requiring oxygen. All TGA samples had dimensions of 4 mm x 4 mm x 8 mm. One of the faces was free from SiC coating applied during processing while all the other surfaces had a 5 µm coating. During the experiment the surface without the coating was the only one that underwent a reaction. Finally all the rates obtained for each sample were normalized with respect to the area of the uncoated surface as these varied slightly between samples.

RESULTS AND DISCUSSION

Displacement/Time Data

From the four point bend experimental data, plots of deflection as a function of time were constructed. Fig. 1 shows an example of the raw data for the composite material held at a constant load at 1000°C and tested at different oxygen partial pressures. Note that the initial deflection is different for the different samples due to slight variations between samples (geometry, load). In order to account for these slight differences the deflection needs to be normalized.

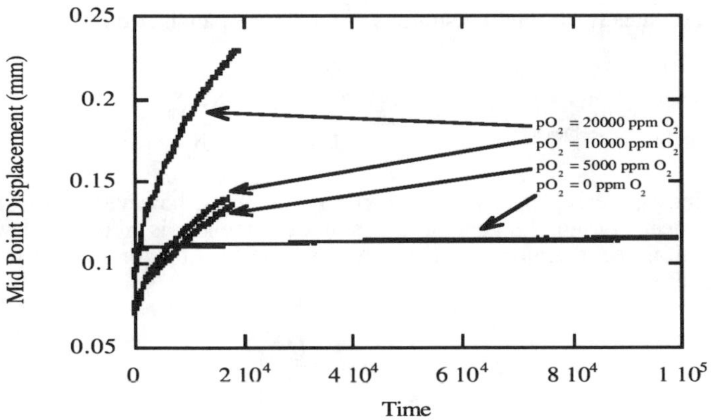

Fig.1: The Deflection behavior of the BN Interphase Composite for varying oxygen partial pressures at 1000°C. This data is not normalized.

The creep rate of a material is influenced by many variables including temperature, T, applied stress, σ, creep strain, ε, elastic modulus, G and grain size, d. In general the creep rate may be given by:

$$\dot{\varepsilon} = f(T,\sigma,\varepsilon,G,d) \tag{1}$$

Di Carlo[8] investigated the creep behavior of several ceramic fibers and found that the fiber creep followed the standard expression for transient creep:

$$\varepsilon = A\sigma^n \left[t.\exp\left(\frac{-Q}{RT}\right)\right]^P \tag{2}$$

where A, P and n are empirically determined creep parameters. The creep parameter, P, is obtained by plotting ln (ε) vs ln (t) at a constant temperature and the activation energy is obtained from a plot of ln (ε) vs 1/T.

In this study the creep deformation of composites was investigated using the above approach, however, initially it was necessary to normalize the deflection data. The deflection was normalized using the following approach. Let the mid point deflection = δ, and the initial mid point deflection = δ_o (at time t = 0). Then we define normalized deflection as follows:

$$\delta^* = \frac{\delta - \delta_0}{\delta_0} \tag{3}$$

Using this normalization, small variations in load and sample dimensions were taken into account allowing different samples to be compared. The following equation relates creep strain to midpoint displacement for the geometry used in this study:

$$\varepsilon = \frac{4\delta h}{s^2} \tag{4}$$

where ε is the outer fiber bending strain, s is the inner span, h is the sample height. Thus:

$$\varepsilon_{norm} = \frac{\varepsilon - \varepsilon_0}{\varepsilon_0} = \frac{4h\delta - 4h\delta_0}{4h\delta_0} = \frac{\delta - \delta_0}{\delta_0} = \delta^* \tag{5}$$

Eqn. 2 describes the creep behavior in inert environments, however in this study the pO_2 has also been varied. In general the pO_2 dependence is a power law for an oxidation reaction. We therefore use Eqn. 6a as a modification for Eqn. 2 to account for oxidizing environments.

$$\varepsilon = A_o pO_2{}^m \sigma^n \left[t.\exp\left(\frac{-Q}{RT}\right)\right]^P \tag{6a}$$

Finally rewriting equation (6a) in terms of normalized deflection (5)

$$\delta^* = \frac{\varepsilon - \varepsilon_0}{\varepsilon_0} = A_1 pO_2^{\,m}\left[t.\exp\left(\frac{-Q}{RT}\right)\right]^p - 1 \qquad (6b)$$

where A_1 is an empirical parameter. Eqn. 6b assumes that the stress dependence of deflection at zero time and as a function of time is the same. It should be noted that in this study, all the samples were tested at approximately the same outer fiber stress. Thus, even if this assumption is not correct, its effect on the result is small.

An example of normalized deflection vs time is given in Fig 2. From this plot it was observed that the deflection increased with increasing oxygen partial pressure at a fixed temperature. At a fixed pO_2, the deflection also increased as temperature increased. Taking the natural log of both sides of Equation 6b the creep parameter, P, is obtained by plotting ln (δ^*) vs ln (t) at a fixed temperature and pO_2, and the activation energy is obtained from a plot of ln (δ^*) vs 1/T at a constant pO_2 and time.

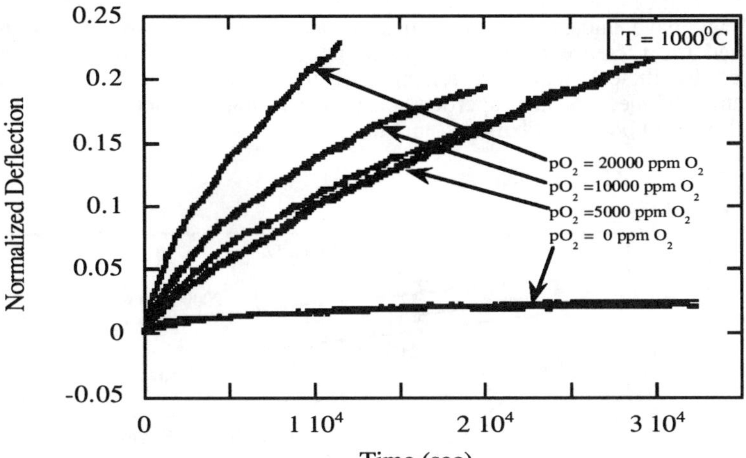

Figure 2: The Normalized Deflection vs Time of the BN Interphase Composite for varying oxygen partial pressures at 1000°C.

The empirical creep parameter, P, was obtained for the different oxygen partial pressures and temperatures. P was relatively constant at about 0.4 in the inert environment and this is the same as the P value obtained by DiCarlo[9] for Nicalon fibers tested in the same temperature range. P increased from 0.4 with increasing oxygen partial pressures for the BN interphase composite. There is no clear physical significance of P. It can be viewed as a measure of deviation from steady state. Note P = 1 for steady state creep and less than 1 for primary creep. Once the creep parameter P has been obtained the activation energy Q can be obtained in the following manner. Normalized deflection was plotted as a function of time for different temperatures at a fixed pO_2. The normalized deflection at a fixed time was obtained and using Eqn. (6b), a plot of ln δ^* vs. 1/T at constant time and pO_2 was used to obtain the activation energy. Such a plot is shown in Fig. 3. In the Ar atmosphere the activation energy for composite deformation, Q, was observed to be constant at 550 ± 20 kJ/mol (see Fig. 3). The activation energy for Nicalon fiber creep is estimated at 600 kJ/mol[9] and this value is very similar to Q obtained for deformation in these experiments which suggests that in an inert environment fiber creep controls composite deformation. In addition, for testing in Ar environments very long times to failure were observed.

Oxidizing environments of various partial pressures of O_2 were chosen to investigate effects of oxidation on the composite material deformation. A value of ~ 60 ± 10 kJ/mol was observed for all oxidizing environments for composite deformation (see Fig. 3). Note, the activation energy was independent of pO_2. However, the deformation rate increased as the pO_2 increased (see Fig. 2). For the BN interface material SEM micrographs (Fig. 4) reveal that the interface formed a reaction product and the amount of reaction product increased with increase in oxygen partial pressure. Also at lower temperatures and pO_2's there is evidence that the interface is being removed faster than the interfacial reaction product formed. Similar observations have been made by other groups[4-6]. The product formed at the interface at the higher oxygen partial pressures is probably a borosilicate glass which may have undesirable properties as an interphase material. As already mentioned the presence of the interphase is critical for the superior properties displayed by these composites. Reaction of the interphase may lead to severe degradation of mechanical properties since the reaction product is a brittle phase that will bond to both the fibers and matrix.

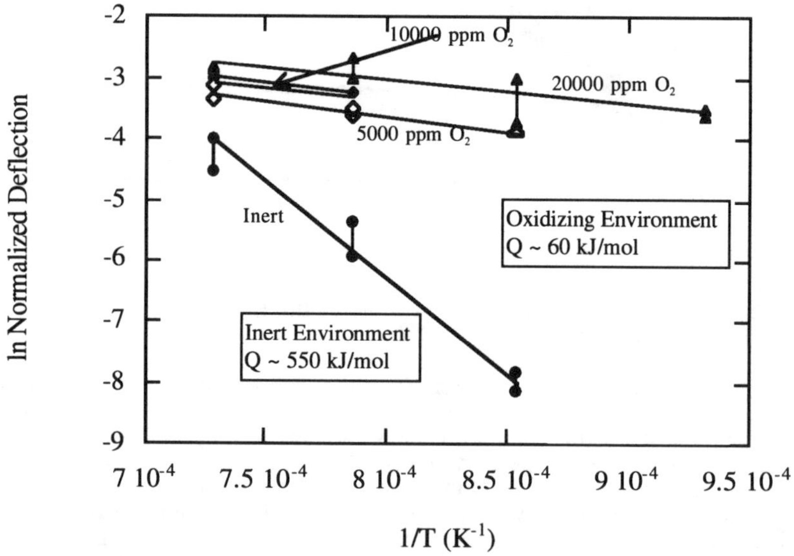

Figure 3. Activation Energies obtained for Composite Deformation at different pO_2's.

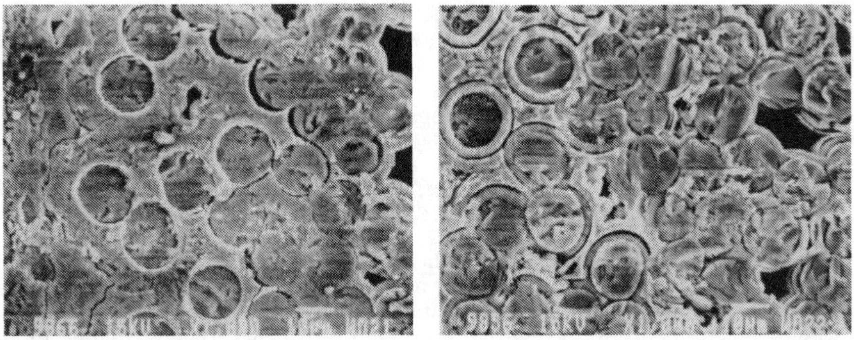

Figure 4. SEM Micrograph of BN Interphase material after testing at 1100°C with 5,000 and 20,000 ppm oxygen environment. Increased pO_2 leads to increased reaction product at the interface.

TGA DATA

From the TGA data, plots of weight loss as a function of time were obtained (Fig. 5.). All the mass loss vs time data is normalized with respect to the exact exposed area as this varied slightly from sample to sample. The weight loss increased with increasing temperature and increasing oxygen partial pressure. Parabolic weight loss kinetics is observed for the BN interphase composite. It was observed that at the higher pO_2's the highest temperature sample (1100°C) lost less weight than the sample tested at 1000°C. Fig 5 shows that at longer times the higher temperature sample stops losing weight. A possible explanation is the pinching off of the interface at higher temperatures due to the decreasing viscosity of the product oxidation layer that is forming. This low viscosity product layer effectively seals off the interface preventing further reaction.

For the BN interphase the TGA data exhibits parabolic kinetics and the rate is related to the weight loss as follows:

$$k_l = \frac{\Delta M^2}{t}\frac{1}{\rho^2 A^2} \tag{7}$$

where ΔM is the mass loss (g), t is time (sec), ρ is the density of BN = 2.25 g/cm^3, A is the calculated area of exposed surface that is interphase ($8.8 \pm 0.5 \times 10^{-3}$ cm^2). It is possible to get an activation energy for the mass loss from:

$$Rate = \left(\frac{\Delta M^2}{t}\right)\frac{1}{\rho^2 A^2} = k_o \frac{xP}{RT}\exp\left(\frac{-Q}{RT}\right) \tag{8}$$

where P is pressure (Pa), x is the fraction defining the partial pressure of oxygen, k_o is a constant, R is the gas constant (J mol^{-1} K^{-1}), T is temperature (K) and Q is the activation energy (J/mol).

If pO_2 is held constant then equation 8 has the following form:

$$k_l T = Const.\exp\left(\frac{-Q}{RT}\right) \tag{9}$$

A plot of ln (k_lT) vs 1/T will yield a straight line with a slope of -Q/R. From this it is possible to obtain the activation energy for reaction of the BN interphase. This plot is presented in Fig. 6 and from this the activation energy has been calculated to be 65 ± 10 kJ/mol independent of the partial pressure of oxygen. This is similar to the activation energy obtained for composite deformation in the previous sub-section (50 ± 10 kJ/mol).

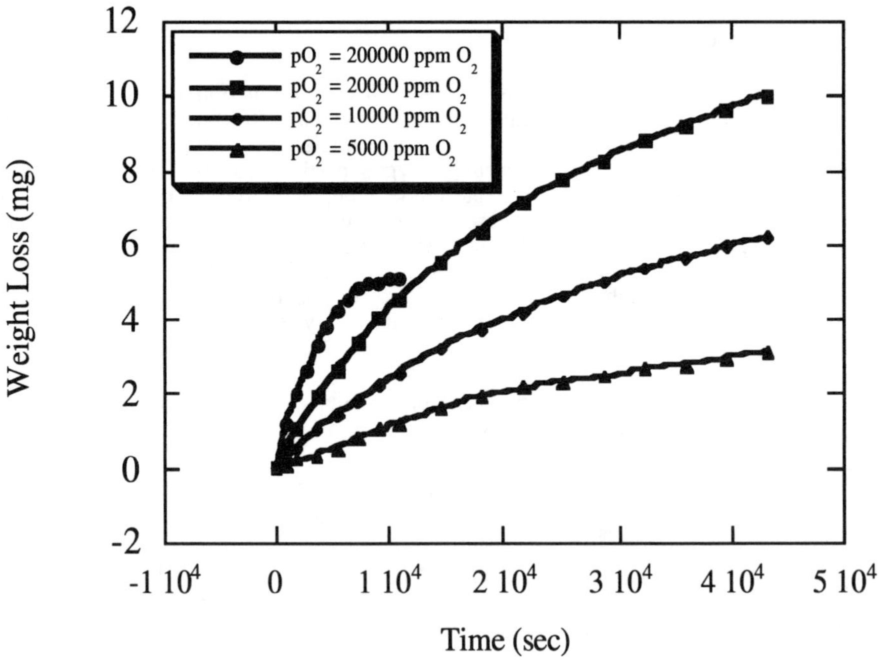

Figure 5. Weight Loss vs Time for varying oxygen partial pressures at 1000°C

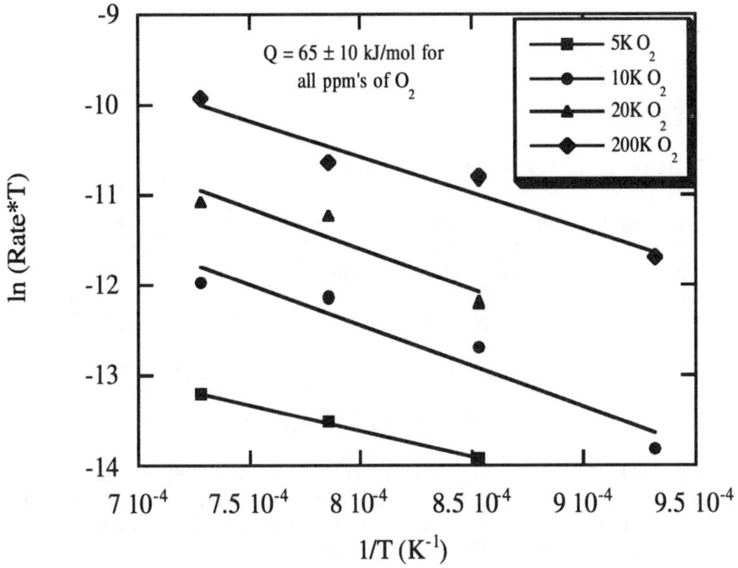

Fig. 6: Activation Energies obtained for the Carbon Interphase Recession Rates

SUMMARY AND CONCLUSIONS

The time dependent deformation under constant flexural load of SiC/SiC composites with a BN interphase has been obtained as a function of pO_2 and temperature. The samples tested were notched beams. It is shown that the presence of oxygen accelerates sample deformation and the effect depends on both pO_2 and temperature. Further, it is shown that the normalized deformation behavior can be analyzed in terms of the formalism developed for fiber creep. From these results activation energies for deformation were obtained in both inert and oxidizing environments. In inert environment the activation energy observed for time dependent deformation (550 ± 20 kJ/mol) is similar to the activation energy for Nicalon® fiber creep. This suggests that composite deformation is controlled by fiber creep in this environment. In oxidizing environments the activation energy for time dependent deformation decreases significantly to 65 kJ/mol. It was also shown that the activation energy for time dependent deformation in oxidizing environments and the activation energy obtained for interphase reaction from the TGA data are similar suggesting that the reaction of oxygen with the interface controls time dependent deformation in oxidizing environments. Thus, based on the approximately equal values of the activation energy, it can be concluded that the kinetics of deformation in an inert environment is controlled by the creep of fibers. Further, in an oxidizing environment the kinetics of deformation is controlled by the rate of interface oxidization.

REFERENCES

1. A.G. Evans and D.B. Marshall,. *Acta Metall.*, **37**, [10], 2567, (1989)
2. C.H. Henager Jr. and R.H. Jones, *Ceramic Eng. Sci. Proc.*, **13**, 411 (1992)
3. C.H. Henager Jr. and R.H. Jones, *Ceramic Eng. Sci. Proc.*, **14**, 7 (1993)
4. G.N. Morscher, *J. Am. Ceram. Soc.*, **80**, [8], 2029, (1997)
5. E.Y. Sun, S.R. Nutt, J.J. Brennan., *J. Am. Ceram. Soc.*, **79**, [6], 1521, (1996)
6. E.Y. Sun, H.T. Lin, J.J. Brennan.,, *J. Am. Ceram. Soc.*, **80**, [3], 609, (1997)
7. C.H. Henager Jr. and R.H. Jones, *Mat. Sci. and Eng.*, **A166**, [1-2], 211, (1993)
8. J.A. DiCarlo, *Comp. Sci. and Tech.* **51**, [2], 213, (1994)

ACKNOWLEDGMENTS

PNNL is operated for the U.S. Department of Energy by Battelle Memorial Institute under contract DE-AC06-76RLO 1830. The research at The University of Washington was supported by the National Science Foundation under Grant Number DMR-NYI-9257827 and by the Du Pont Young Professor Award.

ENVIRONMENTALLY-INDUCED FAILURE MECHANISM MAPPING FOR CONTINUOUS-FIBER, CERAMIC COMPOSITES.

C. A. Lewinsohn*, C. H. Henager, Jr., and R. H. Jones
Pacific Northwest National Laboratory*,
PO Box 999,
Richland, WA 99352,
USA.

ABSTRACT

Ceramic matrix composites (CMCs) are considered for a variety of applications because they possess tolerance to flaws and they do not fail catastrophically. Crack propogation in these materials is controlled by the presence of fibers that bridge crack faces. Removal of the fiber bridging stresses can occur by a variety of environmentally induced mechanisms. These mechanisms, and the effects that they have on subcritical crack growth and residual toughness of CMCs, will be discussed. The results of an experimental investigation of the conditions under which interface removal or oxidation embrittlement of the bridging fibers controls the rate of subcritical crack growth will be used to construct a failure mechanism map. The failure mechanism maps illustrate regions of temperature and oxygen concentration where each crack growth mechanism operates. It will be shown that temperature and oxygen concentration are key parameters in determining the crack growth mechanism.

INTRODUCTION

To obtain the desirable properties of CMCs debonding must occur between fibers and the matrix as cracks propogate through the matrix., leaving fibers bridging the matrix crack. Debonding is usually obtained by addition of a constituent at the interface between the matrix and the fiber that is prone to Mode II failure. Due to the additional requirements of high-temperature durability, carbon, boron nitride, and silicon carbide are commonly used as the interphase material prone to debonding. Subsequent to matrix cracking, the interphase and fibers that bridge cracks are exposed to the ambient environment. Previous studies [1-3] have identified four different elevated temperature, subcritical crack growth mechanisms in materials reinforced by silicon carbide fibers in a silicon carbide matrix. (SiC_f/SiC_m). The occurrence of each mechanism is determined by environmental conditions and microstructural characteristics of the materials.

It has been proposed that the formation of a brittle oxidation product on the fibers creates Griffith type flaws and decreases the strength of the fibers [4]. As the fibers progressively weaken they fail in the wake of the crack tip. Fiber failure decreases the bridging stresses and increases the crack tip stress intensity leading to additional crack growth. Since the the peak load that can be supported by a CMC is governed by the strength of the fibers, CMCs that have undergone oxidation embrittlement can support lower peak loads than unexposed material. In addition, the formation of a brittle oxidation product between the fiber and the matrix can prohibit matrix sliding and strain accommodation during fracture [1,5]. Without matrix sliding and strain

* Operated for the U.S. DOE by Battelle Memorial Institute under Contract DE-AC06-76RLO 1830.

To the extent authorized under the laws of the United States of America, all copyright interests in this publication are the property of The American Ceramic Society. Any duplication, reproduction, or republication of this publication or any part thereof, without the express written consent of The American Ceramic Society or fee paid to the Copyright Clearance Center, is prohibited.

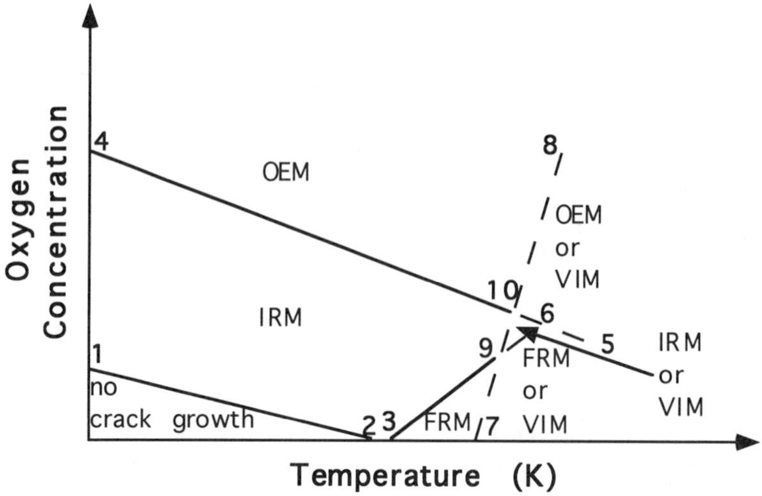

Figure 1. Schematic Failure Mechanism Map.

accommodation stress builds up in the fibers and they fracture in the plane of the matrix crack leading to subcritical crack growth as described. Since both of these processes may occur simultaneously, these mechanisms will be referred to as the oxidation embrittlement mechanism (OEM). Clearly, fibers with lower oxidation rates will extend the lifetime of composites undergoing oxygen embrittlement.

A second subcritical crack growth mechanism [6], referred to as the interface recession mechanism (IRM), involves removal of the interphase by oxidation (subsequent to matrix cracking) which increases the crack opening displacement and causes additional crack growth. In addition, during IRM, fiber creep occurs. Since the strength of the fibers is not affected by interphase recession, exposed materials can support similar peak loads to unexposed materials. In some cases, a transition from IRM to OEM can occur. This transition depends on the oxidation kinetics of the interphase , the fiber, and the matrix; the temperature and oxygen concentration; and the thickness of the interphase.

Failure mechanism maps are a graphical guide to determining the environmental conditions under which each of the four crack growth mechanisms described above occur. A schematic representation of a failure mechanism map is shown in Figure 1. This type of map plots the range of conditions under which each mechanism controls crack growth. Since accurate models of all these mechanisms do not exist, and because the values of required parameters for a specific material may not be known, experimental methods must be employed to allow construction of failure mechanism maps.

Although explicit solutions for the boundaries shown in the schematic failure mechanism map shown in Figure 1, the slope and relative positions of the boundaries can be intuitively determined. The failure mechanism map shown in Figure 1 is not drawn to scale and other parameters, such as interface thickness and gas flow rate, may change the relative positions of the boundaries. The line between points 1 and 2 defines the lower limits of measurable interphase oxidation. Once interface oxidation causes recession of debonded interfaces, IRM controlled crack growth can occur. As the temperature is increased, the rate of silica formation increases. At low to intermediate temperatures carbon oxidation is faster than silica growth, because of a larger rate constant, but at higher temperatures silica growth rates exceed those of carbon oxidation, because of a higher activation energy for silica growth. Once sufficient silica grows on fibers to create a critical flaw, or to bond the fibers to the matrix, OEM controls crack growth: this condition is defined by the area above the line from points 4 to 5. As the temperature increases, so does the fiber creep rate. Although carbon oxidation will always begin at lower temperatures than fiber creep, the creep rate of the fibers has a larger activation energy than that of carbon oxidation and it is possible that the rate of crack growth can be controlled by relaxation of the fiber bridging stresses due to fiber creep at high enough temperatures [3]. This mechanism is referred to as the fiber relaxation mechanism (FRM). The region where FRM controlled crack growth occurs is defined by the area to the right of the line between points 3 to 6. Finally, it has been postulated that, at high temperatures, an oxidation product can form between the fiber and the matrix that has sufficiently low viscosity to flow rather than fracture in a brittle manner [2]. Viscous flow of such an oxide layer would also cause the fiber bridging stresses to relax and lead to subcritical crack growth. This mechanism is referred to as the viscous interface mechanism (VIM) and it is defined by areas to the right of the dashed line between points 7 to 8. Since VIM has not been observed in the CMCs investigated in this work, the boundary defining its occurrence is shown by a dashed line and the regions to right are labelled as OEM or VIM, IRM or VIM, and FRM or VIM. In addition, we have not determined if crack growth controlled by FRM could occur at a faster rate than that controlled by OEM, hence the hypothetical boundary line (which assumes that FRM does not occur faster than OEM) from point 6 to 5 is also shown by a dashed line.

EXPERIMENTAL METHOD

The specimens were fabricated by chemical vapor deposition of silicon carbide on a fibrous preform (CVI). The fibrous preform consisted of approximately eight layers of 2-D plain weave (0/90°) "ceramic-grade" Nicalon®* fibers. Prior to deposition of the matrix, a thin layer of carbon was deposited on the fibers. According to the manufacturer** the thickness of this layer is 150 nm. The resulting composites contained approximately 40 volume-percent fibers and 10-15% porosity. Single-edge notched beam specimens (SENB) were cut from extra compact-tension specimens. Thus two surfaces of the SENB specimens (the tensile and compressive surfaces) were not coated with the protective, CVD silicon carbide layer.

Subcritical crack growth experiments were conducted to investigate the environmentally induced failure mechanisms of the CFCC material studied. All experiments were performed by flexural loading of single-edge-notched beam specimens (SENB) in 1/4 four-point bending in a fully articulated silicon-carbide fixture contained within a mullite tube. The atmosphere inside the mullite tube was controllable. The deflection of the specimen midpoint was measured by an alumina pushrod, also containing a thermocouple, attached to a strain-gauge extensometer. The displacements were corrected for differences between the load-point and mid-point and for the compliance of the test apparatus. Since, the subcritical crack growth rate was determined from an indirect measurement, which was based on calculating the instantaneous compliance of the specimen, it is also referred to as the effective crack velocity. The final crack length was measured for several specimens by sectioning the specimens, polishing with diamond pastes, and observing the crack with optical microscopes. Agreement between the crack lengths determined optically and via the compliance measurement was within 25%. The crack length calculated from the compliance measurement was consistently 20-25% higher than the optical measurements. Scanning electron microscopy (SEM) was used to investigate fracture surfaces. In addition, the residual toughness, after subcritical crack growth, was measured for some specimens. Further details of the testing methods can be found elsewhere [3].

RESULTS AND DISCUSSION

Three experimental techniques were used to investigate the subcritical crack growth behavior of these materials during exposure to different combinations of temperature and oxygen concentration. The first was observation of the specimen displacement as a function of time during crack growth experiments and investigation of the relationship between the effective crack

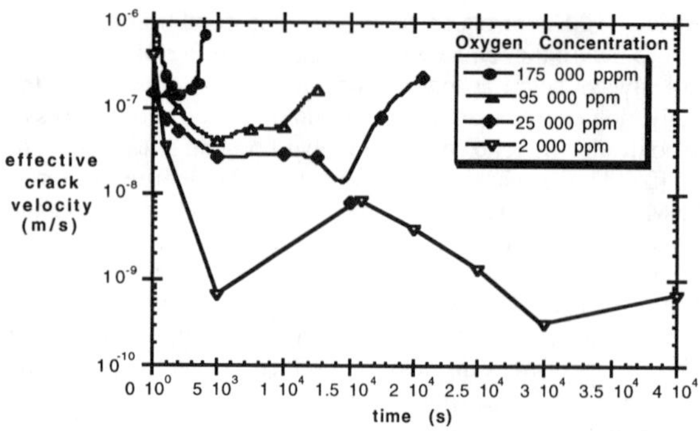

Figure 2. Rate of subcritical crack growth, at 1073 K, as a function of oxygen concentration.

velocity and time. The second was comparison of the fracture toughness, calculated from the peak load of SENB specimens tested in flexure, after various treatments, to the intrinsic fracture toughness of the same material. The third involved investigation of the fracture surfaces via SEM.

Subcritical Crack Growth

The subcritical crack growth rate, at 1073 K, of specimens with 150 nm C interphases, in atmospheres containing different concentrations of oxygen, is shown in Figure 2. Subcritical crack growth in gettered argon (< 20 ppm O_2) is characterized by a continuously decreasing crack velocity with time. This behavior has been explained by considering relaxation of crack-bridging tractions due to time-dependent creep of fibers: this mechanism is referred to as the fiber relaxation mechanism (FRM). Predicted crack velocities from a micromechanical model describing this behavior agree well with experimental measurents [3]. In addition, the activation energy for crack growth in argon [7] matches that reported for fiber creep [8].

The specimens tested in environments containing oxygen exhibited significantly different subcritical crack growth behavior than those tested in gettered argon. The effective crack velocity initially decreased, but then it increased by several orders of magnitude . Evidence of a glassy phase on the fibers or at the fiber-to-matrix interface was not obtained from observations, using optical and scanning electron microscopy, of specimens after they had been tested. In addition, TGA measurements (between1073-1373 K and 2 000 - 200 000 ppm O_2) on the same materials indicated only weight loss [9]. If the material underwent passive oxidation (formation of silica) a weight gain would be observed. The activation energy for subcritical crack growth, prior to the rapid increase in velocity, was measured as 47 kJ/mol. The activation energy for removal of the carbon interphase was measured as 50 kJ/mol using thermogravimetric analysis (TGA) [9].

The effect of interphase recession on fiber push-in behavior was measured on specimens of a similar composite material, reinforced with Hi-Nicalon™ fibers with a 1 μm-thick carbon interphase, that had been oxidized in air for increasing amounts of time [10]. The oxidation treatments were conducted at 1073 K to minimize thermally induced microstructural changes in the fibers. Complimentary examination of the oxidized specimens verified that the length of interphase recession increased with increasing oxidation time. The typical loading behavior of fibers during push-in testing of fibers after different amounts of oxidation is shown in Figure 3 As the interphase recession distance increases, the slope of the load versus displacement curve decreases. This decrease in the slope of load versus displacement during fiber push in is likely due to the decreased resistance to sliding caused by removal of the interphase. Hence, interphase

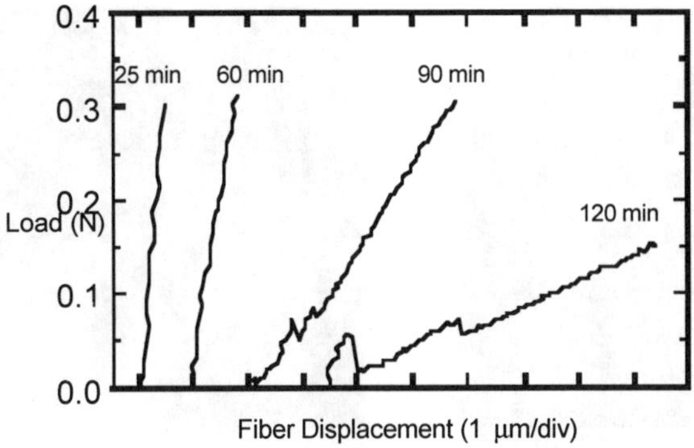

Figure 3. Typical fiber push-in initial loading curves for various oxidation times at 1073 K in air.

recession leads to a more compliant bridging zone. As interphase recession proceeds the crack opening displacement will increase leading to additional crack growth.

As shown in Figure 2, the time at which a rapid increase in effective crack velocity occurs decreases as the oxygen concentration increases. For experiments performed in atmospheres containing greater than 2 000 ppm O_2, the time-to-failure decreased linearly with oxygen concentration (Figure 4). Although other investigators have derived models relating the growth rate of an oxide layer on the surface of bridging fibers to the failure time of a composite, they have only done so for constant oxygen concentrations. The results shown in Figure 4 illustrate that the oxygen concentration clearly plays a role in determining the failure time. The accepted models of oxide growth on silicon carbide predict a linear dependence of the growth rate on oxygen concentration. Thus, the time to form a fixed thickness of an oxide layer would be inversely proportional to the oxygen concentration. This suggests that composite failure is likely controlled by the time required for an oxide layer to reach a fixed thickness, such that flaws of a critical size

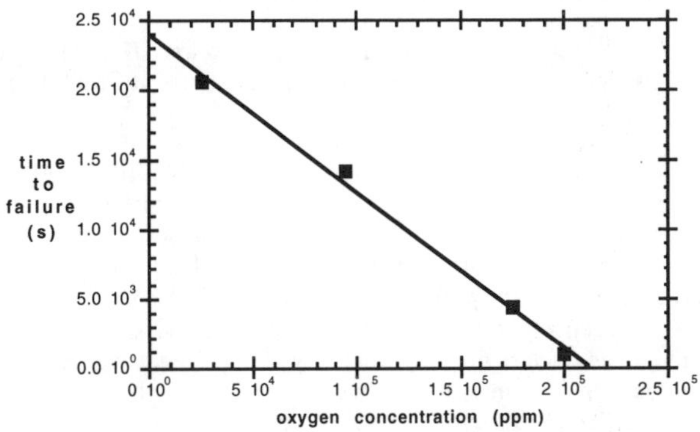

Figure 4. Dependence of failure time on oxygen concentration at 1073 K.

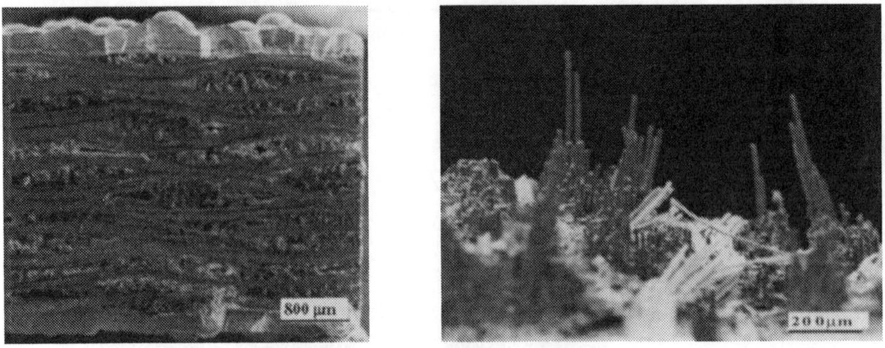

Figure 5. The fracture surface of a specimen broken at 1073K in an inert environment.

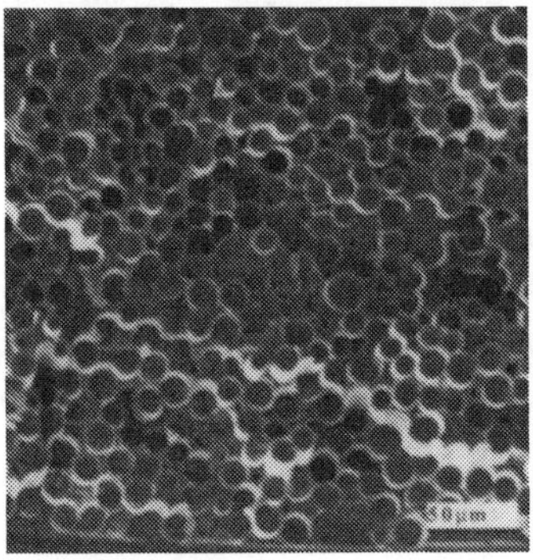

Figure 6. The fracture surface of a specimen broken at 1073 K after oxidation with 50% O_2.

become present or bonding across the fiber/matrix interface occurs. This is in agreement with other models of the effect of oxidation on the failure of these types of CMCs [4,11].

Residual Fracture Toughness and Fractography

The intrinsic fracture toughness of the CFCC was tested at 293, 1073, and 1373 K, in gettered argon (<20 ppm O_2). At all three temperatures the fracture behavior was similar. The material behaved linearly elastic until failure. Despite the inability to support more than approximately 5% of the peak load after failure, the specimens remained intact and did not fail catastrophically. The toughness values calculated from the peak load were 17.5 ± 0.5, 17.3, and 17.1 MPa-$m^{1/2}$ at 293, 1073, and 1373 K, respectively. The fracture surfaces of these specimens were also similar. Extensive fiber pullout and fracture markings on many of the fibers were readily apparent (Figure 5). The fracture toughness values of specimens subject to subcritical crack growth in an environment containing 2% of O_2 at 1373 K (IRM conditions) that were measured at 1373 and 1073 K were 16.4 ± 0.5 MPa-$m^{1/2}$ and 17.5 ± 0.5 MPa-$m^{1/2}$, respectively. On the other hand, the fracture toughness of a specimen that was subject to subcritical crack growth in an environment containing 2% of O_2 at 1373 K (IRM conditions) followed by exposure at 1073 K to 500 000 ppm O_2 for 45 minutes without stress, was measured at 1073 K as 15.0 ± 0.5 MPa-$m^{1/2}$. Limited fiber pullout was seen on the fracture surfaces of the latter specimen (Figure 6). Thus, since IRM does not affect the fiber strength the residual fracture toughness and pullout behavior is similar to the intrinsic value. Since OEM leads to fiber strength degradation it increases the probability of fiber fracture near the matrix crack and it also leads to a decrease in residual toughness.

Failure Mechanism Mapping

The results of subcritical crack growth, residual fracture toughness, and microscopy were combined to identify the subcritical crack growth mechanisms under a variety of conditions. These results are shown as a failure mechanism map in Figure 7. Since there are large ranges of values for which experiments have yet to be conducted, this failure mechanism map is by no means definitive. This map serves as a basis for choosing critical experiments to determine with more precision the conditions under which each mechanism operates. In addition, the data serves as a guideline for future modeling efforts.

Results reported by other investigators [8,9,11-16] were combined with the results of this and previous studies in our laboratory (using a CFCC with Hi-Nicalon fibers and a 1-μm-thick carbon interphase) to construct another failure mechanism map (Figure 8). Although the failure mechanism map that is presented contains data gathered from slightly different materials, the relationships among the boundaries separating distinct failure mechanisms is expected to be similar for CFCCs with similar constituents. On the failure mechanism map presented, the range of conditions for which interface recession, without signs of oxidation, was observed in TGA experiments is outlined with a dashed box. Data from critical experiments, using fast fracture tests or subcritical crack growth tests, are shown by symbols. Several of the data were reported for experiments where specimens had been treated at elevated temperatures, but evaluated at room-temperature. Since the oxidation product for all the materials included in Figure 8 is expected to be silica or a boron containing silica-glass, the elastic and fracture properties of the reaction product may be extremely temperature dependent over the range of temperatures investigated. At room temperature, the reaction product would be expected to have a very low failure strain and, hence, promote embrittlement. On the other hand, at elevated temperatures, the oxide layer may posses sufficiently low viscosity to inhibit severe embrittlement.

SUMMARY AND CONCLUSIONS

Subcritical crack growth mechanisms in non-oxide CMCs were investigated by analysis of the dependence of the subcritical crack growth rate, measurement of residual toughness, and examination of fracture surfaces. IRM was characterized by an initially decreasing subcritical crack growth rate followed by a rapid rate of growth, no change in residual toughness, moderate amounts of fiber pullout, and clearly visible fracture markings on broken fibers. OEM was characterized by a rapid rate of subcritical crack growth, decreased residual toughness, and nearly flat fracture surfaces. The predominance of OEM or IRM is governed by the competition between interface recession, which is characterized by a low activation energy, and oxidation, which is characterized by a higher activation energy but a faster reaction rate. Therefore, the occurrence of OEM or IRM will be determined by the temperature, oxygen concentration, and interphase thickness. Failure mechanism maps were constructed that identifies conditions under which each mechanism has been observed.

ACKNOWLEDGMENTS

The authors wish to thank Dr. J. Eldridge, NASA Lewis Research Center, for performing the fiber push-in experiments and John Keaveney for assisting with the mechanical testing. The authors are also extremely grateful to Professor Raj Bordia, University of Washington, for fruitful discussions. This work was supported by Basic Energy Sciences under U.S. Department of Energy (DOE) contract DE-AC06-76RLO 1830 with Pacific Northwest National Laboratory, which is operated for DOE by Battelle.

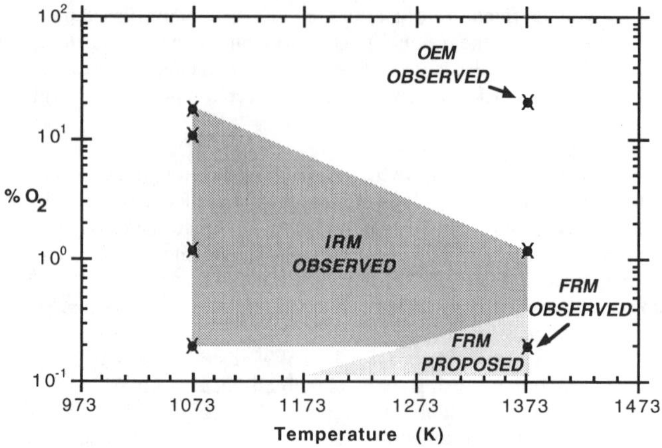

Figure 7. A failure mechanism map identifying mechanisms of subcritical crack growth under the conditions investigated.

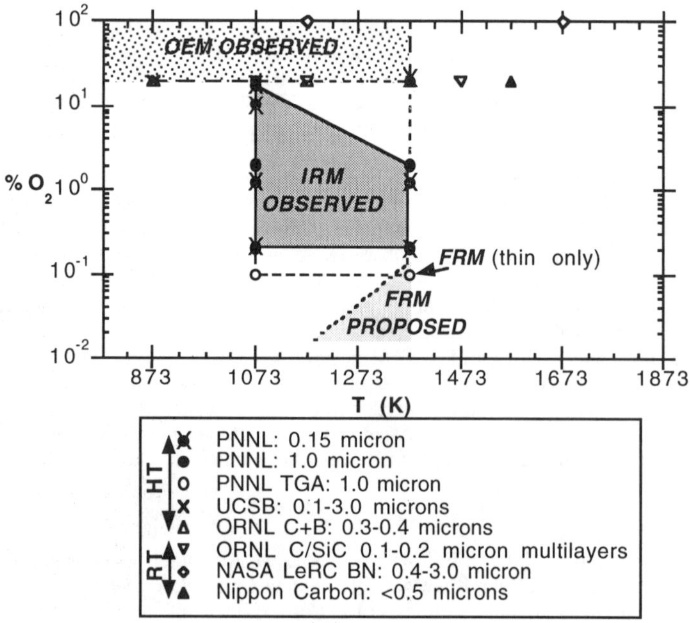

Figure 8. A failure mechanism map compiled from literature data from Nicalon reinforced CMCs.

REFERENCES

1. J.D. Cawley, "Effect of Interphase Carbon Thickness on Environmental Resistance of Continuous Fiber-Reinforced Ceramic Matrix Composites," pp. 377-384 in High-Temperature Ceramic-Matrix Composites II: Manufacturing and Materials Development, Evans A.G. and Naslain R. [Eds.], Ceramic Transactions, vol. 58, The American Ceramic Society, Westerville, OH, 1995.
2. K. Jakus, J.E. Ritter, and R.H. Schwillinski, "Viscous Glass Crack Bridging Forces in a Sintered Glassy Alumina at Elevated Temperatures," J. Am. Ceram. Soc., **76** [1], 33-38 (1993).
3. C.H. Henager Jr., and R.H. Jones, "Subcritical Crack Growth in CVI-SiC Reinforced with Nicalon Fibers: Experiment and Model," J. Am. Ceram. Soc., **77** [9], 2381-2394 (1994).
4. E. Lara-Curzio, "Stress-Rupture of Nicalon/SiC Continuous Fiber Ceramic Composites in Air at 950°C," J. Am. Ceram. Soc., **80** [12], 3268-3272 (1997).
5. G.N. Morscher, "Tensile Stress Rupture of SiC_f/SiC_m Minicomposites with Carbon and Boron Nitride Interphases at Elevtaed Temperatures in Air," J. Am. Ceram. Soc., **80** [8], 2029-2042 (1997).
6. R.H. Jones, C.H. Henager, Jr., and C.F. Windisch, Jr., "High Temperature Corrosion and Growth of SiC-SiC at Variable Oxygen and Partial Pressures," Mat. Sci. & Eng., **A198**, 103-112 (1995).
7. R. Bodet, J. Lamon, N. Jia, R.E. Tressler, "Microstructural Stability and Creep Behavior of Si-C-O (Nicalon) Fibers in Carbon Monoxide and Argon Environments," J. Am. Ceram. Soc., **79** [10], 273-86 (1996).
8. C.A. Lewinsohn, C.H. Henager, Jr., and R.H. Jones, "Effect of Fiber Type on Slow Crack Growth in Continuous-Fiber Composites at Elevated Temperatures," pp. 423-433, Advances in Ceramic Matrix Composites III, Ceramic Transactions v. 74, The American Ceramic Society, Westerville, OH (1996).
9. C.F. Windisch Jr., C.H. Henager Jr., G.D. Springer, and R.H. Jones, "Oxidation of the Carbon Interface in Nicalon-Fiber-reinforced Silicon Carbide Composite," J. Am. Ceram. Soc., **80** [3], 569-574 (1997).
10. C.A. Lewinsohn, J.I. Eldridge, and R.H. Jones, "Techniques for Measuring Interfacial Recssion Distances Using Fiber-Push-in Tests," to appear in Cer. Eng. & Sci. Proc., 1998.
11. A.G. Evans, F.W. Zok, R.M. McMeeking, and Z.Z. Du, "Models of High-Temperature, Environmentally Assisted Embrittlement in Ceramic-Matrix Composites," J. Am. Ceram. Soc., **79** [9], 2345-2352 (1996).
12. F. E. Heredia, J.C. McNulty, F.W. Zok, and A.G. Evans, "Oxidation Embrittlement Probe for Ceramic-Matrix Composites," J. Am. Ceram. Soc., **78** [8], 2097-2100 (1995).
13. G.N. Morscher, "Tensile Stress Rupture of SiC_f/SiC_m Minicomposites with Carbon and Boron Nitride Interphases at Elevtaed Temperatures in Air," J. Am. Ceram. Soc., **80** [8], 2029-2042 (1997).
14. H.T. Lin, and P.F. Becher, "Effect of Fiber Coating on Lifetime of Nicalon Fiber-Silicon Carbide Composites in Air," Mat. Sci. Eng. A., (1997).
15. E. Lara-Curzio and L.L. Snead, Oak Ridge National Laboratory, private communications.
16. T. Ishikawa, "Recent Developments of the SiC Fiber Nicalon and its Composites, Including Properties of the SiC Fiber Hi-Nicalon for Ultra-High Temperature," Comp. Sci. Tech., **51**, 135-144 (1994).

COMPOSITIONAL DEPENDENCE OF THE OXIDATION BEHAVIOR OF $SiC-ZrO_2$-MULLITE COMPOSITES

Cheng-Yuan Tsai and Chien-Cheng Lin*

Department of Materials Science and Engineering
National Chiao Tung University
Hsinchu, 300, Taiwan

ABSTRACT

The oxidation of hot-pressed SiC-particle (SiCp)/zirconia/mullite composites with various zirconia contents, exposed in air isothermally at 1000 and 1200°C respectively for up to 500 hours, was investigated. A clear critical volume fraction of zirconia existed for exposure at either 1000 or 1200°C. The oxidation rate increased dramatically beyond 20 vol% ZrO_2. Additionally, oxidation modes of SiC/zirconia/mullite composites were classified on the basis of the gradient of the silica layer (formed on individual SiC particles) thickness versus depth curve. When the volume fraction of ZrO_2 was below a threshold limit, the composites exhibited Mode I behavior; otherwise, Mode II was observed. For composites with zirconia content above the threshold limit, the formation of zircon, as a result of the reaction between zirconia and oxidation product (i.e. silica), might change the oxidation behavior from Mode II to Mode I.

INTRODUCTION

Many investigators have worked on the oxidation behavior of SiC-reinforced mullite and zirconia/mullite matrix composites at elevated temperatures.[1-8] However, most works did not realize the composites could demonstrate very different oxidation behavior due to the variation of their compositions. Recently, Lin[9] and Zangvil *et al.*[10] have elucidated the oxidation behavior of SiC-reinforced mullite matrix and zirconia/mullite matrix composites after exposure in air at 1000 to 1350°C. They claimed that not only did zirconia content heavily influence the oxidation kinetics, but also that several different oxidation modes appeared to be operating in different composites. Furthermore, they proposed two basic oxidation modes, i.e., Mode I and Mode II, based on the relationship between oxygen diffusivities in the matrix and in the oxide layer.

The effect of zirconia content of SiC/zirconia/ mullite composites on oxidation behavior was conducted by Tsai *et al.*[11] They indicated the oxidation rates of those composites dramatically increased as the zirconia content exceeded a certain threshold volume fraction, while below this limit oxidation rates remained fairly low. They also noted that, at 1200°C, the formation of zircon

To the extent authorized under the laws of the United States of America, all copyright interests in this publication are the property of The American Ceramic Society. Any duplication, reproduction, or republication of this publication or any part thereof, without the express written consent of The American Ceramic Society or fee paid to the Copyright Clearance Center, is prohibited.

due to the reaction between the matrix and the oxidation product could cause a sharp decrease in oxidation rate. In this work, we will study the effect of zirconia content on the oxidation behavior by (1) measuring the weight gain of each composite; and (2) plotting the curves of silica layer thickness with respect to the depth of the corresponding SiC particle. The oxidation modes of SiC/ZrO_2/mullite composites with various ZrO_2 content are then discussed.

EXPERIMENTAL PROCEDURE

Composites of SiCp/zirconia/mullite, all containing 30 vol% of SiC particulate, were fabricated by the hot pressing technique. The starting materials were commercial mullite powder (0.2 μm average, KM-mullite, Kyoritsu Ceramic Materials Co., Nagoya, Japan), 3 mol% Y_2O_3 partially stabilized zirconia (0.3 μm average, TZ-3Y, Toyo Soda Mfg. Co., Tokyo, Japan), and SiC powder (7 μm average, Cerac Inc., Milwaukee, WI). Table I lists the composition and hot pressing conditions, relative density, and XRD phases of each composite. The as hot-pressed composites were ground, then cut into 10x6x3 mm pieces. All specimens were ground and polished with diamond paste, and weighed before and after oxidation using a precision electronic balance (Model R200D, Satorius AG, Goettingen, Germany), with accuracy to 0.01 mg. Composite samples were exposed to air in a box furnace (Model 51333, Lindberg, WI) isothermally at 1000 and 1200°C for up to 500 hours. One specimen for each composition was drawn from the furnace at various intervals for the measurement of weight change. The extent of oxidation was expressed by the weight change per unit surface area or weight gain. In addition to the observation of microstructure, the oxide layer thicknesses of individual SiC particles at various depths were measured by a scanning electron microscope (SEM) and a transmission electron microscope (TEM).

RESULTS AND DISCUSSION

(1) Weight gain vs. zirconia content

Figure 1 shows the relationships between the weight gain and the volume fraction of ZrO_2 for exposure at 1000°C/500 hr and 1200°C/500 hr, respectively. It indicated that the weight gains for the composites containing less than 20 vol% ZrO_2 were comparatively small, while the weight gain increased rapidly when the composites contained more than 20 vol% ZrO_2. It appeared that there was a critical volume fraction of zirconia content required for rapid oxidation. In other words, oxidation rate would be accelerated abruptly beyond this critical zirconia content, whereas the weight gain remained fairly low below the critical value. It is also noted that the weight gain curve for 1200°C slightly declined as the ZrO_2 content was larger than 80 vol%. This is attributed to the formation of zircon as a result of the interaction between zirconia and the oxidation product (SiO_2).

Table I. Designation, Compositions, Hot Pressed Conditions, Densities and XRD Phases of Composites

Sample	Composition[#]	HP Conditions[*]	Relative density (%)	XRD phase[∀]
MZY15/SiC	(Mullite + 15 vol%3Y-PSZ) + 30 vol%SiC	1600°C/45min	99.1	M, α-SiC, t-Z, m-Z
MZY20/SiC	(Mullite + 20 vol%3Y-PSZ) + 30 vol%SiC	1600°C/45min	99.0	M, α-SiC, t-Z, m-Z
MZY25/SiC	(Mullite + 25 vol%3Y-PSZ) + 30 vol%SiC	1600°C/45min	98.3	M, α-SiC, t-Z, m-Z
MZY30/SiC	(Mullite + 30 vol%3Y-PSZ) + 30 vol%SiC	1600°C/45min	98.5	M, α-SiC, t-Z, m-Z
MZY50/SiC	(Mullite + 50 vol%3Y-PSZ) + 30 vol%SiC	1600°C/45min	98.3	M, α-SiC, t-Z, m-Z
MZY80/SiC	(Mullite + 80 vol%3Y-PSZ) + 30 vol%SiC	1600°C/45min	98.0	M, α-SiC, t-Z, m-Z
MZY100/SiC	3Y-PSZ + 30 vol%SiC	1600°C/45min	99.9	α-SiC, t-Z, m-Z

#The volume percentage of the matrix in parentheses is 70% for each composite, the balance being SiC_p. The ZrO_2 content is based on the matrix itself, i.e., vol% of Y-PSZ=volume of Y-PSZ × 100/(volume of mullite + volume of Y-PSZ).

*all under the pressure of 30 MPa and 1 atm Ar atmosphere.

∀ M=mullite; t-Z=t-ZrO_2; m-Z=m-ZrO_2

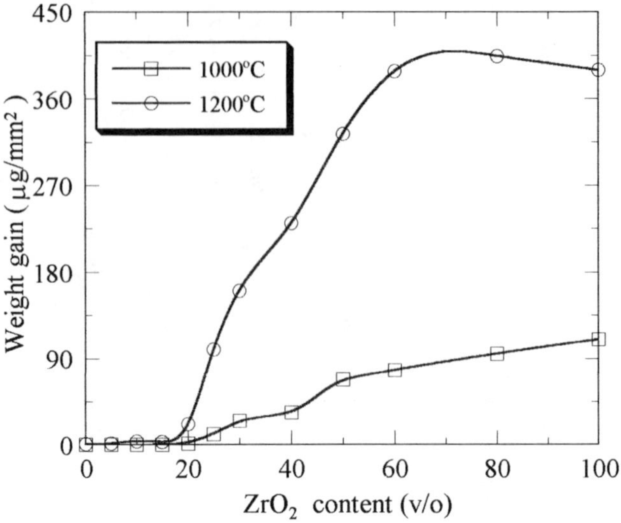

Fig. 1 Curves of weight gain vs. zirconia content at 1000 and 1200°C for 500 hours.

(2) Microstructural observations

Two distinct microstructures viewed from the cross-sectional direction perpendicular to the surfaces of two exposed samples are shown in Figs. 2 and 3. Figure 2 displays an SEM micrograph of MZY25/SiC after exposure at 1200°C/500 hr. The gray irregularly shaped phase encircled by a dark layer is SiC, the bright particles ZrO_2, and the matrix is mullite. The outermost surface is on the left of the micrograph, and to the right is the inner region. During the exposure, oxygen diffused inwards to reach the surface of SiC particles where oxidation took place. SiC particles in the outer region began to be oxidized and then encompassed by a dark layer of oxidation product, which was identified as SiO_2 by the EDS. Figure 2 also reveals a large oxidation zone as well as the variation of the silica layer thickness with respect to the depth of the corresponding SiC particle. SiC particles were partially oxidized all the way from the surface to the depth of more than 370 μm which is beyond the field of view. Experimental results indicated that the silica layer thickness, from the surface to the location at about 120 μm, rapidly decreased with depth but was subsequently followed by slow growth in thickness with depth. Figure 3 displays an SEM cross-sectional view of the oxidation microstructures for MZY15/SiC after exposure at 1200°C/500 hr, indicating a very limited oxidation

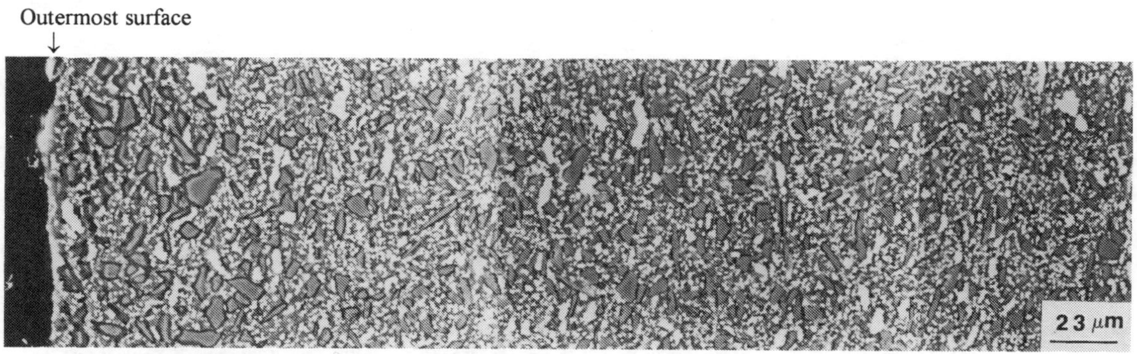

Fig. 2 An SEM cross-sectional view of MZY25/SiC composite after exposure at 1200°C/500 hr, revealing a large oxidation zone as well as the variation of the silica layer thickness with depth. SiC particles were partially oxidized all the way from the surface to the depth of more than 370 μm, which is beyond the field of view.

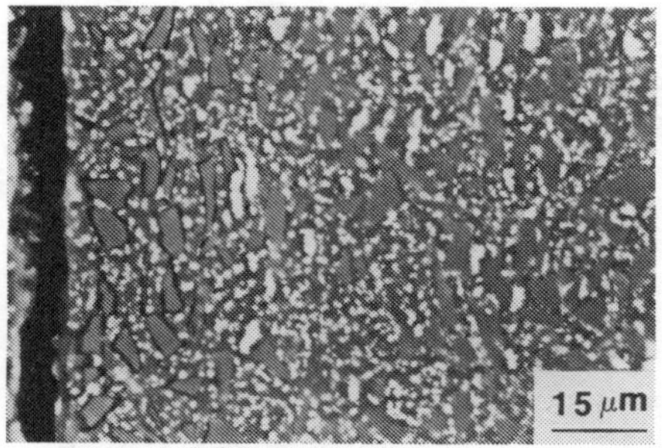

Fig. 3 An SEM cross-sectional view of MZY15/SiC composite after exposure at 1200°C/500 hr, indicating a very limited oxidation zone compared to that shown in Fig. 2.

zone (≈ 40 μm) compared to that shown in Fig. 2.

(3) Silica layer thickness versus depth curves

(a) Oxidation at 1000°C

Figure 4 illustrates the silica layer thicknesses at different depths for various composites after exposure at 1000°C/500 hr/air. SiC particles at the outermost surface for all composites were far from complete oxidation after long-term exposure since the average size of SiC particle was taken at about 7 μm and the largest silica layer in MZY100/SiC was about 1 μm. The curve of MZY20/SiC had a relatively steep slope (ranging from $-8x10^{-3}$ to $-30x10^{-3}$) with a limited oxidized zone of about 40 μm. In contrast, MZY30/SiC, MZY50/SiC, and MZY100/SiC had much shallower slope (ranging from $-0.6x10^{-3}$ to $-4x10^{-3}$) with much larger oxidation depths than that of MZY20/SiC. The difference between MZY20/SiC and MZY30/SiC (or composites containing zirconia content above 30 vol%) implied a dramatic transition in oxidation behavior because of the variation in the zirconia content. The threshold value of zirconia content where this transition occurred was around 20 vol% based on the weight gain data in Fig. 1. With a large slope and a limited oxidation depth, the oxidation behavior of MZY20/SiC at 1000°C was categorized as Mode I. All of the composites containing more zirconia than the threshold value, exhibiting a similar tendency of a small slope and a large oxidation depth, were classified as Mode II.

(b) Oxidation at 1200°C

Figure 5 depicts the curves of silica layer thickness versus depth for various composites after exposure at 1200°C/25 hr. The characteristic of a steep slope ($\approx -13x10^{-3}$) with a small oxidation depth was observed in MZY15/SiC compared with much smaller slopes (ranging from $-0.6x10^{-3}$ to $-6x10^{-3}$) for MZY50/SiC and MZY100/SiC. This observation is analogous to that for MZY20/SiC at 1000°C and, thus, the oxidation behavior of MZY15/SiC belonged to Mode I. However, the oxidation behavior markedly differed among MZY25/SiC, MZY50/SiC, and MZY100/SiC. For MZY25/SiC after exposure at 1200°C/25 hr, the curve could be divided into two regions based on very different slopes: the outer region had a larger negative slope ($\approx -26x10^{-3}$), while the inner region had a smaller negative slope ($\approx -0.4x10^{-3}$) with slightly oxidized SiC. The fact that the behavior of MZY25/SiC at 1200°C/25 hr encompassed two distinctive regions implies that the oxidation mode changed during exposure due to the formation of zircon.[11] The oxidation mode of MZY25/SiC at 1200°C/25 hr was thus categorized as Mode II+I or a mixed mode as defined previously.[9] Interestingly, after exposure at 1200°C for 25 hours, MZY50/SiC rather than MZY100/SiC had the thickest

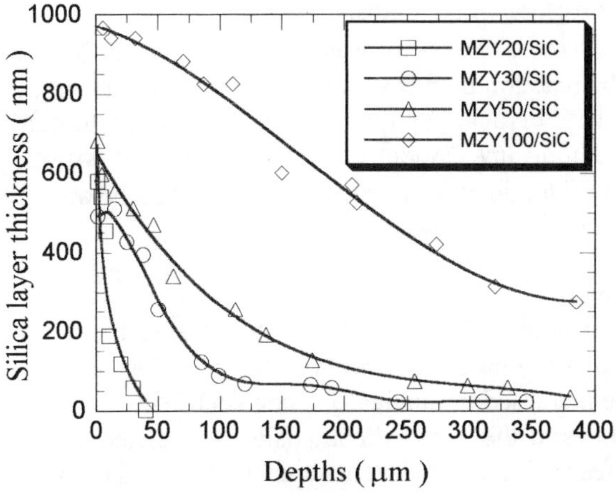

Fig. 4 The silica layer thicknesses at various depths after exposure at 1000°C/500 hr.

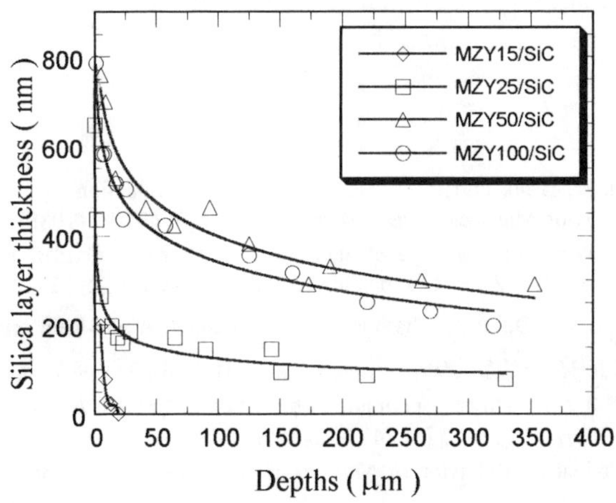

Fig. 5 The silica layer thickness at various depths after exposure at 1200°C/25 hr.

silica layer at corresponding depths. This outcome differed from exposure at 1000°C/500 hr. This observation was also attributed to the presence of zircon, resulting from the reaction between zirconia and silica at 1200°C. More zirconia particles accounted for a larger quantity and an earlier formation of zircon in composites. Without any evident change in slope, but with a larger oxidation depth, the oxidation mode of MZY50/SiC at 1200°C/25 hr was classified as Mode II; MZY100/SiC was as well.

CONCLUSIONS

1. A clear critical volume fraction of zirconia existed for the oxidation of SiC-particle (SiCp)/zirconia/mullite composites at either 1000 or 1200°C. The oxidation rate increased dramatically when $ZrO_2 > 20$ vol%
2. Oxidation modes of the exposed composites are related to zirconia content: (i) Below the threshold limit of zirconia content, the oxidation mode exhibited the behavior of Mode I; (ii) When the zirconia content exceeded the threshold limit, the composites exhibited the behavior of Mode II.
3. For composites with zirconia content above the threshold limit, the formation of zircon, as a result of the reaction between zirconia and oxidation product (i.e. silica), could change the oxidation behavior from Mode II to Mode I.

REFERENCES

[1]M. P. Borom, M. K. Brun, and L. E. Szala, "Kinetics of Oxidation of Carbide and Silicide Dispersed Phases in Oxide Matrices," *Adv. Ceram. Mater.*, **3** [5] 491-97 (1988).

[2]H. Y. Liu, K. L. Weisskopf, M. J. Hoffmann, and G. Petzow, "Oxidation Behavior of SiC Whisker Reinforced Mullite (-ZrO_2) Composites," *J. Eur. Ceram. Soc.*, **5** 123-33 (1989).

[3]K. L. Luthra and H. D. Park, "Oxidation of Silicon Carbide-Reinforced Oxide-Matrix Composites at 1375 to 1575°C," *J. Am. Ceram. Soc.*, **73** [4] 1014-23 (1990).

[4]C. Baudin and J. S. Moya, "Oxidation of Mullite-Zirconia-Alumina-Silicon Carbide Composites," *J. Am. Ceram. Soc.*, **73** [5] 1417-20 (1990).

[5]M. I. Osendi, "Oxidation Behavior of Mullite-SiC Composites," *J. Mater. Sci.*, **25** 3561-65 (1990).

[6]M. B. Ricoult, "Oxidation Behavior of SiC-Whisker-Reinforced Alumina-Zirconia Composites," *J. Am. Ceram. Soc.*, **74** [8] 1793-802 (1991).

[7]B. W. Sheldon, E. Y. Sun, S. R. Nutt, and J. J. Brennan, "Oxidation of BN-Coated SiC Fibers in Ceramic Matrix Composites," *J. Am. Ceram. Soc.*, **79** [2] 539-43 (1996).

[8]K. N. Lee and R. A. Miller, "Oxidation Behavior of Mullite-Coated SiC and SiC/SiC Composites under Thermal Cycling between Room Temperature and 1200-1400°C," *J. Am. Ceram. Soc.*, **79** [3] 620-26 (1996).

[9]C. C. Lin, "Microstructure Development, Phase Stability, and Toughening in Mullite Matrix Composites," Ph.D. Thesis, UIUC, 1991, Chaps. 4, 5, and 6.

[10]A. Zangvil, Y. Xu, and G. Fu, "Oxidation Mechanisms of Mullite/ZrO_2/SiC-Platelet Composites"; pp. 1003-14 in *Ceramic Transactions 46: Advanced Ceramic Composites*, vol. II. Edited by J. P. Singh and N. D. Bansal. The American Ceramic Society, Westerville, OH, 1994.

[11]C. Y. Tsai, C. C. Lin, A. K. Li, and A. Zangvil, "Effect of Zirconia Content on Oxidation Kinetics of Silicon Carbide/Zirconia/Mullite Composites," *accepted by J. Am. Ceram. Soc.*

HIGH-TEMPERATURE MECHANICAL PROPERTIES OF ADVANCED NICALON FIBERS AND THEIR RELATIONSHIP TO COMPOSITE FAILURE

C. A. Lewinsohn*, H. Serizawa**, D. E. Johnston***, G. E. Youngblood*, R. H. Jones*, and A. Kohyama****

* Pacific Northwest National Laboratory, P.O.Box 999, Richland, WA 99352, USA
**Joining and Welding Research Institute, Osaka University, Osaka, Japan
*** Associated Western Universities/University of Washington, Seattle, WA, USA
**** Institute of Advanced Energy, Kyoto University, Kyoto, Japan

ABSTRACT

Ceramic matrix composites, reinforced by small diameter (<20 μm) ceramic fibers, posses a combination of flaw tolerance and non-catastrophic failure modes, in addition to desirable (thermal) properties of monolithic ceramics. Not only the fast-fracture strength, but also environmentally-induced subcritical crack growth may be controlled by the strength of the reinforcing fibers. Under certain conditions of temperature and environment, oxidation of fibers that bridge matrix cracks can lead to embrittlement and subcritical crack growth. Extensive research and development has lead to small diameter silicon carbide fibers with low oxygen content and nearly stoichiometric compositions. In this research, to examine the effects of oxidation on the strength of one such fiber, Hi-Nicalon Type S, for reinforcement of ceramic matrix composites, fiber tensile tests at high temperature and fiber oxidation tests at 1373 K were carried out. Hi-Nicalon fibers were also tested to compare with Hi-Nicalon type S. From the results of research for effects of gauge length on fiber strength, the strength-gauge length relationship of Hi-Nicalon type S was found to follow the weakest-link rule. The results of tensile tests at high temperature indicated that the strength of Hi-Nicalon type S slightly increased with increasing test temperature. Subsequent to oxidation, there was a change in strength-limiting flaw distribution and fiber strength, between fibers tested at room temperature and 1373 K, in Hi-Nicalon Type-S fibers.

INTRODUCTION

Silicon carbide based fibers with high tensile strength, high elastic modulus and good thermal stability are promising candidates for the reinforcement of ceramic matrix composites (CMCs) [1-3]. Many types of SiC-based fibers have been produced industrially: for example, Nicalon of Nippon Carbon Co., Ltd. and Tyranno Lox-M of UBE Co., Ltd.. At high temperature, however, microstructural changes in those SiC-based fibers occur because of surplus oxygen in the fibers [4-6], thus degrading the properties of CMCs. Therefore SiC-based fibers with a low oxygen content have been developed: Hi-Nicalon of Nippon Carbon Co., Ltd. and Tyranno Lox-E of UBE Co. Ltd.. Recently, based on research on the effect of C/Si ratio on thermomechanical properties of fibers, the near-stoichiometric and highly crystalline SiC fiber, which was named Hi-Nicalon type S, has been developed by Nippon Carbon Co., Ltd. [7].

To obtain the desirable properties of CMCs, the bonding between the fiber and matrix must be such that debonding occurs between fibers and the matrix as cracks propagate through the matrix, leaving fibers bridging the matrix crack. Subsequent to matrix cracking, fibers that

To the extent authorized under the laws of the United States of America, all copyright interests in this publication are the property of The American Ceramic Society. Any duplication, reproduction, or republication of this publication or any part thereof, without the express written consent of The American Ceramic Society or fee paid to the Copyright Clearance Center, is prohibited.

Table I Characteristic properties of fibers used

	Hi-Nicalon type S	Hi-Nicalon
Diameter	12 μm	14 μm
Density	3.20 Mg/m^3	2.73 Mg/m^3
Tensile Strength	2.45 GPa	3.02 GPa
Tensile Modulus	390 GPa	273 GPa
Oxygen	0.8 wt%	0.6 wt%

bridge cracks are exposed to the ambient environment. Non-oxide CMCs offer desirable combinations of mechanical and thermal properties for a variety of applications, however their constituents are prone to oxidation. In some cases, the composite can be designed to offer passive oxidation protection, but usually only at temperatures high enough for the growth rate of a passive oxide layer to seal off cracks prior to extensive damage. Oxidation of the fiber or the matrix, after matrix cracking has occurred, may lead to oxidation embrittlement [8-12]. It has been proposed that the formation of a brittle oxidation product on the fibers creates Griffith type flaws and decreases the strength of the fibers [8,9]. Since the oxidation product has a composition similar to silica glass, the viscosity of the oxide layer is likely to be temperature dependent. The oxidation product would be expected to be most brittle at low temperatures, and have a lower viscosity at high temperatures. Hence, oxidation embrittlement, due to the growth of an oxide layers on the fibers, would be more severe at lower temperatures. On the other hand, the formation of a brittle oxidation product between the fiber and the matrix would prohibit matrix sliding and strain accommodation during fracture [9,12]. Without matrix sliding and strain accommodation, stress builds up in the fibers and they fracture near the plane of the matrix crack. Due to the temperature dependent properties of the oxide reaction layer, oxidation embrittlement, due to the growth of an oxidation product between the fibers and the matrix, would be more severe at lower temperatures.

In this study, to investigate the capability of Hi-Nicalon type S fibers for CMCs, fiber tensile tests at high temperature were carried out with 190 mm gauge length. In general, fiber strength is usually measured according to ASTM standards in USA or JIS standards in Japan where the gauge length is 1 inch or 25 mm, respectively. So the effects of gauge length on the fiber strength was examined. In addition, since fiber oxidation can seriously degrade the mechanical behavior of CMCs, measurements of oxide layer thickness, as a function of time, and fiber strength tests after oxidation were carried out. Fiber strength tests, after oxidation, were performed at room temperature and 1373 K to investigate whether the temperature dependent properties of the oxidation products influenced the resulting fiber strength.

EXPERIMENTAL

Materials used for the fiber tensile tests were Hi-Nicalon type S and Hi-Nicalon, manufactured by Nippon Carbon Co., Ltd., Tokyo, Japan. Characteristic properties of these fibers (from the same batches as used in this study), as reported by the manufacturer, are shown in Table I where the tensile strength at room temperature (R.T.) was measured with a 25 mm gauge length according to JIS standards (JIS R 7601). In this study a 190 mm gauge length was used to ensure that the ends of the fibers, where the load was applied, were cold during high-temperature testing. The fiber tensile tests were carried out with a Micropull Tensile Test Machine (Micropull Science) with a constant crosshead rate of 5.5 μm/s. Hi-Nicalon type S fibers were tested at R.T., 1373 K, and 1673 K, in air. Hi-Nicalon fibers were tested at R.T. and 1373 K, in air. In addition, Hi-Nicalon fibers that were annealed at 1373 K in argon for 2 hours and fibers that were oxidized at 1373 K in air for 2 hours also were examined. To study the oxidation behavior of Hi-Nicalon type S fibers, fibers were oxidized at 1373 K for 10, 50, 100 and 210 hours, in air. The oxide layer thickness of fibers was measured directly with SEM observations. Hi-Nicalon type S fibers were also oxidized in air, at 1373 K for 10 h and their strength measured at 1373 and R.T..

To evaluate the creep resistance of the fibers, the bend stress relaxation technique [13] was used. Fibers were wound around a silicon carbide rod that was placed inside a hollow silicon carbide cylinder with a slightly larger inner diameter than the outer diameter of the rod. The rod, fibers, and outer sleeve were then placed in a furnace and held at various temperatures for 100 h in flowing argon. After these treatments, the fibers were removed and the residual curvature was measured. A parameter related to the amount of strain relaxation, presumably due to fiber creep, was calculated from the initial radius of the rod and fibers, r_0, and the radius of curvature of the fibers after heat-treatment, r_f,

$$m_{BSR} = 1 - r_0/ r_f \tag{1}$$

Thus, a value of 1 for m_{BSR} represents no creep and a value of 0 for m_{BSR} represents complete relaxation.

RESULTS AND DISCUSSION

Effect of Gauge Length

The strength, σ, distribution of brittle fibers like SiC, glass or carbon fibers usually follows a classical Weibull distribution described by the following equation [14],

$$P(\sigma) = \exp\left[-\left(\frac{\sigma}{\sigma_0}\right)^m\right] \tag{2}$$

where $P(\sigma)$, m, and σ_0 are probability of survival, shape parameter (Weibull modulus), and scale parameter (σ_0), respectively. The results of fiber tensile tests of Hi-Nicalon type S with 190 mm gauge length at R.T. were analyzed according to this equation and the strength of the fiber, σ_0, was 1.33 GPa according to the classical Weibull distribution. The fiber diameter, 12.7 μm, was determined from SEM measurements of Hi-Nicalon type S fibers.

Equation (2) does not explicitly include the effects of gauge length. To predict the strength at a different gauge length, the results were analyzed by using the following modified equation, which was proposed by Watson and Smith [15] and Gutans and Tamuzs [16],

$$P(\sigma) = \exp\left[-\left(\frac{l}{l_0}\right)^\alpha\left(\frac{\sigma}{\sigma_0}\right)^m\right] \tag{3}$$

where l and l_0 are the different gauge lengths ($l > l_0$), and α is a parameter between 0 and 1. Phoenix et al. [17] suggested that $\alpha = 0.60$ for Kevlar 49 fibers, and Watson and Smith [15] obtained $\alpha = 0.90$ for carbon fibers. In the case of fibers that obey the weakest-link rule, α becomes 1.0 and Wu and Netravali [18] reported that $\alpha = 1.0$ for single Nicalon fibers. In contrast, recent data for ultra high-strength polyethylene fibers yield α near zero [19]. To compare with Nippon Carbon's results shown in Table I, 190 and 25 mm were used for l and l_0, respectively, and α was assumed to be 1.0. Then, from the analysis of the fiber tensile tests according to Equation 3, the strength of Hi-Nicalon type S with 25 mm gauge length was calculated to be 2.09 GPa. If the fiber diameter was assumed to be 12.0 μm according to Table I as reported by the manufacturer, the strength of Hi-Nicalon type S with 25 mm gauge length was calculated to be 2.35 GPa, which is in agreement with Nippon Carbon's results considering that the typical value for the standard deviation is 0.36 GPa and the Weibull modulus is relatively low. Since the standard deviation of the fiber diameter was 0.78 and the accuracy of SEM observation was considered to be 10%, the strength-gauge length relationship of Hi-Nicalon type S was found to follow the weakest-link rule.

Table II Strength of Fibers

Temperature	Hi-Nicalon type-S RT	Hi-Nicalon type-S 1373 K	Hi-Nicalon type-S 1673 K	Hi-Nicalon RT	Hi-Nicalon 1373 K
Weibull Strength (σ_0)	1.33 GPa	1.60 GPa	1.74 GPa	2.88 GPa	2.94 GPa
Weibull modulus, m	4.5	4.1	6.1	5.1	6.6
# specimens	31	30	32	34	30

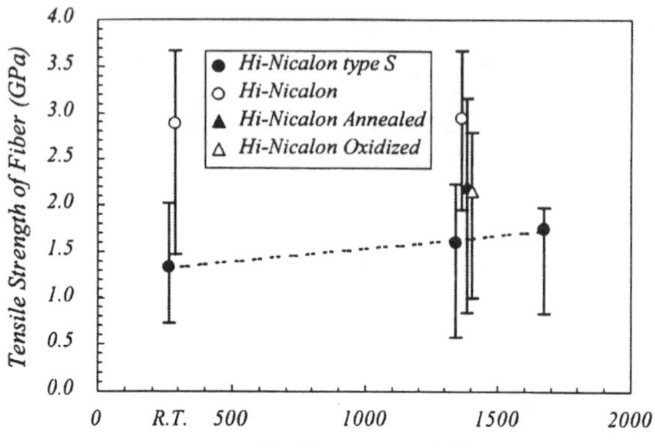

Figure 1. Tensile strength of advanced Nicalon fibers as a function of temperature.

Fiber Tensile Tests at High Temperature

The results of fiber tensile tests at R.T. and high temperature, for Hi-Nicalon type S fibers, are shown in Table II. The strength values were based on fiber diameters measured from SEM observations of the tested fibers. The strength of fibers was calculated from the classical Weibull distribution analysis, where at least 30 fibers were tested. The effect of test temperature on the tensile strength values of the fibers are shown in Figure 1. In the case of Hi-Nicalon, the strength of the fibers scarcely changed with increasing test temperature. On the other hand, for Hi-Nicalon type S, the strength increased slightly with increasing test temperature up to 1673 K, in air. One possible explanation for this apparent increase in strength is the formation of a thin layer of oxide that blunts, or heals, pre-existing flaws in the fibers. Since Takeda *et al.* [20] has measured a decrease in the strength of Hi-Nicalon Type-S with increasing heat-treatment temperature, it is likely that simultaneous changes in the fiber microstructure occur during high temperature testing. Since the initial grain size of the Hi-Nicalon type S is larger than that of the Hi-Nicalon fibers, if grain growth occurred in both fibers at a similar rate (i.e., controlled by the same diffusional process) then the strength of the Hi-Nicalon type-S fibers would decrease more slowly than the Hi-Nicalon fibers. If the decrease in strength occurred at a slow enough rate then another process, such as flaw healing due to the growth of a surface oxide, may be capable of mitigating the strength loss temporarily.

Fiber Oxidation Behavior

The effects of oxidation time on oxide layer thickness of Hi-Nicalon type S are shown in Figure 2. The thickness was measured from SEM observation of the cross section of the fibers normal to their axis. The thickness gradually increased with increasing oxidation time. To analyze the oxidation results the Deal/Grove linear-parabolic model [21], which describes the growth of silica (SiO_2) on silicon, was applied. This model describes the growth of silica,

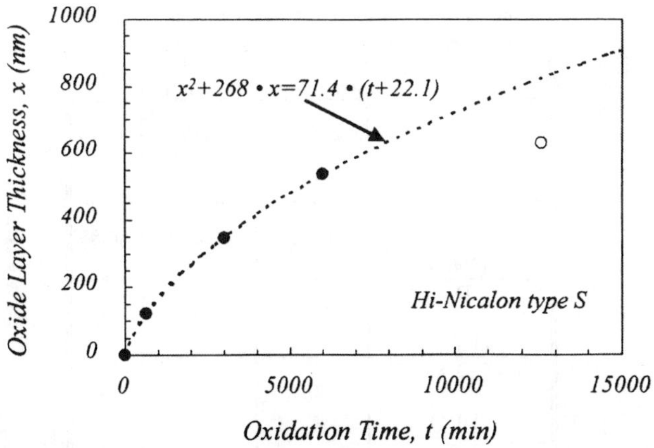

Figure 2. Oxide layer thickness, as a function of time, on Hi-Nicalon type-S fibers.

SiO_2, using a mixed-control model, in which the diffusion of oxygen through the oxide layer and the chemical reaction at the SiC/SiO_2 interface act in series. These two processes are coupled by the concentration of the diffusing species at the interface. The Deal/Grove model predicts that, for thin oxide layers, the chemical reaction at the SiC/SiO_2 interface is rate controlling. As the oxide thickens, the growth rate is limited by the coupled control between the interfacial reaction and the diffusion of oxygen through the SiO_2. Finally, for thick oxides, the growth rate is diffusion controlled. For fast gaseous diffusion, their model is,

$$x^2 + Ax = B(t+\tau) \tag{4}$$

where x and t are oxide thickness and time, respectively. The time constant, t, allows the origin (x = 0) of the model to adjust to allow for an initial oxide layer or regime not described by the model. B is the parabolic rate constant. B/A is the linear rate constant.

The oxidation behavior of silicon and SiC has been well described by the Deal/Grove model [21]. Ramberg et al. [22] examined the oxidation behavior of single-crystal SiC (α-SiC) and CVD-SiC (β-SiC), and reported the results of these rate constants as shown in Table III. In this research, the Deal/Grove model was applied to the oxidation results for 0, 10, 50 and 100 hours. The results fitted this model very well, and the parabolic rate constant, B, was calculated to be 71 nm^2/min. This value is within the range reported by Ramberg et al. [22]. Thus the oxidation behavior of Hi-Nicalon type S was found to fit the Deal/Grove model. The oxide layer growth rate is lower than that observed for earlier Nicalon fibers. Thus, Hi-Nicalon type-S fibers would be expected to delay the onset of oxidation embrittlement relative to more oxidation prone fibers.

Recently, Zhu et al. [23] have solved for the oxide layer thickness on cylindrical specimens using the same approach as Deal and Grove. They found that at long times, their solution diverged from the Deal -Grove solution, Equation 4, and predicted smaller oxide layer thickness. This may be why the measured oxide layer thickness after 210 hours did not fit the Deal-Grove model. Another explanation, however, was that the oxide layer thickness was not clearly observed by SEM because not only the edges of fibers but also the surface of fibers were oxidized. So the measured oxide layer thickness for 210 h oxidation may not be accurate. Further analysis will be performed to determine if the data fit the model proposed by Zhu et al. [23].

Table III Rate constants for Deal/Grove model [22].

	B (nm^2/min)	B/A (nm^2/min)	τ
Single-crystal SiC (000$\bar{1}$)	239.10	0.1678	4.573
Single-crystal SiC (0001)	11.15	0.0199	86.85
CVD SiC ($\bar{1}\,\bar{1}\,\bar{1}$)	225.50	0.1841	0
CVD SiC (1 1 1)	12.98	2.807	0
Hi-Nicalon type S	71.4	0.266	22.1

Table IV Strength of Hi-Nicalon fibers after annealing or oxidation

	RT	1373 K	1373 K argon anneal	1373 K air oxidation
Weibull strength (σ_0)	2.88 GPa	2.94 GPa	2.20 GPa	2.16 GPa
Weibull modulus, m	5.1	6.6	3.7	5.4
# specimens	34	30	31	31

Fiber Tensile Tests after Oxidation

Preliminary experiments were performed on Hi-Nicalon fibers that were annealed in argon or oxidized in air for 2 h at 1373 K. The Weibull distribution parameters (σ_0, m) of the strength of these fibers is shown in Table IV. After either of these treatments, the strength decreased significantly although the diameter did not change. This decrease in strength is probably due to microstructural development in the fiber, such as grain growth. Takeda et al. [20] has also reported a decreasing strength with increasing temperature for Hi-Nicalon fibers annealed in argon. It is also possible that these results are due to damage to the fiber surfaces caused by removal of the protective sizing, during the oxidation and annealing treatments, and subsequent handling steps.

In subsequent experiments, Hi-Nicalon type S fibers were oxidized for 10 h at 1373 K, in air. The results are shown in Figure 3 and Table V. Fibers that were oxidized and then tested at 1373 K were slightly weaker than those that were initially unoxidized and tested at 1373 K. Fibers that were oxidized at 1373 K and tested at room temperature were weaker than those tested at 1373 K. These results indicate that the growth of an oxide layer on the fibers may have created strength limiting flaws that were larger than the intrinsic flaws in the fiber. In both cases, the fibers that were oxidized at 1373 K for 10 hours were cooled to room temperature prior to mechanical testing. Although the oxidized fibers that were tested at 1373 K were held for 15 minutes prior to loading, the flaw size in the oxide layer should have been the same at both test temperatures. Therefore, since the strength of the fibers was lower at room temperature, than at 1373 K, the toughness of the oxide layer appears to be temperature dependent. Whether this temperature dependence of the fiber strength, due to the oxide layer, is responsible for oxygen embrittlement in silicon carbide composites is the subject of ongoing studies.

Table V Strength of Hi-Nicalon type-S fibers after oxidation

	untreated		oxidized	
	RT	1373 K	1373 K	RT
Weibull Strength (σ_0)	1.33 GPa	1.60 GPa	1.41 GPa	1.1 GPa
Weibull modulus, m	4.5	4.1	4.5	5.6
# specimens	31	30	31	21

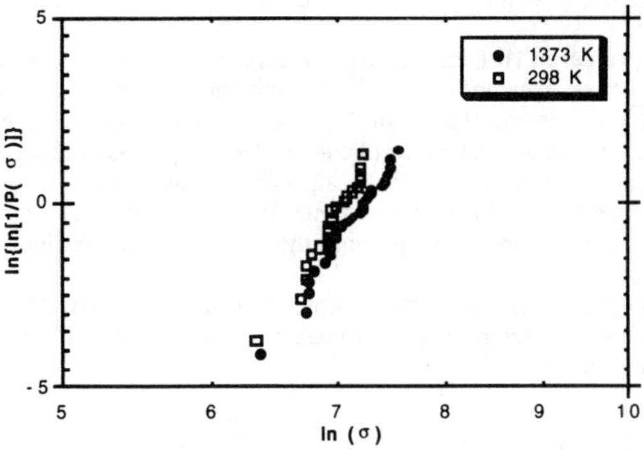

Figure 3. Weibull plot of the strength of Hi-Nicalon type-S fibers after oxidation, in air, at 1373 K.

Fiber Bend Stress Relaxation

To evaluate the resistance of the newly developed Hi-Nicalon type S fibers to creep, bend stress relaxation tests were performed. A comparison of the bend stress relaxation parameter, m_{BSR}, is shown in Figure 4. At a given temperature, a higher value of m_{BSR}, indicates a greater resistance to creep. As can be seen from the results shown in Figure 4, Hi-Nicalon type S fibers were more creep resistant than Hi-Nicalon fibers. Lewinsohn et al. have shown that fibers with higher creep resistances inhibit subcritical crack growth in CMCs at elevated temperatures [24].

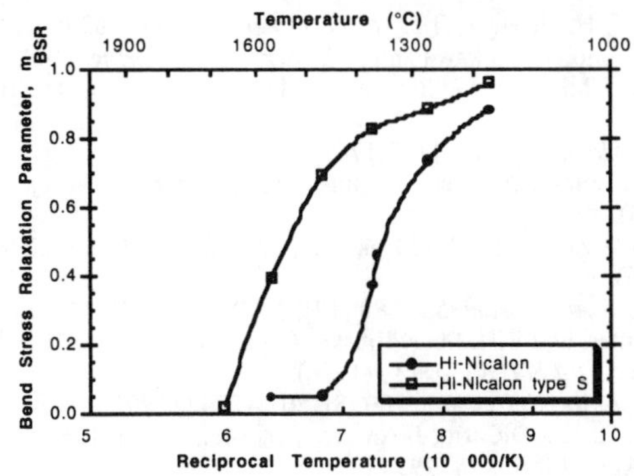

Figure 4. Bend stress relaxation of Hi-Nicalon and Hi-Nicalon type S fibers as a function of temperature.

CONCLUSIONS

To estimate the capability of newly developed advanced SiC fibers (Hi-Nicalon type S) for reinforcement of ceramic matrix composites, fiber tensile tests at high temperature with 190 mm gauge length and fiber oxidation tests at 1373 K were carried out. The conclusions can be summarized as follows:

(1) The relationship between the strength of Hi-Nicalon type S at room temperature and the gauge length was found to follow the weakest-link rule, described by the Weibull distribution.

(2) The strength of Hi-Nicalon type S slightly increased with increasing test temperature up to 1673 K. On the other hand, the strength of Hi-Nicalon scarcely changed and decreased after annealing in argon for 2 hours or oxidation in air for 2 hours at 1373 K. These results suggest that microstructural development and surface oxidation occur concurrently.

(3) The strength values of Hi-Nicalon type S fibers that were oxidized at 1373 K for 10 h, in air, were lower than unoxidized fibers, and the strength values were lower when measured at room temperature than 1373 K. These results suggest that the fiber strength is controlled by the temperature dependent properties of the surface oxide layer.

(3) The oxidation behavior, at 1373 K in air, of Hi-Nicalon type S was found to fit the Deal/Grove linear-parabolic model for times as long as 100 h. The oxidation resistance of Hi-Nicalon type S fibers may prolong the lifetime of composites during oxidation embrittlement.

(4) Hi-Nicalon type S fibers were more resistant to bend stress relaxation than Hi-Nicalon fibers. The higher creep resistance of these fibers would also inhibit subcritical crack growth due to fiber creep.

ACKNOWLEDGMENTS

The authors would like to express their sincere appreciation to Dr. Y. Imai and Mr. M. Takeda (Nippon Carbon Co., Ltd.) for providing the materials. This work was maintained by National Institute for Fusion Science under Japan-USA Program of Irradiation Test for Fusion Research. This work was also supported by Basic Energy Science under U.S. Department of Energy (DOE) contract DE-AC06-76RLO 1830 with Pacific Northwest National Laboratory, which is operated for DOE by Battelle.

REFERENCES

1. P. J. Lamicq, G. A. Bernhart, M. M. Dauchier and J. G. Mace, *Am. Ceram. Soc. Bull.*, **65**[2], 336(1986).
2. J. R. Strife, J. J. Brennan and K. M. Prewo, *Ceram. Eng. Sci. Proc.*, **11**[7-8], 871(1990).
3. R. H. Jones and C. H. Henager, Jr., *J. Nucl. Mater.*, **212-215**, 830(1994).
4. K. Okamura, M. Sato, T. Ishikawa and T. Yamamura, *Sintering '87*, **Vol.1**, 61(1987).
5. T. Yamamura, T. Ishikawa, M. Shibuya, T. Hisayuki and K. Okamura, *J. Mater. Sci.*, **23**, 2589(1988).
6. Y. Hasegawa, *J. Mater. Sci.*, **24**, 1177(1989).
7. M. Takeda, J. Sakamoto, A. Saeki, Y. Imai and H. Ichikawa, *Ceram. Eng. Sci. Proc.*, **16**[4-5], 37(1995).
8. A.G. Evans, F.W. Zok, R.M. McMeeking, and Z.Z. Du, *J. Am. Ceram. Soc.*, **79** [9], 2345-2352 (1996).
9. E. Lara-Curzio, , *J. Am. Ceram. Soc.*, **80** [12], 3268-3272 (1997).
10. K. Jakus, J.E. Ritter, and R.H. Schwillinski, *J. Am. Ceram. Soc.*, **76** [1], 33-38 (1993).
11. M.R. Begley, *Int. J. of Sol. & Struct.*, (1996).
12. G.N. Morscher, *J. Am. Ceram. Soc.*, **80** [8], 2029-2042 (1997).
13. G.N. Morscher, and J.A. DiCarlo, J. Am. Ceram. Soc., **75** [1], 136-140 (1992).
14. W. Weibull, J. Appl. Mech., **18**, 293 (1951).
15. A. S. Watson and R. L. Smith, *J. Mater. Sci.*, **20**, 3260(1985).
16. J. Gutans and V. Tamuzs, *Mech. Compos. Mater.*, **20**, 1107(1984).

17. S. L. Phoenix, P. Schwartz and H. H. Robinson IV, *Compos. Sci. Technol.*, **32**, 81(1988).
18. H. F. Wu and A. N. Netravali, *J. Mater. Sci.*, **27**, 3318(1992).
19. P. Schwartz, A. N. Netravali and S. Sembach, *Textile Res. J.*, **56**, 502(1986).
20. M. Takeda, A. Urano, J. Sakamoto and Y. Imai, *J. Nucl. Mater.*, in press.
21. B. E. Deal and A. S. Grove, *J. Appl. Phys.*, **36**[12], 3770(1965).
22. C. E. Ramberg, G. Cruciani, K. E. Spear, R. E. Tressler and C. F. Ramberg Jr., *J. Am. Ceram. Soc.*, **79**[11], 2897(1996).
23. Y.T. Zhu, S.T. Taylor, M.G. Stout, D.P. Butt, and T.C. Lowe, *J. Am. Ceram. Soc.*, **81** [3], 655-660 (1998).
24. C.A. Lewinsohn, C.H. Henager, Jr., and R.H. Jones, pp. 423-433, Advances in Ceramic Matrix Composites III, Ceramic Transactions v. 74, The American Ceramic Society, Westerville, OH (1996).

FLEXURAL CREEP AND CREEP RUPTURE OF MONOLITHIC ZIRCON AND ZIRCON SILICON CARBIDE FIBER COMPOSITES

Umashankar Anandakumar and Raj N. Singh
Department of Materials Science and Engineering
University of Cincinnati, Cincinnati, Ohio 45220-0012.

ABSTRACT

The creep and creep rupture behaviors of the monolithic zircon and zircon matrix silicon carbide fiber unidirectional composites is studied in bending mode. The composites exhibited higher load bearing capability and lower steady state strain rates than the monolithic zircon. Creep studies are conducted on zircon-SCS-6 silicon carbide fiber composites at two different temperatures of 1350°C and 1400°C and at stresses of 100 MPa, 150 MPa and 200 MPa for a maximum duration of 120 hours, and in monolithic zircon at 1300°C and at stress levels of 50 MPa, 100 MPa, and 200 MPa. These stresses are above and below the matrix cracking stress of the composites at these temperatures. Apparent steady state creep is observed at all the stress values at both the temperatures in composites. The fracture surfaces and tensile surfaces of crept specimens are observed using optical and scanning electron microscopes to study the creep mechanisms and the observations are correlated with experimental results. The study showed that composites exhibited steady state creep with a stress dependence similar to the monolithic materials. The microstructural observations indicated that steady state creep by creep mechanisms occurred at lower stress and temperature levels, while at higher stresses creep cavitation played a dominant role in rupture of the samples. Linkage of creep cavities resulted in the accelerated creep and creep rupture in composite samples at stresses above the matrix cracking.

INTRODUCTION

Ceramic composites are candidate materials for high temperature applications such as gas turbines, heat exchangers and other applications. Such applications would require the material to withstand stresses at high temperatures for an extended period of time. Hence it is important to study the creep behavior of the monolithic material and ceramic composites in order to understand the roles

To the extent authorized under the laws of the United States of America, all copyright interests in this publication are the property of The American Ceramic Society. Any duplication, reproduction, or republication of this publication or any part thereof, without the express written consent of The American Ceramic Society or fee paid to the Copyright Clearance Center, is prohibited.

of matrix and fiber, and their contribution to the overall creep behavior of the material. Creep behavior in ceramic composites is a complex phenomenon which depends on individual creep resistances of the matrix and fiber [1-3]. Also, load is continuously transferred between the fiber and the matrix depending on the difference in their individual creep rates in order to achieve equal strain in the matrix and fiber [4]. The microstructure of the composite plays a strong role in determining the overall creep resistance. When the matrix has a lower creep resistance than that of the reinforcing fiber (as is usually the case), composite damage occurs in the form of periodic fiber fracture since most of the load is carried by the fiber. However, when the matrix has a higher creep resistance, periodic matrix cracking occurs. Such a situation is more detrimental since it exposes the fibers to the reaction environment and results in degradation of the interface and decreases the intrinsic strength of the fiber. A comprehensive review of the creep behavior in ceramic composites has been given by Holmes. [2]

In the present study, a zircon matrix silicon carbide fiber composite is subjected to flexural creep at 1350°C and 1400°C at stresses of 100 MPa, 150 MPa, and 200 MPa. The stress level of 200 MPa was above the matrix cracking stress of the composite. Under these conditions, the matrix has a much lower creep resistance than that of the fiber. The stresses of 100 MPa and 150 MPa were chosen to avoid inelastic deformation mechanisms such as matrix cracking, debonding, and interfacial sliding during the initial stage of creep. Similar studies were conducted on the monolithic zircon at 1300°C and at stress levels of 50 MPa, 100 MPa, and 200 MPa. The tests on the monolithic zircon could not be conducted at these stress levels at higher temperatures due to its high creep rate and low rupture strength. The tests were conducted till steady state creep conditions were achieved and for a maximum duration of 110 hours.

Microstructural examination was done on fractured surfaces of samples, and on the tensile surface of flexural bars for samples which did not fail during the creep test. The nature of the interface, fiber pullout, matrix cracking, void nucleation and linkage were studied. The microstructural observations were correlated with the creep behavior of the composites.

EXPERIMENTAL

The composite samples used in this study consisted of a zircon matrix uniaxially reinforced with SCS-6™ silicon carbide fibers. The details of the processing technique have been described elsewhere[5]. It involved tape casting of zircon powder, followed by lamination of tapes and fiber mats, binder burnout and hot pressing at 1640°C in a nitrogen atmosphere at a stress of 10 MPa. The hot pressed samples were ground to remove the excess layer of zircon on either side of the composite and the final composite had a fiber fraction of approximately 25 volume percent. Monolithic zircon was processed using a similar route but by hot pressing at 1600°C. The samples had free silica (20 volume percent) and zirconia due to decomposition of the zircon during hot pressing. The silica phase was present as a continuous phase along the grain boundaries. Ground samples were cut to dimensions of 4.0mm x 1.35mm x 25 mm. The composite and monolithic samples were polished using silicon carbide paper and diamond paste of progressively finer size. Final polishing was done using a vibratory polisher and an alumina slurry of particle size 0.3 μm.

Creep tests were performed in three point bending mode on zircon-silicon carbide composite samples at two different temperatures of 1350°C and 1400°C and at three different stress values of 100 MPa, 150 MPa, and 200 MPa and on the monolithic zircon at 1300°C at stress levels of 50 MPa, 100 MPa, and 200 MPa. The tests were conducted in a flowing nitrogen atmosphere in order to avoid the effects of oxidation and interface degradation on the creep behavior. The tests were performed using a MTS servohydraulic mechanical testing machine using fixtures made of thoriated tungsten. The furnace was heated by two carbon elements and the temperature was monitored and controlled using two tungsten - rhenium thermocouples. The strain in the sample was measured using a silicon carbide contact extensometer. The extensometer was capable of measuring displacement of the order of one μm. The extensometer and the furnace chamber were water cooled to prevent fluctuations in strain and temperature. The tests were carried out for a maximum period of 110 hours. The samples were loaded to the creep stress at a rate of 100 MPa/sec. Prior to the creep tests, the matrix cracking stress and failure stress of the composite and monolithic samples were measured. The stresses chosen for creep tests were below(100 MPa and 150 MPa) and above(200 MPa) the matrix cracking stress of the composite in order to study the effects of inelastic phenomenon on the creep behavior of the composite.

Crept samples were polished to expose the fiber layers on the tensile surface of the composites. Microstructural observations of the tensile surfaces of crept monolithic and composite samples and of the fracture surface of samples that failed during creep were conducted using a Jeol-3000 scanning electron microscope. Features such as fiber pullout, periodic fiber and matrix fracture, void nucleation in the matrix, linkage of cavities, and creep crack growth were studied.

RESULTS AND DISCUSSION

Figures 1 and 2 show the creep curves of zircon silicon carbide samples subjected to creep deformation at 1350°C and 1400°C at three different stress levels of 100 MPa, 150 MPa, and 200 MPa. The stress levels of 200 MPa was above the matrix cracking stress of the composite at both the temperatures. These stress levels were chosen to study the effect of matrix cracking on the rupture life of the composites. The tests were conducted for a maximum duration of 120 hours in the case of the composite sample crept at 1350°C and 100 MPa. The composites exhibited steady state creep at all the different creep conditions. Figure 3 shows the creep behavior of the monolithic zircon measured in four point bending mode at temperature of 1300°C and stresses of 50 MPa, 100 MPa, and 200 MPa. The lower temperature was chosen due to the high creep rate exhibited by monolithic zircon. It can be seen that the minimum strain rate exhibited by composites at a higher temperature is much lower than that of the monolithic zircon. Figure 4 shows the variation of steady state creep rate vs applied stress for composites crept at 1350°C and 1400°C. It can be seen that they exhibit a stress dependence similar to the monolithic polycrystalline materials. The creep stress exponent is nearly the same at both the temperatures, indicating that the creep mechanism does not change significantly between the two temperatures.

From Figure 1, it can be seen that at longer durations, the creep curves exhibit strain jumps. At 1350°C, creep curves at 100 and 150 MPa exhibit strain jumps of different magnitudes. The nature of strain jumps was verified by comparison with the displacement of the actuator, which was measured independently in order to ensure that the strain jumps are not caused by the slippage of the extensometer. Strain jumps have been previously observed in creep of ceramic composites [6]. This might probably be caused due to inelastic deformation mechanisms such as fiber failure or matrix cracking. The sample crept at 200 MPa and 1350°C, exhibited an apparent steady state creep rate for a very short period of time.

Figure 2 shows the composite creep curves at 1400°C together with the steady state creep rates measured from these curves. Strain jumps similar to those measured at 1350°C can be seen in the sample crept at 100 MPa. These strain jumps are accompanied with an sudden change in composite creep rate. The sample crept at 200 MPa and 1400°C failed by creep rupture in less than 0.5 hours. The steady state strain rate at 1400°C was an order of magnitude higher than that measured at 1350°C. The creep rate in monolithic zircon was found to be two orders of magnitude higher than that of zircon- silicon carbide composite measured at 1350°C. This indicates that during creep, the fibers carry a major portion of the applied stress.

MICROSTRUCTURAL OBSERVATIONS

Figure 5 shows the matrix cracks and fiber failure on the tensile surface of the sample which underwent creep at 1350°C and 150 MPa. Very few matrix cracks were seen on the tensile surface of the sample. Though the initial applied stress was below the matrix cracking stress of the composite, fiber rupture at longer durations would result in additional load being transferred from the fiber to the matrix. If this load results in a stress higher than the matrix cracking stress of the composite, it would result in multiple matrix cracking. The sample crept at 1350°C and 100 MPa did not exhibit any fiber failure or matrix cracking as the stress was well below the matrix cracking stress. The strain jump observed during the creep test was most likely caused by fiber failure away from the tensile surface (layer of maximum tensile stress) due to the statistical fiber strength distribution. Creep cavitation, cavity linkage to form creep cracks, and creep crack propagation was seen in the sample crept at 1350°C and 200 MPa. Cavitation damage occurs in the monolithic ceramics as a result of different mechanisms such as grain boundary sliding and viscous hole growth in the glassy second phase [7,8]. The sample also exhibited matrix cracking on the tensile surface of the sample. The initial applied stress was above the matrix cracking stress of the composite and the multiple matrix cracking was caused during the initial loading period.

Figure 6 shows fracture surface of sample that failed by creep rupture at 1400°C and 100 MPa in less than fourteen hours. The fracture surface shows that there is minimal fiber pullout during creep rupture and most of the fibers have failed in the fracture plane. Also, multiple matrix cracks deflected at the fiber matrix interface can be seen on the fracture surface. Fracture was most likely caused by the sudden fiber failure in the composite. Similar to the results at 1350°C, it can be seen that the matrix cracking occurs at stresses below the matrix cracking stress during creep deformation which leads to failure of the composite in an accelerated manner. Figure 7 shows multiple matrix cracking without any fiber failure on the tensile surface of sample crept at 1400°C and 150 MPa, where the test was stopped after ten hours. The absence of fiber failure in this composite indicates that fiber failure occurs towards the later stages of creep, prior to creep rupture. Fiber failure would result in further accelerated creep due to load transfer to matrix which would result in further matrix cracking, till matrix cracking reaches a saturation. Figure 8a and 8b show the formation of cavities in the same sample, at the secondary silica phase in the matrix, the linkage of cavities, and growth of creep cracks. Cavities are likely to be formed due to the viscous flow of silica phase and grain boundary sliding, which then link together to form creep cracks. The fracture surface also exhibits minimal fiber pullout and failure seems to have occurred by the growth of a creep crack to failure.

These results show that even at stresses below the matrix cracking stress of the composite, inelastic deformation such as matrix cracking and fiber failure occur at longer durations, and matrix cracking stress is not an ideal design stress under conditions of creep. At stresses above the matrix cracking stress, rupture of composites by accelerated creep due to cavitation and creep crack growth occurs. The strain jumps observed during creep tests were most likely contributed by fiber failure and matrix cracking in the composite.

CONCLUSIONS

Ceramic matrix composites exhibit steady state creep with a power law behavior similar to polycrystalline monolithic ceramics. The creep resistance of zircon silicon carbide composites is much higher than that of the monolithic zircon, indicating that the major portion of the creep stress is carried by the fibers. At stresses higher than the matrix cracking stress and/or high temperatures, creep rupture occurs in an accelerated manner due to creep cavitation of the matrix, linkage of cavities, and creep crack growth to failure. The fracture surfaces of ruptured samples exhibit minimal fiber pullout indicating that fracture occurred by the failure of fibers in the fracture plane. Even at stresses below the matrix cracking stress of the composite, damage in the form of matrix cracking and fiber failure occur at longer durations. These are indicated by strain jumps observed in the creep curves. The steady state strain rate at 1400°C is one order of magnitude higher than that measured at 1350°C in zircon silicon carbide composites. The nature of creep mechanisms in the matrix and fiber could not be determined from scanning electron microscopy techniques.

ACKNOWLEDGEMENTS

This program has been supported by the US-Department of Energy, Basic Energy Sciences Division under contract number DE-FG02-91ER45459.

REFERENCES.

1. J.W. Holmes, Y. Park and J.W. Jones, "Tensile Creep and Creep Recovery Behavior of a SiC-Fiber Si_3N_4- Matrix Composite", *J. Am. Ceram. Soc.*, 76 [5] 1281-1293 (1993).

2. J.W. Holmes, "Elevated Temperature Deformation and Fatigue Deformation of Fiber Reinforced Ceramic Matrix Composites", in "Elevated Temperature Mechanical Behavior of Fiber-Reinforced Ceramics", *edited by S. Nair and K. Jakus*, Butterworth Hienneman.

3. X. Wu and J.W. Holmes, "Tensile Creep and Creep-Strain Recover Behavior of Silicon Carbide Fiber/Calcium Aluminosilicate Matrix Ceramic Composites", J. Am. Ceram. Soc., 76 [10] 2695-700 (1993).

4. S. Goto and M. McLean, "Role of Interfaces in Creep of Fiber-reinforced Metal-Matrix Composites-I, Continuous Fibers", *Acta metall. Mater.* Vol 39, No.2, 00.153-164,1991.

5. R.N. Singh and A.R. Gaddipati, "A uniaxially Reinforced Zircon-Silicon Carbide Composite", *J. Mater. Sci.*, **26**, 957-962 (1991).

6. J.W. Holmes, "Tensile Creep Behavior of a fiber-reinforced SiC-Si_3N_4 Composite", *J. Mater. Sci.*, 26 (1991) 1808-1814.

7. Rishi Raj, "Fundamental Research in Structural Ceramics for Service near 2000°C", *J. Am. Ceram. Soc.*, 76 [9] 2147-74 (1993).

8. K.S. Chan and R.A. Page, "Creep Damage Development in Structural Ceramics", *J. Am. Ceram. Soc.*, 76 [4] 803-26 (1993).

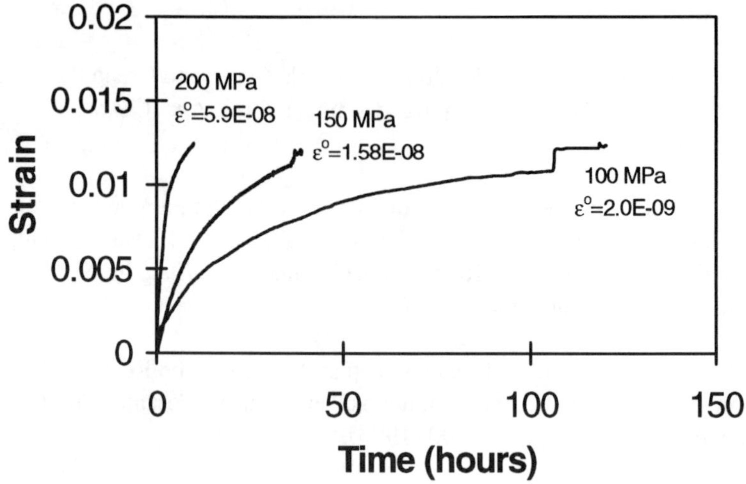

Figure 1. Composite creep curves obtained at 1350°C and stresses of 100MPa, 150 MPa, and 200 MPa. The corresponding steady state rates are also shown in the figure.

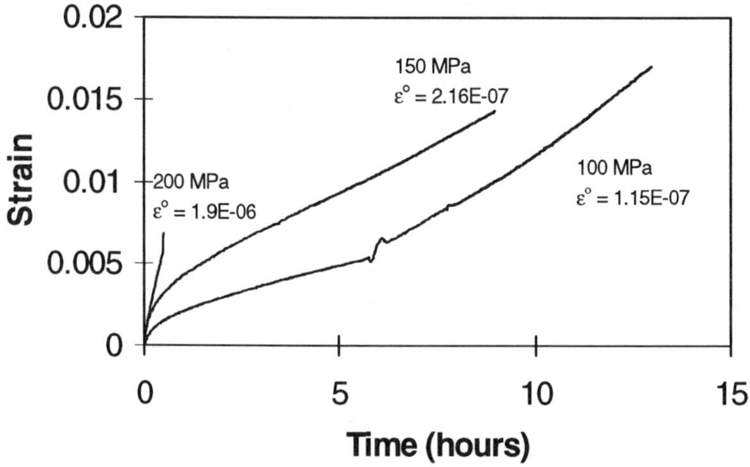

Figure 2. Composite creep curves obtained at 1400°C and stresses of 100 MPa, 150MPa, and 200 MPa. The corresponding steady state rates are also shown in the figure.

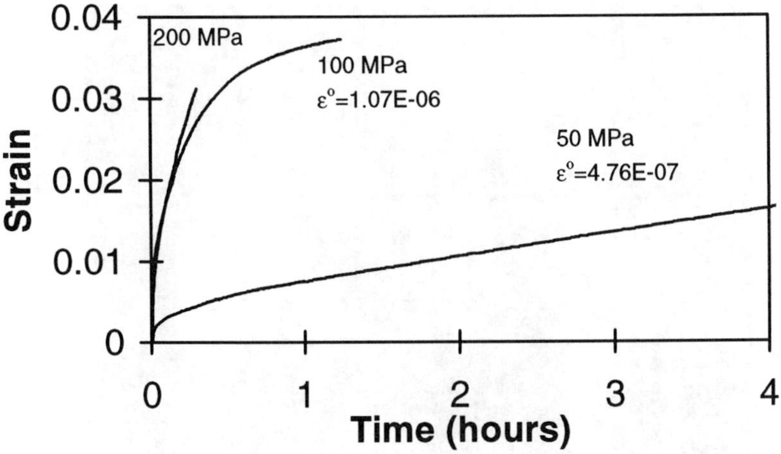

Figure 3. Creep curves of the monolithic zircon obtained at 1300°C and stresses of 50 MPa, 100 MPa, and 200 MPa.

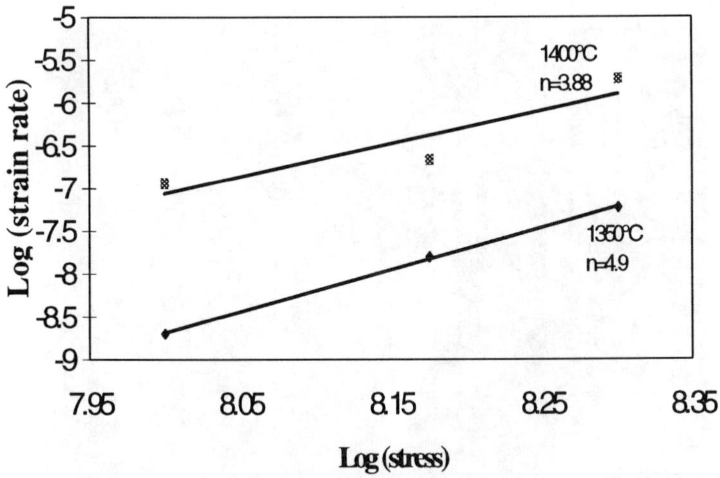

Figure 4. Steady state creep rate vs applied stress for zircon-silicon carbide composites.

Figure 5. Micrograph showing matrix cracking and fiber failure in a composite sample subjected to creep at 1350°C and 150 MPa. Very few matrix cracks are seen on the tensile surface of the sample.

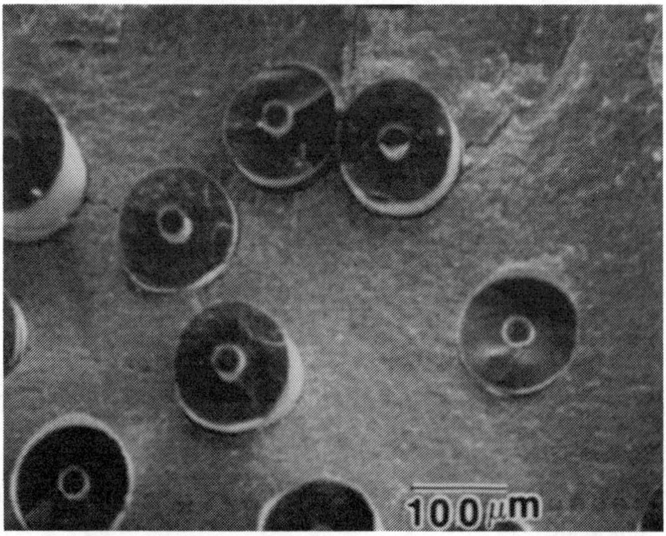

Figure 6. Micrograph showing fracture surface of composite that failed by creep rupture at 1400°C and 100 MPa in less than fourteen hours. The sample exhibits multiple matrix cracking with minimal fiber pullout.

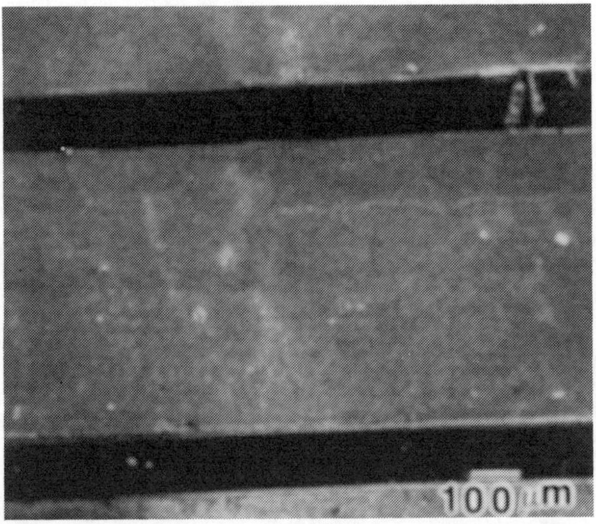

Figure 7. Micrograph showing matrix cracking and crack branching in acomposite crept at 1400°C and 150 MPa without any fiber failure after ten hours. Periodically spaced matrix cracks are seen on the tensile surface

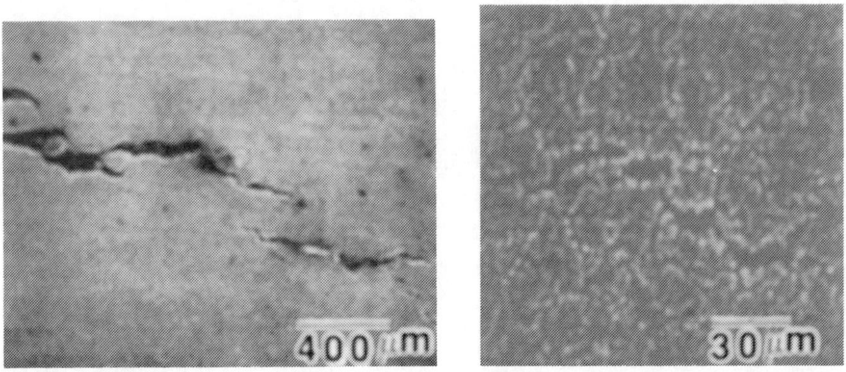

(a) **5** (b) **5**

Figure 8. Tensile surface of a composite which failed by creep rupture at 1400°C and 200 MPa. Cavity formation and linkage of cavities to form creep cracks can be seen. a) 75x b) 1000 x

CREEP DEFORMATION IN SMALL GRAIN SIZE Al_2O_3+TiN COMPOSITE POLYCRYSTALS

Qiu Tai(*) and Alain Mocellin
LSG2M, UMR 7584 - Ecole des Mines
54000 NANCY, France

ABSTRACT

A series of fine grained Al_2O_3+ 10 and 20 vol % TiN composites have been prepared by hot-pressing Al_2O_3 + TiN or AlN + TiO_2 + Al_2O_3 powder mixtures in appropriate amounts. Both these types, i.e. reacted and non-reacted composites were creep tested in compression (10 to 50 MPa) at 1300 and 1350°C in inert atmosphere. The TiN additions are found to enhance the strain rate up to almost one order of magnitude. Other experimental observations show that deformation results from interface and/or grain boundary sliding controlled by interface reaction(s). Somewhat smaller strain rates are observed in the reacted composites and interpreted with reference to the possible presence of TiO_2 rich interface films.

INTRODUCTION

The development of a variety of ceramic/ceramic composites for applications under severe thermo-mechanical conditions has been under active investigation for a number of years, as has the assessment and modelling of their properties. Among the latter the high temperature creep and deformation

---

(*) On leave from Nanjing University of Chemical Technology, Nanjing, P.R. China

To the extent authorized under the laws of the United States of America, all copyright interests in this publication are the property of The American Ceramic Society. Any duplication, reproduction, or republication of this publication or any part thereof, without the express written consent of The American Ceramic Society or fee paid to the Copyright Clearance Center, is prohibited.

behaviour has attracted much attention (see e.g. [1]), particulary so as concerns polycrystalline matrices reinforced with dispersed particles or short fibers such as whiskers. Alumina is one such host which can accommodate both oxide (e.g. ZrO_2) as well as non oxide (e.g. SiC - w) dispersoids (see e.g. [2]). These are frequently observed to reduce its creep rate, except for effects related to impurities at the grain boundaries. It is also known that under appropriate stress and temperature conditions single-phase fine-grained Al_2O_3 can undergo superplastic creep at a gradually decreasing rate as strain develops, because of concurrent grain growth [3,4]. This reduces the potential of Al_2O_3 to be amenable to hot forming operations that could be practically viable.

The question then arises as to whether some dispersed phase could be introduced into fine-grained (i.e. $<\sim$ 1 μm size) polycrystalline Al_2O_3 which would slow down grain growth during superplastic-like deformation, while not adversely affecting other features of the creep mechanism. Such a line of reasoning is well known to metallurgists and has been successfully applied for example to α/β Ti alloys (see e.g. [5]).

In such a context besides the grain growth inhibiting effect which should be expected from the dispersed phase, a major role also may be played by the chemical situation prevailing at grain boundaries and interfaces which results not only from residual impurities in the starting powders but also from details of the processing route which has led to the densified material. More precisely it is conceivable that, everything else being equal, reactively processed composite materials may have grain boundaries and interfaces differing in their compositions and properties from those resulting from the consolidation of non reacting powder mixtures of the same target compositions.

Previous work has shown that fine grained Al_2O_3/TiN composites may be prepared via hot-pressing either non reacting Al_2O_3/TiN or reacting AlN+TiO_2 powders [6,7]. Since Al_2O_3 and TiN are compatible with each other at all temperatures and the corresponding composites may be anticipated to have similar properties to Al_2O_3/TiC which has found applications as cutting tool material, a program was undertaken to investigate the compressive creep behaviour of fine grained aluminas containing ~ 10 and 20 vol % TiN particles. The present paper is a report and discussion of the results thus obtained.

EXPERIMENTAL PROCEDURES

The starting powders used for sample preparation and their chemical analyses are listed in Table I next :

Table I. Chemical analyses of starting powders (from suppliers ; wt %)

	Al	O	C	Fe	Ti	Si	Mg , Ca Na , K	Others
AlN (1)	> 64	2.1	0.06	$7x10^{-3}$	-	-	-	$< 10^{-2}$
TiO_2 (2)	$< 2x10^{-3}$	-	-	$<10^{-2}$	60	$< 10^{-2}$	$< 4x10^{-2}$	$< 3x10^{-2}$
Al_2O_3 (3)	-	-	-	$8x10^{-3}$	-	$3x10^{-2}$	< 0.15	$< 10^{-2}$
TiN (1)	-	1.15	0.15	< 0.1	> 77	-	-	< 0.1

(1) : H.C. Starck Grade C ; (2) Tioxide H.P. ; (3) Alcoa A-16

Four sets of green samples were prepared via colloidal techniques and filter-pressing [8] corresponding to the following initial molar compositions :

$$\left.\begin{array}{l} 1.5\ Ti\ O_2 + 2\ AlN + 5.21\ Al_2O_3 \\ \\ 4.5\ Al_2O_3 + 1\ TiN \end{array}\right\} \rightarrow \quad Al_2O_3 / 9.75 \text{ vol \% TiN}$$

$$\left.\begin{array}{l} 1.5\ Ti\ O_2 + 2\ AlN + 1.76\ Al_2O_3 \\ \\ 1.84\ Al_2O_3 + 1\ TiN \end{array}\right\} \rightarrow \quad Al_2O_3/19.6 \text{ vol \% TiN}$$

After binder removal at moderate temperatures each composition was vacuum hot-pressed at 1673 °K in graphite die under a 45 MPa load applied for one hour. From previous experience [6,7] it was known that such a treatment would yield > 99% dense composites with compositions as indicated above, which hereafter will be refered to as "10T90A" and "20T80A" respectively. Explicit mention of their resulting or not from the AlN + TiO_2 reaction taking place during heating will also be made.

Small 4.5 X 3.5 X 9 mm^3 bars to be used for creep experiments were diamond cut and polished from the as-pressed discs. Figure 1 shows typical examples of sample microstructures before testing, no obvious differences could be detected between reacted and non-reacted materials under such magnifications.

Figure 1 - Examples of composite microstructures (a) fracture surface of a 10T 90A sample ; (b) transmission micrograph of a 20T 80A sample - A and T stand for Al_2O_3 and TiN grains respectively.

In all cases the Al_2O_3 grain sizes estimated as mean intercept lengths were 0.9 ± 0.1 μm whereas the TiN particle sizes determined in a similar way evere found to equal 0.3 ± 0.1 μm with a tendency to take the lower values of such an interval in the reacted materials. As can be noted on Figure 1 the small individual TiN crystals are frequently found to cluster in groups of up to 10 to 15, even at low TiN volume fractions. Some amount of TiN/TiN interface area therefore is also present in such composites besides Al_2O_3 grain boundaries and Al_2O_3/TiN interfaces. No residual phases or significant crystallographic textures were detected with conventional X-ray diffraction techniques.

Creep tests were carried out under uniaxial compression and Ar+He atmosphere in a tungsten resistance heated MTS 810 Material Test System equipment. The temperature was controlled by two W/W-Rh (type C) thermocouples and one pyrometer (IRCON 2λ, 1100-2000°C - USA). The maximum temperature deviation was <2K. The LVDT extensometer setup was capable of measuring sample height changes as small as 0.1 μm. For each test another dummy sample of the same material was placed next to the loaded one in order that possible structural changes resulting from the applied stress could be distinguished from those arising from plain thermal annealing. The temperature and stress levels investigated were 1300 and 1350°C and 10 to 50 MPa repectively. The loads were continuously adjusted during testing in order that constant stress be maintained on the samples. The latter were typically deformed to true strains not exceeding - 0.16 in order to avoid barreling, and after completion of the tests they were rapidly cooled to room temperature. Microstructure observations, density measurements by the Archimedes method and X-ray diffraction analyses with cobalt Kα radiation (λ = 0.17889 nm) were performed after creep testing as described before.

RESULTS AND DISCUSSION

Deformation behaviour.

The variations of the steady state creep rate with applied stress are shown on Figures 2 and 3 for the reacted and non - reacted composites respectively.

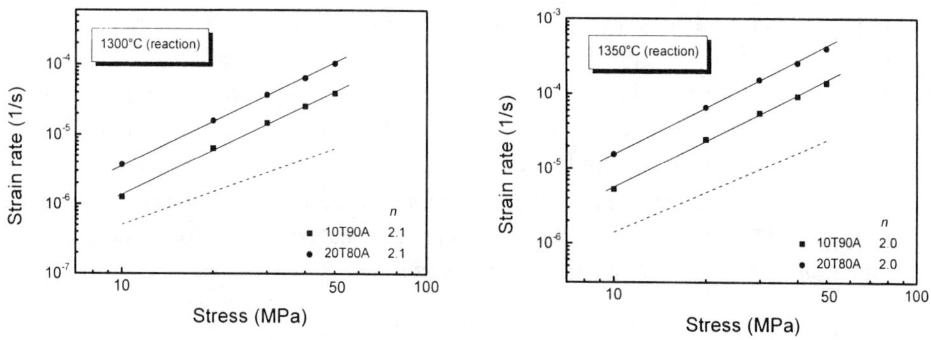

(a) (b)

Fig. 2 - Variation of strain rate with stress for the reaction hot pressed Al_2O_3-TiN composites (a) 1300°C, (b) 1350°C. (broken line = pure Al_2O_3, calculated from the results obtained by Xue and Chen [4]).

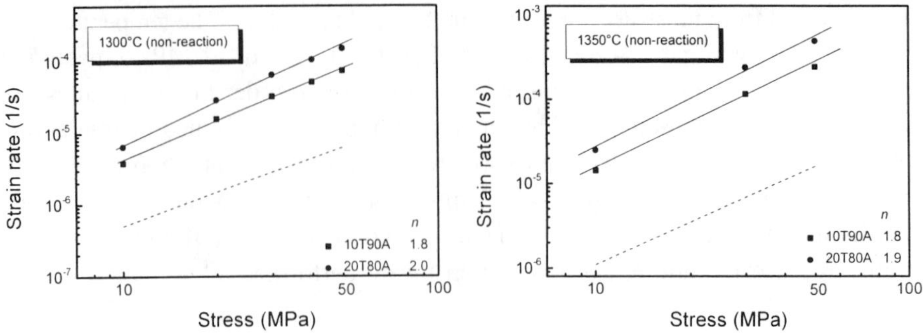

Fig.3 - Variation of strain rate with stress for the non-reaction hot pressed Al_2O_3-TiN composites (a) 1300°C, (b) 1350°C. (broken line = pure Al_2O_3 calculated from [4])

The stress exponent n of the usual creep equation : $\dot{\varepsilon} = A_o \exp\left[-\frac{Q}{RT}\right]\frac{\sigma^n}{d^p}$ was calculated from the slopes of straight lines fitted to the data. In all cases n was found very close to 2, ranging from 1.8 to 2.1. No significant difference in stress exponent is found between the reacted and non-reacted materials, nor between the 10 and 20 vol % TiN composites. The dotted lines on Figures 2 and 3 correspond to experimental observations by Xue and Chen on single phase undoped Al_2O_3 after correction (on the basis of their constitutive equation) for a grain size corresponding to the present 0.9 μm value [4].

In absolute terms therefore it appears that creep rates have been enhanced by TiN additions in either reacted or non reacted materials, and for given stress and temperature conditions the ranking is as follows:

$\dot{\varepsilon}$ (Al_2O_3) < $\dot{\varepsilon}$ (10T 90A) < $\dot{\varepsilon}$ (20T 80A). The differences amount to factors of about 2 or 3.

It is interesting now to assess the effect of the processing route on the strain rate for a given TiN content. Figure 4 shows results obtained with both sets of samples plotted as strain rate in the reacted versus non - reacted material for equivalent stress and temperature conditions. If both were equal the data points would align along the first diagonal (dotted line). A deviation though is clearly apparent and $\dot{\varepsilon}$ (reacted) < $\dot{\varepsilon}$ (non-reacted). Interestingly the difference (again a factor of 2 or 3) tends to decrease and perhaps disappear at the higher strain rates ($> 10^{-4}\ s^{-1}$).

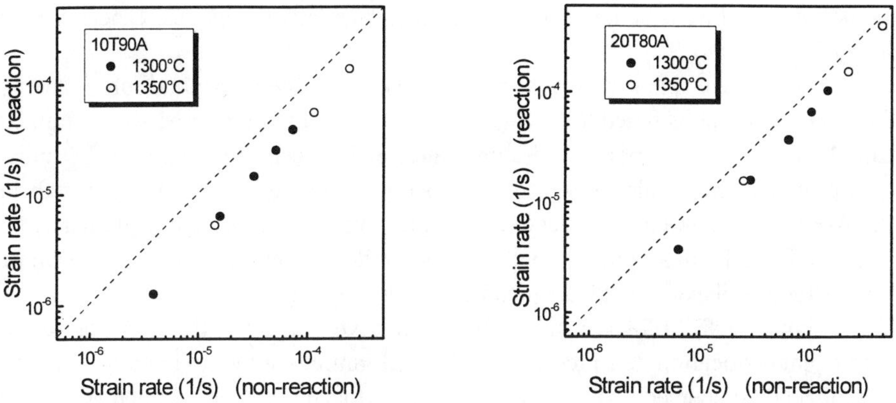

Figure 4 - Effect of processing route on recorded strain rates for identical temperature and stress conditions (a) Al_2O_3/10% TiN, (b) Al_2O_3/20% TiN.

The influence of grain sizes on the creep rates (i.e. the grain size exponent "p" in the constitutive equation) has not been evaluated at this stage. But it is worth mentioning that all recorded creep curves did evidence a well-behaved steady-state after a short initial equilibration transient, indicative of stable microstructures. Finally since so far the creep tests were performed at two different temperatures only, crude estimates of apparent activation energies may be derived from log $\dot{\varepsilon}$ versus 1/T plots at constant stress. Values ranging from 600 to 500 kJ/mole have thus been obtained, with a tendency to decrease under increasing applied stress and/or upon going from the reacted to the non reacted materials respectively. It is emphasized that such data on activation energies ought not be given more significance than to mere orders of magnitude.

Structural observations

Specific gravity measurements showed that a small amount of cavities developed in both types of reacted and non-reacted samples. In fact densities decreased also in the dummy samples, presumably because of gas evolutions during heat treatment. Such variations amounted to about 1.5 to 2.5%, i.e. 0.1 to 0.3% less than in the corresponding deformed samples. Deformation related cavitation therefore is believed not to exceed 0.3% relative densities, especially for low TiN contents.

Grain size measurements by the linear intercept method on scanning electron micrographs failed to reveal significant differences either between strained samples and the corresponding dummy ones, or between the latter and the initial as hot-pressed materials : fine TiN dispersions do indeed inhibit grain growth in the Al_2O_3 matrix at least as effectively as other types of particles do. Similarly, no changes in individual grain aspect ratios could be evidenced : as a whole microstructures have remained equiaxed after deformation.

Finally, TEM observations though not very systematic, did not reveal significant dislocation densities in the deformed samples, either in the matrix grains or in the TiN dispersed phases . Figure 5 shows that no or very limited (possibly resulting from indivual grain rotations) crystallographie texturing could be detected either.

Interpretation and discussion

The experimental evidence just presented on : (i) the amount of strain achieved with a stable equiaxed fine grained microstructure and (ii) the stress exponent of the strain rate in all four types of investigated composites, is considered strong evidence that creep has resulted from grain and/or interface boundary sliding, and that rate control was effected via some interface reaction. The TiN dispersed phase acts as an effective growth inhibitor and microstructure stabilizer of the matrix. The stress accommodating mechanism at multiple grain or phase junctions remains to be identified. But given the lack of signifiant texturing (Figure 5) and of obvious dislocation activities from TEM observations, it may be argued that intragranular plastic deformation mechanisms have not played a signifiant role in the present cases.

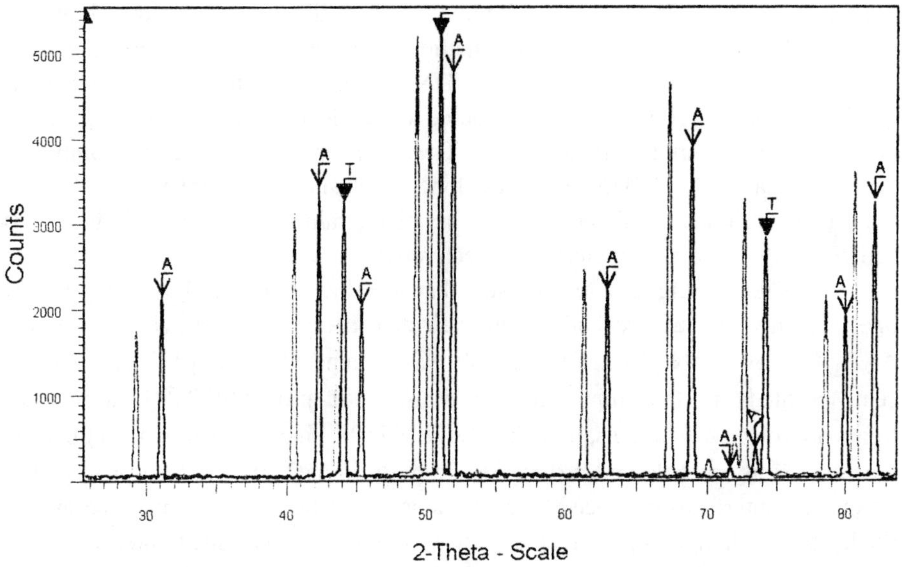

Figure 5 - X-ray diffraction patterns of Al_2O_3/20% TiN composites
Labeled lines (A=Al_2O_3 ; T = TiN) refer to the material after - 0.16 creep strain. Non Labeled lines on their left correspond to the same reflections from the undeformed material. For the sake of clarity on the plot, all angular positions have been shifted : + 2° (2θ) on deformed versus - 1.33° (2θ) on starting materials respectively.

Since it appears plausible that some titanium has entered into solution in the corundum matrix and since Ti has also been reported to enhance the diffusional creep of alumina [9] it is likely that diffusional mechanisms are rapid enough to accommodate localized stress buildups and consequently minimize the formation of cavities. It may also be noted in passing that the present estimates of apparent activation energies are consistent with values found in the literature [2] when interface reaction is concluded to be the rate controlling step of creep.

The specific influence of TiN particles on the rates of deformation now may be understood as follows. Since three different types of interfaces are present in the composites as compared to only grain boundaries in the pure matrix, the possibility exists that under applied stress interface reaction and transport be less easily achieved on one type of interface than on the others. The alumina grain boundaries apparently are fulfilling such a role since the presence of TiN enhances the creep rate. As for the TiN/TiN and Al_2O_3/TiN interfaces the former is not present in large amounts : only a limited fraction of TiN clusters are shown in Figure 1. These behave in first approximation as if they were single crystals since observation has shown that the average TiN particle size remains fairly constant. The TiN phase therefore may be considered as a collection of inert non-interacting particles in a deforming Al_2O_3 matrix. They appear to influence the composite behaviour via the Al_2O_3/TiN interfaces along which sliding is more easily achieved. Such a conclusion is consistent with the relative strain rates recorded in the 0, 10 and 20% TiN containing materials respectively.

It is interesting to note in closing that specific Al_2O_3/ Al_2O_3 and Al_2O_3/TiN interface energies have recently been evaluated in non reactively hot - pressed Al_2O_3 - 5 wt % TiN [10]. A value 0.80 was obtained for the $\gamma AT/\gamma AA$ ratio it being pointed out that thin interface films, most probably TiO_2, were very frequently observed in the sample by HRTEM [10]. Since it is known [11] that Al ions are readily soluble and diffuse very rapidly in TiO_2 it is not surprising that Al_2O_3/TiN interfaces enhanced creep rates as found here. And it may be argued finally that although firm direct evidence is not available, such interface films should be present to a lesser extent in the reactively hot-pressed composites than in the non-reacted ones, because of the different processing. The slower creep rates noted in the reactively pressed composites could conceivably relate to such a difference.

CONCLUSION

The annealing and compressive creep behaviour of Al_2O_3/10 and 20 vol % TiN composites under 10 - 50 MPa at 1573 and 1623 K show that :

- submicron TiN particles effectively inhibit grain growth in the micron size Al_2O_3 matrix and allow deformation with structural stability
- deformation is most likely achieved via interface and grain boundary sliding at a rate which is controlled by interface reaction(s)
- the presence of TiN particles enhances the overall creep rate probably because of easier sliding along Al_2O_3/TiN interfaces than along Al_2O_3 grain boundaries
- the effect is thought to result from TiO_2 based interface films which provide very fast Al diffusion paths
- smaller creep rates observed in reactively versus non-reactively processed composites are consistent with such an assumption.

Work is underway on similar composites with higher TiN contents.

REFERENCES

[1] D.S. Wilkinson, "Creep mechanisms in multiphase ceramic materials", *Journal of the American Ceramic Society*, **81** [2] 275-99 (1998)

[2] Qiu Tai and A. Mocellin, "High temperature deformation of Al_2O_3-based ceramic particle or whisker composites" *Ceramics International*, in press (1998)

[3] J.D. Fridez, C. Carry and A. Mocellin, "Effects of temperature and stress on grain boundary behaviour in fine grained alumina" ; pp. 720-40 in *Advances in Ceramics* vol 10, Structure and properties of MgO and Al_2O_3 Ceramics, Edited by W.D. Kingery. American Ceramic Society, Columbus ; OH 1984

[4] L.A. Xue and I.W. Chen, "Deformation and grain growth of low temperature sintered high purity alumina" *Journal of the American Ceramic Society*, **73** [11] 3518-21 (1990)

[5] J.S. Kim, Y.W Chang and C.S. Lee, "Quantitative analysis on boundary sliding and its accommodation mode during superplastic deformation of two phase Ti-6Al-4V alloy", *Metallurgical and Materials transactions*. Trans **29A** [1] 217-26 (1998)

[6] T. Sperisen, Doctor's Thesis n° 846 *Swiss Federal Institute of Technology.*; Lausanne, Switzerland, 1990

[7] T. Sperisen and A. Mocellin, "Investigation of the displacement reaction in mixed AlN + TiO_2 powders" *Journal of Materials Science* **27** [4] 1121 - 35 (1992)

[8] B. Grobety and A. Mocellin, "Preparation of AlN-TiO_2 powder compacts using colloïdal methods" *Ceramics International* **15** [5] 271-79 (1989)

[9] P.A. Lessing and R.S. Gordon, "Creep of polycrystalline alumina, pure and doped with transition metal impurities" *Journal of Materials Science* **12** [11] 2291-2302 (1977)

[10] J. Jiao, M.L. Jenkins and R.W. Davidge, "Interfacial fracture energy-mechanical behaviour relationship in Al_2O_3/SiC and Al_2O_3/TiN manocomposites" *Acta Materiala* **45** [1] 149-156 (1997)

[11] B. Freudenberg and A. Mocellin "Aluminium titanate formation by solid state reaction of coarse Al_2O_3 and TiO_2 powders" *Journal of the American Ceramic Society* **71** [1] 22-28 (1988)

Damage Evolution, Residual Stress Measurement, and Micromechanical Modeling

MEASURING DAMAGE EVOLUTION IN CERAMIC MATRIX COMPOSITES USING THERMOELASTIC STRESS ANALYSIS

Thomas J. Mackin and Mark C. Roberts

Mechanical and Industrial Engineering
University of Illinois at Urbana-Champaign
Urbana, IL 61801

ABSTRACT

Thermoelastic stress analysis (TSA) was used to monitor damage evolution in several composite systems. A methodology was developed that enables full field surface stress mapping, quantification of stress redistribution, and damage imaging. An entirely new damage imaging method was developed that utilizes infrared images that are out-of-phase with the input load. To explore the general utility of the method, test samples of several ceramic and cement matrix composites were fabricated. It is shown that the TSA method can be used to identify the operative damage mechanism. Model materials were chosen to illustrate the effect of each of three damage mechanisms: a single crack bridged by fibers; multiple matrix cracking; and shear bands. Each mechanism is shown to be effective at redistributing stress and diffusing stress concentrations.

INTRODUCTION

Design guidelines and protocols must be developed in order for ceramic matrix composites (CMCs) to gain widespread acceptance in engineering applications. In particular, it is especially important to quantify the effect of stress concentrations, such as notches and holes, on the allowable design loads. It is known that brittle constituent composites rely on inelastic mechanisms, such as: interface failure, matrix cracking, fiber failure and fiber pullout to enhance ductility and redistribute stress away from locations of stress concentration.[1] The extent of stress redistribution, however, is not well known. Research over the past several years has classified the macroscopic mechanisms of stress redistribution into three general categories.[2,3,4] Class I damage is the development of a single matrix crack bridged by fibers; Class II damage involves the development of

To the extent authorized under the laws of the United States of America, all copyright interests in this publication are the property of The American Ceramic Society. Any duplication, reproduction, or republication of this publication or any part thereof, without the express written consent of The American Ceramic Society or fee paid to the Copyright Clearance Center, is prohibited.

multiple cracks in the matrix; and Class III damage involves the development of a shear damage zone. The operative damage mechanism as well as the extent of stress redistribution in a particular composite will depend upon the composite constituent properties, as well as the properties of the interface.

Brittle constituent composites commonly utilize an interphase to assure interfacial debonding, providing a weak link that enables fiber bridging and pullout. Recently, tough composites have been fabricated that utilize strong interfaces.[5,6] The key difference arises from the toughening mechanism: a class III material need not have a weak interface to develop considerable stress redistribution, local nonlinearity, and enhanced toughness. Furthermore, Aveston, Cooper, and Kelly[7] clearly showed that the addition of strongly bonded fibers to a brittle matrix resulted in a composite with greatly enhanced strength and toughness. The mechanism of stress redistribution was multiple matrix cracking (class II). In any case, the underlying fundamental phenomenon is the development of a non-linear material response that redistributes stress away from regions of stress concentration.

Changes in the fiber, matrix, and interface properties will affect the active damage mechanism as well as the extent of stress redistribution. Fiber and matrix properties are easily measured using standard testing methods,[8,9] while interface properties can be determined using hysteresis[10-12], pull-out, frictional heating[13] or fiber push-out tests[6,14-22]. The macroscopic effect of changes in these properties, however, has not been easy to quantify. We present a non-contacting technique to evaluate stress concentration factors, stress redistribution, damage evolution, classification of damage, and damage tolerance in ceramic matrix composites. The technique utilizes the thermoelastic effect to measure the stress state across the entire surface of a stressed body. This non-contacting method can be used to measure changes in stress distribution and stress concentration factor as a composite is damaged. In addition to quantifying stress redistribution, we show that the damage mechanism can be imaged and identified, thereby assigning a mechanism and damage classification to the measured changes in stress distribution. The results we will present clearly indicate that substantial stress redistribution does occur in ceramic matrix composites, and that each damage mechanism can effectively change a material's notch sensitivity.

Thermoelastic Stress Analysis

Thermoelastic stress analysis utilizes the thermoelastic effect to relate adiabatic changes in the stress state of a material to changes in specimen temperature.[23,24] The relationship between stress and temperature is given by:[23-30]

$$\Delta T = -\kappa \; T_0 \Delta\sigma \qquad (1)$$

where ΔT is the change in temperature brought about by a change in hydrostatic stress $\Delta\sigma$ ($\Delta\sigma = (1/3)*\Delta\sigma_{kk}$) and κ is the thermoelastic constant. In the TSA method a low amplitude sinusoidal stress is applied to the specimen, creating a thermoelastic response that is in-phase with the applied stress. The peak to peak temperature change, ΔT that is in phase with the peak to peak applied dilatational stress is given by [23-25, 31-33]

$$\Delta T = -\kappa T_0 \Delta\sigma \sin \varpi t \tag{2}$$

where ω is the frequency,
$\Delta\sigma$ is the hydrostatic stress amplitude,
and t is time.

A key aspect of this effect is that spatial variations in the applied stress, $\Delta\sigma(x,y)$, will appear as a spatial distribution in temperature, $\Delta T(x,y)$.

In the present tests, a Delta-Therm 1000* TSA system with an infrared array detector was used to measure the surface temperatures of cyclically loaded test samples. Thermoelastic images are collected at loads less than 10% of the proportional limit. This is done to assure that damage is not accumulating during the time required to measure the spatial temperature distribution and to alleviate frictional heating due to dissipation. The Delta-Therm 1000 system allows full field measurements of the surface temperature distribution of standard sized test samples in 30 seconds at a spatial resolution of 120 μm and a temperature resolution of roughly .001K. A sinusoidal load frequency of 10 Hz has been found to result in adiabatic cyclic temperature changes in most materials and was the frequency used in this study.

Materials

To demonstrate the general effectiveness of the TSA method, damage evolution and stress redistribution were evaluated in three different material systems. Each composite exhibited a different damage mechanism. Unidirectional samples of an S-2 glass fiber reinforced CERASET matrix composite and a Cem-fil glass fiber reinforced modified Portland cement were fabricated using a filament winding process.[34] A third specimen was machined from a commercially available industrial cut-off wheel that consists of a phenolic resin matrix embedded with alumina particles. The wheels are available as disks 30 inches in diameter and 0.25 inches thick. Single layers of plain-weave glass

* Manufactured by Stress Photonics, Madison, WI

fiber mat are bonded onto either side of the matrix, resulting in a sandwich structure of matrix between glass mats.

EXPERIMENTAL PROCEDURE

Constitutive Response

Unidirectional tensile specimens measuring 25.4 mm wide and between 15 and 20 cm long were cut from each material. Aluminum tabs were bonded to the ends of each specimen to diffuse gripping stresses during testing. Specimens were tested in an Instron servo-hydraulic load frame using hydraulic grips. A clip-on type extensometer was used to measure elongational strain in the specimens, while load values were recorded using a 22,000 lb load cell. Failure commonly occurred at the grips, providing lower bound estimates to the tensile strength. These values were then used to determine the subsequent damage loads for notched sample testing. Figure 1 shows the longitudinal stress strain response of the remaining material specimens.

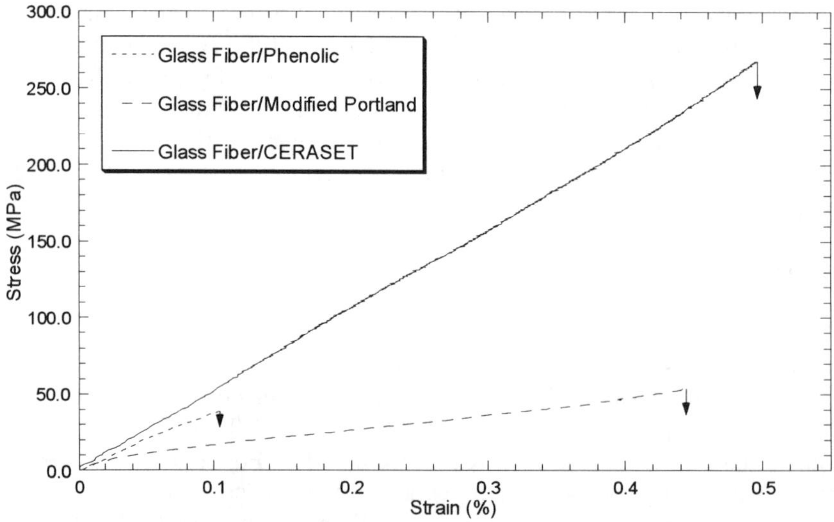

Figure 1: Typical tensile stress strain curves for materials used in damage evolution testing.

Damage Evolution

Double edge-notched samples were fabricated from each material and were used to guarantee damage initiation at a known location. Two coats of commercial flat black spray paint were applied to the specimen to ensure constant emissivity across the specimen. Specimens were allowed to dry for two days prior to testing.

Each material's stress-strain response was used to determine a damage schedule of steadily increasing stresses. Notched samples were loaded to a prescribed fraction of the ultimate strength and then unloaded. A 10 Hz cyclic load at an amplitude that is 10% of the maximum previous load is then applied, and TSA images are acquired. The damage load is incrementally increased, and the cycle repeats until specimen failure. This provides an interrupted picture of the evolution of damage as recorded by the TSA images.

Damage Imaging

Dissipation mechanisms such as fiber sliding can generate frictional heating during cyclic loading[13] that is out of phase with the thermoelastic response of the material. By examining the out-of-phase response, any locations of frictional dissipation can be identified. Regions of relative fiber sliding appear as 'hot' spots in the out-of-phase images and can be used to pinpoint regions of damage. It will be shown that damage in the form of shear bands, and multiple matrix cracks can be identified easily by examining the out-of-phase response.

The cyclic loads applied during acquisition of the in-phase thermoelastic signal are a small fraction of the damage load. However, to facilitate imaging of damage and identifying the specific damage mechanism, the amplitude of the applied cyclic loads was increased while imaging the out-of-phase signals. These higher cyclic load amplitudes enhance dissipation and improve the out-of-phase signal response. Acquisition of these "damage" requires less than one minute resulting in little damage accumulation during signal acquisition.

RESULTS & DISCUSSION

A unidirectional sample of a fiberglass reinforced modified Portland cement composite was notched and tested as described previously. This specimen exhibited substantial stress redistribution, as demonstrated by the in-phase TSA images, Figure 2. The resulting damage images, Figure 3, clearly show the evolution of multiple matrix cracks (class II damage mechanism).

Unidirectional samples of a glass fiber/CERASET matrix composite were notched and tested as described previously. These samples exhibited substantial stress redistribution and complete net section stress failure, Figure 4. The operative damage mechanism was identified as the propagation of shear bands, Figure 5, (class III damage mechanism).

Damage evolution in the glass fiber/phenolic matrix propagated as a single crack bridged by fibers, Figure 6. Double edge-notches resulted in unstable crack propagation, so a single edge-notch was cut perpendicular to the fiber orientation in order to attain stable crack formation. The matrix crack propagated across the entire sample width, leaving an intact bridging zone across the net section. This sample exhibits the class I damage mechanism of a single crack bridged by fibers.

CONCLUSION

Thermoelastic stress analysis was used to document the onset and growth of damage in several brittle constituent composites. In-phase images are used to quantify the stress distribution in the materials, while out-of-phase imaging is used to capture an image that portrays the distribution of the existing damage. The out-of-phase imaging constitutes a new methodology for observing and measuring damage in composites. The operative damage mechanism in each material was identified by examining the out-of-phase thermal signal during cyclic loading. The out-of-phase signal provides a measure of dissipational heating, and provides high thermal contrast in the damage zones. This method presents a new tool for measuring the current damage state in a material. Combined use of the in-phase and out-of-phase images provides a measure of not only the damage, but also its effect on a material's stress distribution and the mechanism of damage evolution. In addition to basic studies of damage, the present methodology can be used as an NDE tool.

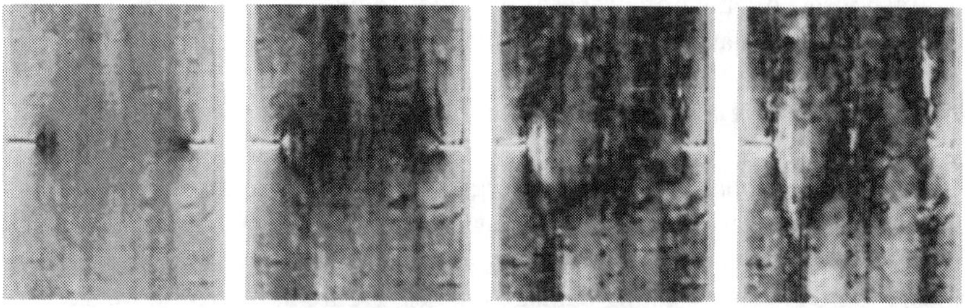

(a) (b) (c) (d)

Figure 2: Representative in-phase TSA images of glass fiber/modified Portland cement after loading to: (a) 5.7 MPa; (b) 8.5 MPa; (c) 11.3 MPa; and (d) 14.2 MPa.

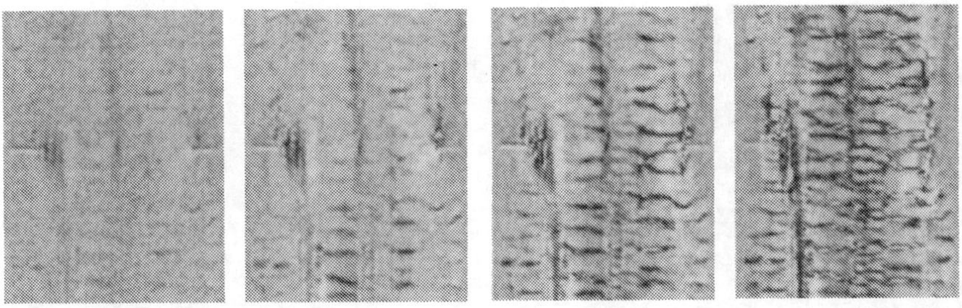

(a) (b) (c) (d)

Figure 3: Representative out-of-phase TSA images of glass fiber/modified Portland after loading to: (a) 5.7 MPa; (b) 8.5 MPa; (c) 11.3 MPa; and (d) 14.2 MPa, illustrating the evolution of multiple matrix cracks.

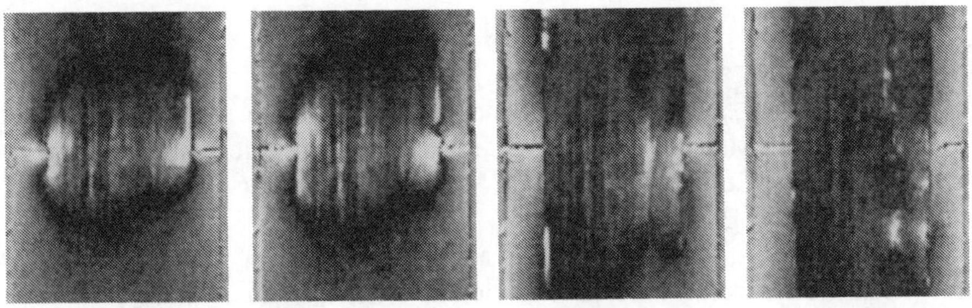

(a) (b) (c) (d)

Figure 4: Representative in-phase TSA images of unidirectional glass fiber/CERASET composite after loading to: (a) 20 MPa; (b) 40 MPa; (c) 60 MPa; and (d) 80 MPa. These images clearly show evolution of shear band damage and load bearing across the net-section.

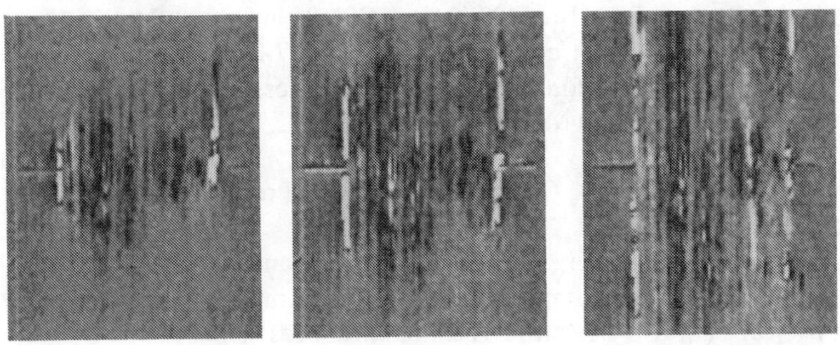

(a) (b) (c)

Figure 5: Representative out-of-phase TSA damage images of unidirectional glass fiber/CERASET composite after loading to: (a) 30 MPa; (b) 40 MPa; and (c) 50 MPa, showing the presence of shear bands as white vertical stripes in the image.

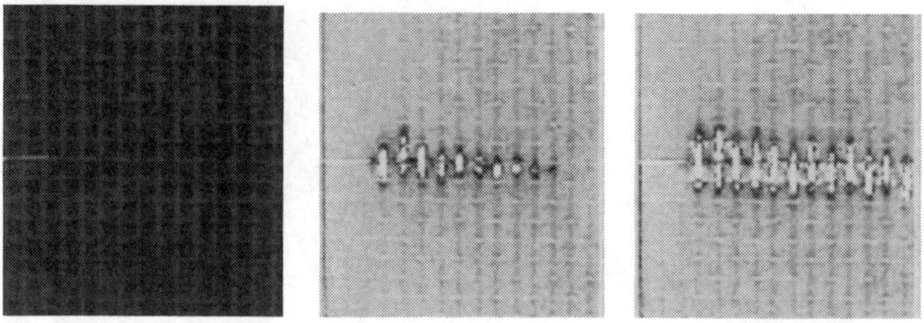

(a) (b) (c)

Figure 6: Images of a glass fiber reinforced phenolic matrix composite. (a) An optical image shows the single edge notch and the glass fabric weave. (b) In-phase TSA image of a bridged crack after loading to 11 MPa where the crack has propagated across most of the sample ligament. (c) After loading to 13 MPa, the crack has traversed the specimen, leaving an intact fiber bridge spanning the specimen.

References

1) Evans, A. G., Zok, F. W., and Mackin, T. J., "The Structural Performance of Ceramic Matrix Composites, High Temperature Mechanical Behavior of Ceramic Composites," Eds. S. V. Nair and K. Jakus, Butterworth-Heinemann, 1995, pp. 3-84.
2) Mackin, T. J., Purcell, T. E., He, M. Y., and Evans, A. G., "Notch Sensitivity and Stress Redistribution in Three Ceramic-Matrix Composites," *J. Am. Ceram. Soc.*, 78[7] 1719-28 (1995).
3) Cady, C., Heredia, F. E., and Evans, A. G., "In-Plane Mechanical Properties of Several Ceramic-Matrix Composites," J. Am. Ceram. Soc. 78(8) (1995) 2065-78
4) Heredia, F. E., Spearing, S. M., He, M. Y., Mackin, T. J., Evans, A. G., Mosher, P., and Brøndsted, P., "Notch Effects in Carbon Matrix Composites," *J. Am. Ceram. Soc.*, 77[11] 2817-27 (1994).
5) Droillard, C. and Lamon, J., "Fracture Toughness of 2-D Woven SiC/SiC CVI-Composites with Multilayered Interphases," *J. Am. Ceram. Soc.*, 79[4], (1996).
6) Faber, K. T., "Ceramic Composite Interfaces: Properties and Design," *Annu. Rev. Mater. Sci.*, 27:499-524 (1997).
7) Aveston, J., Cooper, G. A., Kelly, A., "Single and Multiple Fracture," The Properties of Fiber Composites, Conf. Proc., London, UK (National Physical Lab, IPC Scientific Technological Press), 1971.

8) I. M. Daniel, "Composite Materials"; pp. 812-890 in Handbook on Experimental Mechanics. Edited by A. S. Kobayashi. Society for Experimental Mechanics, Englewood Cliffs, NJ, 1987.
9) R. B. Pipes, R. A. Blake Jr., J. W. Gillespie Jr., L. A. Carlsson, "Test Methods"; Delaware Composites Design Encyclopedia v. 6, Edited by L. A. Carlsson, J. W. Gillespie Jr. Lancaster, PA, 1990.
10) Vagaggini, E., Domergue, J. M., Evans, A. G., "Relationships between hysteresis measurements and the consituent properties of ceramic matrix composites. I. Theory," *J. Am. Ceram. Soc.*, 78[10], (1995).
11) Domergue, J. M., Vagaggini, E., Evans, A. G., "Relationships between hysteresis measurements and the consituent properties of ceramic matrix composites. II. experimental studies on unidirectional materials," *J. Am. Ceram. Soc.*, 78[10], (1995).
12) Domergue, J. M., Heredia, F.E., Evans A. G., "Hysteresis loops and the inelastic deformation of 0/90 ceramic matrix composites," *J. Am. Ceram. Soc.*, 79[1], (1996).
13) Holmes, J. W., Chongdu, Cho, "Experimental observations of frictional heating in fiber-reinforced ceramics," *J. Am. Ceram. Soc.*, 75[4], (1992).
14) Hsueh, Chun-Hway, "Evaluation of interfacial shear strength, residual clamping stress and coefficient of friction for fiber-reinforced ceramic composites," *Acta Metall. Mater.*, 38[3], (1990).
15) Jero, P. D., and Kerans, R. J., "The contribution of interfacial roughness to sliding friction of ceramic fibers in a glass matrix," *Scripta Met. et Mat.*, 24, 2315-18 (1990).
16) Jero, P. D., Kerans, R. J., Parthasarathy, T., A., "Effect of interfacial roughness on the frictional stress measured using push-out tests," *J. Am. Ceram. Soc.,* V. 74, pp 2793-801, Nov. 1981.
17) Carter, W. C., Butler, E. P., Fuller, E. R., "Micro-mechanical aspects of asperity-controlled friction in fiber-toughened ceramic composites," *Scripta Met. et Mat.,* V. 25, pp 579-584, 1991.
18) Warren, P. D., Mackin, T. J., Evans, A. G., "Design, analysis and application of an improved push-through test for measuring interface properties in composites," Acta Metall. Materl, V 40, No. 6, pp 1243-1249, (1992).
19) Kerans, R. J., and Parthasarathy, T. A., "Theoretical analysis of the fiber pull-out and push-out tests," *J. Am. Ceram. Soc.,* 74[7], 1585-96, (1991).
20) Bright, J. D., Danchaivijit, S., Shetty, D. K., "Interfacial sliding friction in silicon carbide-borosilicate glass composites: A comparison of pullout and pushout tests," *J. Am. Ceram. Soc.,* 74[1], 115-22, (1991).

21) Hutchinson, J. W., and Jensen, H. M., "Models of fiber debonding and pullout in brittle composites with friction," Mechanics of Materials, 9, 139, 1990.
22) Gao, Y.-C., Mai, Y.-W., Cotterell, "Fracture of Fiber-Reinforced Materials," J. Appl. Math and Phys. (ZAMP), 39, 550-572, (1988).
23) Enke, N. F., "Thermographic Stress Analysis of Isotropic Materials," Ph.D. Thesis, Department of Engineering Mechanics, The University of Wisconsin-Madison, 1989.
24) Harwood, N., and Cummings, W. M., *Thermoelastic Stress Analysis*, Adam Hilger, IOP publishing, 1991.
25) Wong, A. K., Sparrow, J. G., and Dunn, S. A., "On The Revised Theory of the Thermoelastic Effect," *J. Phys. Chem. Solids,* Vol. 49, No. 4, pp. 395–400, 1988.
26) Wong, A. K., Jones , R., and Sparrow, J. G., "Thermoelastic Constant or Thermoelastic Parameter?" *J. Phys. Chem. Solids,* Vol. 48, No. 8, pp. 749–753, 1987.
27) Budiansky, B., and O'Connell, R. J., "Bulk Thermoelastic Attenuation of Composite Materials," *J. Geophysical Research,* Vol. 88, No. B12, pp. 10, 343–348, December 10, 1983.
28) Budiansky, B., "Thermal and Thermoelastic Properties of Isotropic Composites," *J. Comp. Materials,* Vol. 4, pp. 286–295, July 1970.
29) Zymansky, M. W., *Heat and Thermodynamics,* 5th Edition, McGraw-Hill, 1968.
30) Jackson, E. A., *Equilibrium Statistical Mechanics,* Prentice-Hall International Series in Engineering of the Physical Sciences, 1968.
31) Bakis, C. E., Yih, H. R., Stinchcomb, W. W., and Reifsnider, K. L., "Damage Initiation and Growth in Notched Laminates Under Reversed Cyclic Loading," *ASTM STP 1012,* pp.66–83, 1989.
32) Simonds, R. A., Bakis, C. E., and Stinchcomb, W. W., "Effects of Matrix Toughness on Fatigue Response of Graphite Fiber Composite Laminates," *ASTM STP 1012,* pp. 5–18, 1989.
33) Beyerle, D. S., Spearing, S. M., Zok, F. W., and Evans, A. G., "Damage and Failure in Unidirectional Ceramic-Matrix Composites," *J. Am. Ceram. Soc.*, **75**[10] (1992), pp. 2719–25.
34) Meyer, L. J., Mackin, T. J., Mott, J., "Mechanical Properties of a Filament Wound Glass Fiber Reinforced Portland Cement Composite: Part I. Experiments," to be published in *J. Adv. Cement Based Comp.*

STUDY OF RESIDUAL STRESS IN SAPPHIRE-FIBER REINFORCED GLASS MATRIX COMPOSITE USING FLUORESCENCE SPECTRSOCOPY.

D.Banerjee, and Raj N. Singh, Department of Materials Science and Engineering, and H. Rho, and H. E. Jackson, Department of Physics, University of Cincinnati, Cincinnati, OH, 45221.

ABSTRACT

The well-known shift of optical fluorescence lines of chromium doped alumina or sapphire under pressure is used to measure residual stress on fibers in a sapphire-fiber reinforced glass matrix composite. Residual stress arising from thermal anisotropy is calculated from the observed fluorescence peak shifts of sapphire fiber. The stress is found to vary along the fiber length as well as across fiber cross-section in uncoated fiber embedded in a glass matrix. Modifying the surface of the fibers by applying a carbon coating, which is subsequently burnt out to simulate a weak interface, is observed to reduce the residual stress. No discernible trend in stress variation is observed in modified fibers. The high spatial resolution of the Raman microprobe is demonstrated to be a highly efficient tool to map stresses across a certain section of material.

INTRODUCTION

In recent years spectroscopic techniques have become very popular for measurement of stress and strains in structural ceramics and composites. The chief advantage of the spectroscopy is that, coupled with a microscope, it can achieve spatial resolution in microns. This high resolution enables one to map the stress and strain profiles across a certain area of material, which makes theoretical models more accurate and predictable. The two most important spectroscopic techniques used in this kind of study are Raman and fluorescence spectroscopies. Raman spectroscopy has been used [1-3] to study stresses in polymer matrix composites and in some cases to study polycrystalline ceramics. However, for many structural ceramic materials of importance, the Raman peaks are very weak. Fluorescence spectroscopy, on the other hand, is gaining importance, especially for the study of stress states in sapphire or alumina doped with chromium [4-13]. While the latter is an established structural ceramic, the former has potential to become one as fiber material in ceramic matrix composites.

Study of fluorescence spectra from ruby dates long back because of its use to calibrate pressure in diamond anvil cells. Feher and Sturge [14] reported the effect of stress parallel and perpendicular to the c-axis of alumina doped with Cr^{3+}, V^{2+}, and Mn^{4+} ions on the fluorescence peak shift. The shift was found to have a linear variation with stress. Grabner [4] was the first to derive a relationship between peak shift and applied stress in doped alumina. The shift in peak was related to the applied stress by the following equation:

$$\Delta\nu = (2\Pi_{11} + \Pi_{33})\,\frac{1}{3}\,(\sigma_{11} + \sigma_{22} + \sigma_{33}) + 2(\Pi_{33} - \Pi_{11})(a_{31}\,a_{32}\,\sigma_{12} + a_{31}\,a_{33}\,\sigma_{13} + a_{32}\,a_{33}\,\sigma_{23}) \quad (1)$$

where, $\Delta\nu$ is the shift in peak wave number, Π_{ij} are the piezospectroscopic coefficients, a_{ij} is the orthogonal transformation matrix from crystallographic frame to the reference frame of applied

To the extent authorized under the laws of the United States of America, all copyright interests in this publication are the property of The American Ceramic Society. Any duplication, reproduction, or republication of this publication or any part thereof, without the express written consent of The American Ceramic Society or fee paid to the Copyright Clearance Center, is prohibited.

stress. Ma and Clarke [5] developed a relationship for Cr fluorescence lines which is somewhat different from that by Grabner as shown below:

$$\Delta\nu = \Pi_{11} (\sigma_{11} + \sigma_{22} + \sigma_{33}) + (\Pi_{11} - \Pi_{22})(a_{31}^2 \sigma_{11} + a_{32}^2 \sigma_{22} + a_{33}^2 \sigma_{33}) + 2(\Pi_{33} - \Pi_{11})(a_{31} a_{32} \sigma_{12} + a_{31} a_{33} \sigma_{13} + a_{32} a_{33} \sigma_{23}) \quad (2)$$

The difference is in the additional second term. For the particular case of purely hydrostatic stress this equation simplifies to:

$$\Delta\nu = -(2\Pi_a + \Pi_c) P \quad (3)$$

where, a, and c stands for the respective axes and P is the applied pressure.

Use of fluorescence to measure stress in ceramic matrix composites was first attempted by Ma and Clarke. These authors employed a depth profiling method to measure the residual stress in sapphire fibers embedded in TiAl, and a polycrystalline alumina matrix [6], and a cubic zirconia matrix [7]. He and Clarke [8] calibrated the coefficients of the piezospectroscopic tensor for a 0.05% ruby for uniaxial compression along three principal crystallographic axes at room temperature. The coefficient along the m-direction was found to be different from that along the a-direction. Also, the coefficient along the a-direction was found to vary nonlinearly with stress. The results are explained by the breakdown of the rotational symmetry about the c-axis, because of the relative motion of the Cr^{3+} ion with respect to the oxygen ions. The piezospectroscopic tensors for R_1 and R_2 were also found to be different. Yang and Young [9] determined the dependence of the R_1 and R_2 lines on the fiber strain for alumina-zirconia fiber in a glass matrix. A linear relationship was found between the line shift and fiber strain. These authors calculated the residual strain on fibers in two different types of specimen – fully embedded fiber, and pullout specimen. The fiber ends were found to have a finite value of strain, and residual strain was found to increase from the two ends of the fiber over a certain length of the fiber until reaching a maximum value. From the strain profile measured, shear stress at the interface was calculated using the following relationship:

$$\tau = \frac{E_f r}{2} \frac{de}{dx} \quad (4)$$

where, E_f is the Young's modulus of the fiber, r is the fiber radius, and de/dx is the differential of fiber strain with respect to distance. The shear stress was found to have a peak value near the glass edge for the pullout specimen or at fiber end for the fully embedded fiber. Young and Yang [10] studied the interfacial micromechanics of a model composite consisting of alumina-zirconia fiber, both uncoated and coated with carbon, in a glass matrix using fluorescence spectroscopy. Specimens were subjected to four point bending, and it was found that when matrix strain was low, the fiber strain increases, but at a sufficiently high strain the interface failed and there was no further increase in the fiber strain. The coating led to interface failure at a lower matrix strain indicating that the coating weakened the interface. The maximum shear stress was also found to be lower for the coated fiber.

In this study residual stresses arising from thermal anisotropy in sapphire fiber embedded in a glass matrix have been measured. Unlike other studies, the stresses are measured along fiber length as well as across the fiber cross-section. Effect of a weak interface on the residual stress is also studied.

FUNDAMENTALS OF THE FLUORESCENCE TECHNIQUE

Fluorescence spectra originate from the electronic transition of impurity ions (e.g. Cr, V, Mn) in a crystal. The transition is strongly dependent on the symmetry of the surrounding crystal field. On application of stress, the crystal field around the ion changes and results in a shift in the peak position. The peak shift can be related to the applied stress using the following equation:

$$\Delta\nu = a_{ik}\, a_{ji}\, \Pi_{ij}\, \sigma_{ij}^{c} \qquad (5)$$

where, $\Delta\nu$ is the frequency shift, σ_{ij} is the applied stress, Π_{ij} is the piezospectroscopic coefficient, and a_{ij} is the transformation matrix relating the frame of reference in which the applied stress is defined with respect to the crystallographic axes of the crystal. Π_{ij} is a second rank tensor and due to symmetry of the stress tensor, the Π_{ij} is also symmetric.

In sapphire, the fluorescence arises from the electronic transition of impurity chromium ions (Cr^{3+}) between ^2E to $^4A_2$ states. The chromium ions substitute for aluminum ions in the lattice and are octahedrally coordinated to the neighboring oxygen and aluminum ions (Figure 1). The octahedron is slightly distorted and is elongated in the c direction of the crystal, resulting in splitting of the ^2E levels by 0.004 eV producing R_1 and R_2 lines at 1.790 and 1.794 eV respectively [15].

Except for triclinic and monoclinic systems, the Π matrix is diagonal in the orthogonal frame of reference. Frequency shift, therefore, is insensitive to pure shear stress, resulting in nullifying the off-diagonal terms in the stress matrix. Furthermore, if the crystal belongs to one of the tetragonal, hexagonal, or trigonal systems, $\Pi_{11} = \Pi_{22}$, e.g. in sapphire. Since the Cr^{3+} ion site has 3-fold rotational axis about the c-axis, $\Pi_{11} = \Pi_{22} \neq \Pi_{33}$. In Saphikon fiber the c-axis of the crystal is oriented along the fiber axis. For stresses parallel and perpendicular to the fiber axis, the axial and radial stresses can, therefore, be calculated using the following equations, which are derived from Eq. 2 after simplification [7]:

$$\Delta\nu_1 = 2\Pi_{r1}\,\sigma_r + \Pi_{z1}\,\sigma_z \qquad (6)$$

$$\Delta\nu_2 = 2\Pi_{r2}\,\sigma_r + \Pi_{z2}\,\sigma_z \qquad (7)$$

where, the subscript 1, and 2 refers to R_1 and R_2 lines respectively, and σ_r, σ_z are the radial and axial stresses on the fiber. Calculation of the residual stress, hence, requires a knowledge of the piezospectroscopic coefficients and the shifts for R_1 and R_2 peaks from fibers inside a matrix.

EXPERIMENTAL PROCEDURE

Processing the Composite

The composite system used for studying the residual stress was a sapphire fiber reinforced glass matrix composite. Properties of the glass and fiber are listed in Table I. Composite specimens were prepared following a tape casting, lamination, and sintering route. Glass powder was tape cast to 0.3 mm thickness. Tapes were cut to desired sizes and stacked one on top of the other to a certain thickness. An organic binder was used to keep the tapes in place. Fibers were sandwiched between two such stacks, and the whole unit was laminated at 105°C under pressure. The binder was burnt out by heating in air, and the specimen was then sintered in inert atmosphere at 1000°C. The sintered specimens were ground and polished to optical finish.

The carbon coating on the fiber was achieved by a dip coating technique. A polymeric resin derived from furfuryl alcohol ($C_5H_6O_2$) catalyzed by sulfuric acid is deposited on the fiber.

The fibers coated with resin were pyrolized, which left a coating of carbon on the fiber. Composite specimens with carbon coating were processed, and the coating was subsequently burnt out leaving an air gap between the fiber and matrix, to serve as a weak interface.

Experimental Detail

Two types of specimens were studied for residual stress measurement along the fiber axis:– uncoated fiber completely embedded in the matrix, and modified (i.e. burnt out coating) fiber reaching the ends of the matrix. The fibers lay close to the top surface for maximum clarity, but at a depth more than twice the fiber diameter to retain the residual stress. To measure stress along the radial direction, specimens were cut and polished perpendicular to the fiber length. The piezospectroscopic coefficient along the c axis for the fiber was determined by using a setup shown in Figure 2. The fiber was attached to the metal blocks, one of which was attached to a load cell and the other to a screw mechanism, which would pull or push the block as the screw was turned.

Fluorescence signals were obtained at room temperature in a backscattering geometry using a triple spectrometer in the subtractive mode. Incident light using an argon laser of wavelength 514.5 nm was focused on a surface of the sapphire fiber through an optical microscope with a probe size of 2 μm. Fluorescence spectra were recorded using a liquid nitrogen cooled

Table I.

	Source	Type	E (GPa)	Tensile Strength (Mpa)	Thermal Expn. Coeff. ($X10^{-6}/^{o}C$)
Sapphire	Saphikon[R]	Single crystal Alumina	414	2800-3400	9.0 – c-axis 8.5 – a-axis
Glass	Corning	Borosilicate	60	100	4.0

Charge Coupled Device (CCD). A computer controlled xy-stage moved along the fiber axis for an axial probing, and along the diameter for probing across the cross-section (Figure 3). An input power of 3 – 7 mW was used, and data collection time was kept at 5 – 10 seconds. All spectra were fitted with a Lorentzian shape.

RESULTS AND DISCUSSION:

Since the coefficient of thermal expansion of sapphire is greater than that of the glass in both axial and radial direction the residual stresses on fibers would be tensile in nature. Figure 4 shows the peak shift of uncoated fibers embedded in glass matrix as a function of fiber length. The shift in R_2 is always greater than that for R_1. However, the corresponding piezospectroscopic coefficients are also different. Peak shifts for a fiber subjected to axial tension is shown in Figure 5. The dependence of peak shift on applied stress deviates from linearity at low stress values as shown in Figure 5a, but for most part the variation is linear as shown in Figure 5b, which is a plot of the data in Figure 5a at moderate and high stress values. The piezospectroscopic coefficients along the axial direction calculated from the slope of the lines are:- Π_{z1} = 1.5 cm^{-1} GPa^{-1}, and Π_{z2} = 2.5 $cm^{-1}GPa^{-1}$. Due to experimental difficulties, the coefficients along the radial direction of the fiber couldn't be measured. For calculating the stresses, in this work, the values of the coefficients reported by Feher and Sturge [14] are used instead, which are: Π_{r1} = 3.2 $cm^{-1}GPa^{-1}$, Π_{r2} = 2.7 cm^{-1}

GPa^{-1}, Π_{z1} = 1.4 $cm^{-1}GPa^{-1}$, Π_{z2} = 2.15 $cm^{-1}GPa^{-1}$. Figure 6 shows the variation of stresses along the fiber length calculated from Eq.6 and 7. Radial stress remains more or less the same all through the length, but the axial stress increases gradually from the ends till it reaches a peak saturation value. In case of 7 mm and 2.1 mm long fibers the axial stresses are similar. But for the 0.9 mm long fiber the maximum value is a little less, because the fiber was smaller than the critical length for reaching the saturation stress. The critical length on each side of the fiber is estimated as approximately 0.7 mm. The stresses, as expected, are tensile in both axial and radial directions.

Figure 7 shows the peak shifts along the length of a surface modified (fiber initially coated with carbon and then the coating is burnt out) fiber in glass. The fiber in this case reached the end of the specimen unlike the specimens with uncoated fibers. Figure 8 shows the stresses calculated using Eq. 6 and 7. Radial stress is almost negligible, whereas the axial stress is almost an order of magnitude less than that in the uncoated fiber, indicating the effectiveness of the coating as a weak interface. The variation along the length is also much less abrupt as compared to an uncoated fiber.

Figure 9 shows the peak shifts when the fibers are probed along the cross-section. The peak shifts show opposite trends in uncoated and modified fibers, but the magnitude of the shifts are widely different in the two cases. For the uncoated fiber, the large shift is representative of the stress profile, but in the modified fiber, the shift is too small to derive any interpretation about the stress profile. Also, in modified fiber, the shift in R_1 is higher than that for R_2. R_1 shifts at a rate greater than R_2 for pure radial stress on a fiber i.e. Π_{r1} is greater than Π_{r2} (whereas Π_{z2} is greater than Π_{z1}). In this case, Figure 9b implies that axial and radial stresses are of similar magnitude. In uncoated fiber, however, the axial component of the residual stress is so much larger than the radial component that R_2 shifts more than R_1 even when probed along the fiber diameter. The calculated radial stress (Figure 10) is constant all across the cross-section, but the axial stress for uncoated fiber decreases gradually as a function of distance from the fiber end, reaches a minimum and then again increases to the maximum value at the fiber end, as expected from the peak shift. The maximum value of axial stress as a function of diameter across the cross-section is similar to the minimum value when probed along the length in a completely embedded fiber. This suggests that the axial stress measured along the length of the fiber is limited to the surface of the fiber. If this stress is a surface stress, it would decay as we go deeper inside the fiber till the effect from the other end takes over, which is exactly what we see in Figure10. The radial stress remains uniform all along the cross-section, which is expected for a uniform circumferential pull. For the coated fiber, the stresses are very small and the same trend as above is also exhibited here, though at a much smaller scale.

A theoretical model to estimate the residual stress for a similar situation is used to verify the validity of the stress values. In the existing models, one uncertainty is selecting the ΔT value as it is not known at what temperature stresses began to set in. A theoretical shear lag model by Budiansky et. al [15] is used here to verify the axial stress value as given below:

$$\sigma_{ax} = [\lambda_2' E_m E_f (1-V_f) \Delta\alpha \Delta T]/[\lambda_1' E_c (1-\nu)] \tag{8}$$

where, E_f, E_m, E_c are the elastic modulii of the fiber, matrix and the composite respectively, ν is the Poisson ratio of the matrix and ν_f is that of the fiber, $\Delta\alpha$ is the thermal expansion mismatch, and,

$\lambda_1' = 1- [(1-2\nu)(E_f - E_c)/2E_f(1-\nu)]$
$\lambda_2' = (E_f + E_c)/2E_f$.

Fiber volume fraction in the specimens studied was negligible so that, $E_c \approx E_m$, and $V_f \approx 0$. ΔT is taken as 750^0C, which means stresses began to set in from the softening point of glass (~ 800^0C), and ν is taken as 0.2. The remaining values are taken from Table I. Axial stress is calculated as 1080 MPa, which is similar to the values measured.

The radial stress may be calculated using the following relationship:

$$\sigma_r = -E_f (\alpha_m - \alpha_f) \Delta T [(1/(1+ E_f / E_m))] \quad (9)$$

which gives the radial stress to be 135 MPa, which is also very close to the measured values.

CONCLUSION:

Fluorescence spectroscopy is found to be capable of characterizing the residual stresses on fibers in a model ceramic matrix composite. The piezospectroscopic coefficients of Saphikon fiber along the axial direction are measured. Fiber ends are observed to have finite amount of residual stress. The axial component of the residual stress on fibers arising from thermal anisotropy is observed to vary along the fiber length, as well as along the fiber cross section. The axial component is found to be maximum at the fiber surface, decreasing gradually as a function of fiber diameter until reaching a minimum. The radial component of stress remains more or less the same along the length as well as across the cross-section. The dramatic effect of coating on reducing the residual stress on fibers is demonstrated. The measured stress values are in good agreement with the values calculated from theoretical models.

REFERENCE:

1. M. R. Gallas, Y. C. Chu and G. J. Piermarini, "Calibration of the Raman Effect in α-Al_2O_3 Ceramic for Residual Stress Measurement," *Journal of Materials Research,* **10** [11] 2817-22 (1995).
2. N. Sato, N. Tatsuda and T. Kurauchi, "Application of Laser Raman Spectroscopy to the Analysis of Stress Distribution of Fibers in Composites," *Journal of Materials Science Letters*, **11** 365-66 (1992).
3. J. D. Belnap, J. F. Tsai and D. K. Shetty, "Direct Measurement of Crack Shielding in Ceramics by the Application of Raman Microprobe Spectroscopy," *Journal of Materials Research*, **9** [12] 3183-93 (1994).
4. L. Grabner, "Spectroscopic Technique for the Measurement of Residual Stress in Polycrystalline Alumina," *Journal of Applied Physics*, **49** [2] 580-583 (1978).
5. Q. Ma and D. R. Clarke, "Stress Measurement in Single-Crystal and Polycrystalline Ceramics Using Their Optical Fluorescence," *Journal of the American Ceramic Society,* **76** [6] 1433-40 (1993).
6. Q. Ma and D. R. Clarke, "Measurement of Residual Stresses in Sapphire Fiber Composites Using Optical Fluorescence," *Acta Metallurgica et Materiala,*" **41** [6] 1817-23 (1993).
7. Q. Ma and D. R. Clarke, "Stress Measurement Using Optical Fluorescence"; pp 27-37 in *Experiments in Smart Materials and Structures*, Edited by K. S. Kim, AMD-Vol. 181, Book No. H00888 (1993).
8. J. He and D. R. Clarke, "Determination of the Piezospectroscopic coefficients for Chromium Doped Sapphire," *Journal of the American Ceramic Society,* **78** [5] 1347-53 (1995).

9. R. J. Young and X. Yang, "Interfacial Failure in Ceramic Fiber/Glass Composites," *Composites Part A*, **27A** 737-41 (1996).
10. X. Yang and R. J. Young, "Determination of Residual Strains in Ceramic-Fiber Reinforced Composites Using Fluorescence Spectroscopy," *Acta Metallurgica et Materiala*," **43** [6] 2407-16 (1995).
11. S. E. Molis and D. R. Clarke, "Measurement of Stresses Using Fluorescence in an Optical Microprobe: Stresses around Indentations in a Chromium Doped Sapphire," *Journal of the American Ceramic Society,* **73** [11] 3189-94 (1990).
12. Q. Ma and D. R. Clarke, "Piezospectroscopic Determination of Residual Stresses in Polycrystalline Alumina," *Journal of the American Ceramic Society,* **77** [2] 298-302 (1994).
13. Q. Ma, M. C. Shaw, M. Y. He, B. J. Dalgleish, D. R. Clarke, and A. G. Evans, "Stress Redistribution in Ceramic/Metal Multilayers Containing Cracks," *Acta Metallurgica et Materiala*," **43** [6] 2137-42 (1995).
14. E. Feher and M. D. Sturge, "Effect of Stress on the Trigonal Splitting of d^3 Ions in Sapphire (α-Al_2O_3)," *Physical Review*, **172** [2] 244-249 (1968).
15. J. H. Eggert, K. A. Goettel, and I. F. Silvera, "Ruby at High Pressure I. Optical Line Shifts to 156 Gpa," *Physical Review*, **40** [8] 5724-32 (1989).
16. B. Budiansky, J. W. Hutchinson and A. G. Evans, "Matrix Fracture in Fiber Reinforced Ceramics," *Journal of Mechanics Physics and Solids*," **34** [2] 164-89 (1986).

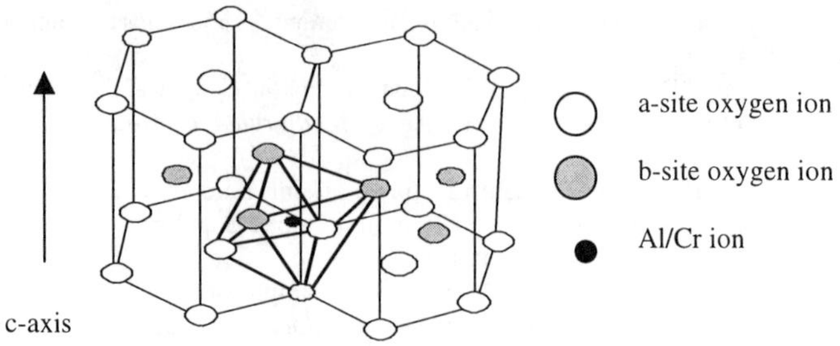

Figure 1. Structure of alumina

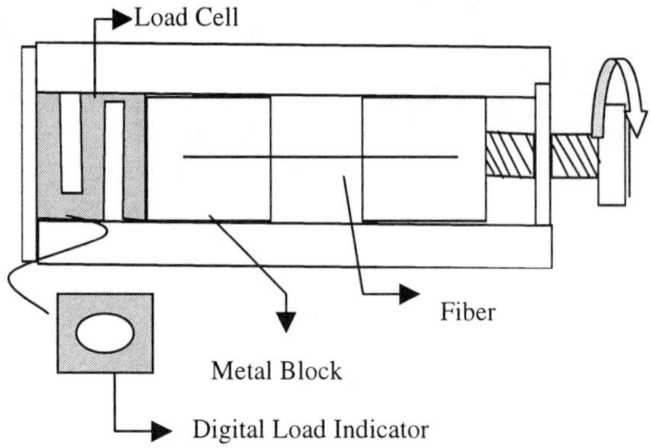

Figure 2. Fixture for pulling fiber along the axis.

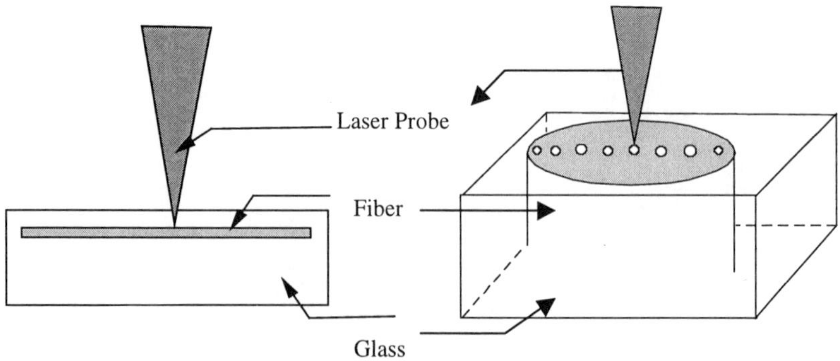

Figure 3. Schematic of fiber probing geometry.

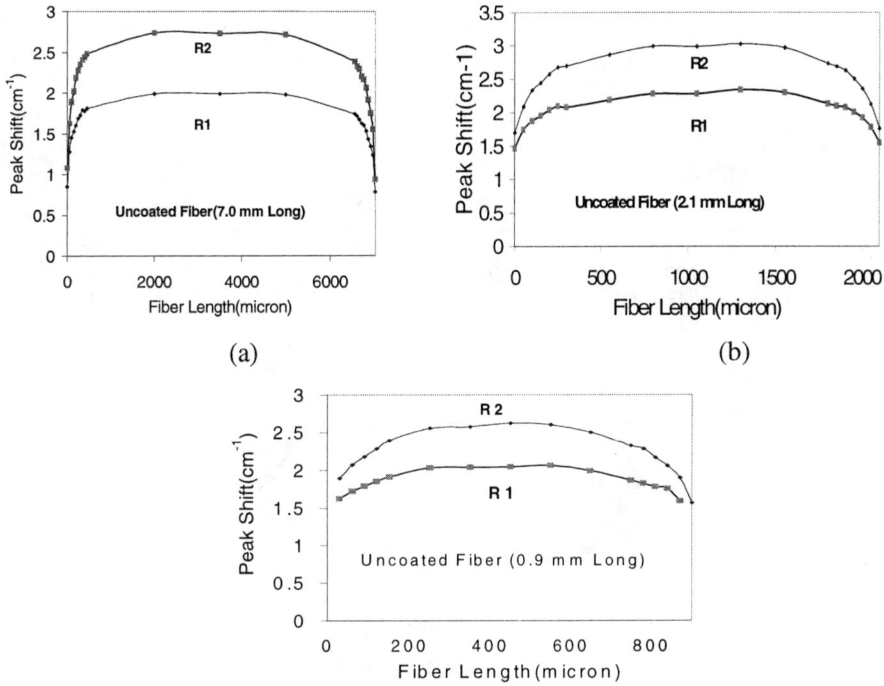

(c)

Figure 4. Peak shifts along the lengths of uncoated fibers embedded in a glass matrix.

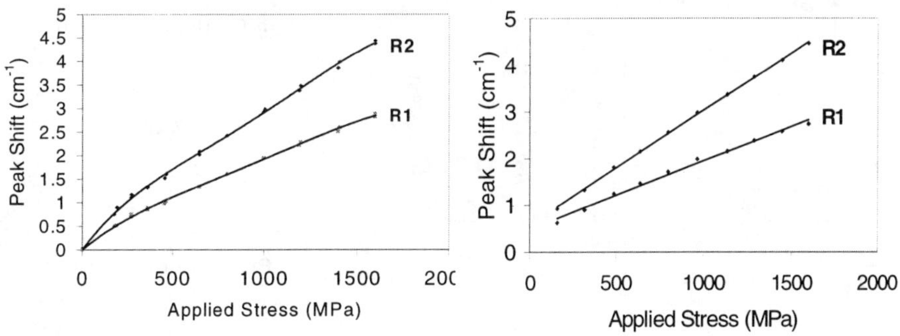

(a) (b)

Figure 5. Peak shifts in a sapphire fiber subjected to axial tension.

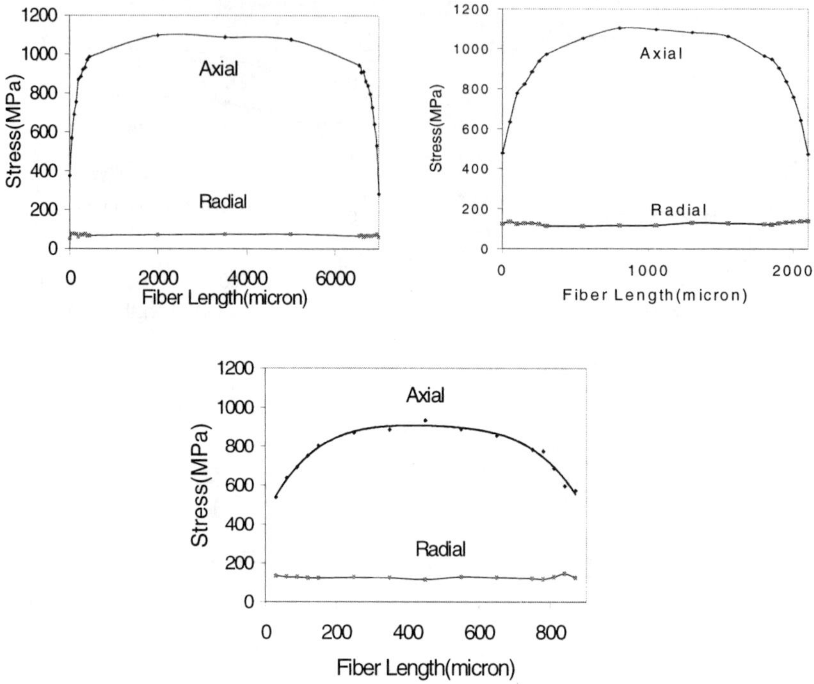

Figure 6. Residual stress calculated for uncoated fiber in glass.

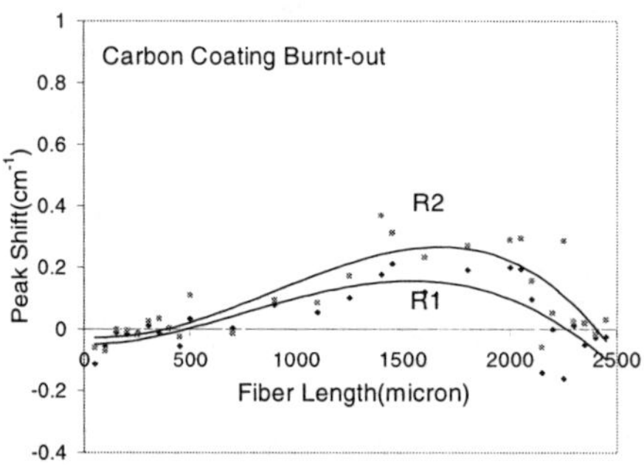

Figure 7. Peak shift along the length of modified fiber.

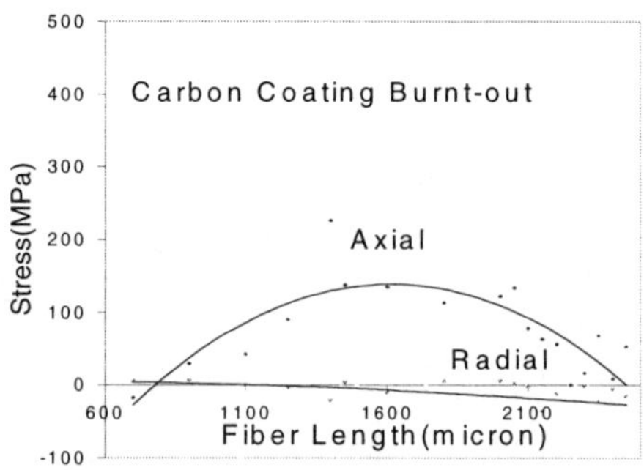

Figure 8. Calculated stresses on modified fiber.

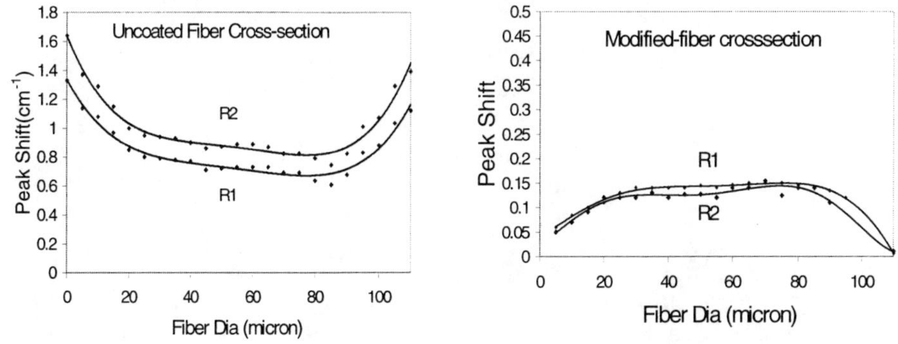

Figure 9. Peak shifts across cross-section for (a) uncoated fiber, and (b) modified fiber.

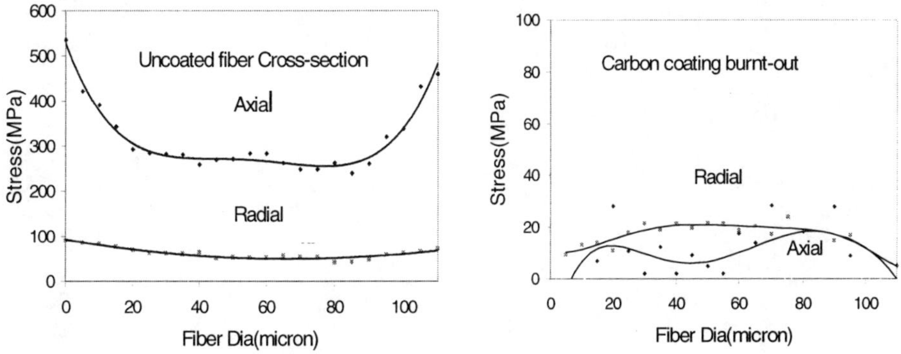

Figure 10. Calculated stresses for (a) uncoated fiber, and (b) modified fiber.

A NOVEL APPROACH FOR THE FIBER BRIDGING ANALYSIS IN CERAMIC MATRIX COMPOSITES

Yongjian Sun and Raj N. Singh
Department of Materials Science and Engineering
University of Cincinnati, P. O. Box 210012
Cincinnati, Ohio 45221-0012

ABSTRACT

Debond Length Measurement (DLM) is introduced as a new approach to measure the fiber bridging stress in ceramic matrix composites. In situ measurements of the debond length in fiber-bridged cracks are performed in a transparent glass matrix composite. From the debond length profile along fibers bridging a matrix crack, the fiber bridging stresses are determined based on both the force and energy balance models. As an independent approach, the crack opening displacements (COD) of the matrix cracks are measured in situ, from which the fiber bridging stresses are also calculated. The distribution of fiber bridging stresses determined from DLM is compared with the results obtained from COD measurement. The results from DLM and COD techniques are found to be in good agreement.

INTRODUCTION

The most important advantage of fiber-reinforced ceramic matrix composites (FCMC) over the monolithic ceramics is their higher fracture toughness. The greatly enhanced toughness of FCMC is largely governed by the bridging (traction) behavior of fibers at the cracked matrix surfaces [1, 2] which is also associated with the processes of fiber-matrix interfacial debonding and sliding [3, 4]. The fracture toughness introduced by fiber traction is dependent on the fiber bridging stress distribution T(x) [2]. Therefore, the determination of in situ distribution of the fiber bridging stress T(x) along a matrix crack is very important for the evaluation of the toughness.

To the extent authorized under the laws of the United States of America, all copyright interests in this publication are the property of The American Ceramic Society. Any duplication, reproduction, or republication of this publication or any part thereof, without the express written consent of The American Ceramic Society or fee paid to the Copyright Clearance Center, is prohibited.

A conventional method for determining the fiber bridging stress is the crack opening displacement (COD) measurement. The relationship between T(x) and u(x) was expressed by Marshall and Cox (MC [2]) as,

$$T(x) = 2\left[u(x)\tau_f V_f^2 E_f (1+\eta)^2 / r\right]^{1/2} \tag{1}$$

where η is a constant which is equal to $E_f V_f / E_m V_m$, V and E denote the volume fraction and elastic modulus, respectively, the subscripts m, f, and c of V and E denote the matrix, fiber, and composite, respectively, τ_f is the interfacial frictional shear stress at the fiber/matrix interface, and r is the radius of the fiber. Using this technique, the COD profile u(x) can be measured and related to the fiber bridging stress profile. The fiber bridging toughness can be calculated using the following expression [1],

$$G = \int T(u)du \tag{2}$$

where T(u) is the fiber bridging stress as a function of the COD u(x).

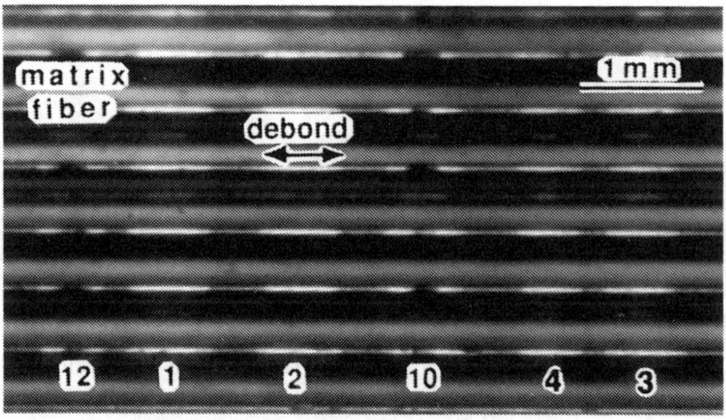

Figure 1. An optical micrograph showing the matrix cracking and interfacial debonding on the tensile surface of a transparent composite

The debond length is another physical parameter which is more directly associated with both the process of fiber bridging and interfacial debonding/sliding.

The relationships among the externally applied stress, interfacial properties, and debond length were derived from both the force [3] and energy balance [4] approaches. However, these relationships had not been experimentally examined in FCMCs until the recent work done by Sun and Singh [5, 6]. In this study, the fiber-matrix interfacial debonding phenomenon was directly observed as indicated by white bands with matrix cracks running across in the middle of these bands and is shown in Fig. 1 [5,6]. The observation and measurement of debond length at the fiber/matrix interface of this study was facilitated by carefully fabricating a transparent composite system [7]. Using this new approach, the experimental relationship between the debond length and applied stress was obtained as shown in Fig. 2 for the first time.

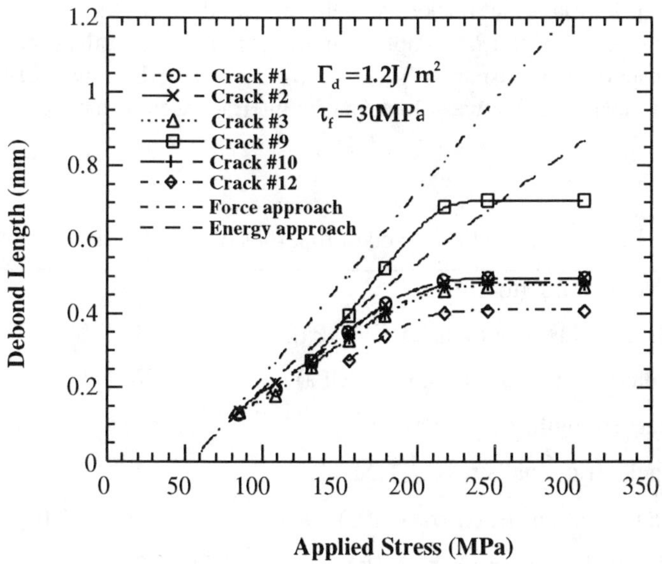

Figure 2. The experimental and theoretical relationships between the Debond length and applied stress.

This research focuses on the development of a novel technique for measuring the fiber bridging stress by debond length measurement (DLM). Using this technique and comparing it with the conventional COD method, the fiber bridging stress distribution along a matrix crack will be determined, and a new bridging law will also be established.

EXPERIMENTAL PROCEDURE

Borosilicate F glass was used as a matrix material, and SiC SCS-6 fibers were used as a reinforcement material. These materials were selected to have similar coefficient of thermal expansion (CTE) in order to minimize the thermal residual stresses upon composite fabrication. Transparent glass composites with a close-to-full density were fabricated by Tape Casting/Binary Sintering (TCBS) method [7]. The composites made for this study had 7% fibers by volume. After grinding, cutting and polishing, the specimens with dimensions of 1.95x5.0x50 mm^3 were prepared for four point flexure tests. The DLM and COD measurements were performed in situ on the same specimen at the same time. Thin slices of sample with 0.5-1.0mm in thickness were prepared from the same batch of composites as the COD/DLM samples for measuring interfacial properties. The interfacial properties, as listed in Table 1, were characterized by fiber pushout tests. The composites with a weak fiber/matrix interface were achieved.

Table 1. Mechanical and Interfacial Properties of SiC_f-Glass Composites

Property	Value
Fiber volume fraction, V_f	7%
Elastic modulus of composite, E_c (GPa)	80 [7]
First matrix cracking stress, σ_{cr} (MPa)	74 [7]
Ultimate strength, σ_u (MPa)	273 [7]
Interfacial frictional stress, τ_f (MPa)	7.5 ± 1.5 [6]
Debond initiation stress, σ_d (MPa)	16 ± 2.1 [6]
Interfacial shear strength, τ_f (MPa)	20 ± 3 [6]
Debond energy, Γ_d (J/m^2)	0.85 ± 0.2 [6]

COD Measurement

The COD measurements were made on a composite loaded in the four point bending mode using a micro measure machine equipped with CCD camera as shown in Fig. 3. The highest magnification of this microscope is 2000X. An image freezing function was used to achieve the highest resolution of 0.4 μm. The COD images were recorded in a computer with respect to the position of each measurement. The crack length was also measured at a certain load/stress level using a lower magnification of about10x. The COD measurements were made under the same load/stress level as the DLM in order to obtain data under identical conditions.

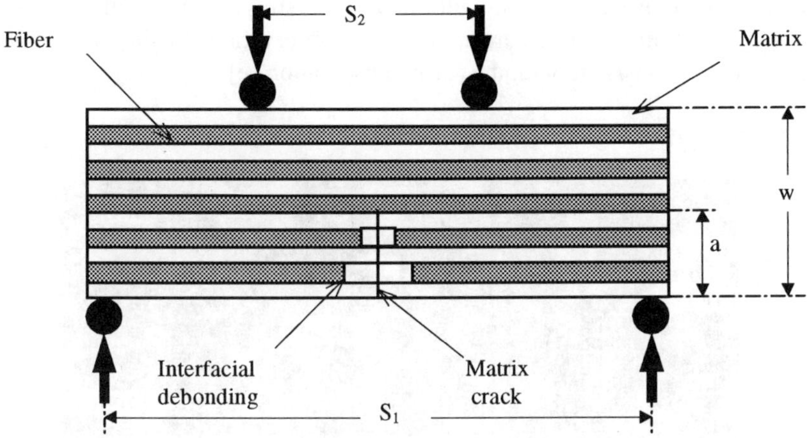

Figure 3. The schematic diagram showing specimen configuration for debond length and crack opening displacement measurements.

Debond Length Measurement (DLM)

The debond length measurements were made on the same sample at the same time as COD was measured. Matrix cracking occurred on the tensile surface of a flexure bar when the applied load increased beyond the matrix cracking stress. Accompanying matrix cracking, interfacial debonding was also created because of the weakly-bonded interface. The debond length was measured in situ using the photographing and video recording methods without unloading the specimen. The magnitude of the debond length was taken as half of the length of the white band in Fig. 1. The debond length for each fiber layer along a matrix crack was obtained with respect to the fiber positions along the crack path. Therefore, both the debond length and COD profiles were obtained for each matrix crack under the same load/stress level.

RESULTS AND DISCUSSION

Debond Length and COD Profiles

The debond profile along the matrix crack propagation direction is observed for the first time in this study. Figure 4 is an optical micrograph from a sample loaded to 135N or 85MPa stress. The debond length profiles of the four different matrix cracks are plotted in Fig. 5. The variation in the debond length values is found to be small because of the uniformity of the stress distribution created upon loading in the four point flexure mode. This observation on the debond length measurement is consistent with the earlier observation [6].

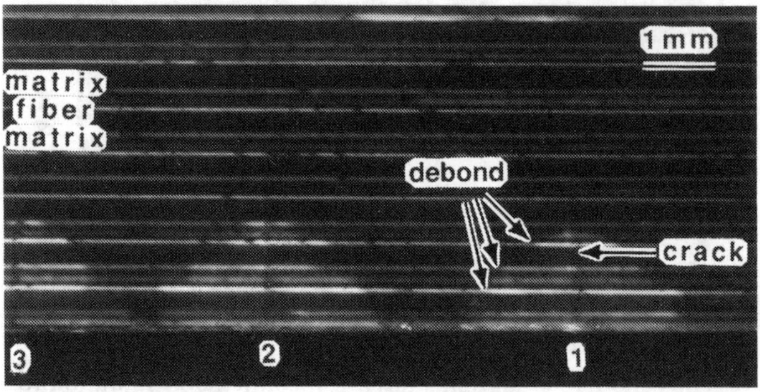

Figure 4. An optical micrograph of the interfacial debond profile in a transparent composite subjected to a force of 135N in the four point flexure mode.

The COD profiles measured for the four matrix cracks are shown in Fig. 6. The crack lengths are also measured and found to be very close to each other, as shown in Fig. 6. At a stress of 85MPa, the matrix cracks do not reach the neutral axis of the flexure specimen, which is about 2.5mm from the ends of the sample.

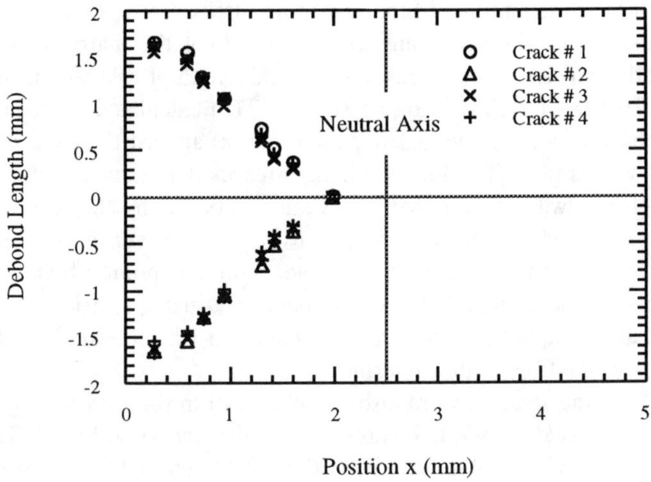

Figure 5. The debond length profiles for the four matrix cracks subjected to a force of 135N.

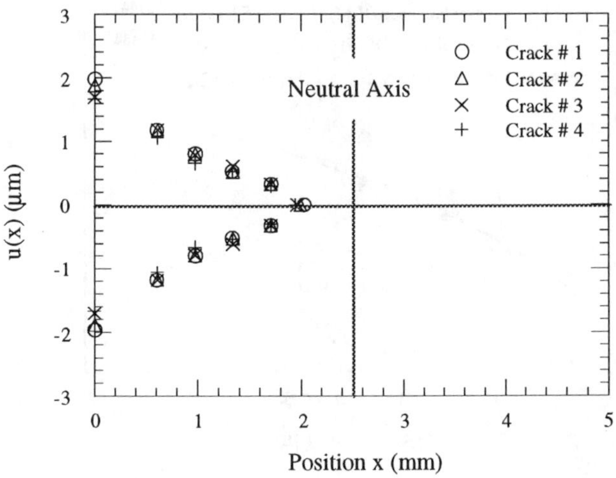

Figure 6. The COD profiles for the four matrix cracks subjected to a force of 135N.

Determination of the Distribution of Fiber Bridging Stress

The distribution of fiber bridging stress T(x) along the matrix crack #2 is calculated from the experimentally measured debond length profile and using DLM models. Figure 7 shows the fiber bridging stresses T(x) calculated from the DLM (both the force balance and energy balance approaches) against the experimentally measured COD values u(x). The fiber bridging stresses determined by the energy balance approach is always larger than the values obtained by the force balance approach. This result is not only expected from the difference in their fiber bridging stress expressions but also reasonable from the point of view of the fracture mechanics. The energy balance approach is usually a stricter approach than the force balance approach for crack propagation (the interfacial debond is one kind of crack at the fiber/matrix interface).

The fiber bridging stresses were also calculated from the measured COD for crack # 2 using Eq. 1 (MC Model). The results are also shown in Fig. 7. The fiber bridging stress distributions obtained from the DLM and COD measurements displayed the same trend. But, the differences in the fiber bridging stress values between DLM and COD measurements still cannot be ignored.

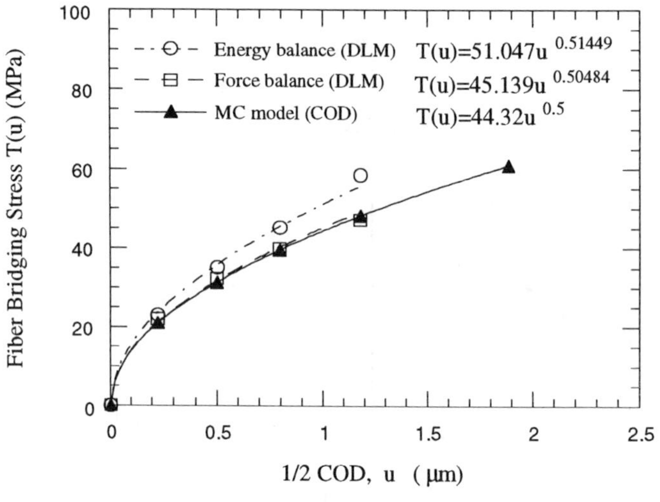

Figure 7. A comparison of the fiber bridging stresses calculated from the DLM and COD measurements for Crack # 2 subjected to a force of 135N.

Curve fitting of the data of T(x) vs u(x) using a simple power law was done which gave an exponent value of 0.514 for the energy balance approach and 0.505 for the force balance approach, which are close to 0.5 for MC [2] (COD). The fiber bridging stress determined from the COD is very close to the stress values obtained from the force balance approach of the DLM. This is because the COD measurement relies on the same force balance criterion as is used for the force balance approach of the DLM except for an extra stress term, σ_d, which become insignificant with the increase of debond length and COD.

SUMMARY

A new approach for determining the distribution of fiber bridging stress based on the debond length measurement (DLM) is developed. The debond length profiles along fibers bridging a matrix crack are observed and measured in situ in this study. The fiber bridging stresses are calculated from the debond lengths using both the force balance and energy balance models. The fiber bridging stresses are also determined from the independently measured COD. Both the DLM and COD measurements provided similar estimates of the distribution of fiber bridging stresses in this study.

The discrepancies in the measured values of the fiber bridging stresses between the DLM and COD methods are also found. It is seen that the MC model of COD measurement is acceptable in the case of a very weak interface. But, for a stronger interface, it is expected that the energy balance approach of DLM should provide a better estimate of the fiber bridging stress.

ACKNOWLEDGMENT

Authors are grateful to Dr. James Webb for designing the flexure fixture. This research was supported by National Science Foundation.

REFERENCES

[1]A.G. Evans, "Perspective on the Development of High-Toughness Ceramics", *J. Am. Ceram. Soc.* **73** [2] 187-206, (1990).

[2]D.B. Marshall and B.N. Cox, "A J-Integral Method for Calculating Steady-State Matrix Cracking Stresses in Composites", *Mechanics of Materials* **7**, 127-33 (1988).

[3]B. Budiansky, A.G. Evans and J.W. Hutchinson, "Fiber-Matrix Debonding Effects on Cracking Aligned Fiber Ceramic Composites", *Int. J. Solids Structures* **32** [3/4], 315-28 (1995).

[4]S. Li, S.P. Shah, Z. Li and T. Mura, "Micromechanical Analysis of Multiple Fracture and Evaluation of Debonding Behavior for Fiber-Reinforced Composites", *Int. J. Solids Structures* **30** [11], 1429-59 (1993).

[5]Y. Sun and R.N. Singh, "The Generation of Multiple Matrix Cracking and Fiber-Matrix Interfacial Debonding in a Glass Composite", *Acta Metall. Mater.* **65**, 1657-1673 (1998)

[6]Y. Sun and R.N. Singh, "A Novel Approach for Determining Fiber Bridging Stress in Ceramic Matrix Composites by Debond Length Measurement", to be submitted for publication.

[7]Y. Sun and R.N. Singh, "The Fabrication of a Toughened and Transparent Glass Composite", *Advances in Ceramic-Matrix composites III, Ceramic Transactions* **74**, 141-151 (1996).

A 2-D MODEL FOR CRACK DEFLECTION IN THERMOELASTICALLY LOADED INCLUSION CONTAINING CERAMICS

Marc Genevrier and Alain Mocellin
LSG2M, UMR 7584 - Ecole des Mines
54000 NANCY, France

ABSTRACT

Toughening via crack deflection by second phase particles or fibers has attracted much attention for a number of years. Experimental results recorded with dispersion reinforced ceramic/ceramic composites however have remained below theoretical expectations. The latter were for the most based on a priori assumed crack paths. In this paper such a restriction is relaxed. A single elliptical inclusion separated from the matrix by a thin interphase film is considered. A crack under mode I loading is assumed to propagate toward the inclusion quasi-statically. The local stress state resulting from thermoelastic mismatches and external loading is first calculated by the method of complex potentials. Then the trajectories of cracks in such an environment, are determined under the assumption that their tips are assigned to follow lines of zero shear stress. It is argued in conclusion that crack deflection in itself can only provide modest toughening in ceramic composites.

INTRODUCTION

Crack deflection by second phase particles or fibers has long been recognized as a possible toughening mechanism in brittle matrix composite materials. Attempts at quantifying the effect usually refer to the classical paper by Faber and Evans [1] and refinements brought by subsequent anthors [e.g. 2,3] An important feature of such models is that some geometry is a priori postulated for the crack as a sequence of links from one inclusion to another along the main

To the extent authorized under the laws of the United States of America, all copyright interests in this publication are the property of The American Ceramic Society. Any duplication, reproduction, or republication of this publication or any part thereof, without the express written consent of The American Ceramic Society or fee paid to the Copyright Clearance Center, is prohibited.

direction of fracture, and calculations are carried out therefrom. Such an approach, although not beyond criticism in itself, may be useful to model composites when the matrix is more dilatant than inclusions (e.g. Al_2O_3 reinforced with SiC) in which the latter are reached by the propagating crack. It is however unsatisfactory in reverse cases, i.e when the matrix thermal expansion is less than that of the inclusions and that cracks therefore run in the matrix only. Examples of such systems may be found in the literature [e.g.4]. On the other hand, making a priori assumptions on the crack path in a heterogeneous medium may overlook the contributions of residual stress fields, such as arising from thermal expansion differences from inclusions and matrix. Such residual stresses in themselves may also yield some toughening. Models to quantify such an effect have been devised, again with some simplifying assumptions, for example that the residual stress field may be approximated as being periodical [5]. More recently some refinements have also been brought to such models [e.g. 6, 7] making more explicit use of the Eshelby formalism [8] and accounting for finite volume fractions of the particulates. They have generally found encouraging agreement with experimental measurements. But to the present authors' knowledge most if not all such theoretical models have not attempted in explicit detail to either predict trajectories to be followed by cracks in such heterogneous stress fields or confirm that they would run as a priori hypothesized.

In a neighbouring field of research however such an issue has been considered for some time (see e.g. [9] and references therein). In particular several examples of elastic interactions between a (macro)-crack with (micro)-defects such as circular hole, dilational inclusion or microcrack have been treated [10]. To that end the stress fields were represented by complex potentials ϕ (z) and Ψ (z) [11] which of course restricted the applicability of the argument to 2D geometries only. But among others such an approach also has the merit of bringing to light that crack deflection per se does not only and/or always lead to significant toughening in the corresponding medium [12]. In the present paper the method of complex potentials [11] is also used to investigate some aspects of crack deflection in two dimensions which do not seem to have attracted much theoretical attention so far. Our argument will be based on some rather drastic assumptions. It is nonetheless hoped that some of the findings may be of practical interest, at least qualitatively. The presentation consists of two parts :

- the determination of the residual stress state around a single inclusion within a remotely loaded matrix (only the case of a more dilational inclusion will be considered).
- the derivation of crack paths in the matrix under the simple minded assumption that crack tips are assigned to follow lines of zero shear stress. In conclusion comments will be made on the amount of toughening which may result in such a context. Because of space limitations only the principles and major steps of the derivation are presented here. More detail can be found in reference [13] which can be made available.

THE RESIDUAL STRESS STATE

Figure 1 depicts the geometry and loading which are considered here : an elliptical inclusion (major and minor axes : a and b respectively) at angle α with the direction of remote loading of the matrix. A thin continuous film of a third phase is assumed to separate the inclusion from the matrix. Two cartesian coordinate systems with origins at the inclusion center will be used, Ox_0y_0 related to the direction of matrix load (having intensity σ^a), and Oxy bound to the inclusion geometry respectively. A crack initiating far from the inclusion and running parallel to and at a distance X_i from Oy_0 will eventually near the inclusion.

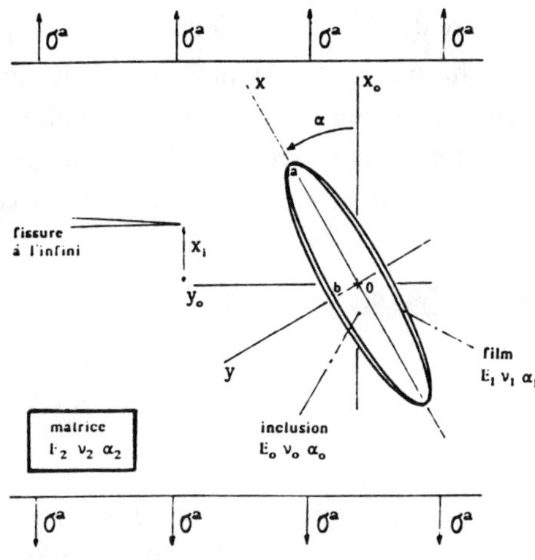

Figure 1 - Assumed geometry of a sheathed elliptical inclusion being approached by a crack in an infinite 2-D matrix under Mode I load.

The elastic and thermal expansion properties of the three media will be assumed to be constants noted E_i, ν_i and α_i respectively, where i = 0, 1, 2 refer to inclusion, film and matrix respectively. Also used in equations to follow will be ; $\mu = \frac{E}{2(1+\nu)}$ and $\kappa = 3 - 4\nu$ or $\kappa = \frac{3-\nu}{1+\nu}$ for plane strain or plane stress conditions respectively.

Assume now that cooling from a high temperature stress-free condition to a lower temperature ΔT apart has produced a residual stress field due to thermoelastic mismatches, and to which σ^a may also be applied. When the complex variable z = x + iy is used to represent positions in the inclusion cordinate system, it is known that the state of stress wihin the system can be determined from the Kolosov-Muskhelishvili potentials $\phi(z)$ and $\Psi(z)$ as :

$$\sigma_x + \sigma_y = 4 \cdot \mathrm{Re}\,(\phi'(z))$$

$$\sigma_y - \sigma_x + 2i\,\tau_{xy} = 2\,[\bar{z}\,\phi''(z) + \Psi'(z)] \qquad (1)$$

$$2\mu\,(u + iv) = \kappa\,\phi(z) - z\,\overline{\phi'(z)} - \overline{\Psi'(z)}'$$

Where u and v are displacements along x and y respectively and Re (z), $\bar{z}$ and f ' (z) represent the real part, complex conjugate and derivative of f with respect to z respectively. Three such potential couples pertaining to the three media in presence obviously are needed to solve the problem. When assuming ideally rigid interfaces, continuity of deformations between matrix and film requires :

$$\Phi_1(z) + z\,\Phi_1'(z) + \Psi_1(z) = \Phi_2(z) + z\,\Phi_2'(z) + \Psi_2(z)$$

$$\mu_2/\mu_1\,[\kappa_1\,\Phi_1(z) - z\,\Phi_1'(z) - \Psi_1'(z)] + 2\mu_2\,(\alpha_1 - \alpha_2)\,\Delta T \cdot z = \qquad (2)$$

$$\kappa_2\,\Phi_2(z) - z\,\Phi_2'(z) - \Psi_2'(z)$$

and a similar system (3) not explicited here between film and inclusion. Other conditions are : the stress tensor applied to the matrix from infinity (i.e.σ^a) is known, and within the inclusion the stress state is uniform. Thus :

$$\Phi_2(z) = 1/4\,(\sigma_x^a + \sigma_y^a)\cdot z + O(1/z)$$

$$\Psi_2(z) = 1/2\,(\sigma_y^a - \sigma_x^a + 2i\tau_{xy}^a)\cdot z + O(1/z) \qquad (4)$$

$$\Phi_0(z) = 1/4\,(\sigma_x^o + \sigma_y^o + i\Omega)\cdot z$$

$$\Psi_0(z) = 1/2\,(\sigma_y^o - \sigma_x^o + 2i\tau_{xy}^o)\cdot z \qquad (5)$$

With Ω a constant to be detrmined and

$$\sigma_x^a = \frac{\sigma^a}{2}(1 + \mathrm{Cos}\,2\alpha) \quad ; \quad \sigma_y^a = \frac{\sigma^a}{2}(1 - \mathrm{Cos}\,2\alpha) \quad ; \quad \tau_{xy}^a = -\frac{\sigma^a}{2}\sin 2\alpha$$

Now the elliptical geometry can be transformed into a circular one via the following complex variable changes. Letting first : $R_i = \dfrac{a_i + b_i}{2}$ and $m_i = \dfrac{a_i - b_i}{a_i + b_i}$, where a_i and b_i are the major and minor elliptical axes respectively (i = o for the inclusion/film, i = 1 for the film/matrix interfaces) together with the condition : $R_1^2\, m_1 = R_o^2\, m_o$*. New complex variables ξ and ζ now can be introduced such that $z = R_o(\xi + m_o / \xi)$; $Z = R_1(\zeta + m_1 / \zeta)$, and with :

$\xi = \dfrac{R_1}{R_0}\zeta$, and $|\xi| = 1, |\zeta| = 1$ the equations of the inclusion /film and film/matrix interfaces respectively.

The resolution of the system of Equations (1) to (5) is now made easier in the ζ plane with polar coordinates : $\zeta = re^{i\varphi}$ (r and φ obviously are not polar coordinates in the Z plane of Figure 1). The analytical resolution procedure is analogous to that developed by Cherepanov [14] and is relatively straight forward, although rather tedious. First Φ_1 and Ψ_1 are obtained from (3) and (5) as functions of ξ, then Φ_2 and Ψ_2 from (2) as functions of ζ and Ω. The latter is then identified after dividing by z and taking the limit at infinity. Finally Φ_0 and Ψ_0 are obtained, bearing in mind that whenever possible, rational functions of Z are expanded in series, and only the first (i.e. constant) and second (i.e. $1/r^2$) terms are retained. It

---

* This condition prevents the film to be of uniform tickness in the initial geometry of Figure 1. But with such definitions both the inclusion and film are transformed into circles in the ξ and ζ complex planes respectively, wich simplifies the mathematics somehow.

may be noted that the $1/r^2$ dependance of stresses in the matrix is to be expected in two-dimensional problems [15] in contrast with $1/r^3$ in 3-D.

Figure 2 shows an example mapping of iso stress lines relative to σ^a computed with the above scheme, for the case of a sapphirine inclusion in a cordierite matrix with a presumably vitreous film of thickness f ≈ 4% of the mean inclusion size [13].

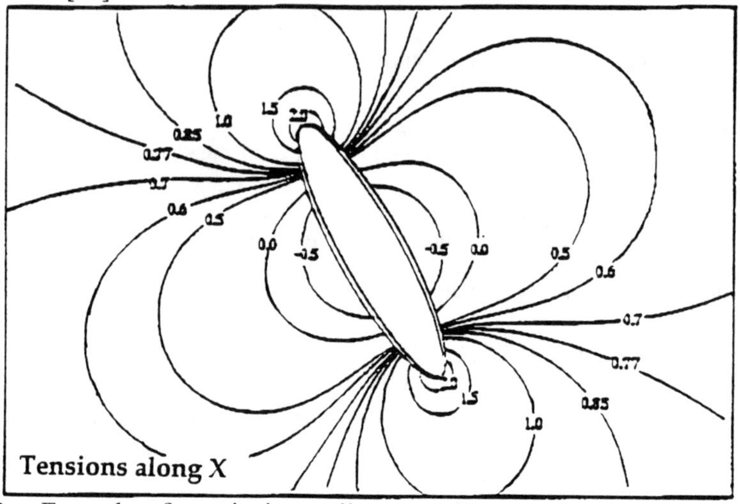

Figure 2 - Example of matrix isotensile stress lines as fractions of σ^a for a sapphirine inclusion in cordierite with inclusion angle with $\sigma^a = \alpha = 30°$, ΔT = 800 K and σ^a = 200 Mpa. Arbitrarily selected film characteristics where : E_l = 70 GPa, $\nu_l = 0.2$ and $\alpha_l = 5 \times 10^{-6}\ K^{-1}$

PREDICTION OF CRACK PATHS

When a crack is introduced as shown on Figure 1 it may be felt not unreasonable, although it has been shown not strictly correct [9], that such a crack will propagate so as to satisfy K_{II} = O at its tip for every position of the latter. In particular such a condition amounts to assuming propagation in a quasi static fashion with no concurrent stress relation.

The calculation is more conveniently carried out in the (O ; ζ) coordinate system before returning to the initial (O ; z) reference coordinates. Thus if $r = r(\varphi)$ represents the parametric equation positioning the crack tip P, the condition of no shear at P implies :

$$-(\sigma_{rr}-\sigma_{\varphi\varphi})\sin 2\vartheta - 2\tau_{r\varphi}\cos 2\vartheta = 0 \tag{6}$$

where ϑ is the angle between the normal to OP and the tangent in P to the crack trajectory :

$$t\,g\theta = \frac{1}{r}\cdot\frac{dr}{d\varphi} = \frac{r'}{r}$$

Substituting in (6) and making use of expressions previously established yields finally :

$$2\frac{r'}{r} = \left[1-\left(\frac{r'}{r}\right)^2\right]\cdot\frac{\sin 2(\varphi+\alpha)-\frac{2S}{r^2}}{\cos 2(\varphi+\alpha)-\frac{2U}{r^2}} \tag{7}$$

where S and U are dimensionless quantities incorporating geometrical and physical parameters of the problem (U varies linearly with $\Delta T/\sigma^a$). Equation (7) which is not easy to integrate can be put in the form :

$$1+\frac{r'}{r}F(r\,;\varphi) = \sqrt{1+F^2(r\,;\varphi)} \tag{8}$$

where F is the ratio sin (...) / cos (...) in (7).
Integration at this stage can be performed numerically to yield the position r (φ) of P, which in turn is converted into (x,y) via :

$$x(P) = R_1 r\cdot\cos(\varphi+\alpha) + \frac{R_1 m_1}{r}\cdot\cos(\varphi-\alpha)$$

$$y(P) = R_1 r\cdot\sin(\varphi+\alpha) - \frac{R_1 m_1}{r}\cdot\sin(\varphi-\alpha) \tag{9}$$

Note that "C" the constant of integration must be defined through the initial position of the crack tip at infinity (Figure 1). One finds : $x = R_1\sqrt{C}$; $y = +\infty$ with reference to the case where $\alpha = o$. Also note that finding whether a crack will reach the inclusion when propagating, amounts to finding φ_i such that r (φ_i) = 1.

Now, bearing in mind that the overall loading of the system is in mode I, the corresponding stress intensity factor when the crack tip is at P, is given by :

$$K_I = \sqrt{\frac{2}{\pi}} \cdot \int_{-\infty}^{\ell} \frac{\sigma_I(s)}{\sqrt{\ell - s}} \cdot ds \qquad (10)$$

where s is a curvilinear abscissa along the crack trajectory (s = ℓ at P)

$\sigma_I = \frac{\sigma_x + \sigma_y}{2} + \sqrt{\left(\frac{\sigma_x + \sigma_y}{2}\right)^2 + \tau_{xy}^2}$ is the local principal stress tending to crack opening in mode I. Equation (10) was selected here as an approximate yet acceptable expression for the stress intensity factor since our ultimate objective is to estimate only toughness variations as they arise from just a short part of the overall crack length.

Figures 3 shows examples of calculated crack trajectories for the case of sapphirine inclusions in a cordierite matrix. Figure 4 shows a similar set of paths for an aspect ratio a/b = 5, together with their corresponding opening stress profiles. It is interesting to note that for cracks initiating close enough to the median axis σ_I first decreases upon approaching the inclusion before going through a maximum and decreasing again when the crack front is beyond the inclusion. In practice therefore : should the crack propagation be stopped, this will occur in front of the inclusion since around and beyond it the crack will be expected to accelerate.

The stress intensity factor tending to crack development can be calculated from σ_I and (10). It is found that for a number of configurations weakening should result rather than toughening because of the σ_I variation (Figure 4). But this effect is overcome, although slightly, by the concurrent increase in crack length which is representative of fracture surface energy. Typical examples of stress intensity variations ΔK_I and net toughening ΔK_{Ic} are again illustrated on Figure 5 for various inclusion aspect ratios and conditions otherwise similar to those of previous Figures. Here K_I^0 and K_{Ic}^0 refer to the case of the cordierite matrix only, i.e. with no inclusion present, under σ^a loading (for K_I^0) and such that a straight crack would propagate through, normally to σ^a and whithout being deflected (for K_{Ic}^0).

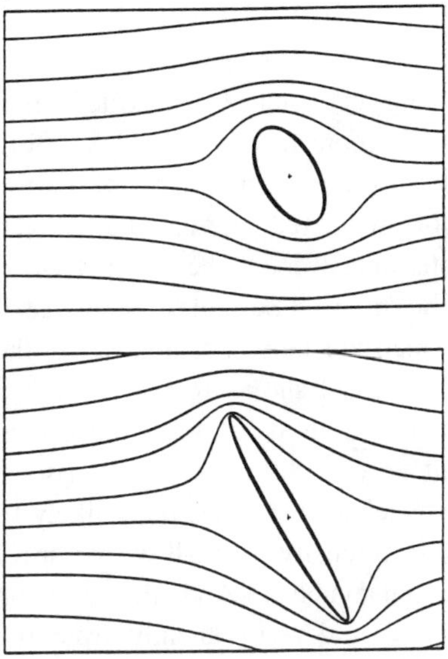

Figure 3 - Effect of aspect ratio a/b on crack paths around a sapphirisne inclusion in a cordierite matrix : a/b = 2 and a/b = 10 respectively. Other conditions as on figure 2.

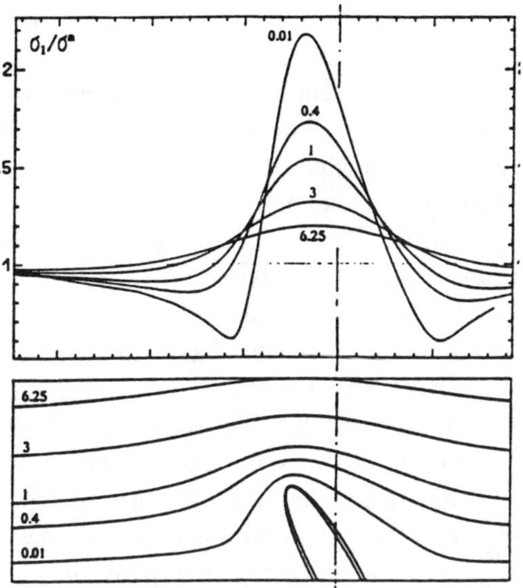

Figure 4 - Examples of crack paths for a/b = 5 and their corresponding crack opening stress profiles σ_I. Other conditions as on figure 2.

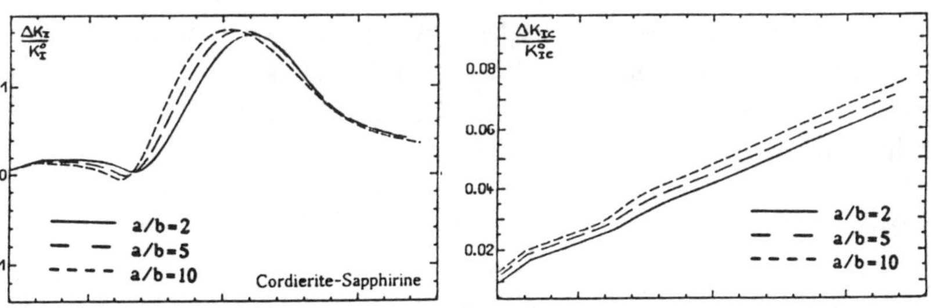

Figure 5 - Estimates of stress intensity factor changes (a) and toughness increases (b) versus position around sapphirine inclusions in cordierite. Crack initiates at Xi = 0.3 (cf. Fig. 1). Other conditions as on previous figures.

CONCLUSION : HOW MUCH TOUGHENING ?

The present model suggests that toughening due solely to crack deflection in brittle matrix composites should be significantly more modest than heretofore claimed or anticipated, when conditions are such that cracks propagate within the matrix only. Other effects of course which are not considered here could alter such a conclusion, for example if matrix-inclusion interfaces were to be reached, or in relation with the polycrystalline nature of the matrix. Several limiting cases therefore may be considered. For inclusions which are more dilational than the matrix and for slow quasistatic propagation, the model shows that a crack will experience a variable driving force along its path around an inclusion. A weakening should result which will be overcompensated but only slightly, by the increase of fracture energy due to the increased crack length. Such a situation is most likely to be encountered when the externally applied stress intensity is small with respect to that resulting from internal thermoelastic mismatches. And the ultimate toughening which may be recorded in practice should result mainly from the average residual stress within the matrix as generated by interaction of stress fields of individual inclusions. Conversely, when a crack propagates rapidly toward a more dilational inclusion, e.g. under high external loading compared to internal mismatches, local residual stress fields are more likely to be ignored and the crack will be marginally deflected. More inclusion/matrix interfaces therefore should be reached and other effects will come into play.

Similarly, for less dilational inclusions than the matrix, cracks will have a tendency to be attracted toward interfaces rather than being repelled. Again other effects such as debonding pullout or fracture of the inclusion will take over and the contribution of crack deflections to the overall toughness will generally remain rather limited.

REFERENCES

[1] K.T. Faber and A.G. Evans, "Crack deflection processes. I Theory" *Acta Metallurgica* 31 [4] 565 - 76 (1983).

[2] Y.C. Chiang and T.W. Chon, "Toughness models of whisker reinforced ceramic matrix composites" ASTM Special Technical Publication 1080 (1990) p. 101.

[3] H. Liu, K.L. Weisskopf and G. Petzow, "Crack deflection process for hot-pressed whisker reinforced ceramic composites". *Journal of the American Ceramic Society* 72 [4] 559 - 63 (1989)

[4] R.W. Davidge and T.J. Green, "The strengh of two-phase ceramic-glass materials"
Journal of Materials Science 3 [6] 629-34 (1968)
[5] R.A. Cutler and A.V. Virkar, "The effect of binder thickness and residual stresses on the fracture toughness of cemented carbides".
Journal of Materials Science 20 [10] 3557-73 (1985)
[6] Z. Li and R.C. Bradt, "Micromechanical stresses in SiC-reinforced Al_2O_3 composites "
Journal of the American Ceramic Society 72 [1] 70-77 (1989)
[7] M. Taya, S. Hayashi, A.S. Kobayashi and Y.S. Yoon, "Toughening of a particulate-reinforced ceramic-matrix composite by residual stress"
Journal of the American Ceramic Society 73 [5] 1382-91 (1990)
[8] J.D. Eshelby, "The determination of elastic fields of an ellipsoïdal inclusion and related problems"
Proceedings of the Royal Society (London) A241 376-96 (1957)
[9] A.A. Rubinstein, "Mechanics of the crack path formation".
International Journal of Fracture 47 [] 291-305 (1991)
[10] A.A. Rubinstein, "Macrocrack-microdefect interaction"
ASME Joural of Applied Mechanics, 53 [] 505-10 (1986)
[11] N.I. Muskhelishvili,
Some basic problems in the mathematical theory of Elasticity, P. Noordhoff Ltd., Holland (1963)
[12] A.A. Rubinstein, "Crack-Path effect on Material Toughness"
ASME Journal of Applied Mechanics 57 [] 97 - 103 (1990)
[13] M. Genevrier, doctor's Thesis
Institut National Polytechnique de Lorraine ; Nancy France, 1992

[14] G.P. Cherepanov,
Mechanics of brittle fracture, R. de Witt and W.C. Cooley Ed., Mc Graw Hill Pub., New York (1979) pp. 698-701
[15] C.H. Hsueh, P.F. Becher and P. Angelini, "Effect of interfacial film on thermal stress in whisker-reinforced ceramics"
Journal of the American Ceramic Society, 71 [11] 929-33 (1988)

THE EFFECT OF FABRIC ORIENTATION ON THE FLEXURAL BEHAVIOR OF CFCCs: EXPERIMENT AND THEORY

N. Miriyala*, P. K. Liaw, C. J. McHargue and A. Selvarathinam
The University of Tennessee
Knoxville, TN 37996

L. L. Snead
Oak Ridge National Laboratory
Oak Ridge, TN 37831

ABSTRACT

Four-point bend tests were performed on a Nicalon™-fabric reinforced alumina-matrix composite fabricated by the directed metal oxidation (DIMOX™) process at room temperature. The flexural strength of the material was significantly higher when the loading direction was parallel to the fabric plane, as compared to loading direction normal to the fabric plane. The difference in flexural strength in the two orientations was related to the low interlaminar shear strength of the material. Analytical and numerical (FEM) modeling of the flexural behavior showed that, when a composite with interlaminar weakness is subjected to flexure loads, specimen failure occurs at a lower load when the loading direction is perpendicular to the fabric plies, relative to the loading direction parallel to the fabric plies.

INTRODUCTION

A majority of the commercially available continuous fiber ceramic composites (CFCCs) developed to date, particularly the non-glass matrix composites, use fiber-reinforcement in the form of woven or braided fabrics. Since there is no reinforcement in the thickness direction, these laminate (2-D) composites suffer from poor interlaminar shear strength, which manifests in the form of interlaminar cracks when they are subjected to loads normal to the fabric plies. Thus, the flexural behavior of a laminate CFCC can be strongly affected by the orientation of the loading axis to the fabric plane. However, till date, only a limited number of studies [1, 2, 3] have addressed the influence of fabric orientation on the flexural behavior of CFCCs. Accordingly, it was the objective of the present study to investigate the effect of fabric orientation on the flexural behavior of a commercially available CFCC, namely, a Nicalon™ fiber-fabric reinforced alumina matrix composite fabricated by the directed molten metal oxidation (DIMOX™) process [4].

---

* Current Address: Solar Turbines Inc., San Diego, CA 92186-5376.

To the extent authorized under the laws of the United States of America, all copyright interests in this publication are the property of The American Ceramic Society. Any duplication, reproduction, or republication of this publication or any part thereof, without the express written consent of The American Ceramic Society or fee paid to the Copyright Clearance Center, is prohibited.

EXPERIMENTAL DETAILS

Materials

Nicalon™/Al_2O_3 composite

The Nicalon™/Al_2O_3 composite was fabricated using the directed metal oxidation (DIMOX™) process (DuPont-Lanxide Corporation, Newark, DE). In this process, eight layers of a 12-harness satin-woven Nicalon™ fiber-fabric were stacked up, with alternate layers being rotated by 90° to achieve a balanced lay-up. The fabric preform was then given a duplex coating of boron nitride (BN) and SiC using chemical vapor deposition (CVD) techniques. The BN acts as the interlayer between the fiber and the matrix, and the overlaying SiC coating protects the fiber and the interlayer during matrix infiltration. At the completion of the duplex coating process, the preform was heated in contact with an aluminum alloy at 900 to 1,200°C in air. The matrix in the composite was formed by the growth of α-Al_2O_3 into the preform, starting from the alloy/preform interface [4]. The nominal fiber and residual-aluminum contents in the composite are approximately 35 % and 1%, respectively, by volume [4]. The average porosity content in the composite plate used in the present study was 15 % [5].

Mechanical Testing

The flexure tests were performed on a commercial servohydraulic load-frame (Model 810, MTS Systems Corporation, Minneapolis, MN) using a semi-articulating bend fixture made of silicon carbide (SiC). To objectively assess the effect of fabric orientation on the flexural behavior of the composite, test specimens of square cross-section were fabricated using diamond tooling. The flexure bars were 50 mm long, 3 mm wide and 3 mm thick. The specimens were subjected to four-point bend loading at room temperature, using the general guidelines of ASTM Standard C 1341 [6]. The loading and supporting spans during the tests were 20 mm and 40 mm, respectively. The flexure tests were conducted under displacement control at a crosshead movement rate of 8.3×10^{-3} mm/s. The load was applied to the specimens either parallel or perpendicular to the fabric plies. Accordingly, the specimen configurations were referred to as edge-on and transverse, depending on whether the load was parallel or perpendicular to the fabric plies, respectively (Figure 1).

RESULTS AND DISCUSSION

Stress-Strain Behavior

Typical stress-strain curves for the Nicalon™/Al_2O_3 composite under flexure loading are shown in Figure 2. It can be seen from Figure 2 that the stress-strain curves are nonlinear in both edge-on and transverse orientations. It is also evident from Figure 2 that the ultimate flexural strength (UFS) and strain-to-failure values are significantly higher in the edge-on orientation, as compared to those in the transverse orientation. The

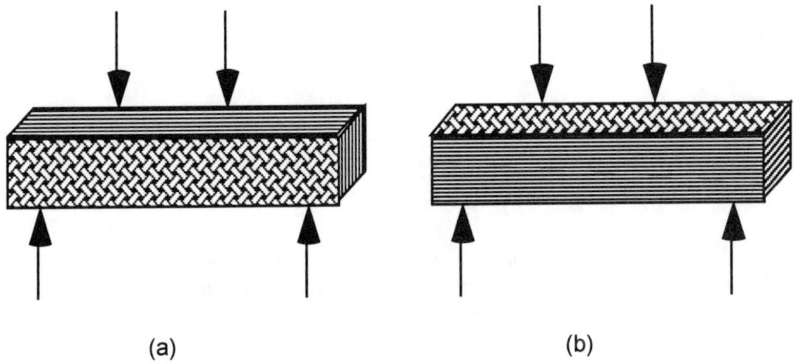

Figure 1. Orientations for flexure-testing laminate composites: (a) edge-on and (b) transverse.

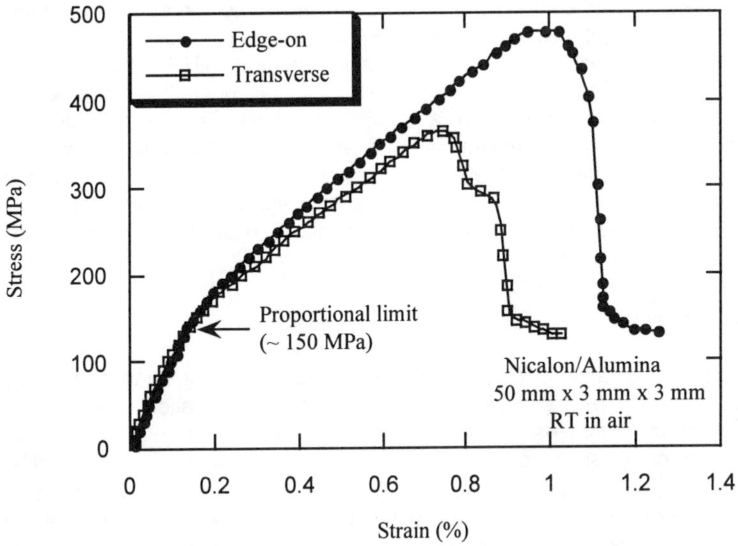

Figure 2. Flexural stress-strain curves for the Nicalon™/Al_2O_3 specimens in the edge-on and transverse orientations.

ultimate flexural strength (UFS) of the composite was 489±21 MPa (mean ± standard deviation from 10 tests) in the edge-on orientation, and 336±34 MPa (mean ± standard deviation from 10 tests) in the transverse orientation. However, the proportional limit (σ_{pl}) was about the 150 MPa in both orientations. The flexural stress (σ) and strain (ε) values were calculated from the following equations based on simple-beam theory [6].

$$\sigma = \frac{3PL}{4bh^2} \qquad (1)$$

and

$$\varepsilon = \frac{4.36\delta h}{L^2} \qquad (2)$$

where σ is the maximum stress in the outer fibers at a given load (P), L is the distance between the supporting pins, b is the specimen width and h the thickness, ε is the maximum strain in the outer fibers at a given load, and δ is the midspan deflection at a given instant of the test.

It is important to note here that the UFS values calculated using Equation 1 are meaningful only when the specimens fail in a predominantly tensile mode. Equation 1 does not consider the shift of the neutral axis to the compression side of the specimen, which takes place in ceramics, once cracking occurs on the tensile side of the specimen [7]. Since the strength of ceramics, in general, is higher in compression than in tension, the shifting of the neutral axis can result in UFS values, as calculated from Equation 1, being significantly higher than the ultimate tensile strength (UTS) of the material. In addition, Equation 1 also does not take into account the propagation of delamination/interlaminar cracking in a specimen. However, simple equations to calculate the UFS from the maximum load in a flexure test, which take into account the neutral axis shift and/or the propagation of interlaminar/delamination cracks, have yet to be developed. Hence, Equation 1 was used to calculate the UFS values in this investigation as a consistent basis to study the influence of fabric orientation on the flexural behavior of the Nicalon™/Al_2O_3 composite.

Damage Mechanisms

Under flexure loading, laminate (2-D) composites in the edge-on orientation are subjected to in-plane shear stresses, while in the transverse orientation they experience interlaminar shear stresses. The interlaminar shear strength (ILSS) of the Nicalon™/Al_2O_3 composite was 22±4 MPa [7], as measured by subjecting double-notch specimens to compression load using the procedure outlined in ASTM Standard 1292 [8]. Heredia et al. [9] reported a value of 100 MPa as the in-plane shear strength (IPSS) of a Nicalon™/Al_2O_3 composite, which is similar to the material used in the present study.

The specimens tested in the edge-on orientation failed in a predominantly tensile mode. The tensile (bottom) surfaces of an edge-on specimen loaded to failure are presented in Figure 3. It is apparent from Figure 3 that the specimen fractured into two pieces by the failure of the 0° and 90° yarns. In the transverse orientation, however, after the propagation of Mode I cracks, a number of delamination cracks developed in the specimen, leading to the "specimen collapse" rather than fracture into two pieces. Typical cross-section of a specimen monotonically loaded to failure in the transverse orientation is shown in Figure 4. It can be seen from Figure 4 that Mode I cracks are

Figure 3. Tensile (bottom) surfaces of a Nicalon™/Al_2O_3 specimen loaded to failure in the edge-on orientation.

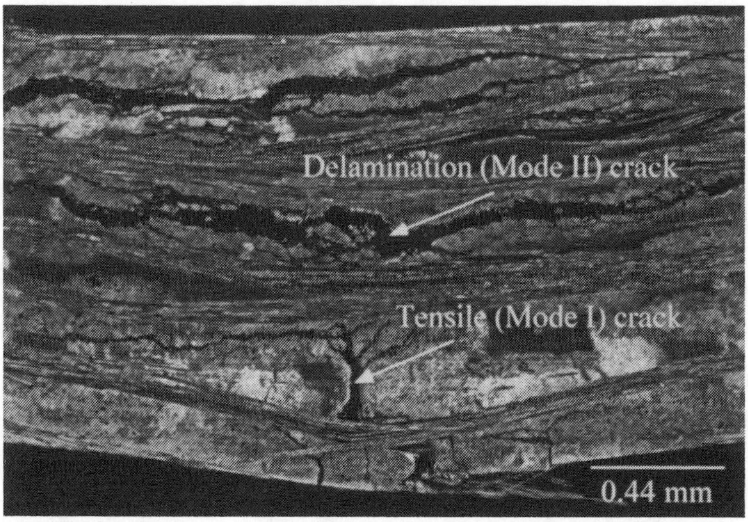

Figure 4. Cross-section of a Nicalon™/Al_2O_3 specimen loaded to failure in the transverse orientation.

present in the specimen portion subjected to tensile stresses, and Mode II (delamination) cracks are mostly confined to the specimen portion subjected to compressive stresses. Thus, the failure of a transversely oriented specimen at a lower load, relative to that in the edge-on orientation, could be attributed to the low ILSS of the material.

To further understand the effect of fabric orientation on the flexural behavior of laminate (2-D) composites, the mechanical response of a composite beam subjected to four-point bend loading was modeled using analytical and numerical (finite element) techniques, with a particular emphasis on the influence of interlaminar weakness on the normal-stress and midspan-deflection values.

MODELING OF FLEXURAL BEHAVIOR

In the analytical and numerical models, it was assumed that the composite material consisted of isotropic plates of two different materials perfectly bonded together. The two materials were considered to have different elastic modulus (E) and Poisson's ratio (ν) values. In the Nicalon^{TM}/Al_2O_3 composite specimens used in the current study, on the average, there were eight layers of fabric reinforcement. Accordingly, it was assumed that the laminate composite used in modeling the fabric orientation effect consisted of eight layers of material 1, each 0.35 mm in thickness. To simulate the interlaminar weakness, one layer of material 2 was introduced between two consecutive layers of material 1. The thickness of each layer of material 2 was considered to be 0.05 mm. Thus, the laminate composite consisted of eight layers of material 1 and seven layers of material 2, with a total thickness of 3.15 mm. The cross-section of the multi-layered beam is shown in Figure 5.

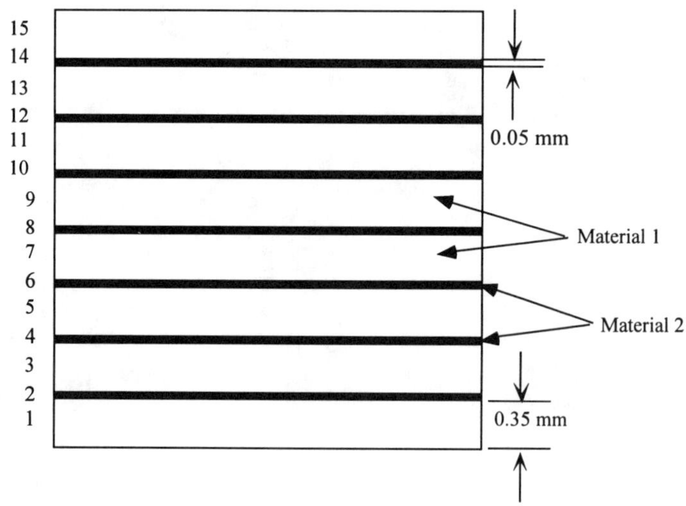

Figure 5. Cross-section of the multi-layered beam used to model the flexural behavior.

In an effort to objectively compare the predictions of the analytical and FEM modes with the experimental results, square cross-sectioned (3.15 mm x 3.15 mm) composite beams under flexure loading were considered in the two models. In both models, a 50 mm long composite beam was subjected to four-point bend loading, with a loading span of 20 mm and a supporting span of 40 mm. The composite beam was subjected to a total load (P) of 500 N (250 N at each loading point). For a load of 500 N, the maximum stress given by Equation 1 is 480 MPa, which is comparable to the average UFS of the Nicalon™/Al_2O_3 composite in the edge-on orientation.

To study the effect of interlaminar weakness on the flexural behavior of the composite beam, the modulus of material 1 (E_1) was taken as 100 GPa, and the modulus of material 2 (E_2) was varied from 1 to 100 GPa, such that the modular ratio η (E_2/E_1) ranged from 0.01 to 1. A modular ratio of 0.01 characterizes the beam with an extreme interlaminar weakness, and a value of 1 for η implies that materials 1 and 2 are equivalent, such that the composite beam behaves like a simple beam of material 1 or 2. To predict the normal-stress and midspan-deflection values, it was assumed that the elastic modulus of the beam was the same in tension and compression.

Analytical Model

In the analytical modeling of the flexure behavior of the multi-layered material, a "transformed-section" method was used [10]. In this method, the cross-section of a multi-layered material is transformed into an equivalent cross-section composed of only one material. The transformed section must have the same bending moment-resistance, and the same location for the neutral axis, if it is to be equivalent. The location of the neutral axis is found by using the condition that the resultant axial force acting on the cross-section is zero. That is,

$$\sum \int_{1} \sigma_{x_1} dA + \int_{2} \sigma_{x_2} dA + \dots\dots + \int_{15} \sigma_{x_{15}} dA = 0 \tag{3}$$

where σ_{x_i} is the longitudinal stress in the ith layer, and A is its cross-sectional area.

The location of the neutral axis will remain unchanged if the area of each layer of material 2 is multiplied by η, provided that the distance from the neutral axis (y) is unchanged. Consequently, we can consider the transformed-section to be consisting of two components: the areas of the thick (0.35 mm) layers unchanged, and the areas of the thin (0.05 mm) layers multiplied by η. Thus, we have a new cross-section consisting of only material 1. The longitudinal stress (σ_x) in a homogeneous beam is given by

$$\sigma_x = \frac{My}{I} \tag{4}$$

where M is the bending moment, and I is the moment of inertia about the neutral axis. Likewise, the longitudinal stress in the beam transformed to material 1 is given by

$$\sigma_x = \frac{My}{I_t} \tag{5}$$

where I_t is the moment of inertia about the neutral axis of the transformed section given by

$$I_t = I_1 + \eta I_2 + I_3 + \eta I_4 \ldots\ldots\ldots\ldots I_{15} \quad (6).$$

When a homogeneous beam is subjected to four-point bending, such that the loading span is half of the supporting span, the midspan deflection (δ) is given by

$$\delta = \frac{11PL^3}{768EI} \quad (7).$$

Accordingly, for the transformed-beam, the midspan deflection δ_t is given by

$$\delta_t = \frac{11PL^3}{768E_1I_t} \quad (8).$$

where E_1 is the modulus of material 1.

Numerical Model

In the numerical modeling of the multi-layered beam under flexure loading, a finite element method (FEM) was used. The PATRAN software (version 5.0) on a Sun workstation was used to mesh the multi-layered beam and generate the finite element code, which was then solved by the ABAQUS software (version 5.1). Eight-node quadrilateral elements were used to mesh the multi-layered beam. The element aspect ratio (ratio of the length to height of the rectangular mesh) used was less than 10. In performing the normal-stress and midspan-deflection calculations, plain strain conditions were used. The Poisson's ratios of materials 1 and 2 were considered to be 0.2 and 0.1, respectively.

Effect of Interlaminar Weakness on Normal-Stress and Midspan-Deflection Values

The results of the analytical and FEM models indicated that the longitudinal stresses in a multi-layered beam with interlaminar weakness could be significantly higher than those predicted by the simple beam theory. The maximum stress values predicted by the analytical and FEM models are plotted as a function of the modular ratio in Figure 6. Both the models predicted the same maximum stress for all modular ratios. For a modular ratio of 0.01, which is the extreme case of interlaminar weakness considered, the two models predicted that the maximum longitudinal stress (tensile or compressive) in the beam will be 525 MPa. Thus, the maximum stress predicted by both models is approximately 10 % higher than that calculated from Equation 1.

In a material with low ILSS, the UFS calculated from the maximum load during a flexure test, using Equation 1 (based on simple-beam theory) does not accurately represent the maximum stress actually experienced by the specimen. This results in a lower UFS value in the transverse orientation than in the edge-on orientation. The interlaminar weakness can also result in a noticeably higher midspan deflection than that given by

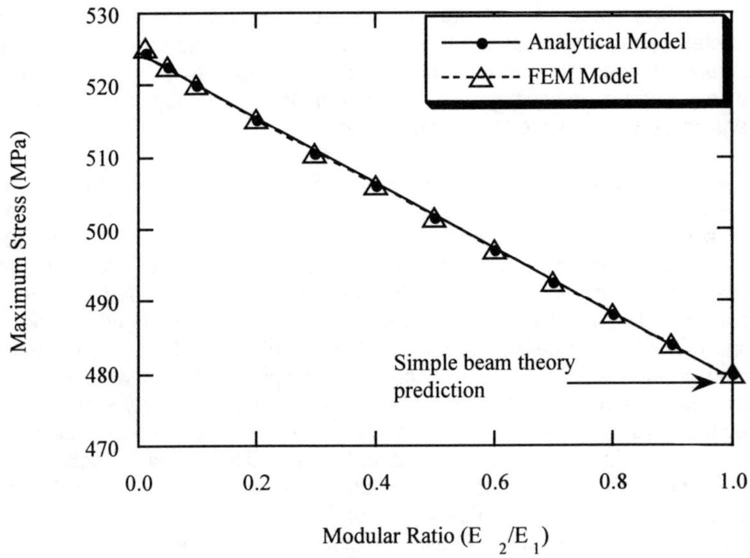

Figure 6. Maximum stress values predicted by the analytical and FEM models for 50 mm long multi-layered beams of different modular ratios subjected to a load of 500 N.

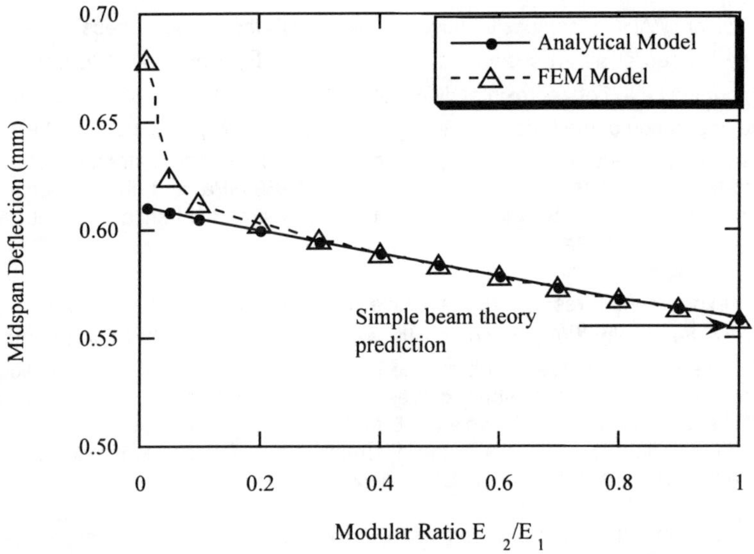

Figure 7. Midspan deflection values predicted by the analytical and FEM models for 50 mm long multi-layered beams of different modular ratios subjected to a load of 500 N.

Equation 7, which is applicable only for homogeneous beams. The midspan deflection values predicted by the analytical and FEM models are plotted as a function of modular ratio in Figure 7. It is apparent from Figure 7 that the midspan deflection of a composite beam weak in shear can be considerably higher than that of a homogeneous beam of the same dimensions, and subjected to the same types of loads.

The midspan deflection values predicted by the FEM model are comparable with the analytical results at higher modular ratios, but are significantly higher at lower modular ratios. This is due to the fact that Equation 8 used in the analytical model to calculate the midspan deflection of the multi-layered beam considered only the effect of bending deformations in the beam. An additional deflection is produced in the beam by the shear deformations [9]. The FEM model considered the effect of both bending and shear deformations to predict the deflection values. The contribution of shear deformation to the deflection of the beam can be significant for composite beams with considerable interlaminar weakness (low modular ratios), while they can be negligibly small for beams that are relatively strong in shear (high modular ratios). Hence, the deflection values predicted by the FEM model are comparable to the analytical model predictions at high modular ratios, but significantly higher at low modular ratios. For a modular ratio of 0.01, the FEM model predicted a midspan deflection of 0.68 mm, which is much higher than a value of 0.56 mm given by Equation 7 based on simple-beam theory. Thus, the midspan-deflection enhancement (the difference between the deflections predicted by the FEM model and the value given by Equation 7) is about 20 % for a modular ratio of 0.01.

Prediction of Flexural Stress-Strain Curves

For the Nicalon™/Al_2O_3 composite used in the study, the tensile stress-strain curve obtained by Fareed et al. [4] showed an initial modulus E_{C0} of about 160 GPa, and a post-σ_{pl} modulus (E_{C1}) of approximately 40 GPa. The flexural stress-strain curves in the transverse orientation of the Nicalon™/Al_2O_3 composite can be predicted from the stress-strain curves in the edge-on orientation, assuming that (i) the stress-strain curves in the both orientations are bi-linear with an initial modulus of 160 GPa, and (ii) the specimen in the edge-on orientation behaves like a simple beam (no stress enhancement), but with a post-σ_{pl} modulus of 40 GPa.

From the experimental results, the average σ_{pl} and UFS values in the edge-on orientation of the Nicalon™/Al_2O_3 composite were 150 MPa and 489 MPa, respectively (Figure 2). Using these values to obtain the bi-linear stress-strain curve in the edge-on orientation, the predicted stress-strain curves in the transverse orientation are shown in Figure 8 for stress enhancement levels of 10 to 40 %. The UFS values were calculated using Equation 1, taking into account the different levels of stress enhancement. The strain (ε) values were calculated using Equation 2.

The σ_{pl} is unchanged in all the stress-strain curves in Figure 8, as the experimentally determined stress-strain curves indicated that the fabric orientation affects only the post-σ_{pl} flexural behavior. For a stress enhancement level of 40 %, the predicted values of UFS and strain-to-failure in the transverse-orientation are 349 MPa and 0.87 %, which compare favorably with the experimentally determined UFS and strain-to-failure values of 336 MPa and ~ 0.80 % for the Nicalon™/Al_2O_3 composite (Figure 2).

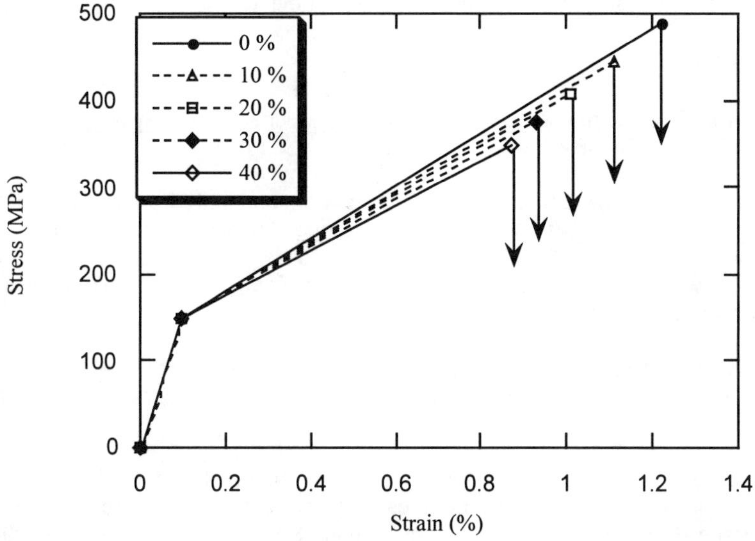

Figure 8. Predicted stress-strain curves in the transverse orientation for different stress levels of stress enhancements (0 % refers to the stress-strain curve in the edge-on orientation).

CONCLUSIONS

1. The ultimate flexural strength of the Nicalon™/Al_2O_3 composite was lower in the transverse orientation, than in the edge-on orientation, due to low interlaminar shear strength of the material.

2. When a composite specimen with interlaminar weakness is subjected to flexure loads, there can be significant increases in the normal-stress and midspan-deflection values, resulting in specimen failure occurring at a lower load in the transverse orientation than in the edge-on orientation.

ACKNOWLEDGMENT

This research was supported by the Department of Energy under Contract No. Martin Marietta 11X-SV483V to the University of Tennessee. P. K. Liaw was also supported by the National Science Foundation under Contract No. EEC-9527527 to the University of Tennessee. The authors are grateful to E. E. Bloom and A. F. Rowcliffe of the Oak Ridge National Laboratory for their help and support. Appreciation is due to A. S. Fareed of

Lanxide Corporation, Newark, DE, and J. A. Morrison of Westinghouse Electric Corporation, Orlando, FL, for providing the composite materials used in this study. They are also thankful to T. A. Long, G. L. Jones and D. E. Fielden of the University of Tennessee, and A. M. Williams, J. W. Jones and D. W. Coffey of the Oak Ridge National Laboratory for their assistance in performing the experimental work.

REFERENCES

1. P. K. Liaw, Fiber-Reinforced CMCs: Processing, Mechanical Behavior, and Modeling, JOM, 47 (10), 1995, 38-44.

2. N. Chawla, P. K. Liaw, E. Lara-Curzio, R. A. Lowden and M. K. Ferber, Effect of Fiber Fabric Orientation on the Monotonic and Fatigue Behavior of a Continuous Fiber Ceramic Composite, in High Performance Composites - Commonalty of Phenomena, K. K. Chawla, P. K. Liaw and S. G. Fishman (editors), The Minerals, Metals and Materials Society, Warrendale, PA, 1994, 291-304.

3. N. Chawla, Effect of Fiber Fabric Orientation on the Mechanical Behavior of Continuous Fiber Ceramic Composites, M.S. Thesis, The University of Tennessee, Knoxville, 1994.

4. A. S. Fareed, D. J. Landini, T. A. Johnson, A. N. Patel and P. A. Craig, High Temperature Ceramic Matrix Composites by the Directed Metal Oxidation Process, SME Paper EM92-216, Society of Manufacturing Engineers, Dearborn, MI, 1992, 1-21.

5. N. Miriyala, Fabric Orientation Effects on the Monotonic and Fatigue Behavior of Continuous Fiber-Reinforced Ceramic-Matrix Composites, Doctoral Dissertation, The University of Tennessee, Knoxville, 1997.

6. C 1341, Flexural Properties of Continuous Fiber-Reinforced Advanced Ceramic Composites, Annual Book of ASTM Standards, American Society for Testing and Materials, W. Conshohocken, PA, 1997.

7. G. D. Quinn and R. Morrell, Design Data for Engineering Ceramics: a Review of the Flexure Test, Journal of the American Ceramic Society, 74 (9), 1991, 2037-2066.

8. C 1292, Shear Strength of Continuous Fiber-Reinforced Advanced Ceramics at Ambient Temperature, Annual Book of ASTM Standards, American Society for Testing and Materials, W. Conshohocken, PA, 1996.

9. F. E. Heredia, A. G. Evans and C. A. Anderson, Tensile and Shear Properties of Continuous Fiber-Reinforced SiC/Al_2O_3 Composites Processed by Melt Oxidation, Journal of the American Ceramic Society, 78 (10), 1995, 2790-2800.

10. J. M. Gere and S. P. Timoshenko, Mechanics of Materials, Second Edition, Wadsworth Incorporated, Belmont, CA, 1984.

KEYWORD AND AUTHOR INDEX

Abnormal grain growth, 245
Akatsu, Takashi, 271, 285, 295, 315
Alumina, 51
Anandakumar, Umashankar, 381
Antenna window materials, 219
Applications, 113
Attrition milling, 231

Balmori-Ramírez, H., 231
Bamba, Noriko, 197
Banerjee, D., 417
Bansal, Narottam P., 3, 135
Belardinelli, R., 161
Bender, Barry A., 89, 101
Blackglas™, 149
Bordia, Rajendra K., 341
Browning, Scott, 89
Bruce, Calvin, 77, 123

Cantonwine, Paul E., 51
Celsian composites, 3, 135
Ceramic fiber composites, 101
Ceramic matrix composites, 51, 77, 113, 341, 407, 429
  flexure test methods, 161
  Nextel™/Blackglas™, 161
  polymer-derived, 161
Ceraset™, 149
CFCC, 451
Choa, Yong-Ho, 185, 197, 209
Coatings
  environmental barrier, 27
  Ti_3SiC_2, 39
COD, 429
Components, 113
Composites, 89, 149, 237, 351, 361, 371, 381
  2-phase particle, 393
  α/β-SiAlON, 307
  mullite-ZrO_2, 231
  Nicalon™/Al_2O_3, 451
  oxide/oxide, 51
  SiC-SiC, 123
  SiO_2-Si_3N_4, 219
  YAG-spinel-alumina, 271
Corrosion, 27
Crack deflection, 439
Crain, S., 39
Creep, 381, 393
CVD, 39, 77
CVI, 77, 123

Damage, 407
  mechanisms, 451
  tolerance, 173
De la Torre, S.D., 231
Debond length, 429
Densification, 257
Divorced eutectic, 315
Durability, 17

Easler, T.E., 113
EDS, 39
Electrical properties, 209
Electrophoretic deposition, 63
Equilibrium, 329
Eutectic composites, 271
  YAG-spinel, 285, 295, 315

Faber, K.T., 149
Fabrication, 113
Failure, 351
Fang, Y., 257
Fibers, 149, 173, 371
  bridging stress, 429
Flexural strength, 451
Fluorescence spectroscopy, 417
Fractography, 51
Fracture strength, 197
Fracture toughness, 237

Genevrier, Marc, 439
Glass matrix composites, 417
Gonczy, S.T., 149
Grain boundary, 307
  sliding, 393

Hardness, 237
Hayashi, Y., 209
Henager, Charles H., Jr., 341, 351
Hermetic seals, 77
Hi-Nicalon™ fibers, 3
High-temperature deformation, 341
Hirschfeld, Deidre A., 27
Huang, Xueren, 219

Impulse excitation of vibration, 161
Interfacial oxidation, 341
Ionic conductivity, 197
Jackson, H.E., 417
Jeong, Y.K., 185

Jessen, Todd L., 89, 101
Johnsson, M., 237
Johnston, D.E., 371
Jones, Russell H., 341, 351, 371

Kohyama, A., 371
Kolaya, Lynne E., 77
Kowbel, Witold, 77, 123

Lackey, W.J., 39
Laminates, 63, 173
Lee, Kang N., 17
Lewinsohn, C.A., 351, 371
Lewis, Nathan, 77
Li, Baoshun, 219
Liaw, P.K., 451
Lin, Chien-Cheng, 361
Liu, Siang-Chun, 329
Liu, Yin, 77

Mackin, Thomas J., 407
Magnesia, 231
Matrix cracking, 381
McHargue, C.J., 451
Mechanical behavior, 101
Mechanical properties, 161, 185, 209, 219, 371
Metastable precipitation, 271, 285, 315
Microstructural modification, 245
Microstructure, 3, 101, 185, 197, 271, 285, 295, 381
Miller, Robert A., 17
Mini-tow, 149
Miriyala, N., 451
Mocellin, Alain, 393, 439
Modeling, 451
  2-D, 439
Mullite, 173

Nagarajan, N., 257
Nanocomposites, 185, 197, 209
Nicalon™, 89, 149
Niihara, Koichi, 185, 197, 209
Nygren, M., 237
NZP, 27

Oxidation, 351, 361
Oxidation resistance, 307
Oxides, 173

Pauletti, S.L., 161
Petrak, D.R., 113
Pettersson, P., 237
Phase stability, 329
Phosphate, 89
Pickering, E., 39
PIP, 77, 123
Porous layers, 63
Porous matrix composites, 51
Preceramic polymers, 113
Pressureless sintering, 185, 245
Processing, 89
Properties, 89

Radsick, Timothy, 173
Reaction bonding, 231
Residual stress, 417
Rho, H., 417
Roberts, Mark C., 407
Rocha-Rangel, E., 231

Sapphire, 417
Saruhan, B., 173
Schneider, H., 173
Segregation, 307
Sekino, Tohru, 185, 197
Selvarathinam, A., 451
Serizawa, H., 371
Shear strength, 135
Sheu, Tzer-Shin, 307
Silicon carbide, 63, 77, 351, 381
  matrix composites, 27
  mullite/YSZ-coated, 17
Silicon nitride, 245, 257, 307
Singh, J.P., 209
Singh, Raj N., 381, 417, 429
Sintering, 51
  additives, 185
Snead, L.L., 451
Stackpoole, Margaret M., 341
Step sintering, 257
Stinton, David P., 27
Strength, 257, 371
Sun, Yongjian, 429
SYLRAMIC™, 113
Systems
  Ti-Si-C, 39
Szweda, A., 113

Tai, Qiu, 393
Tanabe, Yasuhiro, 271, 285, 295, 315
Tensile strength, 3, 135, 149
Thermal conductivity, 123
Thermal shock resistance, 219
Thermodynamics, 329
Thermoelastic stress analysis, 407

Threshold, 361
Time-dependent failure, 101
Titania, 231
Toughening, 439
  mechanisms, 219
Toughness, 197, 257
Transformation, 257
Tsai, Cheng-Yuan, 361
Tsou, K.T., 123

Ünal, Ö., 135
Unidirectional composites, 135

Van Der Biest, Omer O., 63
Vandeperre, Luc J., 63
Vaporization, 329
Vaubert, Virginie M., 27

Wadley, Haydn N.G., 51
Wang, Shuqiang, 271, 285, 295, 315
Water vapor, 17
Whiskers, 237
White, Kenneth W., 245, 257
Withers, James C., 77, 123

XRD, 39

$YAlO_3$, 271, 285
Yao, Junjie, 219
Yasuda, Eiichi, 271, 285, 295, 315
Youngblood, G.E., 371
Yu, Feng, 245, 257

Zhang, Yufeng, 219
Zircon, 381
Zirconia content, 361